The Volcano Adventure Guide

Have you ever wondered what it would be like to stare down into the bubbling crater of an active volcano?

If so, this book is for you! It contains vital information for anyone wishing to visit, explore, and photograph active volcanoes safely and enjoyably. The book begins by introducing readers to the eruption styles of different types of volcanoes and explains the physical settings on Earth where they are typically found. It describes how to prepare for a volcano trip, and how to avoid the dangers associated with being on or near active volcanoes. The author draws on her own experience of working on active volcanoes to explain what is safe and what is foolish, when to watch an eruption and when to stay away. She believes that volcanoes can be visited and enjoyed by all, and includes several examples of volcanoes that can be easily explored by people of all ages, and all levels of fitness and expertise.

The second part of the book provides a comprehensive travel guide to 20 volcanoes around the world that are among nature's most spectacular examples. It also gives short guides to 22 additional volcanoes located in the same regions, that could be visited during the same trips. This section is packed full of practical information including tour itineraries, maps, transportation details, and warnings of possible non-volcanic dangers such as unfriendly wildlife. The two appendices at the end of the book direct the reader to a wealth of further volcano resources. These include websites with up-to-date listings of volcanic activity across the globe, and a list of learned societies and commercial holiday companies offering volcano tours. There is also an extensive bibliography that refers readers to more detailed geological and practical information.

The Volcano Adventure Guide is the first book of its type. Aimed at non-specialist readers who wish to explore volcanoes without being foolhardy, it will fascinate amateur enthusiasts and professional volcanologists alike. The stunning color photographs throughout the book will delight armchair travelers as well as inspire the adventurous to get out and explore volcanoes for themselves.

ROSALY LOPES is an expert in planetary volcanism at NASA's Jet Propulsion Laboratory in Pasadena, California, where she studies volcanism on Earth, as well as other planets and moons. Using data returned by the Galileo spacecraft, Dr. Lopes was responsible for the discovery of 71 previously unknown volcanoes on Io, one of Jupiter's moons. She currently works on the Cassini mission, which is making observations of Saturn and its moons.

Rosaly's field work on Earth has taken her to many active volcanoes, starting with Mount Etna in Sicily (as a member of the UK's Volcanic Eruption Surveillance Team). She has made many trips to active volcanoes around the world and has given lectures to the public in many countries. During these lectures she was often asked "How can I visit an active volcano?" and this inspired her to write this book on volcano adventures. She has also written many scientific papers, encyclopedia articles, and book chapters, and has been featured on two Discovery channel television documentaries. She has won several awards from JPL and NASA, and was chosen by GEMS Television, Miami, as the GEMS Woman of the Year in Science and Technology, 1997.

The Volcano Adventure Guide

ROSALY LOPES

CAMBRIDGE
UNIVERSITY PRESS

PUBLISHED BY THE PRESS SYNDICATE OF THE UNIVERSITY OF CAMBRIDGE
The Pitt Building, Trumpington Street, Cambridge, United Kingdom

CAMBRIDGE UNIVERSITY PRESS
The Edinburgh Building, Cambridge CB2 2RU, UK
40 West 20th Street, New York, NY 10011–4211, USA
477 Williamstown Road, Port Melbourne, VIC 3207, Australia
Ruiz de Alarcón 13, 28014 Madrid, Spain
Dock House, The Waterfront, Cape Town 8001, South Africa

http://www.cambridge.org

First published 2005

Printed in the United Kingdom at the University Press, Cambridge

Typeface ITC Giovanni 10/13 pt *System* QuarkXPress™ [SE]

A catalogue record for this book is available from the British Library

ISBN 0 521 55453 5 hardback

Contents

Preface

The purpose of this book is to introduce its readers to the wonderful world of volcanoes and to help them visit, explore, photograph, and, above all, appreciate volcanoes both in eruption and in repose. Volcanoes have shaped the Earth's surface and are nature's most awesome manifestation of the power within our planet. One of the surprising facts about volcanoes is that they are among the most scenic places on Earth yet only a few of them attract a significant number of visitors. Volcanic eruptions are undoubtedly one of nature's most spectacular events, but relatively few people can claim to have witnessed one first-hand, and most of them did not do so by choice. The increase in adventure travel over the past two decades has not yet reached most of the world's volcanoes. Travel companies are quite willing to take tourists all over the globe to meet gorillas, canoe in piranha-infested waters, or dive with sharks, but not, it seems, to watch a volcano erupt.

What should people do if they want to visit a volcano, particularly an active one? They can choose a volcano within a well-run national park and rely on information and advice given there. But, if the volcano they want to go to is more isolated, or in a country with fewer resources, travelers will be very much on their own. In practice, this may mean not venturing up the slopes at all or, even worse, ending up in places where experienced volcanologists would fear to tread. In order to explore a volcano in the most sensible and enjoyable way, visitors need to know the dangers they might encounter, as well as the wonders they might see. *The Volcano Adventure Guide* strives to give readers this knowledge.

The idea for this book grew out of the questions that potential volcano visitors often asked me whenever I gave a popular-level talk about volcanoes or revealed in conversation what type of work I do. Some of these questions were: "How can I see an eruption?" "Are all eruptions dangerous?" "Can I go somewhere other than Hawaii to see red lava?" "How do people photograph eruptions?" Less often I'd hear "How can I be sure to visit a volcano when it is not erupting?" I realized that there were no books that could answer these questions in a straightforward way.

Most books about volcanoes assume that readers want to learn what volcanoes are and what makes them erupt, but from the comfort of their armchairs. The *Volcano Adventure Guide* is directed at the people who want to learn about volcanoes by visiting them first-hand. I have tried to provide all the necessary information on how to choose a volcano to go to and, once there, how to make the trip a fulfilling and enjoyable learning experience. These are the themes behind the book's introductory chapters that are followed by field guides. The first five chapters prepare the reader to visit volcanoes in general, while the field guides provide detailed guidance on what do and see on specific volcanoes.

The introductory chapters discuss how volcanoes work and address the practical aspects of planning one's own volcano field trip. A whole chapter is devoted to the most serious concern that travelers have about volcanoes, particularly active ones: safety. The potential visitor has to consider questions such as: "How dangerous is an erupting volcano?" "What precautions can I take?" "What happens if the volcano blows up unexpectedly?" "Are there ways to view or photograph an eruption safely?' Stories of how the curious came to grief are part of the lore surrounding volcanoes and one should certainly be concerned about potential dangers. Even dormant volcanoes can be treacherous and should always be approached with caution.

Obtaining information on volcano safety has been, up to now, rather difficult for the non-professional. Popular books about volcanoes do not discuss how to avoid dangerous situations because most of their readers would not need to know this. Even at the specialist level, very little has been written about safety on volcanoes. In general, budding volcanologists learn about avoiding danger from their more experienced mentors. One of the reasons why few people have attempted to write about volcano safety is that it is rather difficult to stipulate rules. Individual volcanoes often have their own quirks and, besides, people tend to have different comfort levels as far as danger is concerned. I cannot decide for others what level of risk is acceptable to them but I can – and do – attempt to explain fully the dangers a visitor might encounter on a volcano. I have also shared my own safety guidelines and what several colleagues and I have learnt from personal experience. The key message about volcano safety is to learn as much as possible about the dangers of the specific volcano one is planning to go to. Once that is done, all visitors – from the cautious to the daring, the unfit to the athletic – can choose from a wide variety of ways how to make their trip safe, enjoyable, and rewarding.

Apart from the safety issues, the potential visitor to a volcano should know a host of practical details such as when it is best to go to there, where the most interesting locales are, and how to get to them. However, such information about specific volcanoes tends to be hard to come by. With few exceptions, travel books do not devote much space to the local volcanoes. Geologic field trip guides can be invaluable but are generally too technical and available only on a limited basis. The field guides in this book attempt to bridge the gap between specialized geological guidebooks and standard travel guides by providing detailed, informal visiting guides to 20 of the world's most famous volcanoes.

Choosing as few as 20 volcanoes to focus on was not an easy task and I have no doubt that some people will be disappointed that their particular favorite was not selected. However, it was necessary to limit quantity in order to provide enough detail on each volcano to optimize readers' visits, while keeping the book sufficiently short (and light) to be taken along on the trip. Short descriptions of an additional 22 volcanoes and geothermal areas were included when they happened to be easily accessible from one of the volcanoes in the field guides.

I arrived at my "short list" of 20 by using my own personal experience of visiting and working on volcanoes, plus three criteria. The first was that the volcano be classified as active, though some of them have not seen an eruption in several hundred years. Even then, paring down to 20 from the world's approximately 600 active volcanoes was a problem. My second criterion was that the volcano should be easily accessible. For example, I expect that many readers would choose to visit Kilauea in Hawaii, but that most would not be willing (or able) to go to Mount Erebus in Antarctica, even though both volcanoes deserve to be seen. The ease of access, however, varies among the 20: some of the volcanoes are located within national parks where there are excellent roads and facilities, while others are off the beaten path. The summits of some can be reached by road or by an easy hike, while others are for the physically fit only. I have left out volcanoes where technical climbing expertise is called for. My aim was to write about accessible volcanoes that would suit the different tastes and expectations of a wide range of potential visitors.

The final criterion in narrowing down choices was variety: volcanoes come in several different types and their eruptions come in a variety of danger levels. The volcanoes selected range from mildly explosive types such as Kilauea in Hawaii to potentially very violent volcanoes such as Vesuvius in Italy. This requirement for variety also takes into account how often a volcano erupts, as eruption frequency is a factor of major importance to many visitors. There are those whose main reason to go to a volcano is to see some activity, while probably just as many would rather not go if an eruption is likely to happen. The volcanoes selected range from those that have been persistently active over the last few years to those that are not likely to erupt again in the near future.

I am hopeful that all readers will find at least one volcano in this book that they will feel inspired to go and see for themselves. I strongly believe that volcanoes should be visited and enjoyed by everyone – be they young or old, frail or fit, cautious or bold. The only requirements for a volcano traveler are curiosity about nature and a sense of adventure. The way – how to choose a volcano and how best to explore it – is what this book is all about.

Acknowledgments

Many friends and colleagues provided invaluable help and support for this book, which included teaching me about "their" volcanoes in the field, reviewing chapters, providing photographs and figures, answering questions, and giving much-needed encouragement. My sincere thanks to Guillermo Alvarado, Robert Carlson, Tim Druitt, Stephen Floyd, Charles Frankel, Henrietta Hendrix, Lucas Kamp, Susan Kieffer, Chris Kilburn, Gudrun Larsen, Adriana Ocampo, Scott Rowland, Stuart Malin, Bill Smythe, and Chuck Wood. Special thanks to Nick Gautier, who commented on the manuscript from a non-geologist's point of view and generously provided many of his photographs. I cannot thank Charlie Bluehawk enough for drawing so many of the figures in his spare time. I am truly grateful to all others who contributed photographs and helped with figures, including Elsa Abbott, Mike Abrams, William Aspinal, Pierre-Yves Burgi, Kathy Cashman, Frederico Chavarria, John Eichelberger, Jim Garvin, Magnús Guðmundsson, John Guest, Tom Mommary, Tom Pfeiffer, Vince Realmuto, Armando Ricci, Oddur Sigurðsson, Eysteinn Tryggvason, Rodolfo van der Laat, Ralph White, and Simon Young.

Simon Mitton, Susan Francis, Jayne Aldhouse, and Anna Hodson, my editors at Cambridge, guided this book through completion.

Last I thank my family: my son and great field assistant Tommy, my parents Atir and Walmir, and my sister Rosane, for their never-failing support.

This book is dedicated to my late friend and mentor Dr. Jon Darius, whose love for life and science remain my constant source of inspiration.

1 Volcanoes of the world

Why go to a volcano?

Active volcanoes are the ultimate adventure destination. Those who have been lucky enough to witness a fire-fountain spouting red lava high into the sky or billowing clouds of steam rising as lava pours into the ocean will testify that volcanoes provide one of nature's most awesome spectacles. Volcanoes are impressive, spectacular, and have a profound effect on life on Earth. Active volcanoes allow us to experience the thrill of hearing a loud explosion which can make the ground shake at our feet, to gaze at the cracking and shifting of a lava lake's surface, and to smell the strangely appealing sulfur odor. Even when they are dormant, volcanoes offer a wide variety of beautiful and strange sights, ranging from the majesty of snow-capped cones to the barren lava wastelands that have often been compared to Hell.

Volcanoes are definitely a destination for those of us who are adventurous travelers. We are no longer a peculiar group made up of daring eccentrics, but a growing number of intelligent people who are not content with spending our vacations in quaint seaside towns. Exotic and unusual destinations are increasingly within our reach. In recent years, the traveling public's growing interest in adventure as well as in ecology has led to a wave of "nature tours" ranging from safaris and whale-watching to shark diving and trekking in the Amazon forest. This trend should eventually make volcanoes one of the most popular destinations on Earth. A trip to a volcano has the potential to be a memorable adventure as well as a first-class lesson in how our planet works and even, in many cases, how history was made.

A trip to a volcano is a challenge to the mind because it gives us the opportunity to see a major geologic process at work. Out of the four fundamental geologic processes that shape the Earth's surface – volcanism, erosion, tectonism, and meteorite impacts – volcanism is the only one we can easily witness making rapid changes in the landscape. Volcanoes can be very cooperative: they can erupt gently for long periods of time, making it possible for us to plan a trip specifically to see the action. For example, the Italian volcano Strómboli has been nearly continuously active for centuries, while Kilauea in Hawaii is currently on its third decade of delighting visitors with an exceptionally "watchable" eruption.

Volcanoes shape not only the Earth's surface but also the course of human history. Eruptions have contributed to the downfall of civilizations, changed the course of wars, and, more frequently, destroyed whole cities killing thousands of inhabitants. On the positive side, volcanoes make fertile lands that are the source of livelihood for numerous people all around the globe. One of the most interesting aspects of visiting a volcano is learning how its eruptions have affected the local people and their culture. Equally fascinating is to find out how the current population views the volcano: feelings run from pride to terror, depending largely on the frequency and character of the predominant eruptions.

Even those who live far away from active volcanoes are vulnerable to their effects. Large eruptions, such as that of the Philippines' Mt. Pinatubo in 1991, can lower temperatures around the world. These eruptions inject large amounts of sulfur gases into the stratosphere, where they combine with moisture to produce a thin aerosol cloud. The cloud causes some sunlight to be deflected, causing slight decreases in average surface temperatures around the world. Although such changes are small (very rarely as much as 1 °C) and do not lead to major climatic effects, it is possible that large eruptions may have more profound effects on the Earth. There are indications that the amount of chlorine injected into the stratosphere by these eruptions may contribute to the depletion of the Earth's ozone

Fig. 1.1. Volcanic eruptions are some of nature's most awe-inspiring events and some, like this one from Kilauea volcano in Hawaii's Big Island, can be watched and photographed safely. (Photograph by the author.)

layer. This problem is still being studied and raises some serious concerns, because we cannot stop eruptions happening.

Given the major importance of volcanoes on the past, present, and future history of our planet it is not surprising that so many people become interested in them, often from a young age. Many children are as interested in volcanoes as they are in dinosaurs and space. Volcanoes have the distinct advantage of being neither extinct nor unattainable to most people. They are all over the Earth to be visited and explored. For those still wondering whether volcanoes are worthwhile traveling destinations I offer the following reasons: volcanoes deepen our understanding of how the Earth evolves and how humans interact with nature's forces. Volcanoes are magnificent in repose and thrilling in action, when they allow us to experience the sounds and sights of nature at work. Volcanoes appeal to our intellect, our sense of adventure, our appreciation of natural wonders, and our fascination with danger (Fig. 1.1). As many youngsters would agree, a visit to an active volcano would rank second in excitement only to a trip into space – or maybe one to Jurassic Park.

Volcanic tours

What are the world's most interesting active volcanoes and where are they located? Our planet has many volcanoes that are considered active – about 600 on land and many more under the sea. On average, about 50 volcanoes erupt each year and about a dozen or more may be active in any particular month. Most people don't hear about these eruptions, either because they are small or because they occur in isolated places and do not have a significant local or global impact. The eruptions that grab the headlines are those which cause loss of life or major economic disasters. If you are determined to see an erupting volcano, you will need to choose either a persistently active volcano (such as Strómboli in Italy) or one that has just begun erupting, and hope that the activity will last until you arrive. Chapter 5 discusses how you can find out about current activity and how to choose a volcano to go to. For now, I'll borrow an idea from my late friend Peter Francis and take us on an imaginary journey to the Earth's most volcanic regions. We will follow the plate tectonic boundaries (discussed in Chapter 2) and break the journey into four volcano world tours: the Ring of Fire, the mid-Atlantic, Africa, and the Mediterranean. Time and money are no object; we will imagine ourselves to be wealthy volcanologists on sabbatical leave.

The Ring of Fire and the Pacific Ocean

Volcanically speaking, this is the Earth's busiest side. The volcanoes that mark the Ring of Fire – over 1,000 of them – are located in no fewer than four continents and span a wide variety of scenarios, climates, and cultures (Fig. 1.2). A good starting point for our journey is the scenic North Island of New Zealand, one of the world's prime volcanic areas. This is where the most violent eruption in historical times took place: the Taupo eruption of AD 186. What we know about this cataclysmic event has been pieced together from geologic studies of the immense ash flow deposits, as no historical records exist. If people lived in the island at

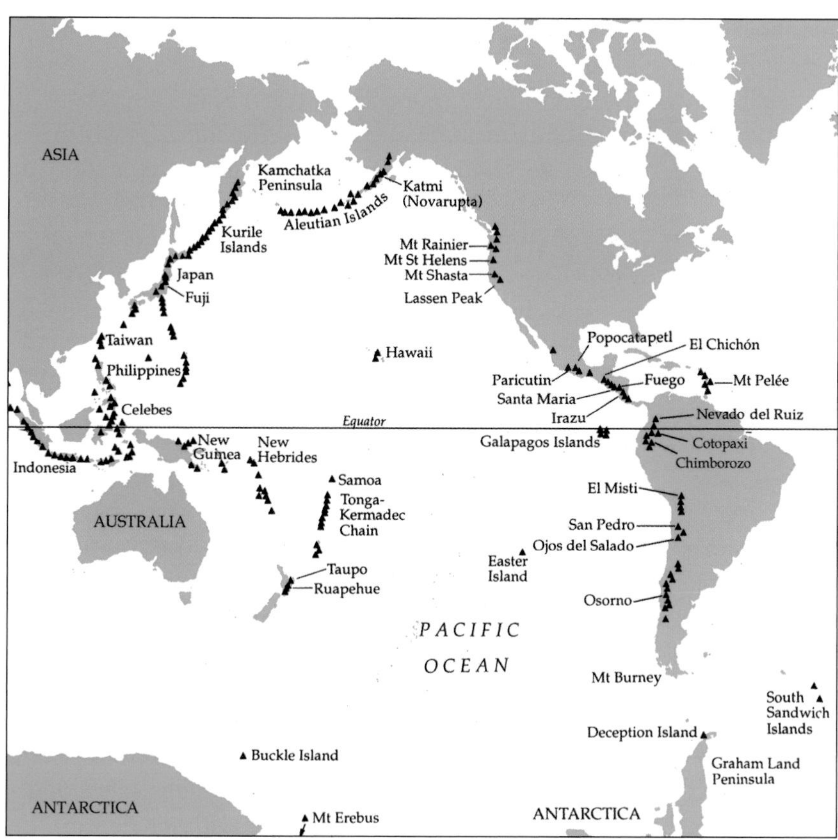

Fig. 1.2. The distribution of volcanoes round the Pacific Ring of Fire. These include some of the deadliest volcanoes known, such as Pinatubo, Krakatau, and Mt. St. Helens. (Modified from Francis, 1993.)

the time, they didn't survive to tell their story. These days the Taupo–Wairakei area offers thermal baths and geothermal power stations, and Lake Taupo Caldera as reminder of the eruption. Aside from the record-setting Taupo, North Island has other volcanic areas that have been active more recently, if rather more sedately. These include Ngauruhoe, a frequently restless and nearly perfectly shaped volcano and Ruapehu, whose latest activity was a small mudpool eruption in 2001. Ruapehu has hiccups every few years, sometimes sending mudflows (also known as lahars, an Indonesian word) down the slopes, but this has not stopped New Zealanders from building ski runs on the side of the volcano. Both Ngauruhoe and Ruapehu are located inside the Tongariro National Park which has good facilities for visitors. The Rotorua–Tarawera area offers a variety of volcanic landscapes, The Tarawera complex of rhyolite domes erupted spectacularly in 1886, burying three villages. This, however, was a minor event compared with what could happen again in the island. The Waiotapu thermal area is a well-known tourist attraction, and the town of Rotorua is within a large geothermal area. For the more adventurous, a visit to White Island is recommended. You can get there by boat or helicopter, but get some local information first and a guide. White Island is very active, there are mudpools and fumaroles, and new craters form often (there is one crater named Donald Duck). A strong and unexpected explosion in July 2000 covered half the island with a thick layer of ash and pumice fragments, but luckily nobody was visiting the island at the time. With so many interesting choices, New Zealand is a great country for volcano tours.

North of New Zealand the Ring of Fire starts curving around towards Asia, along the Tonga–Kermadec island chain, Samoa, and the New Hebrides. These are the South Seas volcanoes whose fantasy versions show up now and then in Hollywood movies such as *South Pacific*, *Bird of Paradise*, and, more recently, *Joe versus the Volcano*. The reality is somewhat different: the inhabitants of these islands are quite friendly and don't throw themselves or visiting volunteers into fuming craters. In fact, many of these volcanoes are nowadays on the quiet side and in no need of appeasing. Travelers who want to see some action should head to Yasur volcano in Tanna Island, part of the nation of Vanuatu. Yasur has been in almost constant but fairly mild activity since its discovery in 1774 and you can – with caution – climb right up to the top and look down at the erupting crater, and up to see the volcanic fireworks. Just don't try to take home rocks as souvenirs: native folks believe that every rock from their active volcano has spiritual significance and have confronted visitors who tried to take them away. Vanuatu has other interesting volcanoes, such as Ambryn, and it is a great destination for the adventurous, as the islands are well off the beaten tourist path. While in this part of the world, visitors may hear about the fascinating Falcon volcano in the Tonga Islands. This undersea volcano is famous for its "disappearing" islands, which are small ash cones formed during eruptions that are quickly washed away after the end of the activity. Volcanoes that come and go are not so uncommon in this part of the world. In May 2000, a lucky group of scientists came to the dormant underwater volcano of Kavachi and much to their delight found it spewing ash, forming a new, temporary beach that is truly far away from it all.

Continuing along the Ring we come to exotic Papua New Guinea, which boasts one of the world's great calderas, Rabaul. This is actually a group of small volcanoes clustered around the rim of a caldera bay. Rabaul had a large eruption in 1937 which generated disastrous tsunamis, killing 500 people. In 1994, Rabaul woke up again with a spectacular eruption from Vulcan and Tavurvur volcanoes. The town of Rabaul was greatly damaged and over 52,000 people had to be evacuated. Luckily only a few deaths occurred. Much more tragic circumstances resulted from the 1951 eruption of New Guinea's other notable volcano, Mt. Lamington, when glowing avalanches (*nuées ardentes*, French for "glowing clouds") devastated 230 km^2 (90 square miles) of land and killed about 3,000 people.

Further west from New Guinea we come to the volcanic wonderland of Indonesia, a country made up of over 13,000 islands and home to 76 historically active volcanoes. Some of the world's most notorious volcanoes are here: Krakatau, Tambora, Merapi (Fig. 1.3), Agung, Semeru, and Galunggung. These are big killers – past eruptions have claimed many thousands of lives. The Krakatau eruption of 1883 is thought by many to be the largest recent historical eruption but, in fact, the blast from Tambora in 1815 holds the record: it produced some 40 km^3 (10 cubic miles) of ash and magma fragments. These fell over thousands of square kilometers, killing crops and causing widespread famine. Krakatau, however, remains in the public's mind as one of the world's most infamous volcanoes and, for this reason, many visitors like to make their way there. Although the original island of Krakatau was destroyed by the 1883 eruption, a much smaller eruption in 1927 generated a new island – Anak Krakatau, meaning the "child of Krakatau" – which one hopes will not live up to its parent's reputation. It is not easy to travel in Indonesia at present because of the often volatile political situation and potential terrorist attacks, but this does not deter every traveler.

The safest place to go to is probably Bali, despite an isolated, recent terrorist attack. Bali is an idyllic island with interesting volcanoes such as Gunung Agung and Gunung Batur. Lombok is another popular tourist destination and it has Gunung Rinjani. Anak Krakatau is potentially dangerous, both because of its location and its mild-to-moderate activity in recent years. It is possible to get there by boat from beach resorts on the west coast of Java, but access has been limited in the last few years since the death of a tourist in 1996. The highest volcano in Java is Semeru, which rises to 3,676 m (12,060 feet) elevation. It is a very active and dangerous volcano; in August 2000 it killed two volcanologists from the Volcanological Survey of Indonesia and injured six others who were on a tour for professional scientists. This is not a place to go without a local guide and, even then, the risk is high. The same could be said about other majestic Indonesian volcanoes such as Merapi, one of the country's most active, which stands over the densely populated city of Yogyakarta. Among the dangers posed by Indonesian volcanoes are lahars. The explosive nature of the volcanoes creates steep flanks and deposits of fine ash; heavy rainfall can lead to devastating and deadly lahars.

The next notable volcano country along the Ring of Fire is the Philippines, site of several extremely dangerous types such as Taal, Mayon, and Pinatubo. The 1991 eruption of Mt. Pinatubo was the world's third largest eruption in the twentieth century and made many international headlines. Mayon is the country's most active volcano. It is known for its symmetrical shape, but the placid beauty hides a rather restless interior. Explosive activity and the growth of a lava dome led to evacuation of people from areas adjacent to the volcano in 2000 and 2001. Taal is a volcanic caldera filled with a lake, with the small but very active Volcano Island in the center. The eruptions have caused disastrous lake tsunamis. Taal is easily accessible from Manila and resorts on the crater rim provide good tourist facilities. Although not the best country from a tourism point of view, the Philippines are worth a stop because of these and other notorious volcanoes.

Fig. 1.3. Merapi in Java is one of Indonesia's most active and dangerous volcanoes. The steep-sided stratovolcano dominates the landscape of one of Java's major cities, Yogyakarta. Merapi's eruptions have caused many fatalities and devastated agricultural land. This type of volcano, part of the Earth's Ring of Fire, is best visited while in repose. (Photograph courtesy of Vincent Realmuto.)

Next comes Japan, another country where the volcanoes are numerous and restless, with Unzen and Sakurajima being the most frequently active these days. Unzen claimed the lives of famous volcano chasers Maurice and Katia Kraff in 1991 (see Chapter 4). A less dangerous highlight for a volcano tour is Mt. Fuji, undoubtedly the most scenic Japanese volcano and famous throughout the world for its postcard-perfect beauty. Fuji seems to be resting since its last eruption in 1707, but it has a history of being restless – it is known to have erupted at least 13 times in the last thousand years. It is considered by the Japanese to be a sacred mountain and each year many pilgrims make their way to the summit. Go in the summer if you want to join them, in the spring if you would like to take your own postcard-style photograph of the volcano with cherry blossoms in the foreground. Asama volcano, in central Japan, tends to have small eruptions, but in 1783 it sent out deadly *nuées ardentes*. Visitors should beware of Aso volcano, where small explosive eruptions have occasionally killed tourists standing on the rim at the wrong time. Go to

Sakurajima for a chance to see some action; the volcano often has small eruptions that can be seen from the base of the mountain. For hot springs, Hakone volcano is the place to visit. It has a beautiful crater lake, varied volcanic features, and plenty of tourist facilities.

The Ring of Fire then stretches along the Kurile Islands and the Kamchatka Peninsula in Russia. These places are not renowned for their tourist facilities, but Kamchatka is rapidly changing. This volcanic peninsula was closed to Westerners until the fall of the Soviet regime but since then it has started to become a popular destination for adventure tours. The major volcanoes in the peninsula are Bezimianny, Karymsky, Kliuchevskoi (Fig. 1.4), and Tolbachik. All are quite active, erupting on average more than once a decade. Bezimianny (in Russian "the nameless one") was considered rather insignificant until 1955, when it woke up with one of the most violent eruptions of the century, sending clouds of ash 45 km (28 miles) above the ground and giant *nuées ardentes* which devastated more than 60 km^2 (23 square miles).

The Ring of Fire continues along the Aleutian Islands, where there are active volcanoes such as Kanaga, but these islands are not easy to get to, and the tourist facilities are limited to say the least. The next group of volcanoes that are reasonably easy to reach are those in Alaska. Highlights are Mt. Spurr, Redoubt, Augustine, Pavlof, Veniaminof, and Novarupta. These tall volcanoes can have very violent eruptions that, because of the sparse population, do not usually cause fatalities. Novarupta caldera had a very powerful eruption in 1912. The eruption, the largest of the twentieth century, was known for the huge pyroclastic flows that formed the Valley of the Ten Thousand Smokes. Although almost all of those steam vents (the "smokes") are now gone, the site has not lost its appeal – a classic lava dome can be seen inside the caldera.

Canada represents a gap in the rich eastern side of the Ring of Fire, as its volcanoes have not been historically active. However, the West Coast of the United States more than makes up for the short gap. The majestic Cascade volcanoes attract millions of visitors

Fig. 1.4. Kliuchevskoi is the highest and most active volcano in Russia's Kamchatka peninsula. The 4,835 m (15,863 feet), beautifully symmetrical volcano has produced frequent eruptions, which range from stream and ash explosions to outpourings of lavas. This area photograph shows the 1993 lava flow in the foreground, dark against the snow. Volcanoes in the Kamchatka peninsula are spectacular, but the area was closed to foreign tourists during the Soviet era and is still seldom visited. (Photograph courtesy of Vincent Realmuto.)

to their beautiful, well-run national parks. All of them are worthy of a visit, from Mt. Baker near the Canadian border to Lassen Peak and Mt. Shasta in California. Highlights include the menacing Mt. Rainier, the still notorious Mt. St. Helens, and the breathtaking Crater Lake. The latter is no longer active, but it is the archetypal volcanic caldera and a must for the volcano aficionado. While in California no one should miss a visit to Long Valley Caldera, which is considered a possible site for a catastrophic eruption in the future.

At this point, two detours from the Ring of Fire are called for: one to the Yellowstone geothermal area, where thousands of visitors every year marvel at Old Faithful and other spouting geysers, and another to Hawaii, the world's prime example of "hot spot" volcanism (discussed in Chapter 2). Hawaii is also the prime example of volcano tourism, thanks largely to Kilauea's Pu'u O'o eruption that started in 1983 and is still delighting visitors, even though it has killed a few of the many thousands of tourists who have flocked there since the eruption started. Kilauea is one of the most accessible active volcanoes and the highlight of the Hawaii Volcanoes National Park. Other Hawaiian volcanoes worth a visit are the gigantic Mauna Loa, the slumbering Hualalai, and the magically beautiful Haleakala. All these volcanoes are easily accessible with excellent facilities for visitors.

South of the US mainland the Ring of Fire continues along Mexico, another country that does not lack volcanic activity. Mexico's most threatening volcano is Colima (Fig. 1.5), but Parícutin is far better known because of the circumstances of its birth – it literally sprang up from a cornfield in 1943. Most visitors climb little Parícutin, but only serious climbers tackle Fuego de Colima, Mexico's most restless volcano, or its resting neighbor Nevado de Colima. Tough climbers will want to go to El Pico de Orizaba, Mexico's highest mountain, rising up to 5,700 m (18,700 feet). Other Mexican highlights are Popocatépetl and El Chichón. Popo, as the volcano is known by those who are not comfortable speaking Aztec, is the snow-capped, majestic volcano which dominates the skyline south of Mexico City. Popo woke up from a five-decade slumber in 1994 and has caused a lot of concern. Most of its explosions have been on the small side but a repeat performance of the powerful historic eruption of 1720 is possible. El Chichón was an obscure volcano until 1982, when it woke up briefly but extremely violently, causing 2,500 deaths and significant effects on the world's climate.

The Ring of Fire's surface expressions continue to be plentiful and notorious south of Mexico. One of them

Fig. 1.5. Colima volcano in Mexico, also known as Volcán Fuego or Fuego de Colima, is one of Central America's most active volcanoes. It rises to 3,850 m (12,631 feet) and can be climbed during times of repose, though it is safer to view it by climbing its older, snow-capped neighbor, Nevado de Colima. Fuego can be extremely dangerous. Its historic activity has included violent explosive eruptions, pyroclastic flows, and debris avalanches which have threatened the nearby city of Colima. The city and the surrounding region are rich in history and archeological remains. (Photograph by the author.)

is Guatemala's Santiaguito dome, which has been growing on the flanks of the extremely dangerous Santa María volcano since 1922 (Fig. 1.6). Santiaguito sends out thick, pasty lava flows that move slowly and occasionally some explosions and pyroclastic flows. The volcano is, unfortunately, not an easy or safe place to visit. Pacaya and Fuego, also in Guatemala, are two very active volcanoes that have mildly explosive (Strombolian) eruptions and are better choices for a visit. Eruptions from Pacaya can often be seen from Guatemala City, the country's capital. El Salvador has Izalco, which was known as the Lighthouse of the Pacific until the nearly continuous activity stopped in

Fig. 1.6. Santa María volcano in Guatemala (right) erupted violently in 1902, devastating a large area of the country. Since 1922, the Santiaguito lava dome (left) has been growing at the base of the 1902 eruption crater. The growth of the dome is marked by almost continuous minor explosions and often by lava extrusions and more violent events. Santiaguito is considered very dangerous, but depending on conditions at the time, the volcano can be viewed safely from a distance. In 1929, a large pyroclastic flow from Santiaguito killed at least several hundred people; some report as many as 5,000. Guatemala is home to several other active volcanoes. (Photograph courtesy of Vincent Realmuto.)

1966 – just as a volcano hotel was being built. Nicaragua is home to Cerro Negro and Masaya, two very active volcanoes that would be very attractive destinations if the country became more politically stable. Although maidens are no longer thrown into the lava lake at Masaya, visitors should remember that some Central American countries can be as volatile as their volcanoes.

An exception to the above caution about Central America is the tiny, ecologically-conscious nation of Costa Rica. This is an easy country to travel in, as its politics are peaceful and its tourism well developed. Costa Rica was one of the countries that pioneered the concept of "eco-tours" and has become a popular vacation spot destination for environmentalists. The country's most active volcanoes are the very active (and dangerous) Arenal, the spectacular Poás, the easily accessible Irazú, and the little-known Turrialba, located in an exceptionally scenic region.

At this point a Caribbean detour is called for. Many volcanoes rise along the Lesser Antilles arc, a result of subduction of the North Atlantic ocean floor beneath the Caribbean plate. Many people have heard of the infamous Mt. Pelée in Martinique and its tragic 1902 eruption, but there are several less well-known but equally interesting volcanoes in the region. Among them are three that share the name La Soufrière ("the sulfur producer"). The homes of the three Soufrières are the charming islands of St. Vincent, St. Lucia, and Guadeloupe, while Dominica has the Grand Soufrière. The volcanoes are easily accessible and the islands receive many visitors, but most of them do not venture up the volcanic slopes. The volcano that made the news during the last years of the twentieth century and continues to do so in the new millenium is another "sulfur producer": Soufriere Hills in Montserrat. Once a haven for visitors wanting to see an unspoiled Caribbean island, Montserrat has been devastated by the continuing eruption. It is hard to tell when tourism will return to the island, but while the volcano is still active it is an interesting choice for adventurous types who can forgo the usual tourist facilities. It may come as a surprise that the most active volcano of the Caribbean is a submarine one which bears the unusual name Kick-'em-Jenny. This volcano, near the island of Granada, was first spotted in 1939, when a black column of ash and steam rose almost 300 m (1,000 feet) out of a boiling sea. It has been growing and should form a new island in the near future. Keep watching the news and, when the volcano emerges, it's time to plan a visit.

The Ring of Fire continues down South America and we come next to Colombia, another volcanically interesting but politically problematic country. Two Colombian volcanoes grabbed the world's attention in recent years by having unexpected and deadly eruptions. Nevado del Ruíz caused one of the major volcanic catastrophes of the twentieth century, while Galeras volcano had a small but ill-timed explosion in 1993 which killed nine volcanologists. Just south of Galeras volcano is Ecuador, a country that owns not only two of the world's most majestic volcanoes – the much-climbed Cotopaxi and the often active Reventador – but also the Galápagos Islands. The Galápagos are all volcanic islands and some of the volcanoes are still very active. One of these is Fernandina, a shallow-sided volcano with a distinctively large summit caldera. Fernandina's last eruption was in early 1995 and another could start any time, as this is a volcano characterized by frequent Hawaiian-style activity. A popular choice for visitors is Isabella Island, home to Sierra Negra (whose summit can be reached on horseback) and Cerro Azul, the latter being one of the most active volcanoes in the Galápagos, but currently off-limits to visitors. Its eruption in 1998 created a serious threat to the island's rare tortoises and some had to be evacuated by helicopter, while others (weighing up to 225 kg, nearly 500 pounds) had to be carried by human rescuers across rugged terrain. Charles Darwin, who made the tortoises famous enough to be worth rescuing, also made important observations about volcanoes and ended up having an active one named after him: Volcan Darwin, also in Isabella Island. The Galápagos are an ideal destination for adventurous types who really want to learn about natural history, but visiting these islands requires some planning. The government of Ecuador has strict measures in force to protect this ecological paradise and the best way to go is to take one of the many organized tours. Only a few places in the islands are open to independent travelers; otherwise you must be in a tour or in the company of a Galápagos naturalist guide.

Continuing down the Andean chain we come to Chile and its imposing volcanoes exemplified by Villarrica and Calbuco, two of the most frequently active. Chile has beautiful national parks and facilities for visitors are very good, though climbing the volcanoes is not easy, as they rise to great heights. The 6,739 m (22,109 feet) high Llullaillaco has the honor of being the world's tallest historically active volcano. Travelers who are particularly adventurous may wish to go down to Cerro Hudson in Patagonia, a volcano that erupted violently in 1991, scattering ash as far away as Australia. The southernmost Andean volcano is Monte Burney, located at the tip of Patagonia, Chile. It is not an easy place to reach but visitors are rewarded with the knowledge that not many make it that far. The Ring of Fire goes on further south to the far-flung South Sandwich Islands, but most travelers would not venture that far. The Ring finally reaches Antarctica at the isolated and grim Deception Island, whose eruption of 1969 destroyed a research base and caused much alarm among the local penguin population.

Volcanoes of the Atlantic

The volcanoes here result from the mid-Atlantic ridge's intense sea-floor spreading activity (Fig. 1.7), a rather different tectonic setting from that of the Ring of Fire. The southernmost active volcano in the mid-Atlantic chain is Norway's Bouvet Island, an uninhabited and almost inaccessible place, probably the most remote volcano in the world that is still considered active. The island, discovered in 1739, had a major eruption about 2,000 years ago, or so magnetic dating tells us. Further north is the isolated island of Tristan da Cunha. It may not get many visitors, but it is known to be a rather interesting place. Its eruption of 1961 forced the evacuation of the small population (a few hundred people) but since then the people have returned, the volcano has remained quiet, and the island has started to appear on one or two cruise brochures. Tristan has a nearby neighboring volcanic island that bears the rather appropriate name of Inaccessible and is, as expected, uninhabited.

Going north, the next volcanic point of civilization is the island of St. Helena, last home of Napoleon and considered a maritime pit stop since it was discovered in 1502. The volcano is no longer active and the island does not offer much to visitors, who must get there by sea, as St. Helena has yet to catch up with the age of air

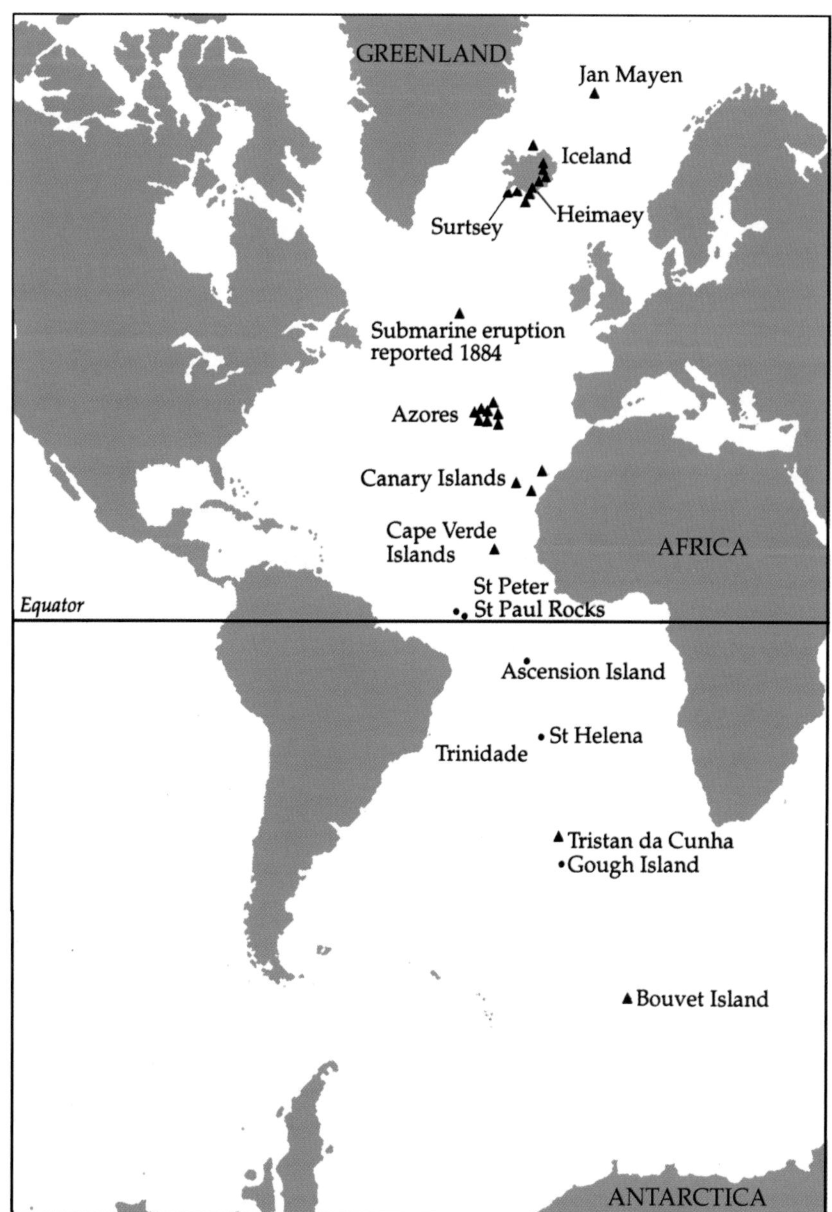

Fig. 1.7. The distribution of volcanoes over the mid-Atlantic ridge. These volcanoes tend to have relatively mild eruptions that are beautiful to watch. Volcanoes in the West Indies and the Mediterranean are not located over the ridge and are discussed in later chapters. (Modified from Francis, 1993.)

travel. Further along the ridge are several volcano remnants such as Ascension Island and the Rocks of St. Peter and St. Paul, none of which is easily accessible.

The island volcanoes get more interesting and the port stops more frequent as the ridge enters the northern hemisphere. There are three rather beautiful groups of islands very popular with visitors: the Cape Verde Islands, the Canaries, and the Azores. The Portuguese Cape Verde Islands are home to Fogo (Fire), a volcano with a magnificent caldera about 8 km (5 miles) wide. Fogo lived up to its name three times in the twentieth century. The last eruption, in 1995, forced the evacuation of 3,000 people, many of who had their homes destroyed by lava flows.

Further north and near the African coast are the Spanish Canaries, a fantastic collection of volcanic islands which have Tenerife, Lanzarote, and La Palma as highlights. El Teide volcano in Tenerife is Spain's highest point at 3,715 m (12,188 feet) and is famous as the volcano that Columbus and his crew possibly saw erupting in 1492, although no one has been able to prove that the "great fire" was indeed an eruption. El Teide erupted in 1909, but it has been quiet since, much to the relief of resort developers. La Palma is a complex volcano with many eruptive centers that are often active, and a spectacular caldera. Teneguia volcano in La Palma erupted as recently as 1971, delighting many visitors. The last eruption in

Fig. 1.8. Iceland's volcanic activity is strongly affected by overlying ice and its meltwater. Grímsvötn, Iceland's most frequently active volcano in historical times, lies largely beneath the vast Vatnajökull icecap. This photograph shows the Grímsvötn eruption of 1998 viewed from the air in its fifth day (December 23, 1998). (Photograph courtesy of Magnus Tumi Guðmundsson.)

Lanzarote occurred in 1824, but another could happen any time. Most visitors to the Canaries go there for the sun and fun, but the volcanoes themselves are worth the trip. The popularity of the islands as vacation resorts presents some disadvantages, as many adventurous travelers are rather put off by the overcrowding. It is possible to get away from the crowds, but you have to try hard to do so.

The Azores islands compete with the Canaries as the prime Atlantic destination for those who appreciate both volcanic landscapes and good weather, but they are far less crowded. The Azores have several historically active volcanoes, including Fayal island, off which the submarine eruption of Capelinhos took place in 1958, and the unusually named Agua de Pau ("Tree Log Water") in São Miguel island, last active in 1564. Submarine eruptions off the islands are common and several of the vents have erupted repeatedly. Some of those have been named after local banks (such as Monaco Bank), a rather unusual practice but one that must make sense to the local population.

Aside from the weather, the highlight of the mid-Atlantic ridge is undoubtedly Iceland. This is the Earth's most volcanic country (Fig. 1.8) with no fewer than 22 active volcanoes, including Hekla, Krafla, and Eldfell in the tiny Heimaey island. Iceland is also famous for its spectacular glaciers, eruptions under ice, geysers, and for its overall geology, which is unique. This is where one can actually see evidence of the spreading action of the mid-Atlantic ridge, which is tearing the island in two. Iceland is also known for the resilience of its people in coping with volcanic disasters and even taking advantage of them, such as when a cooling lava flow was used as a source of heat for local homes. Travelers who are seriously interested in volcanoes should not miss a visit to Iceland, though they would be advised to go there during the summer.

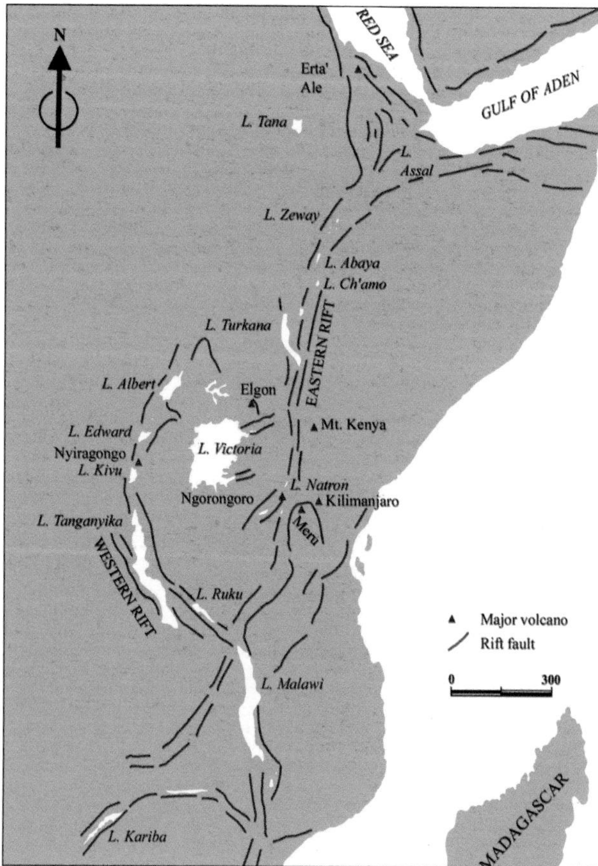

Fig. 1.9. Volcanoes in Africa include the famous Kilimanjaro as well as lesser known volcanoes that are hard to get to, such as Erta Ale. (Modified from Francis, 1993.)

The world's northernmost active volcano is Beerenberg, located in the small and rather bleak island of Jan Mayen. Beerenberg has been active five times since 1633, the last time in 1985. This Arctic island is not a hospitable place, though it is visited by more tourists than one might expect, as its remoteness appeals to adventurous types. Most people, however, would prefer to make a swift return down the ridge to volcanoes in more balmy latitudes.

Volcanoes of Africa

The East Africa Rift Valley, a site of intercontinental volcanism, has some of the world's most fascinating and remote volcanoes (Fig. 1.9). It extends from Ethiopia to Tanzania, crossing zones of major political unrest that can make travel awkward, not to say downright dangerous. It is easy to visit the Rift's northern end, the Red Sea, which is a product of the powerful tectonic forces that have formed the whole valley. However, traveling south from there is another matter. Ethiopia is a country dotted with active volcanoes, including the fascinating Erta Ale (meaning "Fuming Mountain" in the language of the Danakil tribes), a very active volcano whose summit caldera (Fig. 1.10) houses sometimes one, sometimes two lava lakes that may have been active long before they were discovered in 1967. Activity was still going in 2003 (Fig. 1.11), but this is a hard volcano to monitor. It is an even harder volcano to visit. In the early 1990s, a British volcanological expedition to the volcano got turned back by machine-gun-toting tribesmen who didn't seem to

Fig. 1.10. Erta Ale is Ethiopia's most active volcano. The broad, 50 km (30 mile) wide shield volcano rises 613 m (2,011 feet) from below sea level in the barren Danakil depression. Since 1967, or possibly much earlier, its summit crater has housed one, and sometimes two, spectacular lava lakes. Despite the dangers of travelling in the region, the long-term active lava lakes have attracted a number of intrepid visitors. In this view, the black, newly emplaced crust of the lava lake is broken by cracks and a small fire fountain is seen near the crater wall. The lava lake is about 120 m (400 feet) across. (Photograph courtesy of Pierre-Yves Burgi.)

Fig. 1.11. This spectacular nighttime view of the active lava lake in Erta Ale shows that is worth staying awake to observe and photograph volcanic activity. Glowing fresh lava comes out of cracks in the lake's crust and is typically exposed along the crater walls, as the movement of the underlying lava causes the crust to break up as it sloshes against the walls. This characteristic has been used by the author to help identify active lava lakes on Jupiter's moon Io. (Photograph courtesy of Pierre-Yves Burgi.)

Fig. 1.12. Ol Doinyo Lengai in Tanzania is one of the world's most enigmatic volcanoes, the only one that erupts exotic carbonitite lavas. The beautifully symmetrical volcano rises 2,890 m (9,482 feet) above the plains south of Lake Natron in Africa's Rift Valley. Although not an easy place to get to, Ol Doinyo attracts its share of volcano enthusiasts. In recent years, the volcano has hit the big screen – it was portrayed as the Cradle of Life in a Lara Croft movie. (Photograph courtesy of Pierre-Yves Burgi.)

care about science's noble interests. Such discouraging local attitudes might explain why Erta Ale is probably the least studied of the world's persistently active volcanoes, though in recent years several volcanological and other expeditions have gone there. The best way to monitor this volcano's activity is by using remote sensing satellites or scientists and volcano enthusiasts who dare to go. Even if the locals become consistently friendlier in the future, there is still the matter of Erta Ale's setting, the Danakil Depression, which gives a convincing impression of being hotter than the lava lake. The Danakil Depression is, however, a very interesting locale: it has some volcanic vents that are below sea level – the world's lowest land volcanoes in terms of altitude.

Traveling is easier further south, where the Eastern Rift crosses Kenya. Here we find visitors who came to see the wildlife in and around the collection of volcanoes. The most majestic is no longer active: Kirinyaga volcano, formerly known as Mt. Kenya, the second highest point in Africa. The volcano is famous for its African–Alpine environment, where giant lobelias flower and are pollinated by the attractive scarlet-tufted malachite sunbird. The Rift Valley holds soda lakes that are habitats for wildlife and popular with visitors. One example is Kenya's Lake Nakuru, where vast numbers of flamingos come to feed. The Rift Valley continues south and west, around Lake Victoria, into Uganda, Rwanda, and Zaire, and south to Tanzania, which boasts the unique Ol Doinyo Lengai volcano ("Mountain of God" in the Masai language), the world's only active carbonatite volcano (Fig. 1.12). Its lavas have a high proportion of sodium and calcium carbonate – and they erupt to form small flows of very low viscosity and rather low temperature (about 580 °C or 1080 °F). These flows travel only a few meters from their source and are certainly not a threat to life or property, though they have been known to come out quickly at night, threatening unsuspecting visitors who were camping in the caldera. The volcano is often active, and is a great destination for the truly adventurous (such as Lara Croft the Tomb Raider – the volcano was depicted as "the Cradle of Life"). Tanzania is also home to Africa's most famous volcano: the snow-capped Mt. Kilimanjaro, which rises to 5,895 m (19,340 feet), the highest point in Africa. This beautiful mountain was described by Ernest Hemingway in *The Snows of Kilimanjaro.* It is not now active except for steaming fumaroles. It is also the only volcano in the world known to be a birthday gift. The lucky birthday boy was Kaiser Wilhelm II of Germany, who received it from his generous grandmother, Queen Victoria. Still in Tanzania we come to another well-known but no longer active volcano, Ngorongoro, the largest caldera in Africa, with a diameter of some 20 km (12 miles). This spectacular caldera is a worthwhile place to visit because of its collection of wildlife. Not far away is Lake Victoria, the source of the White Nile.

Those wanting to see active lavas should head to Zaire on the western side of the rift, home of Nyiragongo and Nyamuragira, two of the world's most active volcanoes. Both have had sustained lava lake activity and frequent eruptions of fluid lava. Nyiragongo was the cause of many deaths and significant local devastation in 1977, when its persistent lava lake suddenly drained through a series of cracks. In less than 1 hour, 22 million cubic meters (800 million cubic feet) of runny lava rushed down the volcano's flanks at speeds of about 60 km/hour (40 miles/hour), like an incandescent overflow from a dam. About 70 persons were killed and some 800 left homeless. The threat of a repeat performance caused much concern during 1994, when masses of refugees from Uganda were encamped in the area. In 2002, the volcano sent out voluminous lava flows that inundated portions of the city of Goma, causing more fatalities.

It is unfortunate that, aside from a few destinations where eco-tourism has been developed, many of the Rift Valley volcanoes are located in inhospitable and often war-torn countries. However, the situation may improve. Public interest in the Rift Valley has been increased by popular science television shows and books, not only because of the valley's volcanoes but also for its amazing wildlife. Adventure and ecological tours to some of the volcanic areas are becoming easier to find and it is likely that some significant improvements in traveling conditions will be made over the next few years. Some areas in the Rift Valley are becoming major attractions, such as the volcanic Virunga Mountains, habitats of the mountain gorillas made famous by Diane Fossey and later by the movie *Gorillas in the Mist*. The Volcano National Park in Rwanda is a popular destination for international groups, though most people go to see the gorillas rather than geologic formations.

The Rift Valley is not the only place in Africa that has active volcanoes. The western coast has Mt. Cameroon, a massive volcano reaching 4,095 m (13,435 feet), the highest peak in West Africa, as well as the region's most active volcano. This volcano had a fairly conventional lava flow eruption in 1982, just when the movie *Tarzan of the Apes* was being filmed in Cameroon's jungle (lava does not show up in the movie). Cameroon is not exactly a tourism highlight of Africa, but my colleagues who went to study the 1982 eruption found the facilities for visitors adequate, aside from the fact that the best hotel rooms in the capital city, Douala, were occupied by the Tarzan movie crew. A few years later, in 1986, the nearby Oku volcanic field shocked the world when a rare escape of carbon dioxide gas from the bed of Lake Nyos killed 1,700 people as they slept in their villages. The tragedy demonstrated how volcanoes can kill in quiet ways, without fireworks or even disturbance of the landscape. Although toxic gases are known to come out of water lakes on volcanoes, the Nyos disaster was uncommon. Usually, there is sufficient mixing of the lake waters to prevent dangerous build-up of toxic gases near the lake bottom, as happened in Nyos. Why the build-up occurred is not understood. One explanation is that the lake had gone a long time without mixing, and when something (possibly an earthquake or a minor eruption in the lake) caused the waters to begin mixing, the toxic gases rose to the surface and were released as a deadly, ground-hugging cloud. A few years later, in 1984, an explosion and release of poisonous fog happened at the nearby Lake Monoun, killing 37 people, some of whom lost the outer layers of their skin. No one knows if either of these lakes will ever again give rise to such a catastrophe or, indeed, how likely it is that a similar event will occur in a volcanic lake elsewhere.

Although not part of the African continent, a noteworthy volcano worthy of a detour is Piton de La Fournaise ("Furnace's Peak") in the Indian Ocean's island of Réunion. A French island, Réunion can be appropriately described as being very much like Hawaii in its atmosphere, vegetation, and wonderful volcanic scenery, but with one great advantage: its French cuisine. Piton de La Fournaise is an extremely active shield volcano which normally has mild explosive activity and lava effusions every few years. It is an excellent volcano to explore, the only drawback being its isolated location which makes it an expensive place to fly to. However, those who have been there assure me that the expense is very worthwhile.

Volcanoes in the Mediterranean region

Here we find some of the world's better known and most often visited active volcanoes, several of which have been studied since antiquity. At the eastern end of the Mediterranean Sea, Turkey has Nemrut Dagi which erupted in historical times, while Greece has Methana, Nisyros, and the famous Santorini, the site of a cataclysmic eruption that may have destroyed the Minoan civilization.

Italy has a rich collection of active volcanoes including Vesuvius, probably the world's most often photographed and painted volcano. North of Vesuvius is Campi Flegrei (the "Flaming Fields") considered to be potentially even more dangerous than Vesuvius. Between the mainland and Sicily are the volcanic Aeolian Islands which include Strómboli, Lipari, and Vulcano, all very appealing destinations. Sicily has the mighty Mt. Etna, Europe's largest volcano and a very active one (Fig. 1.13). Those who want to go further afield can head for the island of Pantelleria just north of Africa's coast; this volcanic island was last active in 1891.

Fig. 1.13. Mount Etna in Sicily is Europe's largest volcano. Its frequent eruptions often delight visitors, but they can be dangerous. Whether in activity or repose, Etna is one of the world's best volcanoes to visit. This view, taken by the author in March 1981, shows the upper slopes covered with snow. Etna is a popular skiing destination in winter, though eruptions have wrecked havoc with the tourist facilitites. (Photograph by the author.)

The Mediterranean volcanoes offer excellent facilities for visitors, the only drawback being that they are often too popular – it's best to avoid the summer months. Visiting these volcanoes is a lesson in history as well as in geology. It is here that the interaction between volcanoes and people has been documented since antiquity. It was by observing Mediterranean volcanoes that ancient philosophers first started to strive to understand volcanic eruptions. We have the benefit of not only early theories but also records spanning many centuries, an invaluable scientific tool in the search for the truth about how volcanoes work.

Reference

Francis, P. (1993) *Volcanoes: A Planetary Perspective*. Oxford University Press.

2 The basic facts about volcanoes

What are volcanoes and how are they born?

The classic definition of a volcano is an opening on the Earth's surface from which magma emerges. Magma is molten rock containing dissolved gases and crystals, formed deep within the Earth. When a volcano erupts, magma – in the form of lava flows and lava fragments – builds a landform around the opening. This structure, which is often a hill or mountain, is also called a volcano. Most people think of volcanoes as tall mountains, either placid, snow-capped cones like Mt. Fuji in Japan or thundering, violent mountains that belch out towering clouds of ash and cinders, as Mt. St. Helens did in 1980. However, volcanoes come in many sizes and shapes and the character of their eruptions is highly diverse. Sometimes lava just pours out and floods a region, leaving no discernible hill, let alone a mountain. To understand why different types of volcanoes occur, we first have to look at how they relate to the Earth's mosaic of tectonic plates. Plate boundaries are the preferred locations for volcanoes and the types of eruption tend to be similar along each boundary. This is why we find similar volcanoes along each "itinerary" on our world tour.

Tectonic setting of volcanoes

Volcanic eruptions are small manifestations of events taking place deep within our planet. Volcanoes have been called windows into the Earth because they provide many clues to what is going on below the ground. The first clue comes from the distribution of volcanoes around the world, as we saw in our volcano tours. There are some 600 volcanoes around the world that have been active in the last 10,000 years. All these volcanoes are considered active or potentially active, even though some have not erupted for a few thousand years, which is a short interval in geologic time.

Comparing the map of the Earth's tectonic plates (Fig. 2.1) with those from our volcano tours in the last chapter, it becomes obvious that volcanoes are concentrated along several narrow chains which follow the

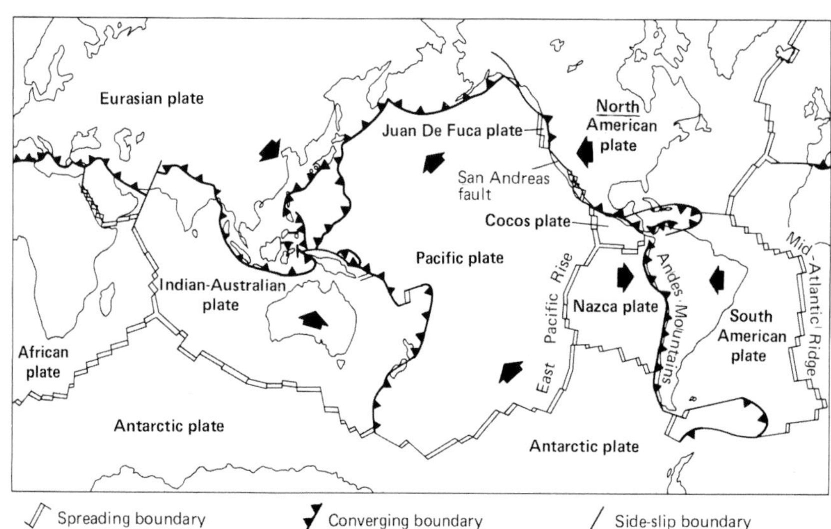

Fig. 2.1. The Earth's tectonic plates. Divergent boundaries, where plates are spreading apart, give birth to less violent volcanoes than convergent boundaries, where subduction zones are located. Arrows show direction of plate movement. (Modified from Decker and Decker, 1991.)

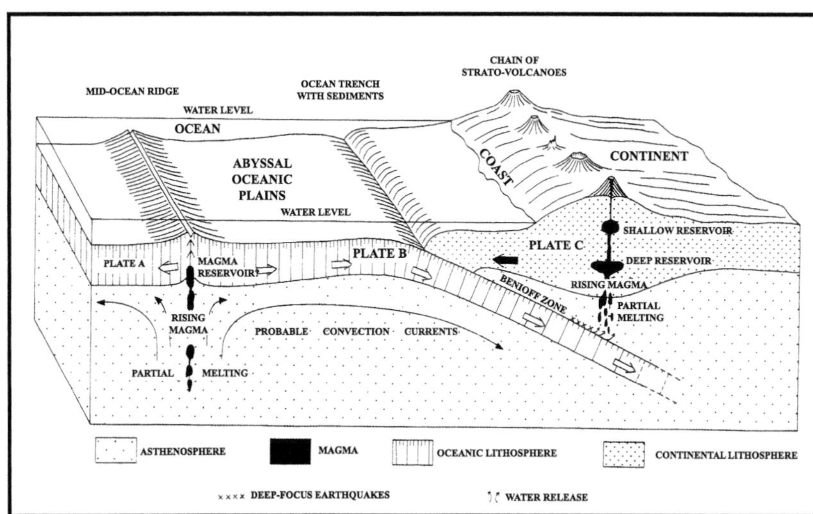

Fig. 2.2. Sketch showing the Earth's mid-ocean ridges and subduction zones. Different types of volcanoes are formed over these tectonic zones. (Modified from Scarth, 1994.)

boundaries of the Earth's tectonic plates. Over 94% of known historic eruptions have taken place along these boundaries. It is also not a coincidence that earthquakes also tend to occur along the same chains, because volcanoes and earthquakes are intimately linked to plate tectonics. They occur in places where the Earth's moving plates are being created and destroyed. In technical terms these places are called, respectively, mid-oceanic ridges and subduction zones (Fig. 2.2).

Tectonic plates form the Earth's cool, rigid crust, which overlies the hot interior. The zone beneath the crust is called the mantle. Heat from the interior causes slow circulation to happen in the solid mantle, which flows extremely slowly, like ice in a glacier. The mantle, flowing upwards, eventually reaches a level – at about 70 km (40 miles) below the Earth's surface – where the temperature is high enough and the pressure low enough to allow a small proportion of the rock to melt. The molten rock (magma) rises away from the surrounding solid denser rock and travels up to the surface through zones of weakness in the crust. Most of these zones are plate boundaries and most volcanoes occur along them – particularly over mid-oceanic ridges and subduction zones. However, the Earth is not a simple place and volcanoes also occur over regions named "hot spots" and in the middle of plates (called intracontinental).

Mid-oceanic ridges

The volcanoes on our "Volcanoes of the Atlantic" tour are islands that have grown tall enough to poke out of the ocean's surface. These and the multitude of submarine volcanoes are born when magma rising along the mid-oceanic ridge creates new crust as it is erupted. Mid-oceanic ridges are zones where two tectonic plates are slowly separating and new oceanic crust is being created by the extruded magma. This is by far the most dominant style of volcanism on Earth but most of it does not represent a threat to human life or property. The downside is that, unless you have access to a submarine, you cannot visit most of these volcanoes. Iceland, the Canaries, and other islands discussed in our tour are some of the few places where the oceanic rift system – in this case the mid-Atlantic ridge – emerges above sea level. Studies of Icelandic volcanism have been especially important for understanding of what goes on deep down in the mid-Atlantic ridge.

In recent years, research submarines have revealed much about the nature of deep-sea volcanic activity, including what the vents look like. In 1979, the manned submarine *Alvin* took the first photograph of an active volcanic vent on the sea floor, about 2.5 km (1½ miles) down on a ridge located off the Mexican coast. *Alvin* literally bumped into what looked like a smoking chimney, breaking off a piece of the column. *Alvin*'s pilot, Dudley B. Foster, attempted to measure the temperature of the vent using the submarine's plastic temperature probe, but it quickly went off scale and began to melt. The instrument had been designed to withstand temperatures up to 330 °C (626 °F) that was slightly below that of the smoking chimney.

Subsequent exploration by *Alvin* and other submarines have uncovered an amazing volcanic world of "black smokers" (Fig. 2.3), as the vents became known, which emit mineral-rich water at a speed of several meters per second. The "smoke" comes from the precipitation of the hot geyser's dissolved minerals

Fig. 2.3. "Black smoker" vents, such as these on the East Pacific Rise, serve as a home for varied species of marine life including the crabs and clams seen here. (Photograph courtesy of Ralph White.)

into the surrounding cold water. Cooler vents are called "white smokers" and they serve as a home for crabs as well as for some species of barnacles and other marine life. Marine biology has taken a new direction since the discovery of these amazing colonies. Some scientists think that life on Earth started in these submarine hydrothermal areas along mid-ocean ridges. These studies also have implications for other planets. Jupiter's moon Europa is thought to contain a liquid ocean underneath its icy crust. There are signs of perhaps recent volcanic activity on this moon's surface and the combination of heat plus ocean has led to much speculation about the possibility of life.

Deep-sea volcanoes on Earth pour out basaltic magma in a relatively quiet, gentle manner. Photographs taken by submarines have shown vast expanses of lava and numerous volcanoes on the ocean floor. These volcanoes outnumber land volcanoes by a considerable amount. It is possible that in the future we may be able to take submarine excursions to see these fascinating areas but, for now, we must be content with exploring the few which rise above sea level. The mid-Atlantic ridge is well endowed with volcanoes above water as was discussed in our "tour," from Jan Mayen at the north to Bouvet Island near Antarctica. Like their undersea neighbors, these island volcanoes erupt basaltic magma, the dark, fine-grained rock that forms the vast expanses of lava common in Hawaii and Iceland. Basaltic volcanoes are, in general, less dangerous than those that are composed of andesites and other more viscous magmas, which are found along subduction zones.

Subduction zones

It is clear that if the Earth's plates are spreading apart in the mid-ocean ridges, they must be either destroyed or pushed back down elsewhere to maintain equilibrium. The zones of plate destruction are called subduction zones and they occur where oceanic and continental plates are pushed against one another, such as along the Ring of Fire in the Pacific. What happens in these zones is that oceanic crust is pushed underneath continental crust and is carried steeply down into the mantle, together with sediments and water. This process generates a tremendous amount of heat. Portions of the oceanic plate and of the lower part of the continental plate are melted and move up towards the surface. These magmas tend to be more "evolved" (meaning modified) than basalt. It is not difficult to see why: magma from subduction zones can be considered second-hand, since it is produced by melting of crust that contains sediments and other materials. Although some of the resulting magma is still basaltic, other types are often produced, such as andesitic magma. The name andesite comes, logically enough, from the Andes, which form one edge of the Ring of Fire. Andesites have more silica (SiO_2) in their composition than basalts. Still higher in silica are rhyolitic and dacitic magmas, which are also found on subduction zone volcanoes. One effect of the higher silica content is to make these magmas more viscous than basalt. Instead of erupting as long, fluid lava flows as basalts tend to do, these magmas either form thick, pasty flows or erupt explosively. Krakatau and Mt. St. Helens come to mind as examples of explosive volcanoes from the Ring of Fire. Because of their explosive nature, eruptions from subduction zone volcanoes have the potential to be catastrophic. They are also the most common type of eruption we see from land volcanoes, a fact that may be to blame for the popular belief that volcanoes are always a threat to human life and should only be viewed from afar.

Mid-oceanic ridges and subduction zones are the key players in the story of how volcanoes form, but they don't explain everything. Some volcanoes are found far away from plate margins, either in the middle of an oceanic plate or far inland on a continental plate. Although they are fewer in number, these volcanoes are no less important to our understanding of what is going on deep down in the Earth.

Hot spots and flood basalts

Hot spots are thought to be long-lived mantle plumes. They originate deep in the mantle and feed basaltic magma to the surface through an overlying plate, away from plate margins. Hot spots can occur underneath either a continental or an oceanic plate. It is thought that the first outpouring of magma from a hot spot is its largest and can produce enormous quantities of runny basaltic lavas. All the known land examples of these eruptions, called flood basalts, occurred millions of years ago and formed vast lava plains such as the Columbia plateau in the northwestern USA. When a hot spot is located underneath an oceanic plate, repeated submarine eruptions occur until, after many millions of years, the pile of lava breaks through the ocean surface forming a volcanic island. The prime example of hot spot volcanism building islands is the Hawaiian chain. All the Hawaiian islands have been formed by a hot spot that has remained underneath the same location for at least the last 70 million years. At the same time, the Pacific plate has been moving towards the northwest. The result is that the hot spot did not form a single volcano but a whole chain in which the islands get progressively younger towards the southeast. Volcanoes in the Big Island of Hawaii, at the younger end of the present chain, are currently being fed by magma from the hot spot, Kilauea being the most active. However, a new island, Loihi, is slowly being built to the southeast and one day will take Kilauea's place as the most active site on these islands.

Intracontinental volcanoes

These eruptions are a bit of a mixed case: they also occur away from plate margins and some are caused by hot spots, but others are the result of rifting. Intracontinental is the least-understood volcano type because the process of magma coming up through thick continental crust is particularly complex. Small areas of intracontinental volcanism occur in the western USA, notably in eastern California and Arizona, where the tectonic setting is unusual because the crust is thinner than would be expected. However, the most dramatic example of intracontinental volcanism is the East Africa Rift Valley, which extends from the Red Sea, through Ethiopia and all the way down to Mozambique (Fig. 2.4). The Rift, which we talked about in the last chapter, is a long, narrow valley bounded on either side by geologic faults. The floor of the valley has dropped down along the bounding straight faults and is in some places 1,000 m (3,300 feet) deep. The volcanoes in the Rift Valley are mostly formed by alkali basaltic magmas. The term alkali denotes that the magma has more sodium and potassium than the more common ocean basalts. One of the best-known volcanoes in the East Africa Rift is Mt. Kilimanjaro in Tanzania; other volcanoes in the Rift Valley, discussed in our African volcano tour, are some of the world's most fascinating and least-understood active volcanoes.

How volcanoes erupt

Volcanoes are not all born equal. The tectonic setting of the volcano influences what type of magma and what type of eruption are produced. But although the characteristics of individual eruptions can vary widely,

Fig. 2.4. Part of Africa's Rift Valley is shown in this false-color radar image of Central Africa taken from the shuttle *Endeavour* by Spaceborne Imaging Radar–C/X-band Synthetic Aperture Radar (SIR–C/X–SAR). The mountains are the Virunga volcanic chain that extends along the borders of Rwanda, Zaire, and Uganda. The large volcano in the center of the image is Mt. Karisimbi. (Image courtesy of NASA/JPL.)

the basic principles of how volcanoes erupt are the same.

Magma comes up from the mantle along plate boundaries and other zones of weakness. Most volcanoes sit atop what is called a magma chamber, which is essentially a stopping and storage place. Some magma chambers are relatively shallow, just a few kilometers below the surface, while others are tens of kilometers deep. The amount of time magma remains in the chamber varies and can be several years. A variety of different factors, such as earthquakes, are thought to trigger eruptions, though the mechanisms are not well understood. What is known is that the molten magma is less dense than the surrounding rock and it eventually travels up towards the surface.

Magma contains dissolved gases, such as water vapor, carbon dioxide, and the pungent sulfur dioxide which gives volcanoes a characteristic smell. As the magma reaches lower temperatures and pressures, the dissolved gases begin to escape, much like the bubbles in a bottle of champagne as it is uncorked. The key factor that determines how explosive the eruption is the way the gases come out of solution. If the magma is very fluid and allows the gas to escape easily, the result is a relatively quiet eruption. However, if the magma is viscous and the gases cannot exolve easily they will eventually escape explosively by ripping the magma to shreds.

What makes a big bang?

How much gas is dissolved in the magma and how easily it can come out are the key factors determining whether an eruption will be explosive or not. If the magma is viscous but also gas-poor it can erupt quietly as pasty lavas forming stubby flows tens of meters thick. Extrusion of these high-viscosity magmas is often likened to toothpaste being squeezed out of a tube and volcanologists sometimes use the expression "toothpaste lava" to refer to some of these flows. Sometimes the lava is pushed upwards to form domes or spines (Fig. 2.5).

Because gases tend to migrate towards the top of a volcano's magma chamber, it is not uncommon for the initial phases of an eruption to be much more explosive than the later stages. In fact, quiet extrusions of pasty lava sometimes follow very violent eruptions. The most famous example is the tall spine that was formed after the fateful 1902 eruption of Mt. Pelée in Martinique. The spine reached over 300 meters (1,000 feet) above the volcano's summit and became known as the Tower of Pelée (see Chapter 13).

Magmas that are gas-poor and have low viscosity produce the least explosive eruptions, but that does not necessarily mean without some volcanic fanfare. Hawaiian magmas are typically of this kind, yet they produce some of the most spectacular eruptions on Earth. As the magma nears the surface, pressure from the rapidly expanding gases can spray the lava high into the air, creating the spectacular fire-fountains that are such a major tourist draw (Fig. 2.6).

Magmas that have low viscosity but are gas-rich have the potential to produce more explosive eruptions; however, the escape of gas can be very gradual and the magma can come out quietly. Some eruptions can start

Fig. 2.5. View of the Novarupta Dome, Katmai, Alaska, a classic volcanic dome. Novarupta was the vent for the 1912 eruption of Katmai volcano, the most violent eruption in North America in the last 100 years. The rhyolite dome, about 65 m (210 feet) high, was extruded into the vent after the eruption. (Photograph courtesy of John C. Eichelberger.)

off with only gas emission, which is later followed by gentle outpouring of degassed lava. Other eruptions can produce successive bursts of escaping gas that hurl lava fragments high above the ground. Despite that, eruptions of low-viscosity magmas are rarely violent and are usually the safest to watch and the most photogenic.

What makes a really big bang? The most lethal type of magma has the combination of high gas content and high viscosity. The high viscosity inhibits the release of dissolved gases until eventually the sudden release of pressure allows them to boil out explosively. Hot fragments, gases, and ash are violently ejected from the vent, producing the most destructive types of eruptions (Fig. 2.7).

The different types of magma

We have mentioned how the viscosity of the magma and the amount of gas dissolved in it determine how explosive an eruption is. But what makes magma gas-rich or gas-poor? What makes magma so viscous that it can builds a stubby lava flow tens of meters high, or so fluid that it can slosh up against the side of a steep cone?

There are several factors that determine magma viscosity but chemical composition, temperature, the amount of dissolved gases, and the proportion of solid materials (such as crystals) are the most important.

Starting with chemical composition, the key factor affecting viscosity is the amount of silica (which is found as SiO_2) in the magma. Basalts, often referred to as basic or mafic rocks, have a relatively low percentage of SiO_2. Basaltic magma comes from the upper regions in the mantle and erupts in relatively primitive, or unaltered, condition. Basalts are usually dark in color because of their component minerals, which are mostly feldspar and pyroxene, with smaller amounts of the green mineral olivine.

Other types of magmas – basaltic andesites, andesites, dacites, and rhyolites – contain progressively higher amounts of silica and they are often referred to as silicic, acidic, or evolved magmas. They are usually much lighter in color than basalts and contain mostly quartz and feldspar (Fig. 2.8).

Fig. 2.6. Fire-fountains are produced by eruptions of low-viscosity Hawaiian magma. These are spectacular events that can usually be viewed and photographed safely. This example is from Kilauea volcano in Hawaii. (Photograph by J. D. Griggs, courtesy of US Geological Survey.)

Silica content, together with alkali content, forms the basis of classification of volcanic rocks, as shown in Fig. 2.9. The silica content (SiO_2 percentage) is plotted in one axis and the sum of the amount of sodium and potassium is plotted as a percentage on the other axis. Most of these names will not be referred to in this book, so readers who are starting to find these explanation too detailed need not worry. It is, however, useful to know that volcanic rocks and lavas are grouped in this way.

How does the silica content of the magma affect its viscosity? Silica and oxygen atoms have strong bonds that form groups of interlinked tetrahedra, a phenomenon known as polymerization. These strong bonds make the magma more viscous or, in other words, less likely to deform or flow. The greater the silica content of the magma, the more polymerized it can be and, consequently, the more viscous. The presence of other minerals and gases can also affect viscosity either by encouraging polymerization or by

Fig. 2.7. Pinatubo is an example of a volcano that can erupt violently. (Photograph by Jon Major, courtesy of US Geological Survey.)

Fig. 2.8. Samples of common types of volcanic rock. Basalt (bottom right) is the darkest. Lighter in color is an andesite sample (hornblende andesite, bottom left). Still lighter in color are dacite (top left) and rhyolite (top right). (Photograph courtesy of Stephen Floyd.)

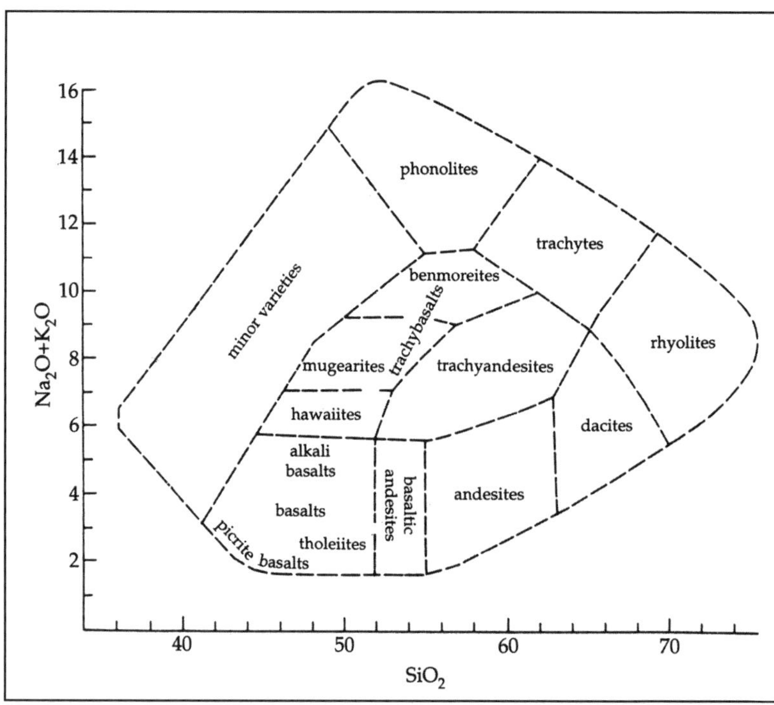

Fig. 2.9. Diagram showing the classification of volcanic rocks in terms of silica content (SiO_2) and alkali content. Basalts contain less silica than other common volcanic rocks such as andesites, dacites, and rhyolites. (Modified from Francis (1993), after an original by Cox *et al.* (1979).)

breaking down the bonds between silica and oxygen. For example, water in the magma lowers viscosity but carbon dioxide increases it. More alkaline magmas tend to be less polymerized and, therefore, flow more easily.

The amount and size of solid crystals also have an effect on viscosity: the greater the percentage of solid crystals per unit volume, the more viscous the magma. In basaltic lavas, crystals (mostly olivine and pyroxene) typically take up only a few percent of the volume, but in dacites the crystals (mostly feldspar) can form as much as 40% of the total volume. We can compare these to cake batter: basaltic magma would be like a white, fluffy cake batter, while a crystal-rich dacite would resemble a heavy fruit-cake mixture. The white cake mixture will be much easier to stir with a spoon and to pour into a baking pan.

Most people are aware that temperature influences the viscosity of fluids. This is why people who drive cars in cold climates use a different grade of oil in winter than in summer. The viscosity of magmas also changes with temperature: the hotter the magma, the more easily it can flow. How hot is magma? This, once again, depends on the composition, as shown in Table 2.1. Basalts erupt at hotter temperatures than the more silicic magmas. While basalt will be mostly solid below 1000 °C (1800 °F), the hottest rhyolites never reach that temperature.

Table 2.1. *Relation between composition and temperature of different magmas*

Magma composition	Temperature (°C)
Rhyolite	700–900
Dacite	800–1100
Andesite	950–1200
Basalt	1000–1250

Since viscosity decreases with temperature, the hotter basalts (those around 1200 °C, 2200 °F) will be more fluid than the cooler ones. If you have noticed that the temperature ranges in Table 2.1 overlap, you may wonder what happens if, for example, a basaltic and an andesitic magma are erupted at the same temperature (say 1000 °C). Will they have the same viscosity? The answer is no, because the more silicic magmas are always more viscous than the less silicic ones at a given temperature: a 1000 °C andesite will flow less easily than a 1000 °C basalt.

Lastly, let's consider gas content. Without a significant amount of gas, even the most viscous magma will not make a big bang, but just ooze out slowly. The word "gas" is used here in a loose way. Volcanologists tend to refer to gases as volatiles – a term that also includes water which will come out of the magma as steam. The amount of volatiles in the magma is linked to composition: basalts typically have less than 1% of volatiles per unit volume, while silicic magmas have more – rhyolites may contain as much as 5%. Thus it is not hard to see why silicic magmas are potentially more explosive than basalt: they have more gas and are more viscous.

Looking again at the distribution of volcanoes around the world, we can better understand why explosive volcanoes are clustered along the Ring of Fire, while those with gentler eruptions are clustered along mid-oceanic ridges and hot spots. A volcano's explosivity is linked to the composition of its magma. The magma composition, in turn, is dependent on the volcano's tectonic setting: mid-oceanic ridge volcanoes erupt basaltic magma, while more silicic magmas are erupted along subduction zones. There are some exceptions to this general trend. Some subduction zone volcanoes are not particularly explosive. One example is Masaya in Nicaragua, which had predominantly gentle basaltic activity, though a sudden explosion in 2001 sent tourists scrambling (luckily, nobody was hurt). In contrast, some mid-oceanic ridge volcanoes, such as Hekla and Katla in Iceland, have had violent explosive eruptions in the past. In general, however, volcanoes sharing the same tectonic setting erupt in a similar way. As we will discuss in the next chapter, volcanoes that are siblings in terms of tectonics will tend to grow up to look alike and to show many similarities in terms of their eruptive personalities.

References

Cox, K.G., J.D. Bell, and R.J. Pankhurst (1979) *The Interpretation of Igneous Rocks*. Allen and Unwin.

Decker, R. W. and B. B. Decker (1991) *Mountains of Fire: The Nature of Volcanoes*. Cambridge University Press.

Francis, P. (1993) *Volcanoes: A Planetary Perspective*. Oxford University Press.

Scarth, A. (1994) *Volcanoes*. Texas A&M University Press.

3 Volcanic eruptions

The different types of volcanic eruptions

The term "volcanic eruption" can describe a wide variety of phenomena, from the gentle oozings of lava that sometimes go on for years to the catastrophic explosions that are quickly over but which impact the region for many years to come. Volcanologists tend to sort different volcanic eruptions by their "character": this is a loose classification that reflects how explosive the eruptions are. It was pioneered by G. Mercalli, who is best known for his scale of seismic intensities. The classifications have been revised many times and volcanologists still disagree on how many types of eruptions there should be and on what these should be called. The only point that everyone seems to agree on is that classifying eruptions by their "character" is not very useful scientifically, but that it is quite handy for the purposes of expedient description. Terms such as "Hawaiian eruption" and "Plinian eruption" immediately bring to mind a picture of how explosive and dangerous the activity in question is. The differences are even clear from the simple diagram in Fig. 3.1.

Mercalli started the tradition of naming the different types of eruptions after volcanoes where they occur often. Hence we have, in order from least to most explosive: Hawaiian, Strombolian, Vulcanian, Peléean, Plinian, and Ultraplinian types. The last two are named after Pliny the Younger, the Roman who observed and described the Vesuvius eruption of AD 79. A separate, special category is that of hydromagmatic eruptions, which happen when water comes into contact with hot volcanic materials.

Before discussing these various groups, it is useful to know that a more quantitative measure of the explosivity of eruptions has been developed. It is called the volcanic explosivity index, or VEI for short. The index was born out of a desire by volcanologists to measure the "bigness" of eruptions using numbers rather than relying on an observer's description (Fig. 3.2). The VEI attempts to minimize unreliable individual perceptions, such as one observer's "huge" eruption being another's "moderate." Such perceptions may depend on the individual's experience, level of fear, and other factors.

The VEI is, however, far from a precise number, as its determination still has to rely on some qualitative descriptions. Unlike in the situation of earthquakes, it is not possible to use instruments to measure the magnitude of an eruption. The index is based on information about the total volume and type of eruption products, the height of the eruption's cloud, descriptive

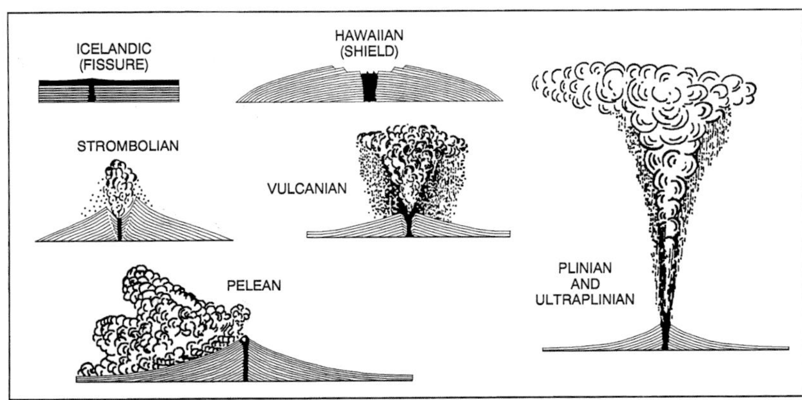

Fig. 3.1. Diagram of the different types of eruption. Icelandic and Hawaiian eruptions are the least dangerous to watch, while Peléean and Plinian are extremely destructive. (Modified from Simkin and Siebert, 1994.)

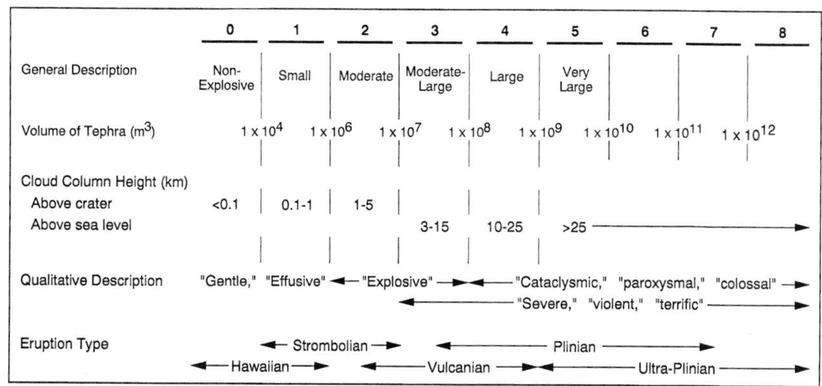

Fig. 3.2. Criteria for the volcanic explosivity index (VEI), expressed in the numbers 0 to 8 along the top. "Tephra" is a generic term for all volcanic fragments that are explosively ejected. (Modified from Simkin and Siebert, 1994.)

terms, and other factors (as illustrated in Fig. 3.2) which are combined to give a simple 0 to 8 number. At the lower extreme are the gentle effusions of lava which are assigned a VEI of 0 and at the upper end are cataclysmic eruptions which have a VEI above 5. Luckily for the Earth's inhabitants, the most common type of eruption has a VEI of 2, while eruptions having a VEI of 6 and above are quite rare. Only one historical eruption is known to have reached a VEI as high as 7: that of the Indonesian volcano Tambora, in 1815.

Use of the volcanic explosivity index complements rather than replaces Mercalli's classification, particularly because two eruptions very different in character can have the same value of VEI. Anyone seriously considering going to active volcanoes should know about Mercalli's major types of eruptions and, in particular, what tends to happen during each of them.

Hawaiian eruptions

Hawaiian eruptions are, in general, the safest and most beautiful eruptions to watch. This type of activity takes its name from, of course, the Hawaiian volcanoes, but it happens all over the world. The magma erupted is high-temperature, low-viscosity basalt, with a low gas (volatile) content. In other words, the magma is a typical oceanic basalt. The resulting eruption is largely non-explosive and rates most often 0, sometimes 1, in the volcanic explosivity index. Such apparent mildness can convey an unwarranted sense of security. One of my British colleagues used to joke that Hawaiian volcanoes were not for "real men" until he visited Kilauea during the early years of the eruption of Pu'u O'o. As our footsteps crunched the tops of the recently cooled, characteristic thin-crusted "shelly" lava, we were reminded by a more experienced colleague that hot, active tubes full of fresh magma were just below us and that we might actually step through the top of one. The thought of unpleasant burns was very sobering to my colleague who proceeded to pick his way rather daintily across the lava. I have not heard him underrate the dangers of Hawaiian volcanoes since.

What can one expect from a Hawaiian eruption? Very long lava flows, often reaching tens of kilometers, are one of the hallmarks of this type of eruption. The flows tend to be thin – a couple of meters or less in height – and very fluid, so they can easily cover tens of square kilometers. Flows often come out of long fissures rather than from a single "hole" or vent and this is one reason why the lava tends to spread over a vast area.

When a large fissure-produced flow is the major product of the activity the eruption is sometimes called "Icelandic," after what is probably the best historical example of this type of activity: the 1783 Laki eruption in south Iceland. This enormous lava field poured out from a series of fissures 24 km (15 miles) long and covered a total of 565 km^2 (218 square miles). In one day, the lava advanced 14 km (9 miles) – an astonishing rate for a lava flow.

Lavas from Hawaiian-type eruptions are typically far more sluggish than the Laki flow. The most common types of lavas on volcanoes are pahoehoe and aa, both of which are Hawaiian words. Pahoehoe is smoother and easier to walk on (that is, in fact, the translation of the Hawaiian word). Aa (a'a in Hawaiian) means a kind of Hawaiian "ooouch," named because it is difficult to walk on. A third type of lava, called blocky, is found on more silicic volcanoes. Pahoehoe lava (Fig. 3.3a) tends to form ropes on its crust as it flows and cools. Aa lava is more broken up (Fig. 3.3b). Pahoehoe lava can change itself into aa if the flow goes over a steep break in slope, such as that shown in Fig. 3.4.

For several years now, visitors to Kilauea in Hawaii have enjoyed watching flows slowly pour into the ocean, courtesy of the long-lived Pu'u O'o eruption. Pu'u O'o, a small cone on the side of Kilauea, has also

Fig. 3.3. The two most common types of lava flows: pahoehoe and aa. (left) The pahoehoe flow shows a characteristic ropy structure. (right) The aa flow has a more broken-up aspect. Both these active flows are from the Pu'u O'o eruption in Hawaii. (Photographs courtesy of Scott Rowland.)

Fig. 3.4. Pahoehoe lava commonly changes into aa when the flow goes over a steep break in a slope such as this one. Note the different textures of the flow at the top (pahoehoe) and bottom (aa). The flow is from the Pu'u O'o eruption. (Photograph courtesy of Vincent Realmuto.)

Fig. 3.5. The Kupaianaha lava pond on Kilauea volcano, Big Island, Hawaii. The photo was taken in 1987, a few years after the start of the Pu'u O'o eruption. The lava pond has now ceased to be active. (Photograph courtesy of Scott Rowland.)

produced the other two great sights of Hawaiian eruptions: lava lakes and fire-fountains.

Lava lakes are one of the most spectacular sights a volcano can provide. This is particularly true at night when lava glows bright red in cracks between constantly shifting cool slabs. Unlike many other volcanic attractions, lava lakes tend to be long-lived. For example, the Halemaumau crater atop Kilauea housed a lava lake for over a century: the lake was discovered by European missionaries in 1823 and it was active nearly continuously until 1924. It has been sporadic since then, but smaller lakes have appeared elsewhere on Kilauea, such as the Pu'u O'o eruption's lava "pond" of Kupaianaha (Fig. 3.5).

Lava lakes and lava flows are enough to make a visit to a volcano in the midst of a Hawaiian-type eruption an unforgettable experience. However, most observers agree that the crème de la crème of volcanic phenomena is fire-fountaining. Magnificent jets of red-hot lava are gushed up into the air like water fountains, often reaching 100 m (330 feet) high. The tallest fire-fountain on record occurred in 1986, not in Hawaii but on the Japanese island of Oshima – the lava reached 1,600 m – a mile high!

What mechanism can cause lava to be spurted up so high? Fire-fountains are not the result of explosions but, rather, of the rapid expansion of gases in the magma. Although basaltic magmas in Hawaiian eruptions are generally not gas-rich, gases often migrate to the top of the chamber and concentrate there prior to an eruption. This is why fire-fountains often happen at the beginning of Hawaiian-type activity, when the gas-rich magma accumulated towards the top of the chamber is erupted. The rapidly expanding gases easily fragment the fluid magma and the gas-and-fragments mixture is shot up into the air. Fire-fountains can last hours or even days, though they often fluctuate in intensity. Some Hawaiian-type eruptions start with a whole row of fire-fountains along a fissure – the so-called "curtain of fire" – a truly majestic sight for those lucky enough to see it.

Hawaiian-type eruptions are in general the most spectacular of all eruption types and also the least dangerous. Hawaiian-type activity rarely kills anyone and most of the known deaths could have been easily avoided. The worse recent disaster involving a Hawaiian-type eruption was the 1977 sudden breaching and draining of the lava lake atop Nyiragongo volcano in Zaire (see Chapter 1). Most of the time, however, Hawaiian-type activity can be watched more safely than any other type of eruption.

Where can one go to see Hawaiian-type activity? The Hawaiian volcanoes are, of course, prime locations and they draw thousands of visitors a year. Other good choices are volcanoes with lava lakes such as Erta Ale in Ethiopia (if you are quite adventurous) and mid-oceanic ridge volcanoes such as those in Iceland. Many of the eruptions from Réunion Island's Piton de La Fournaise and from the Galápagos volcanoes are Hawaiian-type, as are those from the African volcanoes Nyamuragira and Nyiragongo.

Strombolian eruptions

This type of eruption is named after the Italian volcano Strómboli, a tiny island in the Aeolian Sea. Strombolian eruptions have mild to moderate

Fig. 3.6. Strombolian-type eruption from a small cone on Mt. Etna, Sicily, in the winter of 1974. The snow on the ground reflects the glow of lava bombs ejected by the activity. (Photograph courtesy of John Guest.)

explosions punctuated by quiet "rest" periods. The explosions can eject magma fragments hundreds of meters into the air, though they are often less violent than that. Rest periods range from less than a minute to half an hour or more. This cyclic rhythm can last for years or even centuries – Strómboli has been erupting this way for over 2,500 years.

Strombolian eruptions (Fig. 3.6) are spectacular to watch and to photograph, particularly at night when trails of glowing red lava become visible. Like Hawaiian-type eruptions, they share the two traits volcano watchers want the most: spectacular to look at and unlikely to kill anyone. The many generations of Strómboli residents who have lived only a few kilometers downhill from the ever-active vent have rarely needed to worry about their safety. In fact, the island's economy is largely dependent on the tourists who keep coming to see the persistent fireworks.

The Strombolian type of eruption is generally fueled by basaltic magmas, but of a more viscous nature than those involved in Hawaiian-type eruptions. Instead of gushing fire-fountains, Strombolian activity is characterized by a series of explosive bursts that shoot magma high above the ground, usually accompanied by loud bangs. Some volcano watchers are suitably impressed by the noise, which makes them feel they are on a "real" volcano. The VEI of most eruptions in this class is 1 or 2. Lava flows often occur but they tend to be shorter and thicker than those in Hawaiian-type eruptions.

Although Strombolian activity can be watched in relative safety, it is important to remember that many of the fragments ejected by the explosions can be large enough to cause serious damage should they hit someone. Volcanologists have named these chunks "bombs" and, even though they don't explode, they can be quite lethal. Watching a Strombolian eruption safely requires a great deal of distant observation to determine the rhythm and pattern of the exploding vent and the range of the fragments. This is important for two reasons: first, one must be fairly sure that the explosions are not likely to get stronger and, second, one must stay out of the range reached by the falling fragments and bombs. Another note of caution is that the common, persistent type of mild Strombolian activity tends to be interrupted at intervals of months to years by brief periods of more violent activity, in which bombs may be thrown several kilometers away from the vent. It is not a good idea to visit a volcano during one of these violent bursts though it may be possible to watch some of the activity from a safe distance.

Despite these brief and usually unpredictable angry outbursts volcanoes having predominantly Strombolian activity are great destinations for the eruption-seekers. Luckily for us, these eruptions are quite common: they happen on basaltic volcanoes literally around the world, including Mt. Etna in Sicily, Arenal in Costa Rica, La Palma in the Canary Islands, Pacaya and Fuego in Guatemala, and Yasur in the island nation of Vanuatu. Fuji in Japan has had Strombolian activity, and so have Iceland's Heimaey and Askja. Travelers who are eruption-shy can visit Parícutin and Sunset Crater, both fine examples of volcanoes formed by short-lived episodes of Strombolian activity that is unlikely to start again. In spite of the wide choice, few would argue that the ever-thundering Strómboli is still the top volcano in this category. It is the ideal destination for those who simply must see a volcano in action – and photograph those myriads of red trails against the night sky.

Vulcanian eruptions

This type of activity is named for the Italian volcano appropriately called Vulcano. This mountain named not only one type of eruptive activity but also all of the world's volcanoes. Vulcano, a neighbor of Strómboli's in the Aeolian Sea, is a small island which has been erupting, on and off, for millennia. The ancient Romans believed that the blacksmith god Vulcan had his forge deep down the crater, where he crafted Mars' armor and Jupiter's thunderbolts.

Vulcanian eruptions are a step more violent than the Strombolian type, having a VEI between 2 (considered moderate) and 4 (considered large). The magma involved is more viscous than that in Strombolian eruptions. Basalts may be involved in Vulcanian eruptions but usually the magmas are more silicic: andesites, dacites, trachytes, and, occasionally, rhyolites. Thick, viscous lava flows sometimes form during Vulcanian eruptions, but explosive activity is by far the major product. The explosions are larger and noisier than the Strombolian type and the consequences may be far more severe: not only are the bombs generally larger and able to reach greater distances but these eruptions sometimes can destroy part of the volcano itself.

In a typical Vulcanian eruption, large quantities of ash rise up in a black eruption column which can reach 20 km (12 miles) above ground and can be deposited over a considerable area. Larger fragments of magma reach the ground as angular blocks because they are too viscous to round off during flight. Sometimes the fragments are solid when they leave the vent, because Vulcanian activity can start by blowing off the cap of solid magma which has formed over the molten magma.

Vulcanian eruptions are not particularly suitable for watching because the minimum safe distance may be a long way from the volcano. Sometimes, however, the activity is moderate and the explosions contained. In some eruptions, fine ash is ejected from the vent without any signs of an explosion. A Vulcanian eruption of this type attracted many visitors to Irazú volcano in Costa Rica between 1962 and 1965. The considerable amount of ash produced by the eruption killed the local coffee crops, but did not deter either visitors or locals from driving to the top of the volcano to witness the messy spectacle at close range.

Volcanoes which have produced less watcher-friendly Vulcanian activity include La Soufrière in St. Vincent, Merapi in Java, and Manan in Papua New Guinea. In Japan, the inhabitants of the town of Kagoshima have been getting used to the unpleasant ashfalls caused by the Sakurajima volcano, which has been erupting in Vulcanian fashion since 1955. On the whole, however, it is best to plan to see the results of

Vulcanian activity rather than the eruption itself. Today's visitors to Vulcano can climb to the spectacular summit crater, see hot steam coming out of the ground and even enjoy the dubious health benefits of a sulfurous mud bath heated by the volcano's still active entrails.

Peléean eruptions

This type of activity is named after the infamous 1902 eruption of Mt. Pelée in Martinique where it was first observed. The two hallmarks of Peléean eruptions are lava domes, formed by very pasty magmas, and *nuées ardentes*, the deadly mixtures of hot gases and magma fragments which can rush downhill at speeds of well over 100 km/hour (60 miles/hour). *Nuées ardentes* are also known as pyroclastic ("fire-broken") flows, meaning flows of broken magma fragments, ash, pumice, and gas (in volcanological terms, a *nuée ardente* is a pyroclastic flow of poorly vesiculated magma). This type of volcanic phenomenon has been responsible for much of the death and destruction caused by volcanoes.

Peléean eruptions are big killers because nothing can survive the onslaught of a *nuée ardente*. Many readers may recall the devastation caused by the 1980 eruptions of Mt. St. Helens, which flattened trees for miles and drastically transformed the once-verdant landscape around the volcano. The 1902 *nuée ardente* from Mt. Pelée destroyed the city of St. Pierre, taking 28,000 lives. The most tragic Peléean eruption in recent years was that of Mt. Lamington in Papua New Guinea, in 1951. The *nuée ardente* killed 3,000 people, all of whom were unaware that Mt. Lamington was an active volcano, let alone a lethal one.

What makes glowing avalanches so destructive? Like in Vulcanian eruptions, viscous magma is responsible for the ejection of an eruption cloud. The magma often builds a pasty dome inside the crater beforehand, indicating that trouble may be imminent. The pressure inside the dome builds up and it eventually explodes. Normally the resulting eruption cloud is ejected straight up but sometimes the blast is directed sideways. The cloud, consisting of hot gas, magma fragments, and dust, is propelled downhill under gravity rather than upwards. The gases and fine particles do rise up from the mass, but the heavier solid magma fragments avalanche downhill at great speeds. The temperatures in a glowing avalanche can be high enough to melt glass (700 °C, 1300 °F), and the destructive power is great enough to flatten trees and buildings in the avalanche's path. It is extremely unlikely that a human or animal will live through the onslaught of a *nuée ardente*, though it has happened. Two people are known to have survived the destruction of St. Pierre by virtue of being underground at the time (see Chapter 13). One was in his cellar and the other, ironically enough, was in prison.

Peléean eruptions are not very common, which is fortunate given that they can kill so efficiently. Many of the US Cascade volcanoes, including Mts. Shasta and Hood, have had them in their past and could have them again one day. Peléean eruptions have also occurred on the volcano La Soufrière in St. Vincent, Mayon in the Philippines, Merapi in Java, and Bezimianny in Kamchatka, Russia.

Can Peléean eruptions be watched safely? Although the answer is "sometimes," it is not advisable to try, unless a vantage point can be found which is located on high ground where *nuées ardentes* cannot reach. One possibility would be high up on a neighboring mountain to the volcano. Predicting the path of a *nuée ardente* is not easy, even if a volcano has been spewing them out frequently. *Nuées ardentes* can change path abruptly and even travel some distance uphill. This unpredictable behavior is what caused 42 people to be killed during the 1991 eruption of Unzen in Japan, when a lateral surge broke off unexpectedly from the main flow.

Sometimes the dramatic Peléean eruptions can end quietly with the growth of a lava spine in the crater, the best-known example being the Tower of Pelée (Chapter 13). If the danger of *nuées ardentes* is considered past, viewing a dome or spine in the volcano's crater, preferably from high up on the crater rim, is a not-too-crazy proposition. On the whole, however, it is far preferable to inspect the fascinating effects of Peléean eruptions well after the event. The thousands of annual visitors who come to see the ruins of St. Pierre are rarely disappointed, and none complains about Mt. Pelée's quiet state.

Plinian and Ultraplinian eruptions

Plinian eruptions and their more violent siblings, the Ultraplinian type, have the distinction of being named after a person rather than a volcano. The man in question, Roman citizen Pliny the Younger, was lucky to have been well out of the way of the most infamous eruption of all time: Vesuvius in AD 79. His uncle Pliny the Elder, a keen student of natural sciences, was less fortunate and died closer to the volcano, though it is speculated that his death was caused by a mundane heart attack. The younger Pliny documented the eruption in careful detail in two letters to his friend Tacitus, a historian. These letters are widely considered to be the first scientific description of an eruption and, centuries later, earned Pliny the dubious honor of naming the world's most cataclysmic eruption types.

Plinian eruptions have extremely powerful explosions that eject a great volume of fragmented magma, usually many cubic kilometers, which disperse over hundreds of square kilometers. The eruption cloud, coming out at speeds of hundreds of meters per second, can sometimes reach great heights – as much as 45 km (28 miles) above the ground. These eruption columns can reach the Earth's stratosphere, dispersing vast quantities of ash. Some of the ash falls back, blanketing huge areas, but some stays in the atmosphere causing the scattering of sunlight which makes spectacular red sunsets all over the world. Plinian eruptions can affect the Earth's atmosphere by forming aerosols that remain in the stratosphere for several years. Dust clouds in the atmosphere can also be a serious hazard to aviation, as demonstrated by the 1991 eruption of Pinatubo. In the month following the eruption, nine passenger jet planes had their engines damaged by dust from the eruption and were forced to make emergency landings. Since then, communication between volcanologists and aviation authorities has improved to warn pilots of the presence of eruption clouds.

Plinian eruptions can also cause considerable destruction by releasing pyroclastic flows and surges, but the ash-and-pumice fall is generally the major hazard. The pumice produced in these eruptions is indeed the rock many people have in their bathrooms. It forms when gas-rich magma froths up rapidly, so that it becomes saturated with gas bubbles as it solidifies. Much of the world's bathroom pumice comes from the island of Lipari in the Aeolian Sea, which was the site of violent eruptions in the distant past. Large pumice blocks can float on the sea and are sometimes hazardous to navigation.

Another danger of Plinian eruptions comes from the mixing of the ash deposited by eruptions with water – often rainfall or ice melts – to form thick mudflows, another deadly volcanic product (see discussion on hydromagmatic eruptions, below). Plinian eruptions are so violent that even the volcanoes that produced them often do not escape undamaged: the quantity of magma ejected is so large that it drains the magma reservoir, leaving the top of the volcano unsupported. The result is that this top part caves in, forming what volcanologists call a caldera. The diagram in Fig. 3.7 shows the before-and-after profile of Krakatau, the infamous Indonesian volcano that had a Plinian eruption in 1883. The Krakatau eruption had an estimated VEI of 6 (compared to the eruption of Vesuvius in AD 79 with a VEI of 5), which puts it in the Ultraplinian category. In pre-Roman times, an eruption of comparable magnitude took place in Santorini, Greece. Rather than wiping out towns, this eruption was supposedly responsible for the downfall of the whole Minoan civilization.

It is fortunate that Plinian and Ultraplinian eruptions are very rare. We can expect them a few times per century but where they will happen is anybody's guess. Many volcanoes around the world have had these violent eruptions in the past, including Mt. St. Helens in the USA, Colima in Mexico, Novarupta in Alaska, Fogo in the Azores, Tambora in Indonesia, and El Chichón in Mexico. Crater Lake in the US Cascades is the final product of a huge eruption in prehistoric times, which destroyed the former volcano known as Mt. Mazama.

Although Plinian eruptions occur mainly in volcanoes located over subduction plate margins, they can also happen on oceanic ridge volcanoes. Both Hekla and Askja in Iceland have had Plinian eruptions that were fueled by silicic magma formed under these volcanoes because of special conditions. Basaltic magma can produce a Plinian eruption, though this is extremely rare. One example where this happened was in the Tarawera volcano's eruption in 1886, on New Zealand's North Island. Another volcano in that same island, Taupo, had a cataclysmic Ultraplinian eruption in AD 186 but, luckily, this happened before the first Maori settlers arrived.

This brief description of Plinian eruptions should have made it clear that it would be exceedingly difficult – not to say potentially suicidal – to try to repeat Pliny the Younger's feat of witnessing such a violent eruption. A much more realistic alternative is to visit the volcanoes that have erupted this way in the past and to try to reconstruct in one's mind what must have happened. This is what millions of visitors have done as they stand in the streets of Pompeii and look up at Vesuvius.

Hydromagmatic eruptions

This special type of volcanic eruption happens when large quantities of hot magma and water come into contact with each other, with explosive results. This interaction can take a variety of forms, including eruptions within crater lakes, under ice or snow, or eruptions in shallow water at sea. All these are somewhat alike

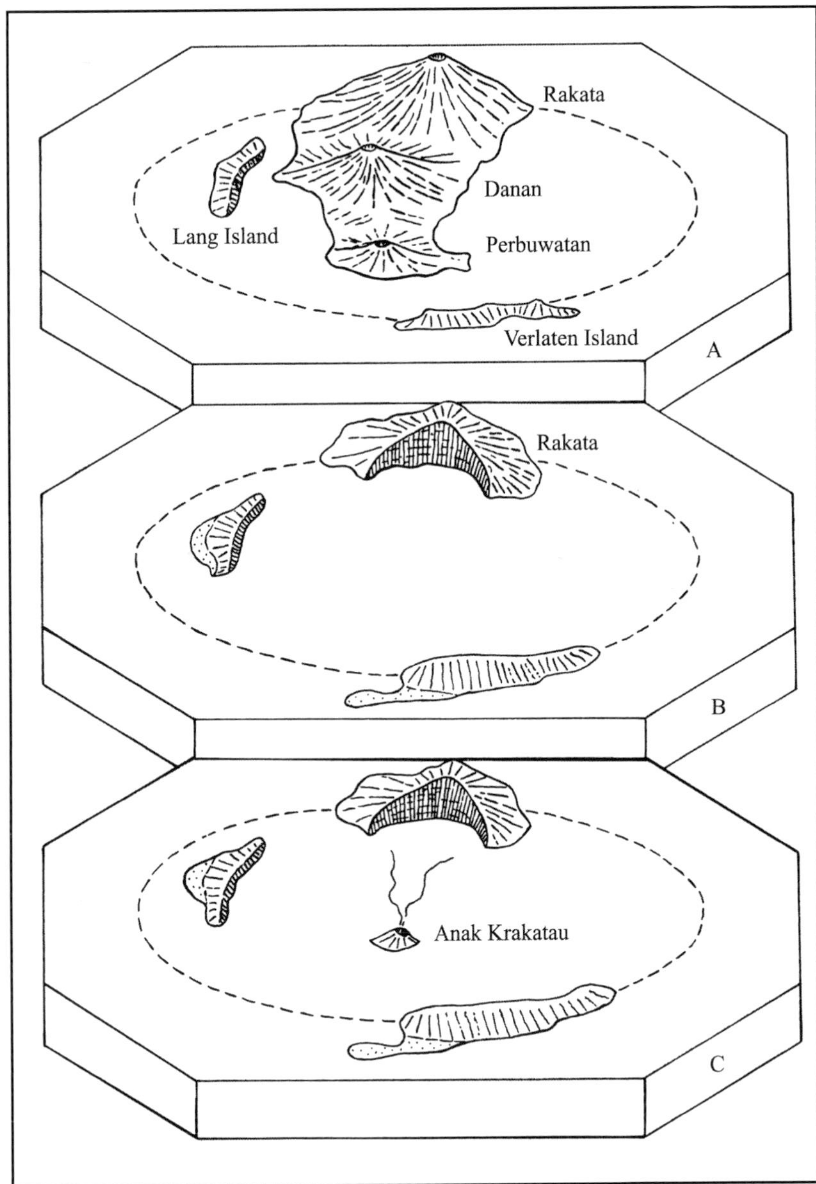

Fig. 3.7. Sketch showing the islands of the Krakatau group before the eruption of 1883 (A), after the eruption (B) and in 1960 (C), showing Anak Krakatau which is still active today. The dashed line indicates the approximate location of the submerged edge of the caldera. (Modified after MacDonald, 1972.)

and are generally referred to as hydromagmatic, hydrovolcanic, or phreatomagmatic eruptions (phreato is a Greek word meaning "water well"). Eruptions that are referred to simply as phreatic are those in which no new magma was involved, meaning that the eruption was triggered by still-hot magma from a previous eruption.

These eruptions can be highly explosive and, therefore, quite dangerous. When hot magma comes into contact with water, the heat of the magma turns the water to steam and the steam expands explosively, tearing the magma apart and producing great quantities of ash. Magma underground can also heat up nearby groundwater until the water flashes to steam, resulting in a violent blast. Most sensible people would stay away from hydromagmatic eruptions. However, now and then some volcano produces an eruption of this type that is suitable for watching from an appropriate distance.

One example was the Poás volcano in Costa Rica during the 1970s. For several years, Poas delighted tourists with small eruptions that took place under the murky waters of its shallow crater lake. The explosions shot plumes of muddy water into the air, sometimes spraying acid water droplets on tourists gathered at the crater rim to watch. Though the steady eruptions no longer happen, the crater lake at Poás still fumes away and small eruptions of mud and sulfur happen occasionally and are viewed by a lucky few.

Some of the most dangerous examples of hydromagmatic eruptions involve the interaction between magma and ice or snow. The effects can be quite devastating because the hot magma can melt large quantities of the frozen water, which then rushes downslope, mixing with ash and debris to produce mudflows. One of the most tragic examples was the 1985 eruption from the Colombian volcano Nevado del Ruíz. A relatively small summit eruption from this white-capped volcano was enough to melt large quantities of snow and ice and to generate a series of mudflows. These were channeled down narrow valleys on the flanks of the volcano and came down during the night. At the mouth of one of the valleys lay the town of Armero. Most of its 22,000 inhabitants never saw daylight again. The few survivors reported that waves of mud overtook the town, carrying away whole houses, cars, and people.

Mudflows are one of the major sources of devastation and tragedy caused by volcanoes. They can be produced by a variety of volcanoes and can be triggered even by minor eruptions. They are most likely to form on volcanoes that are steep-sided and snow-capped and which have large quantities of loose ash and debris on their flanks. Many of the volcanoes in the Cascade range in the USA are capable of producing devastating mudflows: Mt. Rainier has produced enormous mudflows in the past and the danger of repeat performances is very real. The glacier atop Mt. Rainier makes this volcano particularly hazardous, because torrents of meltwater could be released by an eruption.

The dangers of eruptions which cause glaciers and ice-caps to melt is well known to the people of Iceland, who have introduced the term jökulhlaup to volcanology (jökul means "ice-cap" and hlaup "deluge"). Jökulhlaups caused by eruptions from Grímsvötn volcano have released floods of water at outflow rates approaching 40,000 m^3 (about 1.5 million cubic feet) per second – higher than many of the world's greatest rivers. Luckily the glacier-bursts have been confined to the southeastern coast which is largely uninhabited, perhaps because the volcano-wise Icelanders had the good sense not to settle there.

Among Iceland's many volcanic wonders is the island of Surtsey (Fig. 3.8) (see also Chapter 11), another example of water and magma at work. The island emerged from the sea in 1963, a superb example of volcanic eruption under water creating new land. Surtsey's formation was a spectacle hard to duplicate. The eruption was first seen on November 14, 1963, by local fishermen out in a boat. They thought that the dark column of smoke might be a ship on fire. Luckily, they realized their mistake before getting too close for safety, as they saw tall jets of steam and ash rising above the surface of the sea. As is typical for shallow-water eruptions, powerful explosions rip the magma apart, sending ash and fragments hundreds of meters into the air, in a curious pattern known as a cock's-tail plume because of its arcuate shape.

Fig. 3.8. The eruption of Surtsey, Iceland. This photo, taken in 1963, shows clouds of white steam and black volcanic ash formed when hot lava meets seawater. Note the fire-fountain in the island, partly obscured by the clouds. (Photograph courtesy of John Guest.)

Surtsey's growth was rapid: about 24 hours later the new island projected its ashy head above the surface and by November 19 the new elongated volcano was 600 m (2,000 feet) long and had reached 43 m (140 feet) in height. Once the volcano was above sea level, its activity changed character, becoming a typical Strombolian eruption. Five months after its birth, Surtsey was a sizeable cone over 150 m (500 feet) above sea level and 1,700 m (5,600 feet) long. Despite the size of the island, the loose ash could have been easily eaten up by the continuously pounding waves. However, in 1964 a lava lake filled the island's crater and soon lava begun to flow towards the sea. For about

Table 3.1. *General relationships between volcano types, lava composition, eruption style, and common eruption characteristics*

Volcano type	Predominant lava	Eruption style	Common eruption characteristics
Shield	Basaltic, fluid	Generally non-explosive to weakly explosive	Lava fountains, long lava flows, lava lakes, lava pools
Composite	Andesitic, less fluid	Generally explosive but sometimes non-explosive	Shorter lava flows, pyroclastic falls, tephra falls, pyroclastic flows and surges
Composite	Dacitic to rhyolitic, viscous to very viscous	Typically highly explosive, but can be non-explosive	Pyroclastic falls, tephra falls, pyroclastic flows and surges, short lava flows, lava domes

Source: Tilling (1989).

a year, lava flowed out of the crater, covering Surtsey's loose ash slopes and assuring the island's survival. When all activity stopped in 1967, Surtsey had an area of 2.8 km^2 (1 square mile), all of which was to be carefully protected for a unique study to document the evolution of its geology and ecology. The studies go on to this day.

Can you plan to view a hydromagmatic eruption? In general, no. Because of their potential violence, these eruptions are not really suitable for watching. However, opportunities may strike occasionally. Visitors to Poás may be lucky enough to see some manifestations of the volcanic power underneath the lake. Surtsey-type eruptions are rare but, when they happen, they can sometimes be seen from a boat at a safe distance. Indeed, some Icelanders not only saw Surtsey in eruption but even took advantage of the late-stage activity to bathe in warm waters heated up by the lava flows!

The products of hydromagmatic eruptions are rather fascinating to see. Surtsey island still lives and can be viewed from a boat or plane. Iceland has a host of other spectacular features related to hydromagmatic eruptions, such as the products of glacier-bursts. On a smaller scale, visitors to the Big Island of Hawaii can often see a lava flow entering the sea and marvel at the notion that the island is growing before their eyes.

How volcanoes grow

Given that eruptions come in many different types it is not surprising that the volcanoes they build do not all look the same. The "typical" volcano pictured in most people's minds is a steep, convex cone resembling those we see in Japanese postcards or movies set in the South Pacific. This may be why some visitors to the Big Island of Hawaii are surprised when they gaze at the gentle slopes of Mauna Loa, which looks very much like the top half of a UFO. "It doesn't look like a volcano!" is the comment I once heard while standing on the terrace of the famous Volcano House Hotel, from where the unassuming shape of Mauna Loa can be viewed in all its glory. I knew at once that the disappointed tourist could not be a native of Iceland, where many volcanoes also look like upside-down saucers. I also wondered how native Hawaiians and Icelanders react when they see for the first time the impressive loom of a typical Ring of Fire volcano like Mt. Fuji.

The fact is that there is no "right" or "typical" shape for a volcano. What a volcano looks like is largely determined by its predominant type of activity and, in turn, by the composition of the magma and the volcano's location on the Earth's mosaic of tectonic plates (Table 3.1). It is therefore possible to tell a lot about a volcano just by looking at its shape – some people call this "reading the landscape." More formally, this is known as geomorphology.

How do the different shapes of volcanoes come about? Let's consider those Hawaiian and Icelandic volcanoes, like Mauna Loa and Skjaldbreid, which look so unimpressive to some. They are known as shield volcanoes because they look like shields dropped face-up on the ground. The term came, naturally, from Iceland and its past history full of shield-wielding, bloodthirsty warriors. The major products of Hawaiian and Icelandic eruptions are long, fluid lava flows. Imagine what happens when flow after flow, all coming out from a single source or from a line of vents, are layered on top of one another. Eventually these flows will build a concave, shallow-sided mound, as illustrated in Fig. 3.9. Some shield volcanoes are topped by large, flat-bottom craters produced by collapse. These craters can be rather spectacular when they are filled up by lava lakes.

If lava flows are erupted but they are more silicic and

more pasty than basalts, the result will be a lava dome. Domes are formed by the accumulation of thick, pasty flows and they are quite different in shape from shields, as can be seen from the diagram in Fig. 3.1. Domes are much smaller than shields, rarely exceeding 100 m (330 feet) in height. Their flanks are also much steeper, typically 25 to 30 degrees as opposed to the range of 4 to 8 degrees which is typical for shields. Another major difference is that domes are usually formed by a single phase of eruptive activity rather than by the accumulation of many eruptions. In volcanological terms, one-episode features or volcanoes are called monogenetic. They are, predictably, smaller than the polygenetic volcanoes that are the result of many eruptive episodes.

Another common type of monogenetic volcanic feature is the cone formed by magma fragments from Strombolian-type activity. Because the falling fragments of lava will tend to accumulate around the vent, most cones have steep and fairly uniform slopes. Most cones are small (tens of meters high) and grow on the sides of larger volcanoes. However, repeated bursts of explosive activity can produce larger cones, including some which are volcanoes in their own right, such as Parícutin and Sunset Crater. Although the activity in those volcanoes lasted for a considerable time, they are considered monogenetic and are not likely to erupt again.

The world's most common type of volcanoes is called stratovolcano or composite volcano (Fig. 3.9). These usually imposing structures are formed by the accumulation of magma fragments from explosive activity interlayered with lava flows. They tend to have symmetrical shapes with graceful upsweeping slopes and are what many people think of as "typical" volcanoes. In fact, there is a sound basis for such thinking because virtually all continental volcanoes are composite, including some of the world's best known: Vesuvius, Mt. Etna, Mt. Rainier, Mt. St. Helens, Mt. Pelée, and Mt. Fuji. Although they look majestic, these volcanoes are puny compared with oceanic giants like Mauna Loa. For example, Mt. Fuji in Japan has the largest volume of all composite volcanoes: 870 km^3 (210 cubic miles). It rises to 3,700 m (12,140 feet) and has a base extending across 30 km (19 miles). The volume of Mauna Loa, however, is larger than that of Mt. Fuji, even when only the shield's above-water volume is considered.

Because most of this book's volcanoes are composite, it is worth discussing them in a little more detail. Their upsweeping profiles are the result of the evolution of the volcanic activity as the volcano gets older. When these volcanoes are young their eruptions tend to come from the central conduit but, as they age, fractures open on the lower flanks from which lava flows come out. Explosive eruptions still take place at the summit and, gradually, the summit cone gets steeper relative to the lower slopes. Erosion also plays a part in accentuating the shapes of composite volcanoes by removing fine material from the upper slopes and

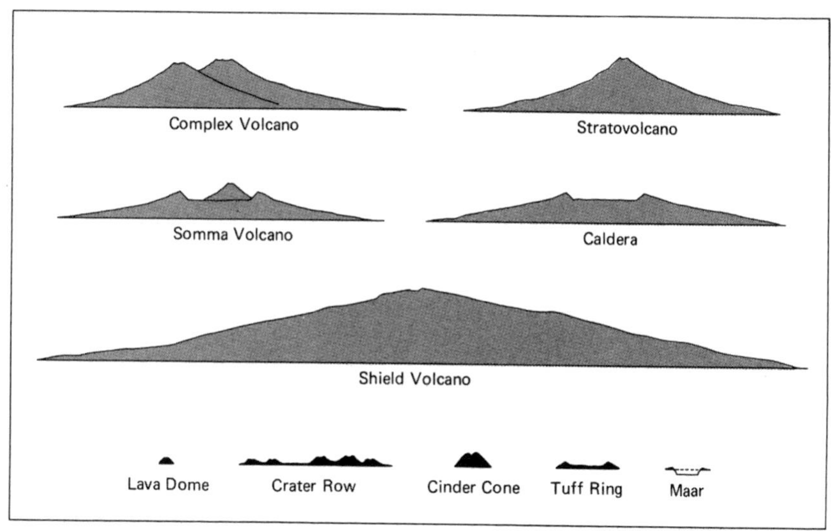

Fig. 3.9. The major types of volcanoes. Schematic profiles are vertically exaggerated by a factor of 2 (types shaded in gray) and by a factor of 4 (types in black). Relative sizes are approximate, as dimensions vary within each group. Somma and complex volcanoes are special types of composite volcanoes. Complex (or compound) volcanoes are multiple, genetically related volcanoes that grow in the same location. Somma volcanoes have collapsed calderas as a result of huge explosions, inside which new composite cones form at a later stage. Volcano types shaded in gray are much longer-lived than the other types. Complex volcanoes generally have lifespans of 1 to 10 million years, somma volcanoes and calderas from 100,000 to 1 million years, and stratovolcanoes from 10,000 to 100,000 years. In contrast, lava domes and cinder cones have lifespans from 1 to 100 years, while the other types generally last less than 1 year. (Modified from Simkin and Siebert, 1994.)

depositing it around the base, where it flattens the lower slopes. In general, the older the volcano and the longer it has been inactive, the more pronounced is its upsweeping profile.

Some composite volcanoes, like Hekla in Iceland, look quite different from the classic Mt. Fuji shape. This is because the eruptions from Hekla are fed from long fissures that cut across the volcano, rather than predominantly from a central vent. The number and location of a volcano's vents (which are places from where magma comes out) exert a major influence on the volcano's shape.

Vents can have the form of pipe conduits, coming out from depth to a single opening at the top, or they can have the form of long cracks or fissures, which can transect the volcano. Although some volcanoes have a single (pipe) vent, most have either a fissure system or else numerous separate vents which are often marked by small cones superimposed on the main volcano.

Single, central vents tend to produce volcanoes with symmetrical profiles, while elongated fissures produce overturned-canoe profiles, elongated in the direction of the fissure. A system of fissures or small vents that cut across the volcano in a well-defined direction is called a rift zone. Mauna Loa is a good example of a volcano with well-defined rift zones. Most of the volcano's eruptions start somewhere along these zones. In plan view, these volcanoes are elongated in the direction of the predominant rift zone.

Some volcanoes have lots of lateral vents, either of the fissure or the pipe kind. Eruptions on these volcanoes can start just about anywhere and, as a result, their flanks are usually dotted with cones. Mount Etna is a very good example of this type of activity. The "plumbing" underneath Etna is very complex and new vents have sprung up all over the mountain in what seems to be a disorderly fashion although, even here, there are some preferred zones. Etna's pattern of eruption is not very reassuring for the folks who live on the volcano's slopes as, in theory, a new eruption could start up right in the middle of town. In fact, numerous Etnean towns have been threatened by lava flows and some have been destroyed. However, this fact should not discourage anyone from visiting Mt. Etna, as volcanoes very rarely go bang in the night (or day) without precursor signs. It is, therefore, quite safe to stay in one of the charming Etnean towns despite the volcano's Mediterranean attitude towards orderly behavior.

Volcanoes come in many shapes, sizes, and forms of behavior. The diversity makes them fascinating and no two are totally alike. However, danger always lurks underneath any volcano considered active. Before setting out to explore volcanoes, it is important to understand the hazards and to know how to keep yourself safe.

References

MacDonald, G.A. (1972) *Volcanoes*. Prentice-Hall.

Simkin, T. and L. Siebert (1994) *Volcanoes of the World*, 2nd edn. Geoscience Press.

Tilling, R.I. (1989) *Short Course in Volcanology*, vol. 1, *Volcanic Hazards*. American Geophysical Union.

4 Visiting volcanoes safely

How dangerous are volcanoes?

Volcanic eruptions are some of the most feared natural disasters, but often they cause no significant damage. Eruptions don't usually make headlines unless they have caused major loss of life or property, or are so large that they have a global impact on the atmosphere. Many people, therefore, do not realize how frequently eruptions occur and how widespread they are on our planet. There are about 600 potentially active volcanoes in the world, plus a whole lot more hidden under the seas. On average, about 50 volcanoes have eruptions each year and a dozen or more may be active in any given month. Some of these not-talked-about eruptions may be spectacular to watch – or too dangerous to go near!

Just how often do volcanoes cause major loss of life? On average, a really catastrophic eruption happens a few times per century, though our statistics are limited by rather incomplete historical records, particularly for volcanoes in the New World and in very remote locations.

Table 4.1 shows the death tolls for the most notable eruptions during the last millennia. We can gather some interesting facts from this information, such as: (1) many of the world's most dangerous volcanoes are located in the Ring of Fire; (2) pyroclastic flows (*nuées ardentes*) and mudflows are by far the most deadly products of a volcanic eruption; and (3) eruptions often kill by indirect means, such as by triggering tsunamis or by destroying vital crops leading to widespread starvation. Ironically, volcanic ash acts as a very effective ground fertilizer and this is why the slopes of many dangerous volcanoes are heavily populated. Once in a while, however, a volcano extracts a high price for its bounty.

Compared with other natural disasters such as floods, hurricanes, and earthquakes, volcanic eruptions affect relatively few people. However, it is estimated that about 10% of the world's population (about 360 million people) live on or near potentially dangerous volcanoes and this number is on the increase. Civil authorities and volcanologists are increasingly working together to try to forecast eruptions, assess their potential hazard, and evacuate the local population if necessary. All of these are extremely complex tasks and are often beyond the means of the developing countries where so many of the world's most dangerous volcanoes are located.

The visitor to a potentially active volcano should ideally find out what the local standards of volcano monitoring and eruption forecasting are, and whether the volcano has shown any activity of late (see Chapter 5 for sources of information). It is also wise to understand what the hazards of that particular volcano are and what its eruptions have been like in the past. If the volcano is showing signs of activity that may be of the explosive, dangerous type, it is worth finding out if there are any plans for evacuation in place, in case the worst happens. It is important to remember that these standards and procedures vary widely from country to country and even from region to region.

The very minimum that a visitor to an active or potentially active volcano should know is what the most common volcanic hazards are and how they can be avoided – and these are the topics discussed in this chapter. Most visitors to volcanoes, including professional volcanologists, will never encounter some of the dangerous situations described here, but it is worth knowing what they are and how they can happen. When dealing with active volcanoes it is ignorance, not curiosity, that kills the most.

How volcanoes kill: two tragic events

Volcanoes can kill people in large numbers or in isolated incidents. Pyroclastic flows, mudflows, debris avalanches, and eruption-induced tsunamis are the usual culprits in the mass killings. Visitors to volcanoes

Table 4.1. *Some notable volcanic disasters since the year AD 1000 involving fatalities (figures rounded off to nearest ten)*

			Primary cause of death				
Volcano	Country	Year	Pyroclastic flow	Debris flow	Lava flow	Post-eruption starvation	Tsunami
Merapi	Indonesia	1006	1,000[a]				
Kelut	Indonesia	1586		10,000			
Vesuvius	Italy	1631			18,000[b]		
Etna	Italy	1669			10,000[b]		
Merapi	Indonesia	1672	300[a]				
Awu	Indonesia	1711		3,200			
Oshima	Japan	1741					1,480
Cotopaxi	Ecuador	1741		1,000			
Makian	Indonesia	1760					
Papadajan	Indonesia	1772	2,960				
Lakagígar	Iceland	1783				9,340	
Asama	Japan	1783	1,150				
Unzen	Japan	1792					15,190
Mayon	Philippines	1814	1,200				
Tambora	Indonesia	1815	12,000			80,000	
Galunggung	Indonesia	1822		4,000			
Nevado del Ruíz	Colombia	1845		1,000			
Awu	Indonesia	1856		3,000			
Cotopaxi	Ecuador	1877		1,000			
Krakatau	Indonesia	1883					36,420
Awu	Indonesia	1892		1,530			
Soufrière	St. Vincent	1902	1,560				
Mt. Pelée	Martinique	1902	29,000				
Santa María	Guatemala	1902	6,000				
Taal	Philippines	1911	1,330				
Kelut	Indonesia	1919		5,110			
Merapi	Indonesia	1951	1,300				
Mt. Lamington	Papua New Guinea	1951	2,940				
Hibok-Hibok	Philippines	1951	500				
Agung	Indonesia	1963	1,900				
Mt. St. Helens	USA	1980	60[c]				
El Chichón	Mexico	1982	>2,000				
Nevado del Ruíz	Colombia	1985		>22,000			
			65,140	53,900	28,000	89,340	53,090

Notes:

[a] Includes deaths from associated mudflows; however, the validity of the 1006 eruption has been questioned.

[b] Includes deaths from associated explosions and/or mudflow activity; estimates are unreliable and probably too high.

[c] Principal causes of deaths were a laterally directed blast and asphyxiation.

Source: Tilling (1989).

will not normally encounter these dangers, because (one hopes) the local authorities will have evacuated the danger area before a catastrophe happens. However, it is important to understand fully that fatal mistakes can be made even by professional volcanologists, as happened in two recent eruptions – those of Unzen in Japan and Galeras in Colombia. These two tragic events are proof that volcanoes do not always behave in ways that can be considered logical and the unexpected sometimes does happen.

Unzen, Japan: June 3, 1991

The volcano had been restless for over 6 months, after a rest period of nearly 200 years. Japanese officials and

Fig. 4.1. Damage to houses from the 1991 pyroclastic flows from Unzen. (Photograph by T.J. Casadevall, US Geological Survey.)

volcanologists around the world were in alert, as this is a particularly dangerous volcano. Located in the island of Kyushu in southern Japan, Unzen (Fig. 4.1) had last erupted in 1792, producing a tidal wave that killed 15,000 people – the worst known volcanic disaster in Japan's history.

The renewed seismic unrest was followed by small ash emissions and, on May 20, by extrusion of pasty lava from the Jigoku-ato crater, forming a lava dome which grew to over 40 m (130 feet) high. On May 24 a large explosion was heard, part of the dome collapsed, and the first in a series of pyroclastic flows came down the flanks, stopping just 2 km (just over 1 mile) from the town of Kamikoba. Heavy rains made the situation more hazardous because of the possibility of mudflows and thousands of people were evacuated from the area, though some were allowed back when the rains ceased. Mudflows continued to be the major worry for the local population, the danger of pyroclastic flows being confined to areas closer to the volcano where a "forbidden zone" was established.

The eruption attracted a number of Japanese journalists and scientists to the area. Amongst the volcanologists were the well-known husband-and-wife team, Maurice and Katia Krafft, who had come in the hope to film the pyroclastic flows. With them was American volcanologist Harry Glicken who, by a curious twist of fate, had missed being killed in the explosion of Mt. St. Helens 11 years earlier. Harry was supposed to man the observation post on Mt. St. Helens on the disastrous day but had to fly out of town for a meeting. His friend and mentor David Johnston replaced him and lost his life when the volcano erupted (see section "Mount St. Helens" in Chapter 8). Ironically, Harry himself would meet the same death at Unzen a decade later when, in his early thirties, he was nearly the same age as Johnston. These two men are the only two American volcanologists to date who have lost their lives in volcanic eruptions.

Harry Glicken, the Kraffts, and a number of Japanese members of the press ventured into the "forbidden zone" at the base of the volcano on June 3. At around 4 p.m. local time, an explosion shook the area, signaling the collapse of a portion of the summit lava dome. This was followed shortly after by a large pyroclastic flow, moving down a river bed at speeds of up to 100 km/hour (60 miles/hour). The flow traveled a distance along the river bed of over 4 km (2½ miles). Pyroclastic flows tend to follow topographic lows and it is likely that the observers, particularly the daring and experienced Kraffts, thought themselves to be relatively safe. However, an unexpected ash cloud surge was apparently detached from the main flow, engulfing the observers and destroying houses and trees, reaching the outskirts of the town of Kamikoba. Unable to outrun the surge, 42 people died including the three volcanologists.

The Unzen tragedy had an element of surprise – the unexpected surge from the main pyroclastic flow – but there is no doubt that the Kraffts and Glicken were well aware of the risks they were taking. The area was considered dangerous and pyroclastic flows are known to be the most deadly of volcanic products. They can attain great speeds, produce unexpected lateral surges and even travel uphill. Maurice and Katia Krafft were as famous for the tremendous risks they took as for the spectacular films and photographs of eruptions that they obtained.

A rather different, but equally tragic, event happened 2 years later in the other side of the world. Unlike Unzen, this volcano had been quiet and its victims did not know that it would soon wake up.

Galeras, Colombia: January 14, 1993

The 4,170 m (13,760 feet) high Galeras is one of the most active volcanoes in South America and the only one to be given the "Decade Volcano" status by scientists in the 1990s, meaning that the volcano was considered dangerous and recommended for intensive study. To help focus the attention of volcanologists on Galeras, an international conference was organized in January 1993. Many of those who attended chose to join a field trip to visit the object of their research. Galeras had been quiet and had given no indication of waking up (although some, with the benefit of hindsight, argue that there were signs of imminent danger). About 30 scientists toured the crater and some actually descended into it to sample volcanic gases. Stan Williams, an American volcanologist, led the party down the summit crater and into the inner crater inside a small volcanic cone. By early afternoon, the group had completed their sampling and started making their way out, with three lingering in the inner crater (one reportedly to smoke a cigarette). The blast came without warning, sending a black column of ash 2.5 km (1½ miles) into the air and showering the area with glowing bombs. Stan Williams and Andrew McFarlane were near the rim of the summit crater at the time – they tried to run but were knocked down by bombs, some still glowing. Dragging themselves, they managed to find some shelter behind rocks and their backpacks. They were amongst the handful who were injured but survived.

The heat inside the crater was so intense that a few of the bodies recovered later were burned beyond recognition. Visibility at the volcano's summit was less than a meter. In spite of that, geologist Martha Calvache and two of her Colombian colleagues, who had been further down the volcano's flanks at the time, rushed back to the crater and descended inside. The three Colombians managed to pull Stan and Andy further towards safety, where they waited several hours to be rescued by helicopter. The blast had killed nine people, including five volcanologists: three Colombian, one British, and one Russian. These five deaths remain the worst accident to have happened to the volcano science community. Colleagues around the world were stunned by the surprise of the event and its ill-timing: had the volcano exploded some 20 minutes later, there wouldn't have been any deaths.

Unexpected blasts are not by any means common, but they often have tragic consequences because people are not prepared for them. The Galeras event was a small one by volcanic standards, as was the 1979 explosion from Mt. Etna that claimed nine lives (see Chapter 9). The Etna explosion did not even throw out new magma; it was a mere volcanic hiccup. Unfortunately, volcanoes can easily cough up fragments that are large enough to kill.

Being showered by bombs is a situation where some knowledge and advance thinking about the situation may save one's life. It is true that nothing could have saved the unfortunate scientists who found themselves close to Galeras' inner crater, from where the blast came. However, in the case of Etna, the people were around the exploding crater where the density of bombs was much lower. It is estimated that there were about 60 people at the site. All of the nine dead and the 20 or so badly injured were tourists; none of the mountain guides was seriously injured though some reported dodging bombs. Although statistical chance surely played a part, some of the guides may have escaped the tourists' fate by knowing in advance what they should do to increase their chances of survival. The next section discusses what to do if a volcano explodes unexpectedly.

Survival rules for exploring volcanoes

Even volcanoes that are not erupting at a particular time should be approached with caution. There are conditions unique to volcanoes, such as certain types of unstable terrain, which make them particularly hazardous. Besides these, many volcanoes present the potential dangers associated with ordinary mountains, such as unreliable weather conditions at high altitude and remoteness. On the whole, visitors who employ common sense should not encounter problems; however, it is worth knowing about specific dangers that one might encounter on active or recently active volcanoes. Very little has been written about personal safety on active volcanoes, and people who regularly visit active volcanoes tend to make up their own rules based on their individual knowledge and the degree of risk they are willing to take. The rules below – for volcanoes and for eruptions – are my own. I have developed them not only from personal experience but also from many conversations with colleagues far more knowledgeable than myself. I hope that they will help to keep volcano visitors safe and to prevent them from making the mistakes, sometimes fatal, that others have made.

General survival rules for volcanoes

Volcano rule 1: Know before you go

Before leaving home, find out as much as possible about the volcano you plan to visit. No two volcanoes are alike, even if they are considered to be of the same type. It is particularly important to find out the following:

Current activity: is the volcano active and, if not, is it likely to become active soon?

Past activity: what were the eruptions like in the past – particularly the last few hundred years?

Monitoring: is the volcano regularly monitored? Are adequate warnings likely to be given before an eruption?

Local contacts: is there a volcano observatory? Is the volcano being studied by scientists from a local university? Are there local experts on the volcano willing to be contacted to give advice?

Services: what are the local emergency and search-and-rescue services? Are there any at all?

Accessibility: are there roads or trails into the volcano and how far up do they go? Are there hiking trails? What parts of the volcano are accessible considering your own level of fitness and physical ability?

Local information: are there adequate maps of the volcano? Are there guidebooks to the region that give some information on the volcano, lodgings, and local transport?

These questions are addressed for the 30 volcanoes highlighted in this book in their individual chapters. However, given that readers may want to visit other active volcanoes, and also the fact that information can easily become out of date (it need only take a new eruption), Chapter 5 discusses how to obtain information about active volcanoes in general. Gathering all the above information does take some effort, but ultimately it saves time, money, and possibly your life.

Volcano rule 2: Bring along safety equipment appropriate for the specific volcano

Rock climbers, mountaineers, and other outdoor-oriented people know the importance of safety equipment and wouldn't dream of leaving home without the appropriate gear. One of the problems with "volcano equipment" is that what is necessary will vary widely according to the type of volcano and the type of trip. A visit to the tourist spots of the Hawaii Volcanoes National Park requires no more than a sturdy pair of shoes, while a hike to the summit of Colima in Mexico calls for a gas mask, a climbing helmet, and all the usual mountaineering emergency supplies. Common sense dictates that the more remote the area, the more emergency equipment one should take.

High volcanoes are, in many respects, no different from other mountains and those planning to attempt difficult climbs should consult mountaineering and rock climbing books for advice on equipment. However, anyone going hiking on an active volcano – even one in a "fumarolic stage" such as Vulcano – should seriously consider taking along at least a climbing helmet and a gas mask.

Climbing helmets are easy to find and are well worth their weight in one's luggage. They offer protection not only in the case of falls, which happen easily enough when walking on rough terrain, but also in the case of falling volcanic bombs (as long as they are small) if an explosion should take place. An alternative to a climbing helmet is a construction workers' hard hat, which some of my colleagues prefer (Fig. 4.2). It has the advantage of leaving one's ears uncovered but does not offer as good overall protection.

Gas masks are not so easily available and most people would not think of taking one on a vacation. Although they are not often necessary, it is sensible to

Fig. 4.2. A well-prepared volcanologist on the edge of Mt. Etna's summit crater, wearing a hard hat, gas mask, and appropriate clothing. (Photograph courtesy of Christopher Kilburn.)

take one if one is visiting a currently active volcano, or one with numerous fumaroles. Volcanic fumes can be unpleasant and, in some instances, overpowering. Details about gas masks are given in Chapter 5.

While on the subject of equipment, I will add that it is a good idea to show respect for active volcanoes by dressing appropriately. Volcanic terrain is often rough and recently cooled lava flows tend to present sharp edges which can cut badly in the case of a fall. Aa flows are particularly unstable, as their top surface is composed of loose rubble. Sandals are not suitable footwear, although I have encountered a number of tourists hiking on Hawaiian volcanoes who seem to think otherwise. Since it is not uncommon for those who are not used to hiking on aa lava to fall down occasionally, it is also sensible to wear long pants rather than shorts and to put on a pair of gloves when hiking on particularly rough stretches. These precautions become much more important in the case of hikes over recently emplaced lava, which in some places can still be very hot. I once had a painful burn when I lost my balance on a days-old aa lava flow and instinctively put my hand down to stop the fall. I thought then, with some horror, of the number of local people I had seen earlier walking about on the same flow in their ordinary clothes – including some women who were wearing dresses! I have learnt my lesson since then, and now carry (and wear) leather gloves.

Volcano rule 3: Beware of falling through unstable terrain

Falling on aa lava may be painful, but other types of volcanic terrain can be far more dangerous. Although ordinary pahoehoe lava is generally quite safe and easy to walk on, shelly pahoehoe is another matter. This type of lava is relatively common on Hawaiian volcanoes and is the subject of many scary stories told by volcanologists who have worked there. One of my colleagues' experience is a good example: he was walking on a long-cooled lava flow on Mauna Loa, unaware that the lava underneath his feet was made up of a thin shell over a void, when the ground suddenly gave way underneath him. He found himself hanging by his elbows over what might have been an ugly drop. Luckily, was able to pull himself up without causing more of the ground around him to collapse. Although most of the time the drop underneath a shelly crust is only a few feet deep, a fall through one can still cause injury. Rugged clothes and footwear help to avoid cuts from the sharp edges of the lava. However, the best precaution is to skirt around areas of shelly lava or, at least, to walk near the edges of flow channels rather than over the middle part. The easiest way to recognize shelly lava is to listen for "crunchy" noises under one's feet and to look for signs of broken lava crusts (Fig. 4.3).

When walking around the edges of craters or lava lakes it is important to be aware that the terrain may be an overhang and potentially unstable. Furthermore, craters sometimes get larger by repeated collapses – look out for fractures around craters that may be a sign of impending collapse and stay on the safe side of them. The same type of danger exists where lava flows are entering the sea, as newly formed benches of lava tend to be unstable and may collapse into the sea without notice. Look for cracks here as well, and stay on the land side of them.

Fig. 4.3. The unstable crust of a cold shelly pahoehoe flow in Hawaii can hide deep holes.

Even more dangerous than shelly lava is the fine-grained, soft, friable terrain often found on geothermal areas and near active fumaroles. In 1991, a Soviet graduate student was killed when he fell into an active fumarole at Mutnovsky volcano. Reports at the time indicated that the ground he was walking on looked "normal", like dried mud, until it collapsed. The student's body was never recovered, only his rock hammer was found. Not long afterwards the collapsed ground regained a solid but equally unstable crust, and looked "normal" again. This type of volcanic terrain is one of the most treacherous there is and this is why one must tread very carefully around fumarolic and geothermal areas. Some of the most visited of these areas, such as Solfatara in Italy and Bumpass Hell in Lassen, California, have signs posted and areas fenced off where the ground is particularly unstable. In more remote places, one has to use common sense and a great deal of caution.

Volcano rule 5: Find out about local laws restricting access to any areas of a volcano; do not go into restricted areas without official permission

During eruptions, local authorities often close off certain areas to the public and usually with very good reason. However, when a volcano is quiet, many people do not realize that they may be wandering into areas where access is restricted, even if danger is not the reason. For example, during the "Kalapana phase" of the Pu'u O'o eruption in Hawaii, the public was not allowed to go into Kalapana, the town which was eventually overrun by lava flows. The reason behind it was to prevent looting from the houses that were still standing. I did hear about a couple of students getting arrested by the local police; apparently, they had been working upslope from the restricted area and wandered into it without realizing where they were.

A related issue is that of private land. Parts of volcanoes that are outside national parks may well be privately owned and visitors may sometimes unknowingly trespass into someone's property. Most of the time, these visitors encounter no problems, but it is wise to ask locally if there are any areas of restricted access or, better still, to employ the services of an experienced local guide. Remember that in many instances, owners of private properties are protecting themselves against lawsuits that injured visitors may bring against them.

Volcano rule 6: Avoid going anywhere remote alone and let someone know where your party is headed to

This is a general common-sense warning that is applicable to other places as well as volcanoes. Potentially active volcanoes have an added dimension of danger and it is wise to err on the side of caution. It is surprising how many people don't abide by this common-sense advice and it is not rare to find lone trekkers on particularly remote and potentially hazardous areas of a volcano. Local guides can usually be found and people traveling alone should seriously consider hiring one.

Volcano rule 7: Find out about specific local dangers

This is another common-sense rule that is applicable to any wild or remote place. Personally, the worst encounters I have ever had on a volcano were, on one occasion, with a pack of unfriendly dogs and, on another, an even more unfriendly-looking, gun-toting farmer. Neither occasion presented real danger; suspicious farmers usually turn quite amicable when you explain that you are studying "their" volcano and are often keen to hear your opinions. I have heard many tales of volcanologists encountering wild animals, including bears and wild pigs, but all escaped unscathed. However, it makes sense to know what to expect and to take appropriate precautions. It is also sometimes useful to know what not to expect. Some years ago, two professors I know greatly amused themselves in Hawaii by warning their students about the "dreaded Hawaiian ahu," a black and very poisonous snake that camouflaged itself among the ropes of congealed pahoehoe lava. The students, who had to walk for miles on the lava, would have been far happier had they known that there are no poisonous snakes in Hawaii.

A different kind of danger is lightning strikes, which can be a problem on some volcanoes. Large expanses of lava-covered areas do not support a single tree and you may be the tallest thing around. One of my colleagues was hit by lightning on the summit of Mt. Etna but, luckily, she was not hurt. Weather on some volcanoes, particularly the summit areas, can change dramatically in a short period of time, and it is worth planning accordingly.

A British colleague probably had the most unusual experience of a potential local danger on a volcano. He happened to be mapping a network of lava tubes in Hawaii. Lava tubes can extend for many miles and they can be large enough in diameter for one to be able to walk comfortably inside them. Sometimes part of the roof over a tube collapses, forming what is known as a skylight. He climbed out of one of these skylights and found himself in a clearing surrounded by dense jungle, in which a neatly tended marijuana crop grew. Local people no doubt used the tube network to travel easily to an otherwise inaccessible place. Fortunately for the volcanologist, the "owners" of the field were not around.

Volcano rule 8: Know your limits

The only person who I ever saw die on a volcano was a German tourist who underestimated what the short but steep climb from the buses' parking place on Vesuvius up to the summit would do to his heart. He collapsed just as he reached the summit. The efforts of a medical doctor, who happened to be also making the ascent, were not enough to save the unfortunate tourist. A quick rescue by helicopter might have saved his life but few places in the world can boast such response to emergencies. In fact, at that time (the late 1980s) the nearest telephone was some 3 km (2 miles) down the road – I could not have imagined that the busy cluster of restaurants and souvenir shops by the parking place would not even have a single emergency radio (this was, of course, before cell phones). As it was, it took an ambulance over 2 hours to arrive, by which time it was far too late. Visitors from countries where emergency response is fast, particularly in busy tourist areas, should not expect similar services abroad. This, of course, applies to places other than volcanoes. However, the added hazards commonly presented by volcanoes, such as high altitude, steep slopes, rough and jagged terrain, and extremes in weather can test many a healthy person even on a dormant volcano.

One of the most common complaints is heat exhaustion, which is can be serious but is easy to treat if caught early. Lots of water, glucose tablets and a good night's rest cured the mild case I had some years ago. Strangely enough, the weather had been quite cool, but I had over-exerted myself walking over a still-hot lava flow. Hot weather and a long field day landed a colleague of mine in a fairly primitive hospital for an unpleasant night's stay. It is not difficult to avoid such mishaps: pacing oneself sensibly and allowing time for acclimatization to local conditions is all that's needed.

Survival rules for eruptions

When a volcano erupts, the dangers multiply, and other precautions are necessary. The type and intensity of the activity determine the level of danger. For example, viewing a mild eruption in Hawaii from sites designated by the National Park can be done safely by almost anyone. In contrast, going anywhere near a volcano during a Peléean-type eruption should not be attempted by anyone who values their life, unless the visitor can arrange to be accompanied by experts who know the location of a safe vantage point. However, since it is possible (though I hope very unlikely) that a visitor might be caught in the midst of a heavy ashfall or a very explosive eruption, I have included some advice for what to do under such circumstances.

Mildly explosive and effusive eruptions

Hawaiian and Strombolian activity, as we have discussed before, can often be watched safely. This does not mean that they are not potentially dangerous, with falling bombs being probably the most common hazard to watch out for. The key safety rule is to view the eruption from an appropriately distant vantage point and to plan a retreat route should the activity suddenly become more violent.

Eruption rule 1: Stay out of the range of bombs

If a volcano is in the midst of Strombolian explosions or fire-fountains, it is very important to watch what is going on from a considerable distance away before carefully approaching the vent. Watch where the fragments being spewed out (bombs; see Fig. 4.4) are landing, and remain out of range. Because intensity of explosions can vary with time, one should watch the vent long enough to feel comfortable about the potential range. "Long enough" is, of course, largely a matter of personal judgment, because although these eruptions tend to have a pattern and some regularity, their intensity can change, particularly in the case of

Fig. 4.4. Freshly emplaced, still red-hot inside, lava bomb in Hawaii. This bomb is less than a foot in diameter, but capable of causing substantial damage. Note how the ash surface has sunk as a result of the impact. (Photograph courtesy of Scott Rowland.)

Strombolian eruptions. This is another situation where the advice of a guide or local expert can be of great help, as they will be more familiar with that particular volcano's pattern. In the absence of that, the sensible thing to do is to watch a number of explosions before approaching the vent, until one can get a "feel" for the pattern. For example, if explosions happen every couple of minutes, I'd watch them for 30 minutes or so before venturing up to a distance equal to the range of bombs plus 30% to 100% or more, depending on the size of the bombs and their density over an area. Explosions that are spewing out boulder-size bombs should be treated far more cautiously than those sending out small pellets. A helmet or hard hat should always be worn, though they only afford protection from the smaller bombs.

Should the intensity of the activity suddenly increase, or should activity start up unexpectedly, one may find oneself within the range of falling bombs. Although the natural reaction is to run away, this is not the best course of action. The recommended procedure is to look up to see the direction and range of the bombs and, if there is one coming right toward you, move sideways. This procedure requires remaining calm but it can be a life-saver. It is very likely that the tourists killed by the unexpected explosion of Mt. Etna in 1979 would have survived had they known this, as the number of bombs emitted by the blast was relatively low.

If an explosion sends out so many bombs that it becomes impracticable to try to dodge them, the best solution is to run for some shelter, such as behind a large boulder, and to cover one's head with something, such as a backpack. This is what the survivors of the Galeras blast did. They were lucky, however, because few things can offer enough protection from the kind of bombs that volcanoes can easily send up. During the 1979 explosion of Mt. Etna, a bomb penetrated the roof of a Land Rover parked near the crater, though it did not go through the vehicle's floor. Since our group had to complete work near that same crater the following day, we devised a plan should it blow up again: we would run for shelter underneath the chassis of our own Land Rover. Fortunately we never had to put the plan to the test.

Running away is, of course, always a last-ditch effort. In 1986, another of Mt. Etna's summit craters caught some people by surprise. This time, however, the people were volcanologists and they could see that the crater was erupting. They placed themselves at a safe distance, but the activity suddenly begun to increase in intensity. One of the members of the party told me later that he had become quite worried and urged the others to leave the site, however, some of his colleagues were busy taking photographs and decided to stay a little longer. This is a good example of a situation when using one's own judgment is better than a majority consensus. My colleague stayed and, minutes later, a formidable explosion showered bombs all around them. This was followed by another, equally large, blast and yet another. The group had no shelter, no nearby vehicle, and the activity showed no signs of waning. The five members of the group ran blindly downhill while bombs fell all around them. They all managed to escape, but realized that they had been very lucky indeed. Even their cameras came through unscathed and the photos were actually quite good – see Fig. 4.5 for one of them.

There is one type of situation where running away is actually the best choice rather than a last resort. This is when you can judge that the bombs coming down are small pellets and that you can very likely get out of range before they land. When I began working on active volcanoes I was not told about this rather common-sense measure. Instead, my thesis advisor, John Guest, drummed into me the importance of never running away from falling bombs. I was thus rather surprised when Mt. Etna, which is always full of surprises, managed to catch the two of us unawares with a last explosion from a cone that supposedly had

Fig. 4.5. The Northeast Crater, Mt. Etna, in eruption in 1986. Photograph taken by Christopher Kilburn shortly before he had to make a quick exit. Note the person in the foreground – he, too, managed to escape.

stopped erupting the day before. We saw the bombs go up in the air and then John shouted "Run!" For a split second, I hesitated – surely this was the wrong thing to do? But soon I decided to defer to his much greater expertise on active volcanoes. He was right and we did get out of range in time. It must be said, however, that judging how far bombs are likely to fall is not for the novice volcano watcher.

One last word on the subject of bombs: sometimes, as a matter of personal judgment, it is worth deliberately going within their range. Many volcanologists, myself included, have done this on occasion, for a sufficiently strong reason such as to get important measurements or samples. Some people may consider a spectacular photo opportunity to be worth the risk and that is a personal decision. However, I would like to stress that going bomb-dodging for macho reasons is likely to anger Madam Pele or other local volcano deities, who may well be merciless in their revenge.

Eruption rule 2: Beware of unstable terrain

The danger of shelly pahoehoe lava (see *Volcano rules 3*) is made much more serious if there is hot lava flowing underneath the apparently solid crust. One must take great care when walking over the surface of recently emplaced lava, particularly shelly pahoehoe, as lava tubes may still be active underneath. The best approach is, of course, to stay away from shelly lava. However, even ordinary pahoehoe can have unstable crust over lava tubes. These types of lava are common in Hawaii and there have been a few cases of people who have got bad burns from falling thorough unstable crust. All survived by being able to extricate themselves quickly. A colleague who lost his balance and sank up to his knees in molten lava survived by recalling the advice of another volcanologist who had a similar experience. He threw himself backward out of the lava flow and then rolled over on the ground to free his legs from the lava.

Lava flows are, however, often quite safe to walk on, even if the solid crust overlies a molten interior. Because the top surface of a lava flow cools very rapidly and then acts as an insulating layer, it is common for a flow to develop a safe and solid crust while, just a foot or so underneath, the magma is still hundreds of degrees hot. In fact, it is even possible for a person to walk over some flows that are still moving, though this is not recommended practice unless in the case of an emergency. The surfaces of slow-moving aa flows, in particular, are generally quite able to support a person's weight. One's boots, however, are likely to suffer! This is another reason why sturdy footwear is a must on active volcanoes – molten boot soles are far preferable to burnt feet.

Eruption rule 3: Plan an escape route

Eruptions can change unexpectedly and a previously safe location may become a hazardous one. Thus it is sensible to have knowledge of the local topography, roads, and paths. A local guide is best; at the very least one should carry a good map. It is also sensible to study the local terrain before approaching the volcano. Note topographic lows and river valleys which can be particularly hazardous places if flows are present, and forested areas that could be set on fire by lava flows.

Eruption rule 4: Avoid breathing potentially toxic fumes

The importance of carrying a gas mask on an active volcano has already been discussed. Volcanic gases are often just unpleasant; however, they can cause death. Visitors who suffer from asthma should be particularly careful. The only casualty of 1973 eruption of Heimaey, Iceland, was a man who reportedly was overcome by toxic fumes. The gases that come out of volcanoes include carbon dioxide, sulfur dioxide and hydrogen sulfide. The rotten-egg smell that I fondly associate with volcanoes is from hydrogen sulfide, a gas many times more lethal than hydrogen cyanide (the toxic of choice for gas chamber executions). Hydrogen sulfide can be particularly dangerous because one's sense of smell quickly becomes insensitive to dangerously high levels of the gas. Although the concentrations of gases are usually not high enough to cause death directly, they may cause dizziness that could have fatal consequences on an active volcano. A colleague from Hawaii tells of a time when he was observing the activity on Pu'u O'o when suddenly the direction of the wind changed. He found himself enveloped in the gas plume, frantically trying to get his gas mask out of his backpack. After this unpleasant experience he started carrying his mask outside his backpack.

A last resort for those who may find themselves in dangerous conditions without a gas mask is to tie a wet cloth over the nose and mouth. Volcanology folklore says that the cloth should be wetted with urine – personally I think it is best always to carry a gas mask.

In addition to gases, eruptions can produce smoke when, for example, lava flows over a paved road or vegetated area. The smoke can be overpowering, so stay at a safe distance from the flows. A different type of danger, though rare, can be caused by simple steam, as when still-hot lava flows in damp weather. The resulting steam can reduce visibility, which again can be a serious predicament to be in during an eruption. It is sensible to move away from any place that is getting steamed up.

Lava that is entering the sea, as is common in Hawaii, can produce large steam clouds that contain hydrochloric acid. Breathing these fumes can cause respiratory problems and unpleasant eye irritation. It can also lead to tragedy. Two hikers visiting the Pu'u O'o eruption in 2000 died after apparently being overcome by fumes near the shore, where a lava flow was entering the ocean. This was, however, a rare event and can be avoided by being sure to stay upwind and sufficiently far away in case the wind changes direction. Often, lava flows pour into the sea slowly without a significant steam cloud forming so they are quite safe to watch, as many visitors to Hawaii can attest. A last but essential warning about "lava shelves" – seashores being created by active lava flows: do not approach the shore. I stay at least 30 m (100 feet) back. The ocean undermines the lava crust (hence the name "shelf"), which suddenly, and often, collapses.

Eruption rule 5: Watch out for steam and methane explosions

Steam explosions result from the rapid contact of lava with snow or significant volume of water, such as when lava enters the sea or even when it flows over marshy ground. There are few reported cases of such explosions hurting people. However, steam explosions by the shore in Hawaii have caused some injuries due to sharp glass fragments being ejected and being blown by the wind into bystanders' skin and eyes. It is always a good idea to wear glasses or goggles when watching volcanic activity.

Methane explosions can occur when lava comes into contact with vegetation. These explosions tend to happen underneath a moving flow as the methane released by heating plant matter ignites. The force of the explosions sometimes throws boulder-size lava fragments into the air. Deep booming sounds originating from a lava flow mean that these explosions are going on and it is wise to keep away. In the case of an unexpected explosion throwing bombs into the air, follow the procedures for evading bombs discussed above and move away from the flow before another one happens.

Strongly explosive eruptions

The best course of action is to get out of the danger area as fast as possible. Watching a strongly explosive eruption is far too dangerous to be recommended as a visitor activity, unless one can be sure to find a safe vantage point on high ground – preferably many miles away from the volcano. The dangers from a Peléean, Vulcanian, or a catastrophic Plinian-type eruption range, depending on distance, from breathing difficulties caused by heavy ashfall to death. The causes of death can be varied but are all very unpleasant. You may die from asphyxiation and/or severe burns caused by pyroclastic flows or from fatal injuries caused by large falling bombs.

There are, however, some volcanoes where sometimes it is possible to watch pyroclastic flows in a reasonably safe manner. Unzen was considered one of these volcanoes, because of its rare combination of steady magma supply that has produced thousands of pyroclastic flows since 1991 and steep, favorable topography that provides some reasonably safe viewing conditions. However, the 1991 tragedy has made it clear that things can go wrong and that the risk is high.

Because viewing strongly explosive eruptions is not recommended, and I have little personal experience with them, I have not attempted to recommend any safety rules. However, there are things one can do to minimize personal risk from these eruptions, should one be in the unfortunate situation of having to, and these are summarized here.

Get to a safe location

A safe place is somewhere far away from the volcano, preferably on high ground if pyroclastic flows or mudflows are likely. How far away will depend on the volcano and the intensity of the activity. This is one of the many good reasons to find out as much as possible about a volcano before venturing near it. If you know that an eruption may happen, plan your escape route carefully in advance, paying very close attention to the topography.

Be aware of the local topography when making your escape

If there is a danger of mudflows (also called lahars, caused by ash mixing with snowmelt, water from a crater lake, or from recent heavy rains) occurring, stay away from low topographic areas, particularly river channels or erosion valleys (Fig. 4.6). Move upslope to a ridge if possible. Do not cross a bridge if a mudflow is moving beneath it, as the bridge is very likely to collapse. Look upstream before crossing a bridge to make sure a mudflow is not coming down. The same precautions should be followed in case of danger from pyroclastic flows; however, these are not as well confined by topography as mudflows.

If mudflows are the major danger, getting away by driving rather than walking is generally preferable, since mudflows can outrun people but not cars. However, if roads are poor and tend to follow low-lying areas, running to high ground is preferable. Heavy ashfall can also make driving very hazardous by severely reducing visibility.

If pyroclastic flows are likely to come down, running for high ground may be preferable to driving, as pyroclastic flows can outrun cars. In case escape from the flow is impossible, a last resort is to lie face down in a small, low area such as a deep narrow ditch or the basement of a house. That's how two men survived the Mt. Pelée eruption of 1902 – one had locked himself in his basement and the other was a prisoner in a nearly subterranean cell. However, chances of surviving a pyroclastic flow are very small.

Seek shelter from ashfalls

If you are away from the danger zones but ashfall is still very heavy, stay indoors or seek shelter such as a car. Use a gas and dust mask to filter out the ash if possible, otherwise breathe through a piece of damp cloth. Keep your eyes closed as much as possible. If you are inside a house that has a flat or low-pitched roof, pay attention to the amount of ash accumulating on it, as the weight of the ash may cause the roof to collapse. Clear the ash as soon as possible, or seek additional protection under a sturdy table. Heavy ashfalls do not usually last longer than a few hours, but during that time total darkness may occur and it may be impossible to use a radio or telephone because of interference.

Be aware of tsunami danger in low coastal areas

A violent eruption of a volcano near the coast or an island volcano may cause a far-reaching tsunami which could devastate low-lying coastal areas even if

Fig. 4.6. The eruption of Mt. Pinatubo in 1991 produced lahars (mudflows) that caused substantial devastation. These homes along the Abalan River near Sapangbato were inundated by a lahar. (Photograph by Jon Major, courtesy of US Geological Survey/CVO.)

they are hundreds of miles away. The effects of tsunamis can be much more catastrophic than those caused more directly by the volcanic products. For example, almost all the 36,000 deaths from the eruption of Krakatau in 1883 resulted from the great tsunami generated when massive pyroclastic flows entered the sea displacing large volumes of water, and when the caldera itself was formed by a violent blast. In some places the waves are estimated to have been 36 m (120 feet) high when they reached shore.

Earthquakes and volcanic eruptions along the Ring of Fire have generated numerous destructive tsunamis and now there are tsunami-warning systems in use in several areas, including Hawaii and Alaska. The only sensible response to one of those warnings is to get away to high ground – or take the next plane out.

References

Tilling, R.I. (1989) *Short Course in Volcanology*, vol. 1, *Volcanic Hazards*. American Geophysical Union.

5 Preparing and planning a volcano adventure

The first question I ask myself is "Will the volcano be active?" Before setting out on a volcano trip I always find out all I can about recent and past activity, whether or not the volcano is monitored on a sporadic, frequent, or continuous basis and, for remote locations, if there have been recent reports from visitors or locals about signs of activity. The current level of activity and the likelihood of an eruption happening are extremely important factors to consider before deciding to set out for a particular volcano. Some volcanoes are best left alone while erupting but many others are disappointing if they are quiet. Non-experts can access much of the necessary information to make a go/no-go decision. However, some background knowledge about volcano monitoring is necessary to make sense of all the information.

Can volcanic eruptions be forecasted?

The successful forecasting of eruptions is one of the most important goals of volcanology, but it is not an easy one to achieve. The key factors are to understand how volcanoes erupt, to know a volcano's long-term pattern of behavior, and to practice volcano monitoring by a variety of means. Eruptions do not come unheralded, but the critical signs that one is about to happen are not always the same for every volcano. If the volcano is not being monitored, these signs may be missed altogether, with tragic results. Eruptions cannot be stopped, but a lot can be done to minimize the death toll and economic havoc that they can cause.

Volcanologists strive to understand and interpret signs of an impending eruption using a wide variety of information, from seismic activity to analysis of gases a volcano is puffing out. Volcanology is a combination of a number of disciplines – geology, physics, chemistry, and mathematics among them – applied to the study of the transport and eruption of magma. An alternative (and more pessimistic) definition of volcanology was given by G. A. M. Taylor who said it was "the Cinderella science which only marches forward on the ashes of catastrophe." Sometimes it does take a tragedy to advance our knowledge or to insure that adequate monitoring programs are started.

Although we are still a long way from being able to tell exactly when and where an eruption will happen next, by monitoring volcanoes in a number of ways we can detect departures from their normal behavior. Changes may be a precursor to an eruption. One difficulty is that some of the signs found to be eruption precursors on one volcano might not happen at all on another. However, any volcano about to erupt will usually exhibit a number of tell-tale signs and the more we know about that particular volcano, and the longer it has been monitored, the greater are our chances of correctly interpreting those warnings.

Monitoring volcanoes involves making frequent measurements of a variety of volcanic phenomena, including earthquakes, ground movement, chemical compositions of the gases emitted by the volcano, and changes in the local electrical, magnetic, and gravity fields as well as in ground temperature. Most measurements are done locally on the volcano, but others can be done remotely by aircraft or even by Earth-orbiting satellites. Remote sensing techniques have the advantage of enabling monitoring of very isolated or otherwise inaccessible areas where ground monitoring might be too expensive or difficult to do.

Changes in a volcano's pattern of behavior which may indicate that an eruption is imminent are usually reported by scientists studying that volcano in a monthly publication called *Bulletin of the Global Volcanism Network* (see next section for details). This is currently the best resource for finding out about current volcanic activity all over the globe. The *Bulletin* gives reports of activity, sometimes with a day-by-day description of eruptions. It is an excellent source for the potential visitor to find out if an eruption has been happening somewhere and if there are signs which indicate that one may be about to start. Some of the information on precursor signs may not be understandable to those without any background on volcano monitoring, so it is worth briefly summarizing what these techniques are and what they can do.

Volcano monitoring techniques

The most common technique involves measurements of ground movement, either in the form of earthquakes or inflation of the ground. Typically, when magma moves up towards the surface and is fed into a reservoir, it causes the ground to inflate as shown in Fig. 5.1. One can imagine the magma as a balloon being inflated. The pressure causes the ground to be pushed upwards and outwards and it changes the slope of the ground and the distance between two points on the surface. These effects can be measured by a variety of instruments, such as tiltmeters, as well as by repeated field topographic surveys (Fig. 5.2). The instruments are becoming more and more automated and precise so that minute changes in ground level can be detected quickly.

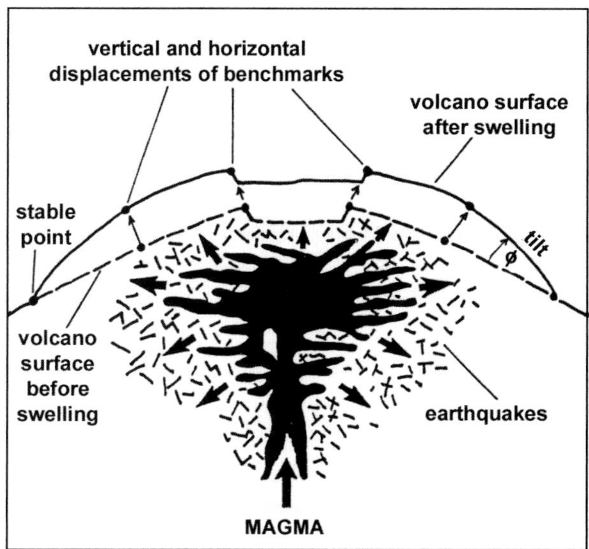

Fig. 5.1. Sketch showing magma moving up towards the surface and causing the ground to inflate (here it is shown exaggerated; the changes are actually very small). The angle ϕ represents the change in slope of the volcano's surface, which can be measured using ground-deformation monitoring. Inflation of the volcano can also trigger seismic activity. (Modified from Tilling, 1989.)

Seismometers are also used to detect magma movement towards the surface, as the inflation of the ground causes fracturing of rock which, in turn, causes earthquakes. A network of seismometers placed on a volcano can easily and accurately provide information on the frequency, location, and magnitude of earthquakes. Measurements of seismic activity and ground deformation are generally the most important for volcano monitoring; however, they cannot by themselves be used to predict eruptions unless the behavior of the volcano is fairly well understood by other means. For example, if the magma chamber is deep, ground deformation at the surface may be too small to serve as a good indicator of magma movement.

Changes in a volcano's geothermal system, the heated waters and fluids that surround the hot magma, can also serve as precursor signs of eruptions. This is because influx of new magma into the chamber, or movement of magma already there, can cause certain gases or fluids to be released. Signs that can be

Fig. 5.2. Leveling survey, Yellowstone National Park. Note bison in the background. (Photograph by US Geological Survey.)

detected at the surface include variations in the temperature, chemical composition, and emission rate of gases and fluids coming out of surface cracks, fumaroles, and hot springs. For example, an increase in the amount of sulfur dioxide gas coming out of fumaroles is generally a sign that new magma is moving towards the surface. At present geochemical monitoring is not as commonly used on volcanoes as ground movement techniques, but as these precursor signs become better understood this is likely to change.

There are other techniques being developed that rely on geophysical effects such as deviations in the local gravitational, geomagnetic, and geoelectric fields. These changes can reflect those in the temperature or mass balance of the volcano's magma, water, gas, and solid rock components. A promising, relatively new way to monitor volcanoes is by using Earth-orbiting satellites. Improvements in the Global Positioning System (GPS) may soon allow the technique to achieve the few parts per million resolution that is needed for volcano monitoring. Satellite observations can also look for anomalous thermal activity on a volcano, which holds great promise as an eruption forecasting technique. Once an eruption starts, satellite data can be very important for monitoring its progress (Fig. 5.3). For example, weather satellites can show volcanic plumes and, in the case of remote volcanoes, the satellite images may be the first indication that an eruption is taking place and the only way to obtain information on the activity. Remote sensing techniques are continuously improving our knowledge of topography, surface changes, and thermal activity at some of the Earth's most remote volcanoes.

Ideally several methods of volcano monitoring are used at the same time and certain combinations of

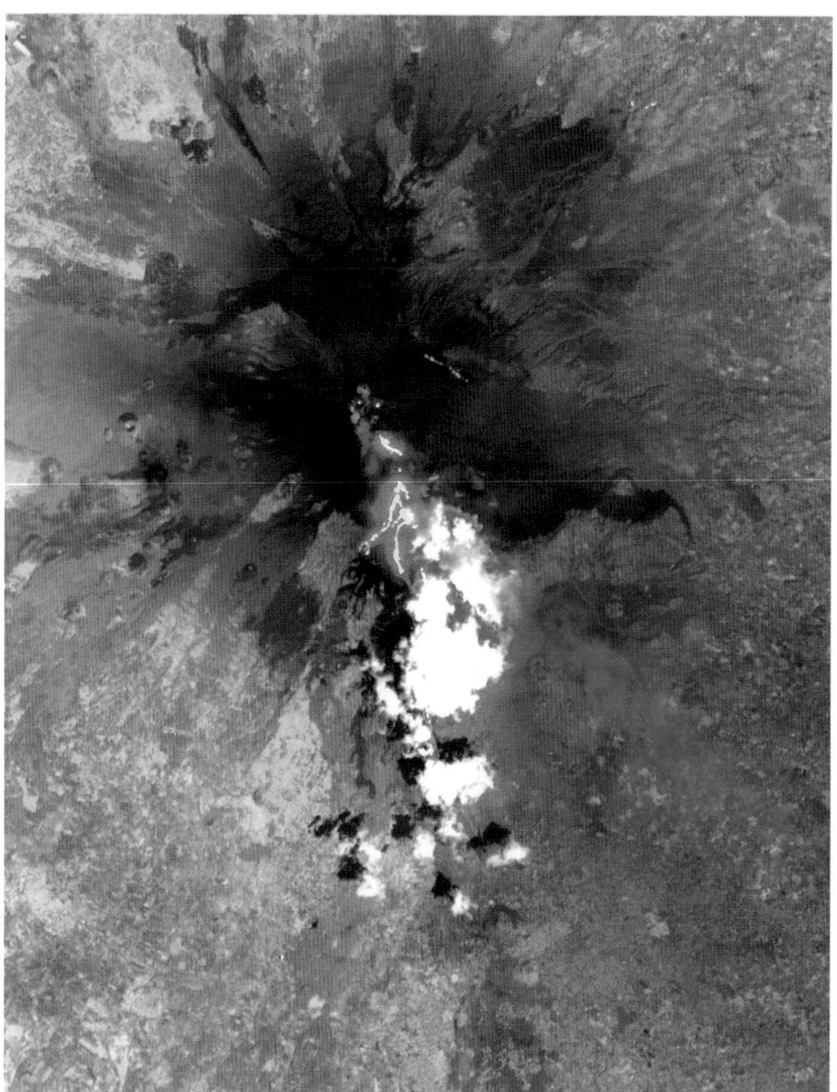

Fig. 5.3. Lava flows, Mt. Etna, Italy. The current eruption of Mt. Etna started on July 17, 2001 and has continued to the present. This image by Advanced Spaceborne Thermal Emission and Reflection Radiometer (ASTER) was acquired on July 29, 2001 and shows advancing lava flows on the southern flank of Mt. Etna above the town of Nicolosi, which was potentially threatened if the eruption increased in magnitude. Also visible are glowing summit craters above the main lava flows, and a small fissure eruption. The bright puffy clouds were formed from water vapor released during the eruption. The image covers an area of 24 × 30 km (15 × 19 miles) (Image courtesy of M. Abrams.)

methods can be powerful ways of forecasting eruptions. Apart from seismic monitoring and ground deformation, which are routinely used together, other promising combinations have been identified. For example, following the Galeras tragic eruption of January 1993, and another 2 months later, survivor Stanley Williams and co-workers noticed a distinct pre-eruption pattern involving two both leaks of sulfur dioxide gas from the caldera and the timing of faint tremors known as long-period quakes. The recorded measurements showed that three phases preceded both the explosions at Galeras. The first was the increased release of sulfur dioxide gas several weeks before the explosions. The second was the decrease in the gas emission, while the energy of the tremors increased, probably because the cracks to conduct the gas to the surface begun to close, causing pressure to build up. In the final phase – when the volcanologists were in the crater – the interval between the tremors went from minutes to seconds and gas was again detected at the surface. If this pattern had been recognized earlier, there is a chance that lives would not have been lost. There has been some controversy surrounding the Galeras eruption and whether it could have been predicted, and I will not go into details here (for those who are interested, see books listed in the Bibliography). I must stress, however, that such a pattern had not been recognized as a precursor of volcanic activity before Galeras, and it is still unclear whether it is present before every Galeras eruption or whether it is applicable to other volcanoes. We know that a similar pattern occurred in Pinatubo in the Philippines, so it may turn out to be a reliable precursor sign for certain types of eruption.

The eruption of Pinatubo in 1991 (Fig. 5.4) can in fact be considered one of the greatest success stories in eruption forecasting. The volcano had been quiet for over 400 years when, on April 2, 1991, it woke up with a series of small explosions. The volcano was known to be highly dangerous, as pyroclastic flows and mudflows had occurred in the past. A number of towns were threatened, plus the US Clark Air Force Base which housed more than 14,000 people. A team of US volcanologists, led by Chris Newhall from the US Geological Survey, came to the area to work with Philippine scientists to understand the volcano's unrest and likely future activity. Their work led to the evacuation of 58,000 people before the volcano's great eruption on June 15, an effort which saved thousands of lives. The US team used the evacuated Clark Air Force Base as their center of operations and left it as the eruption started. They cut it so close that the commanding General, who remained at the base with the team, recounts that he was told by a young volcanologist "Better put some jam in your pocket, General, because we are all going to be toast."

Fig. 5.4. Eruption of Mt. Pinatubo as seen from Clark Air Force Base, June 12, 1991. (Photograph courtesy of Jon Major, US Geological Survey.)

Sources of information on volcanic activity

Once you decide to visit a particular volcano, how do you find out about what its current level of activity is, plus all the other information that, for safety reasons, you should know about? If the volcano is frequently visited by tourists and is monitored, this information will be easier to get than if the volcano is remote and seldom visited. However, a vast amount of information on volcanoes can be easily accessible by the non-specialist and this amount is only likely to increase in this age of fast communication. The listings below are not comprehensive but they give some pointers on how to get started. Addresses and details about the sources are given in the Appendices.

Library sources

These are the "paper products" that can be obtained from libraries or by direct subscriptions, though these are fastly becoming available over the internet. The most useful is the *Bulletin of the Global Volcanism Network*, published by the Smithsonian Institution in the USA (and now available on the Smithsonian's website; see Appendix I). Volcanologists from all over the world send their reports to this monthly bulletin which is a mine of information for the volcano enthusiast. The first 10 years of the bulletin have been compiled into the book *Global Volcanism 1975–1985*. Organized by volcano, this book offers a summary of all the world's volcanic activity during that decade.

To find out quickly whether a volcano is considered active, and what its known activity has been like during the last 10,000 years, look it up in *Volcanoes of the World*, another publication by the Smithsonian group of volcanologists (see Bibliography). The book is organized both by volcano and region and also by date and is an unsurpassed reference work about volcanic activity. The volcano explosivity index (VEI) for each eruption is given wherever it is known or can be estimated; this provides some indication about how dangerous the eruptions from a particular volcano have been in the past.

There are also a number of technical periodicals for those who may wish to research their subject deeper. The International Association of Volcanology and Chemistry of the Earth's Interior publishes the *Bulletin of Volcanology* (with Springer-Verlag) which not only contains technical articles but also news about the society and its various activities. Other technical journals are the *Journal of Volcanology and Geothermal Research* and the American Geophysical Union's *Eos*. Non-technical articles about volcanoes appear from time to time in science and geographical magazines, such as *National Geographic, Discover, Natural History*, and *New Scientist*. Volcanoes are also a popular subject for science and photography books. The Bibliography gives some recommendations about general volcano books and also those about the specific volcanoes included in this book's field guides. Those who plan to visit other volcanoes may be able to find specific books about them through a library search.

Electronic sources

The electronic superhighway can be of enormous use to the volcano enthusiast. One of the best resources is the Volcano Listserv mailing list which has over 800 international subscribers, including a large percentage of professional volcanologists. It acts in a similar way to the Global Volcanism Network, reporting volcanic activity from all over the world. The great advantage of the Volcano Listserv is the fast dissemination of information. In addition, the text (but not figures) of the *Bulletin of the Global Volcanism Network* is relayed over this network once a month. Subscribers to the Listserv often ask for one another's help concerning information about particular volcanoes and the response is usually good. The network, started by volcanologist Jon Fink, has been active for more than a decade and has turned into a major service to volcano researchers, writers, and educators. To subscribe to the Listserv, send an email to volcano@asu.edu.

There are numerous World Wide Web pages devoted to volcanoes. Those listed in Appendix I are likely to remain active in the long term. The Smithsonian Institution's Global Volcanism Program website is particularly useful for finding out about the activity of volcanoes all over the world. In addition to the listed websites, it is worth checking if a particular volcano's country has a website reporting local news and weather. It is also easy to make searches over the network for websites on volcanoes in general.

Local contacts

Local scientists and residents can be invaluable for providing information about a volcano and its state of activity. Suggestions for local contacts are given in this book's field guides. If you are planning to visit other volcanoes than those included here, there are several avenues you can try for getting in touch with local experts. If the volcano is located in a national park, contact the park office asking for information on how to visit the volcano – organized hiking or other tours may be offered and at the very least the office should be able to provide information on hiking and driving on the volcano. The embassy or tourist board of the country the volcano is located in may be able to supply useful guidance. If the volcano is particularly remote or seldom visited, ask the embassy how to get in touch with the local volcano observatory (if there is one) or the geology department of a major university in the country. Be sensitive, however, to the fact that the people working in those places may be far too busy to be able to help you out. A good approach is to ask them if they can recommend a local guide. Failing that, try asking for a guide at your destination's hotel or lodging, as informal arrangements sometimes can be quickly made.

Volcano societies, field trips, and courses

Field trips to volcanoes, led by experts, are often organized by geological or volcanological societies. Although these trips may not appeal to the truly independent traveler, they have the great advantage of expert guides who can often take you to places that might be difficult logistically to reach by yourself. These trips are not generally advertised outside the societies, but in most cases they will welcome non-members who are truly interested. Some of the societies that run field trips to volcanoes are the Geological Society of America (GeoVentures tours) and the Geological Association of Canada. Some museums and specialty natural history tours offer trips that include volcanoes. You also may be able to attend conferences organized by the International Association for Volcanology and Chemistry of the Earth's Interior (IAVCEI) or the International Geological Congress. Both offer a wide variety of field trips associated with their conferences. Some tour companies, particularly those located in places where there are volcanoes nearby, sometimes offer some good options for visiting a volcano and some even employ local geologists as guides. An interesting way of visiting a volcano is to join a research expedition (through Earthwatch or other similar organization) and help a professional volcanologist to carry out field work. Earthwatch often has one or two volcano expeditions a year, which in the past have included Santorini, Kamkatcha, and the Big Bend volcanoes in Oregon.

Essentials for a volcano trip

Most people reading this book are probably seasoned travelers, so I will limit my advice to items that are either essential or particularly desirable to have when one is visiting a volcano. The type and amount of equipment needed depends of course on the nature of the trip and on the area's remoteness and weather conditions. Some more specific hints are offered in the field guides later on, but a short list of essential items will serve you well for most volcano trips. Among the essentials, as discussed in Chapter 4, are protective headgear and gas masks (Fig. 5.5) to be worn on volcanoes that are active or in a fumarolic state (emitting gases). Climbing helmets or hard hats (like those used by construction workers) are easily available. Gas masks can be ordered by mail, over the internet, and sometimes are available at hardware stores (gas masks have become popular in the USA since the terrorist attack on September 11, 2001). Ask the specialty store what types they have to protect against volcanic fumes and ash; note that you don't need something expensive or sophisticated.

Fig. 5.5. Volcanologists in Hawaii wearing gas masks. (Photograph courtesy of Scott Rowland.)

Hiking on recently emplaced lava, even after the lava has cooled, presents special hazards as the surface is often glassy, with sharp fragments which can cause bad cuts in the case of a fall. It is a good idea to protect one's skin by wearing leather gloves, as well as long pants and long-sleeved shirts. Sturdy boots are a must. A less serious problem but one that can cause a lot of discomfort is the fine volcanic ash that is blown around by the wind. Eye protection of some kind is highly recommended, and wearers of contact lenses may like to switch to eyeglasses while hiking volcanoes. Those who, like myself, cannot bear the thought of doing without their contacts should wear sunglasses (preferably with side dust protectors) and take along clear goggles for night-time wear or for when conditions get truly bad.

Geologists usually don't travel without their geologic hammers. However, it must be stressed that rock sampling is not allowed in most national parks, where many volcanoes are located. Some countries, such as Iceland, do not allow sampling at all unless a permit is obtained in advance. In those cases it is best to leave hammers at home.

Volcano photography

Volcanoes are one of nature's most rewarding subjects for photography. There are many books available that show beautiful volcanoes images and readers keen on photography should browse through them for ideas. These books, however, do not reveal how the photographer obtained the shots. A professional photographer could surely write a whole book on how to photograph volcanoes, but this has not yet been done. Hence I felt there was a need to include some general advice on volcano photography, even though I do not claim to be an expert. I have, however, learnt from experience some useful pointers for shooting volcanic landscapes and volcanic activity.

It is best to divide volcano photography into two major types which call for special attention: dormant volcanoes and volcanoes in activity. Dormant volcanoes are not much different from other landscapes as photography subjects, except for the presence of very dark materials such as lava flows and pyroclastics (lava fragments). Since one usually wants the photograph to show both the detail in the lava and also some expanse of bright sky, and sometimes snow, the range of contrast is large. Automatic exposures are useless in this situation, though I have obtained good results by leaving my camera on auto mode and adjusting the exposure by using the spot-exposure option on the darkest part of the frame. When using the camera's manual mode, I will expose for the lava and let sky and snow be overexposed, because I'm generally more interested in the lava as a subject. However, bracketing the exposure is the surest way to obtain a good picture.

Photographing volcanic activity is an entirely different matter. The type of activity I refer to here is Hawaiian or Strombolian, including moving lava flows (Fig. 5.6), where the key subject is red-hot lava (there are many examples of such photos in this book). Safety is the first consideration here. It can be tempting to take unwise risks to get that perfect shot. How much risk to take is a personal call, but at the very least be prepared to grab your equipment and run. Beware of what you put down on the ground, not only because it might delay your escape but also because recently emplaced lava flows are hot and can seriously damage equipment. Tripods are essential for those red lava shots; they should be very sturdy, with steel-spiked tips. I also recommend placing an appropriate filter in front of the camera lens to protect it from damage by ash and, in the case of photographing lava entering the sea, from droplets of acid that can be carried by the wind.

Finding the right place from which to take the best shot is the first consideration, because of both the safety aspect and the crispness of the photo. Hot lava emits heat waves that can make a photo appear out of focus. Make use of prevailing winds and areas of cool

Fig. 5.6. A glowing lava flow photographed at dusk. (Photograph courtesy of Scott Rowland.)

Fig. 5.7. The Pu'u O'o eruption in Hawaii damaged roads and, in some cases, cars. (Photograph by the author.)

lava to avoid this problem. The other major consideration is the time of day (or night). Photographer G. Brad Lewis, who has taken many stunning shots of Hawaiian eruptions, recommends dawn and dusk as being the "magical" times for photographing the "liquid light" of lava trails against the sky. His advice is to pay great attention to the lighting conditions: a few minutes can make the difference between a truly spectacular image and just another lava photo. One interesting condition for a fire-fountain shot is snow on the ground: a shot taken under the right lighting conditions at dawn or dusk can make the whole landscape appear red.

Night-time shots can be truly spectacular, as proven many times by countless shots of red lava trails against a dark sky. In the other hand, lava flows are often best photographed during the day, particularly if it is the damage being caused by a flow that is the major subject for the photo, such as trees or houses catching on fire as they are overrun by lava. Don't overlook the many possibilities for opportunistic, sometimes humorous shots, such as cars and road signs partly buried by lava flows (Fig. 5.7).

Eruptions of more explosive types than Hawaiian and Strombolian should only be photographed from a safe vantage point far away from the active area, or from the air. The rise and expansion of an eruption cloud can make a dramatic sequence of images if you can shoot the sequence quickly enough (here a camera with an auto-winder is an asset). Volcanic clouds are usually dark, mostly brown or gray, so try to expose so that the billowing details within the cloud will come out.

Aerial photographs of volcanoes, whether active or not, are often amongst the most striking that can be taken. Private planes or helicopters are available for hire in many volcanic locations and in some places, like Hawaii, they are a major business. The volcano field guides in this book include information on whether planes or helicopters are available for hire, as well as pointers on some particularly good places on the ground from which to photograph the volcano.

Reference

Tilling, R.I. (1989) *Short Course in Volcanology*, vol. 1, *Volcanic Hazards*. American Geophysical Union.

PART II

Guides to volcanoes

6 Introduction to the field guides

The following chapters consist of field guides to the 20 volcanoes selected for this book. I have focussed on how to make a visit to each volcano instructive, exciting, and enjoyable, rather than teaching the technicalities usually found in geological field guidebooks. Those who wish to read about the geology of each volcano in more detail should consult the Bibliography. Volcanoes are grouped by the regions or countries they are in: Hawaii, the continental US, Italy, Greece, Iceland, Costa Rica, and the West Indies. For each group, there is a discussion of the tectonic/geological setting and of the facilities for travel in each country. At the end of each field guide is a section on other nearby attractions, often other volcanoes. The descriptions of these other volcanoes and geothermal areas are brief but sufficient to help the visitor make up his or her mind on whether to make the side-trip.

Table 6.1. *Classification of volcanoes by type and frequency of eruptions*

	Eruption type		
Frequency	Mildly explosive	Moderately explosive	Violently explosive
Frequently active	Kilauea Mauna Loa	Strómboli Mt. Etna Poás	Arenal
Occasionally active	Krafla	Vulcano Santorini Irazú	Vesuvius Hekla
Infrequently active	Haleakala	Sunset Crater Heimaey	Mt. St. Helens Lassen Peak Montserrat Yellowstone Mt. Pelée

Volcanoes are unique in their ways and it may be hard for visitors to choose between them. Most people will have individual considerations such as the general location of the volcano and expense of getting there, but Table 6.1 and the list below may be of help for those who want to choose based on volcano-specific factors. One important factor is how explosive – and therefore dangerous – the eruptions of particular volcanoes tend to be. Although several different types of eruptions can take place on a single volcano, it is possible to categorize volcanoes in a general way as being mildly, moderately, or violently explosive. The categories here are based on the predominant eruption types and their volcanic explosivity index (VEI) which was discussed in Chapter 2. The VEI ranges from 0 (typical of Hawaiian effusive eruptions) to 7 (for the cataclysmic Tambora eruption, the largest known). Grouping volcanoes by how explosive they tend to be gives the visitor a quick answer to the question of how dangerous the volcano is. The second question in most visitors' minds is how often does a volcano erupt? Although frequency of eruption can vary considerably, it is possible to group volcanoes according to the frequency of their activity in the recent past. Frequently active volcanoes are those which either have been persistently active for the last decade or tend to erupt at least once per decade. Occasionally active volcanoes are those which tend to erupt, on average, at least once per century. Infrequently active volcanoes generally erupt less than once per century and in some cases, such as Sunset Crater, may not erupt again.

If you are still undecided about which volcano to try, consider these less serious (and somewhat prejudiced) factors:

Most frequently active: Strómboli
Most photogenic activity: Kilauea, Strómboli, Yellowstone
Easiest summit access (by road): Kilauea, Krafla, Irazú, Poás
Hardest summit access: Mauna Loa, Arenal
Summit access not allowed: Sunset Crater
Most dangerous summit access (at present): Arenal, Montserrat
Highest elevation summit: Mauna Loa
Most notorious killers: Vesuvius, Mt. Pelée
Most recently born: Eldfell (Heimaey)
Most scenic: Haleakala, Heimaey
Best volcanic sand beaches: Mauna Loa, Vulcano, Santorini
Best laid-out trails: Kilauea, Yellowstone, Sunset Crater
Best local cuisine and wines: any of the Italian volcanoes
Most exotic cuisine: Heimaey and other Icelandic volcanoes
Easiest volcanoes to fly over by private plane or helicopter: Kilauea, Mt. St. Helens, Yellowstone
Best hang-gliding off a volcano (not from personal experience!): Haleakala

7 Volcanoes in Hawaii

Hawaii

All the Hawaiian islands were created by volcanoes. Repeated eruptions for many millennia slowly built up the volcanoes from the bottom of the ocean, until lava broke thorough the surface and island after island was created. At first, there was only black, basaltic lava on the islands. Everything else – the plants, the animals, the people – came to Hawaii from distant lands across the Pacific. The volcanoes are the only original Hawaiians, there long before the coral reefs grew and the first palm tree swayed in the breeze.

Kilauea, Mauna Loa, and East Maui (Haleakala) are three of the youngest and most active Hawaiian volcanoes. Haleakala has erupted only once in recorded history, sometime around 1790, but is still potentially active. Mauna Loa, the largest single mountain on Earth, erupts every 10 years or so, and Kilauea – one of the world's most active volcanoes – has been erupting continuously since 1983.

Hawaii has named the most civilized type of volcanic eruption: it is both spectacular and relatively safe to watch. Few people have died because of Hawaii's volcanic activity, though the lava flows can greedily eat up land and houses, sometimes whole towns. But there is a fruitful side to these frequent eruptions: the long lava flows often reach the ocean, creating new land and yet another black sand beach.

Compared to European volcanoes, Hawaii is a new contributor to the scientific understanding of how the Earth works, because there is no written history of the islands before Captain James Cook came upon them accidentally in 1778. The eruption of Kilauea in 1823 was the first to be written about, by the newly arrived English missionary the Revd William Ellis. The guides who took him to the crater said that it had been active "for the reign of many kings past," probably meaning several centuries. Ellis not only described the activity in the crater but also interviewed locals about two lava flows they had seen in their lifetimes, which he placed at around 1750 and 1790.

In spite of their short scientific history, Hawaiian volcanoes have contributed more to our knowledge of how volcanoes work than almost any others. The frequent and relatively safe eruptions of Kilauea and Mauna Loa have been ideal for a wide variety of studies. Most of the current techniques for volcano monitoring have been pioneered here. The eruptions never fail to spark wonder and curiosity in those who see them, and the science of volcanology owes much to a group of Hawaii residents who founded the Hawaiian Volcanic Research Association in 1911. The Association's seal bears the optimistic words "No more shall the cities be destroyed" and, to achieve this noble purpose, the Association fostered the scientific study of the volcanoes and began to keep invaluable records of volcanic activity on Hawaii and around the Pacific Rim. In 1912, financial support from the Association helped fund the Hawaiian Volcano Observatory, which was to become one of the world's prime institutions for the study of active volcanoes.

Hawaiian volcanoes have also contributed enormously to increasing the public's knowledge and awareness of volcanoes. The Hawaii National Park, created in 1916, deserves a great deal of credit for showcasing the volcanoes, but it is the spectacular eruptions of Kilauea and Mauna Loa that have contributed so much to public education. Most of the eruption footage shown on television or in movies is filmed in Hawaii and many of the photographs of volcanic activity published around the world have been taken here. This popularity has taken some of the mystique away and there are people who consider the overexposure to be something of a turn-off. However, those who consider that these volcanoes offer no opportunity for adventure are very much mistaken. It is true that one's activities within the National Park are restricted and that the well-marked trails can give the ambience of a "tourist volcano." On the other hand, this is what makes Kilauea a great choice for non-experts who want to learn a lot, for parents to take their children to, and for those who do not have the physical ability to go on long hikes. The adventurous types can easily get away from the popular trails and from the crowds. Nobody who has climbed to the summit of Mauna Loa, hiked within the Haleakala crater, or walked over Kilauea's Kau desert can complain about lack of wilderness in Hawaii.

The three Hawaiian volcanoes described in this book are markedly different in what they offer the visitor. Kilauea (meaning "spreading, much spewing"), located on Hawaii's Big Island, is the world's most easily accessible active volcano and is an ideal learning ground for all. Mauna Loa ("Long Mountain") offers a completely different experience. The vast majority of visitors to the Big Island merely photograph this giant volcano from afar. Few venture up its slopes and fewer still reach its 4,169 m (13,677 feet) high summit, from where you can truly feel on top of the world. On the island of Maui is Haleakala ("House of the Sun"), the spectacular crater of East Maui, an older volcano that offers the attraction of an easily accessible summit. Haleakala's remote, untamed slopes offer some of the best hikes in the Hawaiian islands.

There are impressive volcanoes all over Hawaii, but most are no longer active. Mauna Kea ("White Mountain") on the Big Island, named for its frequently snow-capped peak, is famous for its first-class astronomical observatory. Its neighbor Hualalai has the potential to unleash lava flows that would be disastrous for the Kona Coast resorts. Diamond Head is an impressive tuff cone and a much-visited landmark on Oahu. These volcanoes are described briefly in the next chapters. Not only are they fascinating to visit, but they also give us a perspective on their more active neighbors, and a clearer understanding of how the volcanic chain we call Hawaii was born.

Tectonic setting

The Hawaiian islands are the world's best-known example of mid-ocean hot spot volcanism. The

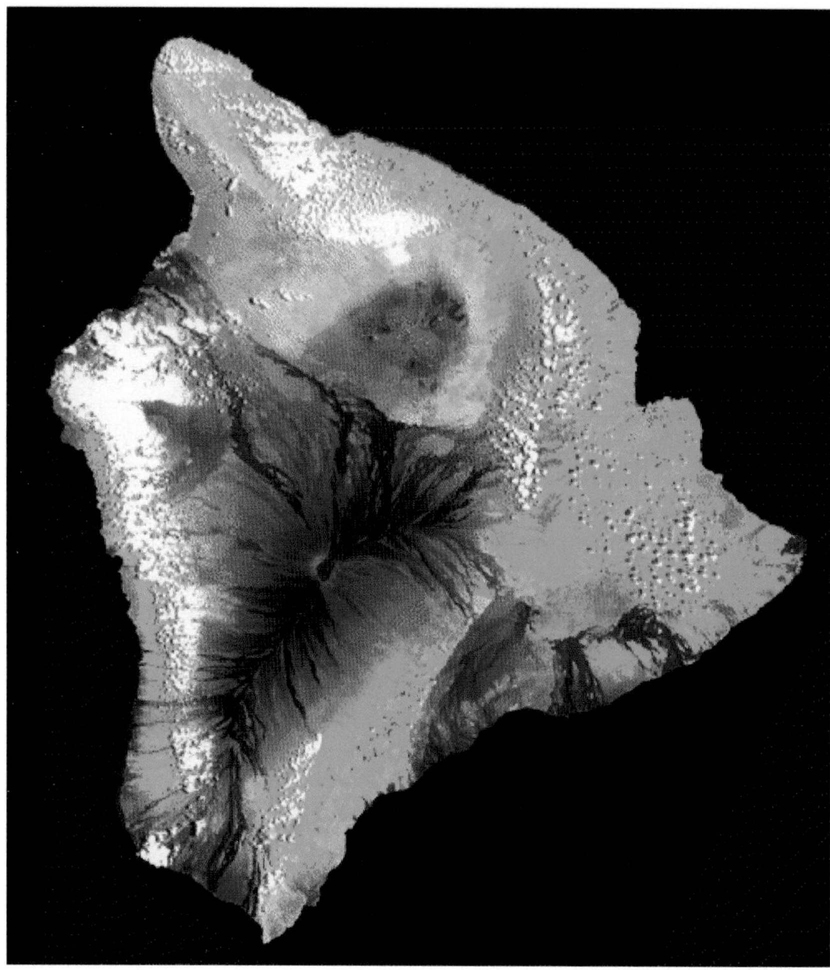

Mosaic of Landsat Multi Spectral Scanner (MSS) images of the Big Island of Hawaii. The caldera and rift zone of Mauna Loa are prominent in the southwestern part of the island. The blue-colored, relatively young flows from Kilauea can be seen in the southeast of the island. The older volcanoes are Hualalai (located to the northwest of Mauna Loa, partly cloud-covered in this image), Mauna Kea (north-central), and Kohala (forming the northwest peninsula).

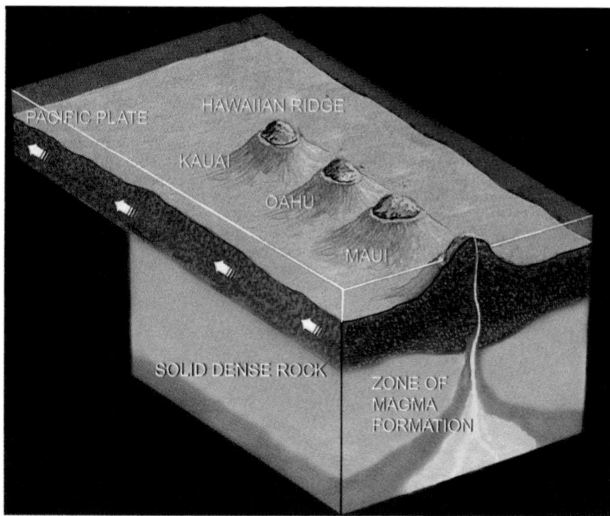

Sketch showing the movement of the Pacific plate over the Hawaiian hot spot. The plate is moving to the northwest and new islands are created over the hot spot. Kauai is the oldest in this picture, Hawaii's Big Island the youngest. (Modified from Tilling *et al.*, 1990.)

progression of volcanism along the island chain is very clear. The lavas of Kauai are 5 to 6 million years old, those of Oahu are 2 to 3 million, Maui's are about one million, and those of the Big Island are youngsters at less than half a million years old. Off the southeast coast of the Big Island, the seamount Loihi is slowly growing – eventually it may surpass Kilauea as Hawaii's most active land volcano.

The Pacific plate is moving in a northwesterly direction over the Hawaii hot spot at a rate of about 10 cm (4 inches) per year, nearly as fast as your hair grows. The Big Island was the last to emerge from the ocean, when lavas from the Kohala volcano broke through the water's surface about 400,000 years ago. Today, Kohala is deeply eroded and forms the northwest corner of the island. The next oldest volcano is the 4,205 m (13,796 feet) high Mauna Kea, which last erupted about 3,600 yars ago. Next down the line is Hualalai, which last erupted in 1801. Mauna Loa has erupted several times this century, the last time in 1984. Kilauea, which is growing against Mauna Loa's southeastern flank, is the youngest and most active volcano on the island.

The evolution of the Hawaiian volcanoes has followed a pattern from frequent eruptions of runny tholeiitic basalt flows during the volcanoes' youth to more explosive, less frequent outbreaks involving more viscous alkaline basalts as the volcanoes age. Hence the volcanoes start out as low shields, like Kilauea, and end up as huge mountains topped by cinder cones and ash layers, like Mauna Kea.

Magma builds the Hawaiian volcanoes but once the supply is exhausted, erosion starts to destroy them. As the volcanoes age and eruptions become less frequent, the power of erosion is felt more and more. Stream valleys create enormous canyons, such as the famous Waimea canyon in Kauai. The actions of the ocean, rain, and wind work together to eventually erode the volcanoes away. Their ultimate fate will be destruction above sea level, with only a cap of coral to mark the site where each great volcano once was. The Pacific plate will continue to carry the old volcanoes to the northwest until, many millions of years from now, they are subducted beneath the Ring of Fire and consumed back into the mantle. However, if the Hawaii mantle plume continues to bring fresh magma to the surface, other Hawaiian islands will be born – and maybe there will always be a piece of paradise in the middle of the Pacific ocean.

Practical information for the visitor

The National Parks

The Hawaii Volcanoes National Park extends from the summit of Mauna Loa to the southeastern coast of Kilauea, preserving a small part of Mauna Loa and most of Kilauea. The Park is the largest historic district in the USA, covering an area of 976 km^2 (377 square miles). On Maui, the Haleakala National Park has two sections: Haleakala Crater, which includes the summit of this spectacular volcano, and Kipahulu District, which extends from the east crater rim down to Maui's coast.

Both parks help to preserve flora and fauna that are unique to Hawaii. The Headquarters of the Hawaii Volcanoes National Park are co-located with the Visitor Center on the edge of Kilauea crater. On Maui, the Haleakala National Park Headquarters are located 1.6 km (1 mile) inside the park's entrance on the northeast flank. Stop at Headquarters on either island to obtain back country permits for overnight hikes (including use of cabins and shelters), as well as detailed maps and a variety of other information. You can call the Hawaii Volcanoes National Park's eruption "hotline" for information on the state of the volcanoes (808-967-7977) or check their website (see Appendix I).

Transportation

A rental car is almost a necessity in Hawaii, though there are bus excursions run by several tour operators based in Hilo and Kailua-Kona in the Big Island and from the major towns on Maui. Public transportation to the Hawaii Volcanoes National Park Headquarters is currently run by the Hele On Bus Company, but the schedules are not particularly convenient. Rental cars are relatively inexpensive in the islands and there are many "fly–drive" deals to Hawaii. Four-wheel-drive vehicles are also available, although they are not necessary unless you want to get to remote locations. Be aware that many rental car companies prohibit the use of vehicles on some narrow roads on the islands, even though these roads are paved. If you are very physically fit, an alternative to rental cars and tourist buses is to rent a bicycle. There are guided bicycle tours on the Big Island and on Maui.

Helicopter tours are popular in Hawaii and widely advertised. There are more safety restrictions on these flights now than there were in the 1980s, when reportedly some pilots would land on still-warm lava. Sightseeing flights using light planes are also offered; they are cheaper but not as popular as helicopter tours.

Lodging

Hawaii does not lack lodgings at all price levels. When visiting Kilauea, it is convenient to stay at one of many bed-and-breakfasts in Volcano village, or at the historic Volcano House hotel in the National Park (see below). The Namakani Paio Campground, nearly 5 km (3 miles) west of the park entrance, is the budget alternative. The Kona Coast offers a wide range of lodgings and is the best place to stay when visiting Mauna Loa, Hualalai, and Mauna Kea. Visitors to Haleakala crater on Maui can find a wide range of lodgings in Kahului (near the airport), or along the island's west coast between Kihei and Wailea, or in bed-and-breakfasts in Kula. A few crater cabins are maintained by the National Park, but advance reservations are a must. Camping within the National Parks is allowed only at designated campgrounds.

Safety and emergency services

Hawaiian emergency services are amongst the best in the world, but rescue services, usually involving helicopters, are extremely expensive. Like everywhere in the USA, you have to pay for medical services, so carrying insurance is a necessity.

Maps

At the entrance to the National Parks in Hawaii and Maui you will be given useful maps showing the major trails and attractions. A wide variety of maps of the Hawaiian islands are available from the US Geological Survey. Many of the maps can be purchased inexpensively at the Visitor Centers on Kilauea and on Haleakala and also at the Jaggar Museum on Kilauea. It is useful to have US Geological Survey 1 : 100,000 topographic maps of the islands plus up-to-date "recreational"-type maps that show the trails on each volcano in detail. Some of the better bookstores stock the U.S. Geological Survey topographic maps at 1:24,000 scale.

The Hawaiian Volcano Observatory (HVO)

Our knowledge of Hawaiian eruptions and volcanism in general has been much advanced by the studies made continuously at the Hawaiian Volcano Observatory, better known as HVO. The Observatory was founded by Dr. Thomas A. Jaggar of the Massachusetts Institute of Technology. Prior to coming to Hawaii, Jaggar was sent by the US government to investigate the disastrous 1902 eruption of Mt. Pelée in Martinique. Jaggar's experiences there convinced him of the need for a US volcano observatory. In 1912, he became HVO's first director and continued until his retirement in 1940. The Observatory is now part of the US Geological Survey and still follows the two major purposes outlined by Jaggar. The first is monitoring the volcanoes, predicting eruptions, and keeping accurate records of the activity. The second is the long-range scientific study of volcanoes that may lead to better monitoring and predicting techniques in the future.

HVO continuously reports the activity of Hawaiian volcanoes to National Park officials, thus making sure that visitors are not allowed to go into dangerous areas. The observatory also posts reports in the *Bulletin of the Global Volcanism Network* (see Bibliography) and on the World Wide Web (Appendix I). The present site of HVO is on the rim of Kilauea crater, at Uwekahuna, next to its former site that now houses the Thomas A. Jaggar museum. The modern building is easily recognized because of its striking glass-enclosed tower, used for observation of activity on Kilauea and Mauna Loa.

Kilauea

The volcano

Everyone who wants to see a volcanic eruption easily and safely should make Kilauea the top choice. Located within a US National Park, this is the world's most visited, most photographed, and probably most studied volcano. Kilauea captures the essence of this

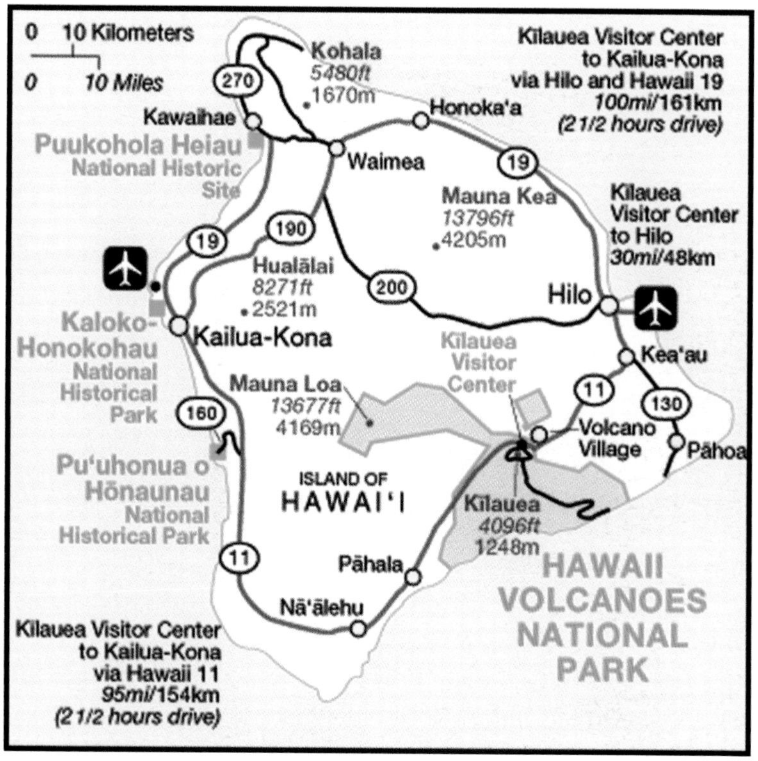

Map of the Big Island of Hawaii showing major roads, towns, and the outline of the Hawaii Volcanoes National Park. (Modified from a drawing by the National Park Service.)

book – that active volcanoes are accessible and can be enjoyed by all. Whenever possible, the National Park allows visitors to see the volcano in action, often constructing temporary trails to safe viewpoints. Kilauea's long-lived Pu'u O'o eruption has spawned numerous helicopter and light plane tours which made eruption viewing possible even for the elderly or disabled. Nobody knows how long Kilauea's Pu'u O'o eruption will last, so my advice is to get there soon. The volcano does have dormant periods that can last several years.

Even if the activity stops, Kilauea will remain one of the best volcanoes to visit and learn from. The trails through the volcanic terrain and into the lava tubes never fail to impress as well as educate those who travel on them. Hiking over the cold lava makes you can feel like an astronaut exploring the lunar landscape, while venturing down the long lava tubes brings to mind Professor Lindenbrock descending to the center of the Earth.

Kilauea is a classic shield volcano, resulting from the eruption of long, fluid lava flows that travel long distances from their vents. Its summit, 1,247 m (4,090 feet) high, is topped by a caldera 4 km (2½ miles) long, 3.2 km (2 miles) wide and up to 120 m (400 feet) deep, within which is the smaller crater Halema'uma'u. The shield's vents are located at the summit and along the two rift zones that radiate from it. The East Rift Zone cuts the coastline at Kapoho and continues for about 56 km (35 miles) offshore, while the Southwest Rift Zone extends to Palima Point and ends a short distance beyond.

The mere sight of the Kilauea caldera makes most first-timers gasp. It was often the site of eruptions, so well contained by the steep walls that buildings on the rim never suffered any damage. In fact, the famous Volcano House hotel used to provide its guests with spectacular views of eruptions. You can still order a flaming cocktail at the bar and gaze at the frozen black crust on the lake, wondering what it would be like to sit there, as countless lucky visitors did, watching the dance of the fire-fountains. Even the irreverent Mark Twain was awed by the magnificent sight of an active lava lake within Halema'uma'u crater. Twain wrote in *Roughing It* that Vesuvius was "a mere toy, a child's volcano, a soup-kettle, compared to this . . ."

These days, with the fire in Halema'uma'u temporarily gone, those who visit Kilauea may think at first that it does not look like a volcano. It is true that you can drive all the way to the summit and see no steep slopes or looming peak, but its frequent eruptions have long delighted visitors. However, not all eruptions are mild and safe to watch. In 1924, an explosive eruption took

The Halema'uma'u on Kilauea. (Photograph by D. Weisel)

place at Halema'uma'u, caused by groundwater coming into contact with hot magma. The resulting series of steam explosions hurled large blocks onto the caldera floor, killing a man who had ventured too close.

Kilauea can have long periods of quiescence, such as the 18 years from 1934 to 1952. This was not good for business at Volcano House. It is said that the 1952 eruption started after the hotel's part owner, known as Uncle George, threw a bottle of gin into Halema'uma'u, to encourage the volcano goddess Pele to spring into action. (Gin is, according to some, Pele's favorite beverage.) Pele rewarded Uncle George with a spectacular show in Halema'uma'u that lasted 136 days, a profitable time for the hotel.

One of the most famous eruptions from Kilauea started in 1969 from a vent in the upper East Rift Zone, which became known as Mauna Ulu ("Growing Mountain"). Great volumes of lava flowed down towards the southern coast, covering parts of the Chain of Craters Road, and fire-fountains as high as 150 m (500 feet) shot from a fissure between the pit craters of Aloi and Alae. A voluminous, rapidly moving lava flow poured out and traveled south, cascading down over the cliffs of the Hilina fault system and forming a broad pool of lava on the flat land below. This eruption was a photographers' delight and the spectacular images of cascading lava and spurting red fountains epitomize Hawaiian activity. The eruption lasted 5 years and became a tourist attraction. Less than a decade later, Kilauea became active again and at the time of writing, the Pu'u O'o–Kupaianaha eruption has been going strong for over two decades.

The Pu'u O'o–Kupaianaha eruption

This long-lived eruption has made such a major impact in Hawaii, and in studies of Hawaiian volcanism, that it is natural for many of us to use it as a time marker. Local folks talk about the island "before the eruption" almost as though volcanic activity was uncommon in Hawaii. This great eruption has, in fact, been singular: the longest and the most voluminous rift eruption in recorded history. Its lava flows have changed part of the island forever, destroying the communities of Kapa'ahu, Royal Gardens, and Kalapana, where many had owned a piece of paradise. On the

The active Pu'u O'o cone in 1984. (Photograph by J.D. Griggs, courtesy of US Geological Survey.)

good side, the Big Island has become bigger – more than a square kilometer (about half a square mile) of land has been added to the coast. For volcanologists, the easy accessibility and long duration of the eruption have provided a golden opportunity to make studies requiring long-term measurements. For photographers, cameramen, and tour operators, the eruption has been an economic boom. Thanks to their work, the public around the world has become more educated about volcanoes – almost anyone with access to a television set has seen the fireworks from Kilauea.

The eruption started in a very ordinary way. Because the magma chamber on Kilauea is shallow, and new magma moving up tends to inflate the volcano like a balloon, eruptions here can often be predicted. Scientists at HVO tracked the magma movement of this one, plotting the earthquake locations on a map and determining approximately where the eruption would start. Not wanting to miss the beginning of the show, a group of scientists decided to camp out near the predicted location in the East Rift Zone. Pele cooperated and did not make them wait long. The next morning, January 3, 1983, the ground cracked open and lava spurted out from the fissures in a long line of splendid fire-fountains.

Over the next four days, a series of fissures opened up, stretching about 8 km (5 miles) along the East Rift Zone. Fire-fountains soon became concentrated along a 1-km section, creating a veritable curtain of fire. From then on, the predicted eruption became unpredictable. All activity stopped after 3 weeks and many thought it was over. In February, the eruption started again, beginning to behave in a pattern of episodes of activity that became typical. Volcanologists now describe the eruption in terms of episodes – to date, there have been more than 50.

By June of 1983, the eruption had settled down to a single vent and there was a regularity to the episodes. Big fire-fountains, fed by a conduit about 15 m (50 feet) in diameter, would gush out for less than a day. The level of magma in the conduit would then drop by over 100 m (330 feet), gradually rising again during the next 3 to 4 weeks' period of rest. Fire-fountains reached 460 m (1,500 feet) and the fallout built a steep cinder cone around the vent. During the first 3 years, the new cone grew to 254 m (835 feet) and became a prominent landmark. Before an official name was chosen, it was known as the "o" vent, because its location coincided exactly with the letter "o" in the word "flow" (of 1965) on the official topographic map. The name began to stick, so when the residents of Kalapana took on the task of naming the cone, they chose the Hawaiian word "O'o", the name for an extinct bird whose yellow feathers adorned the vestments of past Hawaiian kings. Pu'u means hill in Hawaiian, hence the new cone became Pu'u O'o.

In July 1986, after 47 episodes, the eruption changed in a major way. The conduit beneath Pu'u O'o ruptured, and magma begun to erupt from fissures at the base of the cone. Soon a new vent, located 2.9 km (1¾ miles) down the rift from Pu'u O'o, became the main eruption site. Rather than spurting episodic fire-fountains, the new vent erupted quietly and non-stop, and within 3 weeks a small lava shield with a lava pond at its summit was formed. The residents of Kalapana were asked again to choose a name, and they settled on Kupaianaha, meaning "mysterious" or "extraordinary." They did not know that flows from the quiet pond would eventually destroy their community.

Nearly a year later, the walls of the Pu'u O'o conduit, which were no longer supported by a column of magma, began to collapse. By the end of 1988 the collapses had enlarged the original 15 m (50 feet) diameter conduit to an impressive crater over 150 m (500 feet) across. Pu'u O'o is still active and is a spectacular sight from a helicopter.

The smaller, quieter Kupaianaha vent continued erupting continuously, sending out large lava flows within a network of tubes. The flows formed a small shield around the vent, but also traveled far, reaching the ocean, about 11 km (7 miles) away, for the first time at the end of 1986. The small costal community of Kapa'ahu lay on its path and did not survive. By the end of 1990 nearly the whole town of Kalapana would suffer the same fate. The slow-moving lava flows consumed dozens of homes while their owners stood nearby and watched, powerless against the volcano. It had been a long and agonizing wait for many of them. Unlike violent eruptions elsewhere that bring swift destruction, the flows from Kupaianaha did their work slowly and almost meticulously. There was ample time to evacuate, even to move the historic Star of the Sea

The lava pond of the Pu'u O'o eruption. (Photograph courtesy of Scott Rowland.)

Catholic Church, better known as the Painted Church, to a safe location.

The Kupaianaha lava pond attracted many visitors before it shut down in 1992. In 1987, I was one of the many who trekked across still-warm flows to see it. Fire-fountains spurted from the lake, and giant slabs of congealed lava, broken up by red cracks, now and then would sink out of sight.

The eruption has continued, opening new fissures and forming new vents. A new lava shield grew on the western flank of Pu'u O'o, eventually reaching 60 m (200 feet) high. As of 2004, lava still pours into the ocean, and the Park Service makes a special effort to let visitors see the spectacle. At present, no one knows when the activity will end, but many of us hope that it will carry on for a long time, adding new land to the island and inspiring awe and wonder in all those who come to see the real Hawaii.

A personal view: The home of Pele

Mythology is one of my favorite subjects, so not surprisingly I am rather fascinated by Pele, the temperamental volcano goddess who made Kilauea her home. The most interesting aspect of the Pele myth is its endurance in the culture of Hawaii. The gods of the islands' old religion are mostly forgotten, but Pele somehow lives on, perhaps because she has come to symbolize the awe that most of us feel when we see a volcanic eruption – a feeling that can never be described in a scientific manner.

All people living near active volcanoes find their own ways of coping with the constant threat to life and property. Different cultures have relied on a wide variety of coping methods, ranging from human sacrifice to exalted prayer. One can only imagine how the original Hawaiians, who came from tranquil islands in the South Pacific, must have felt in their new home in the Big Island. It is a wonder that they did not take to their boats again, seeking calmer shores without eruptions. Instead, they tried to understand volcanoes and to explain their workings in terms of a rich mythology.

The native Hawaiians believed that all natural forces were gods at work and Pele was assigned to the volcanoes. It is amazing how much these early dwellers grasped about the volcanoes in their new home. They realized that Pele created new land and, although her temperament was suitably volcanic (unpredictable and impulsive), she was never regarded as a force of evil. When her lavas consumed their lands and homes, the belief was (and for some still is) that the goddess was only taking back what is hers, what she had created in the first place.

Pele made her home in Kilauea, but the early Hawaiians knew that she had been elsewhere. Pele supposedly came from their ancient homeland and traveled by canoe to Hawaii, like her people had done. Her first landfall was in the northern islands of the Hawaiian archipelago. Pele needed to make her home in a deep pit, and she moved down the Hawaiian chain digging. However, every time she dug herself a home, her elder sister Na maka o Kaha'i, the goddess of sea and water, flooded the crater with water. According to the legend, Pele had seduced her sister's husband in the old homeland – hence the bitter family feud. The legend of Pele's wanderings tell us that the Hawaiians recognized that the volcanic islands get progressively younger down the chain towards the Big Island, and Kilauea in particular. Here Pele made her home, in the company of her brothers Kane hekili, the spirit of thunder, Ka poho i kahi ola, spirit of explosions, Ke ua a ke po, spirit of rain and fire, and Ke o ahi kama kaua, a spirit that is seen in the "fire spears" thrown out during a Hawaiian eruption.

These legends were carried down as oral history for many generations. It is thought that the original Hawaiians arrived from the South Pacific sometime before AD 450, and that they came from the Marquesa Islands or Tahiti. They had no written language, but a rich knowledge of legends, poetry, and chants. Pele became the most important and revered deity in Hawaii and her exploits – eruptions from the volcanoes – were passed down the generations.

The arrival of Westerners, particularly missionaries, meant that Pele's unchallenged days were over. The worst blow to the old religion was dealt in 1824 by the high chieftess Kapiolani, who had become a Christian. Kapiolani knew that the worship of Pele was the most firmly entrenched in the hearts of Hawaiians, because it was founded on the great and mysterious volcanic forces. Therefore, the best way to earn converts to Christianity was to prove to her people that Pele did not exist. To that purpose, Kapiolani journeyed from Kona to Kilauea, followed by many of her people, who prayed and cried, certain she would die if she dared to challenge Pele. The chieftess marched to the rim of the fiery pit of Halema'uma'u, then went down several hundred feet into the active crater. She said a Christian prayer, then proceeded to defy Pele by throwing stones into the pit and by eating the ohelo berries that are consecrated to the goddess. Kapiolani returned to the crater rim alive, having irreversibly shaken the very foundation of the ancient Hawaiian religion. Her courageous feat was immortalized in the poem "Kapiolani" by Lord Tennyson and described by W. D. Alexander in his *Brief History of the Hawaiian People* as one of the greatest acts of moral courage ever performed.

Pele was beaten but would not be forgotten. In these days of greater religious tolerance, it is common to see worshipers tossing offerings into Halema'uma'u, including sacred berries, coins, and even gin, though true believers do not approve of alcoholic beverages as offerings. Madam Pele, as she is often called, somehow has also managed to coexist with scientific enquiry. Pele's High Priestess is well known to volcanologists and is respectfully allowed to offer prayers when eruptions start. Visitors to the National Park are warned that stealing Pele's rocks will bring them bad luck – a particularly useful myth because it helps to reinforce the Park's ban against collecting rock samples without a permit. Geologists, for the most part, seem to be immune to this curse, though I know of one who had such a spate of bad luck that she sent her samples back to Hawaii. Plenty of visitors to the Park have done the same and their woeful letters are on display at the Visitor Center as a warning to others.

Sightings of Pele are commonly reported, particularly before eruptions, and she is believed to appear in many different female forms, from old crone to beautiful maiden to a white dog. Male visitors to Hawaii should remember that, in principle, any woman could be the goddess in disguise at any time and that it is extremely unwise for any male to anger Pele.

Visiting during repose

Even if Kilauea is active and the hot lava beckons, make time to visit the formations from past eruptions which are conveniently located along the spectacular Crater Rim Drive. This 17.5 km (11 miles) loop road around the caldera is the best possible introduction to the Hawaii Volcanoes National Park. It takes you past cool rainforests and through fields of fresh, barren lava. The highlights are well marked, with overlook stops and trails guiding you along the way. Those willing to hike for a whole day can take the partly paved *Crater Rim Trail* (17.5 km, 11 miles loop) that starts by the Visitor Center and parallels the Drive.

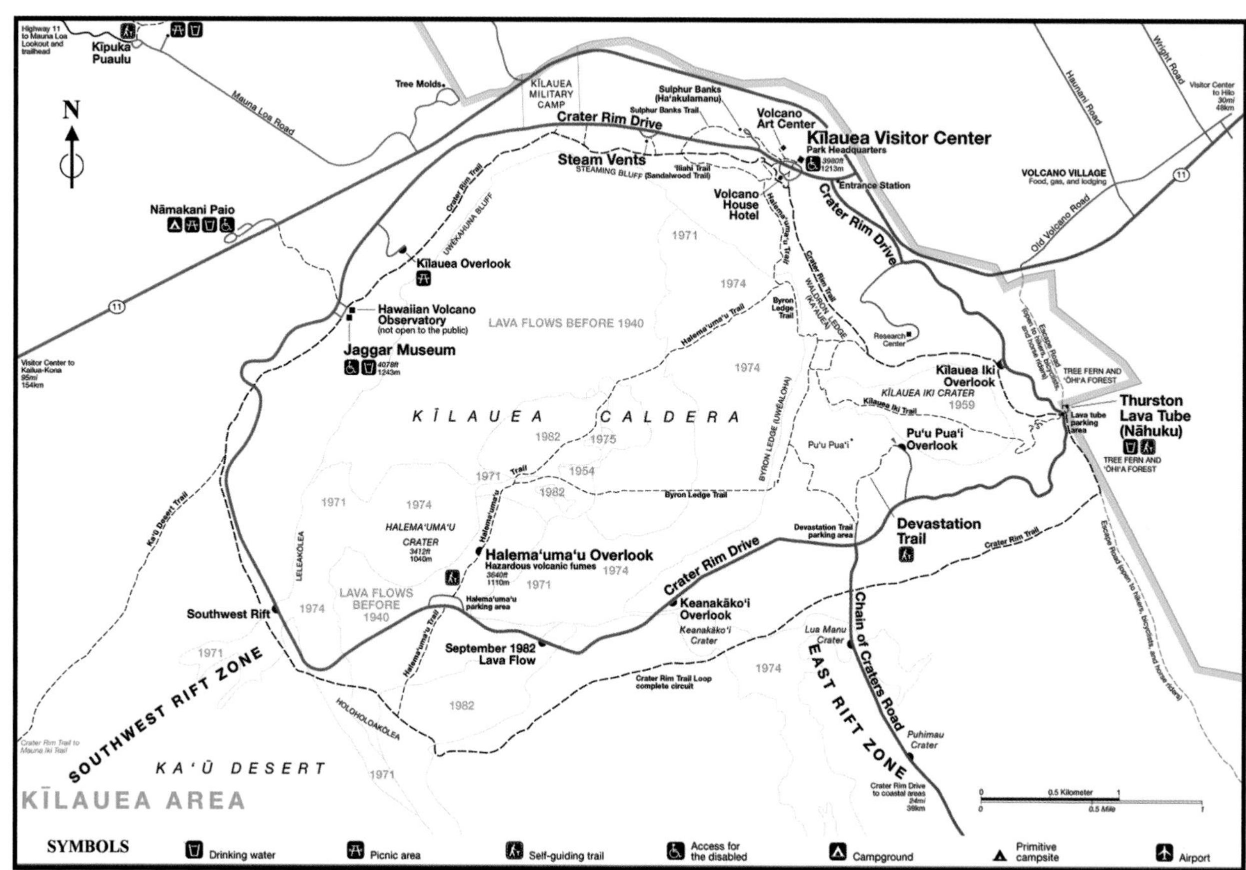

Map of Kilauea caldera showing main trails and the Crater Rim Drive. (Modified from National Park Service figure.)

Kilauea Visitor Center and Park Headquarters

It is best to start your visit at the Visitor Center, located about 400 m (¼ mile) from the Park entrance. The Center has many instructive displays, shows a movie on eruptions, and sells useful trail maps, as well as a variety of books, videos, and posters. You can obtain backcountry camping permits here, and also information about the current volcanic activity, road closures, Ranger-guided walks, and lectures.

The major stops along the Crater Rim Drive are described briefly below – be aware that parts of the road may be closed, so you may have to backtrack. The other major attractions in the Park are located along the Chain of Craters Road, which intersects with the Crater Rim Drive. Lavas from Pu'u O'o have buried parts of this road – inquire at the Visitor Center about its present state.

Volcano House

Located near the Visitor Center, this famous hotel first opened in 1846 as a simple thatched building, which was rebuilt in 1866. Mark Twain and Robert Louis Stevenson stayed here as guests. The 1866 building has survived but is no longer the hotel; it now houses the Volcano Art Center, an art gallery located opposite the modern Volcano House. In 1891, a large, elegant Victorian hotel was built, but it was destroyed by fire in 1940. The present Volcano House was erected in 1941 on the site of the original Hawaiian Volcano Observatory on the crater rim. Drop by and see the 1894 painting by D. Howard Hitchcock in the lobby, showing Halema'uma'u in activity, and the historic fireplace, whose fire the hotel claims has never gone out. Non-guests can come to see an eruption video that is shown twice each evening. The dining room and bar are worth visiting, and looking out of – they both offer magnificent panoramas of the caldera.

The Sandalwood Trail

This easy walk (0.7 km, 1.1 miles one-way) leads you past steaming fumaroles and a sulfur bank. Because more steam condenses in cooler air, the trail is at its most dramatic early in the morning or late in the afternoon. There are a few sandalwood trees left in the upper part of the trail, but they are not easy to spot. Sandalwood trees are now rare in Hawaii; sadly during the nineteenth century they were cut down and the wood sold to foreign merchants. The sandalwood trade is a sad chapter in the history of Hawaii.

The trail starts at Volcano House and takes you down a dense forest of ginger, then along the Steaming Bluff at the edge of the caldera. The steam comes from groundwater heated by hot rocks below. Just a few feet below the surface the temperature reaches 93 °C (200 °F), the boiling temperature of water at this altitude. Tree roots cannot survive in this environment, so only shallow grasses can grow here. Before rainfall started to be collected in rainsheds, puddles condensing from the steam were the only local source of water.

You can loop back to Volcano House or follow the signs to Sulfur Bank. If you take the Sulfur Bank route, you will come to a colorful hillside coated with sulfur, gypsum, and hematite, deposited from the seepage of gases along a major fracture. Hydrogen sulfide fumes provide the rotten-egg smell so characteristic of active volcanoes. The trail then goes uphill and back towards Volcano House. Those who prefer not to make this walk can stop by the Steam Vents overlook on the Drive. There is a short walk from here to the Steaming Bluff.

The Halema'uma'u Trail

This popular trail (5.1 km, 3.2 miles one-way) takes you across the caldera, up to the edge of Halema'uma'u. Starting from Volcano House, the path goes through a tree-fern and ginger forest. The change from the lush vegetation on the rim to the barren lava landscape of the caldera floor is quite striking. Rainfall is abundant on the northeastern rim (about 2400 mm (95 inches) per year), but the prevailing northeastern trade winds become warmer as they pass down into the caldera and rainfall declines rapidly. Halema'uma'u receives half the rainfall of Park Headquarters, which is only about 3.2 km (2 miles) away. The dry, deserted region extends southwest to the Kau desert.

Where the trail reaches the caldera floor, you will see the 1974 pahoehoe lava flow. This eruption rifted the caldera floor and the southeastern rim, producing spectacular fire-fountains about 60 m (200 feet) high. Lava from this eruption covered part of the caldera floor and when it cooled and settled, it left a "bathtub ring" on the walls that can still be seen.

Near the center of the caldera, the trail crosses the 1885 lava flow, on which there are striking orange, white, and yellow sulfur deposits. The trail continues towards the Halema'uma'u overlook, crossing lavas from 1975 and 1954. You will notice that the pahoehoe lava surface is uneven and hilly in places. The mounds are called tumuli; they are formed by hot lava pushing the still-elastic flow crust upwards. Sometimes the crust is fractured and lava tongues spill from the cracks.

Before the trail reaches the Halema'uma'u overlook, it passes close to spatter cones formed by the 1954 eruption. This eruption started out with a spectacular lava fountain on the caldera floor that reached 180 m (600 feet) high. Within minutes, a fissure about 400 m (1/4 mile) long opened, along which a curtain of fire-fountains shot up. The eruption only lasted about 8 hours and few people were lucky enough to see the whole show.

The trail ends near the Halema'uma'u overlook, which is built on lava from 1894. You can go back to Volcano House by following the same trail or choose the slightly longer Byron Ledge trail along the eastern rim.

Thomas A. Jaggar Museum

This small but highly informative museum, considered by many to be the finest within the US National Parks, attracts between 3,000 and 5,000 visitors per day. It was opened in 1987 on the site occupied by the Hawaiian Volcano Observatory from 1948 to 1986. Working seismographs are part of the exhibits, including a portable model that visitors delight in jumping next to, producing their own "earthquakes." On the artistic side, there are murals of Pele done by the renowned Hawaiian artist Herb Kane. The museum's shop sells a variety of books and other volcano-related material. The Hawaiian Volcano Observatory is located next door, but is not open to the public. The site of both buildings is known as Uwekahuna, "the place of the wailing priests". Once there was a temple here, from where ritual sacrifices were offered to appease Pele.

Southwest Rift Zone

The Crater Rim Drive crosses the Kau desert and the upper end of one of the two major rift zones of Kilauea. Young, glistening gray pahoehoe lava covers the land, and cinder cones punctuate the rift. Large fissures, probably formed during the 1868 earthquake and eruption, can be seen near the road. You can walk down the open fissures and see the exposed ash, cinder, and golden pumice from the violent 1790 eruption – these deposits are known as Keanakako'i ash.

Since eruptions began to be recorded in Hawaii, the Southwest Rift Zone has been less active than the East Rift Zone, and only four historical eruptions have taken place from here, the last in 1971. You can follow the Crater Rim Trail across the 1971 lava and the Kau desert, which is not a sand desert but a devastated area of lava flows and fine ash that resembles sand. Gas emissions from the Kilauea caldera are blown into this region by the prevailing winds, and the resulting acid rain guarantees that Kau will remain a stark desert.

Halema'uma'u Crater overlook

This is probably the most visited spot in the Park. The name Halema'uma'u literally means "House of Ferns," an unlikely appellation for a barren and often fiery pit. Hawaiian oral history says that a few centuries ago there were two craters within the Kilauea caldera, one of them filled with ama'u ferns. A violent eruption merged the two craters, presumably destroying the ferns, but the name stuck. Halema'uma'u has been Kilauea's main eruptive vent for at least the last 150 years; its last eruption to date was in 1982. A small patch of 1982 lava can be seen to the right of the overlook. Most of the crater floor is covered by lavas from 1974 and on the far wall there is a prominent dark fissure from that eruption. Below the crater rim is a "bathtub ring" made by the 1967–8 eruption, when lava filled the crater to about 30 m (100 feet) below the rim. There have been times in the past that the crater was so full that it overflowed.

Halema'uma'u is a pit crater, formed by collapse when magma drained away and left the overlying rock unsupported. It is at present nearly 1 km (about 3,000 feet) across and about 83 m (270 feet) deep. Prior to the violent 1924 eruption, the crater was about half its present diameter; immediately after the eruption the depth reached over 350 m (1,200 feet). You may notice that the area around Halema'uma'u is littered with thousands of angular blocks; these were fragments of old lavas thrown out by the 1924 eruption. Although some of the fragments were hot when they were ejected, none of them was from new magma.

You can follow a short trail beyond the overlook to get a closer view of spatter cones and ramparts formed by eruptions in 1954, 1975, and 1982. Small, rather pretty lava stalactites can be seen dangling from spatter overhangs. There are no trails going into Halema'uma'u – even geologists avoid going down into the crater, as the home of Pele is still a place regarded with considerable reverence by many Hawaiians.

Keanakako'i Crater ("Cave of the Adzes")

This crater, about 450 m (1,500 feet) wide and 45 m (150 feet) deep, is famous because ancient Hawaiians used the dense rock exposed in the old funnel-shaped crater floor to make tools. Eruptions in 1877 and 1974 filled the crater and leveled its floor, but the name remains.

Devastation Trail

This short trail (1 km, 0.6 miles one-way) goes through a recovering ohia (*Metrosideros polymorpha*) forest, which is being closely monitored by botanists. The original forest was destroyed in 1959 by the Kilauea Iki eruption, when huge fire-fountains showered the area with ash and spatter fragments. The dead, bleached trees are a dramatic sight. While walking on the trail, look for Pele's tears and Pele's hair. The "tears" are formed by bits of lava thrown up by fire-fountains and cooled rapidly in a drop shape. Because they are frozen very quickly in the air, the "tears" consist of glass, usually jet black in color. Many of the drops draw out behind them thin threads of lava that also freeze in the air. These are Pele's "hair" and some of the "strands" are a striking golden brown. The trail ends at the Pu'u Pua'i ("Gushing Hill") cinder cone overlook into the Kilauea Iki crater. The cone was formed by lava fragments falling downwind from the Kilauea Iki fire-fountaining. Hot gases and steam seeping through Pu'u Pua'i altered the cinders at the its top, turning them yellow.

Thurston Lava Tube

Hawaiian lavas often develop long lava tubes and this one is a particularly fine example. During the eruption, the tube probably extended for tens of kilometers, but now we can visit 137 m (450 feet), the part that has been developed as a tourist attraction. Volcanologists often have to crawl through tubes, but visitors to Thurston can stand up comfortably, as the ceiling reaches 6 m (20 feet) in height. Lava tubes form because the edges of a lava channel move more slowly than its center, cooling and solidifying first. The exposed top surface of the channel also cools and in time a roof is formed. Tubes insulate the lava, allowing it to travel for greater distances than if it was exposed. When the eruption ends, lava may drain away leaving an empty tunnel with a conveniently flat floor. The Thurston tube was formed about 400 years ago by the 'Ai La'au eruption from the East Rift Zone. The eruption also produced a large lava shield – the collapsed top of the shield is the Kilauea Iki crater.

The well-lit trail takes you about 120 m (400 feet) into the tube, entering it through the wall of a pit crater and exiting through one of the natural skylights. Inside you can see numerous lava shelves or benches, which are formed when lava flows at a constant level long enough for its margins to cool and for a roof to start developing. Notice that the walls of the tube have a glazed, glassy surface, typical of the insides of lava tubes.

Between the Thurston Lava Tube and the Kilauea Iki overlook, the road passes through a lush rainforest with many ohia trees. The ohia is usually the first tree to take root on new lava because its small seeds are easily distributed by winds. It is uniquely well adapted to surviving eruptions: as long as the tree does not burn, it can cope with being surrounded by lava. The heat of lava triggers a mechanism that causes the ohia to put out aerial roots. Ohia are easily recognizable because of its red, feathery blossoms, called lehua, which are an important source of nectar for native birds.

Kilauea Iki ("Little Kilauea") overlook

This crater became world famous in 1959, when one of the most spectacular of Hawaii's eruptions started from here. The 5-week show included a 580 m (1,900 feet) fire-fountain, the tallest ever recorded in Hawaii. Cascades of lava half-filled the old crater, forming a layer 111 m (365 feet) deep. You can still see a black, prominent "bathtub ring" left by the receding lava after the eruption ended. Periodic drilling surveys have been made here to find out how fast the lavas are cooling. The last survey, in 1988, found that there is still some molten material in the crater, about 60 m (200 feet) below the cooled crust. The overlook is a superb place from which to photograph the Kilauea caldera with the gentle slope of Mauna Loa in the background.

Kilauea Iki Trail

The trail (4.8 km, 3 mile loop) starts at the Thurston Lava Tube parking lot and goes down 120 m (400 feet) through a forest with a variety of native birds and plants, including wild ground orchids. To traverse the crater, you walk on still-steaming lavas, skirting Pu'u Pua'i on its northern side. The 1959 lava erupted from the reddish vent at the foot of Pu'u Pua'i. On the western end of the crater you can see aa lava full of large, green olivine crystals. It is best to hike on this trail early in the morning or late in the afternoon, when the steam emissions from the crater are more visible.

Chain of Craters Road

This impressive road intersects the Crater Rim Drive between the Keanakako'i crater and the Devastation Trail, and runs for about 32 km (20 miles) towards the

One of the dangers of walking on Hawaiian-type lava flows is well illustrated here. In this case, the cooled crust of the pahoehoe lava flow is thick and there is little danger to the hiker near the "skylight." Skylights are formed when the roof of a lava tube collapses, exposing the hot, molten material flowing underneath. At times, the crust can be dangerously thin and crack under a person's weight. (Photograph courtesy of Scott Rowland.)

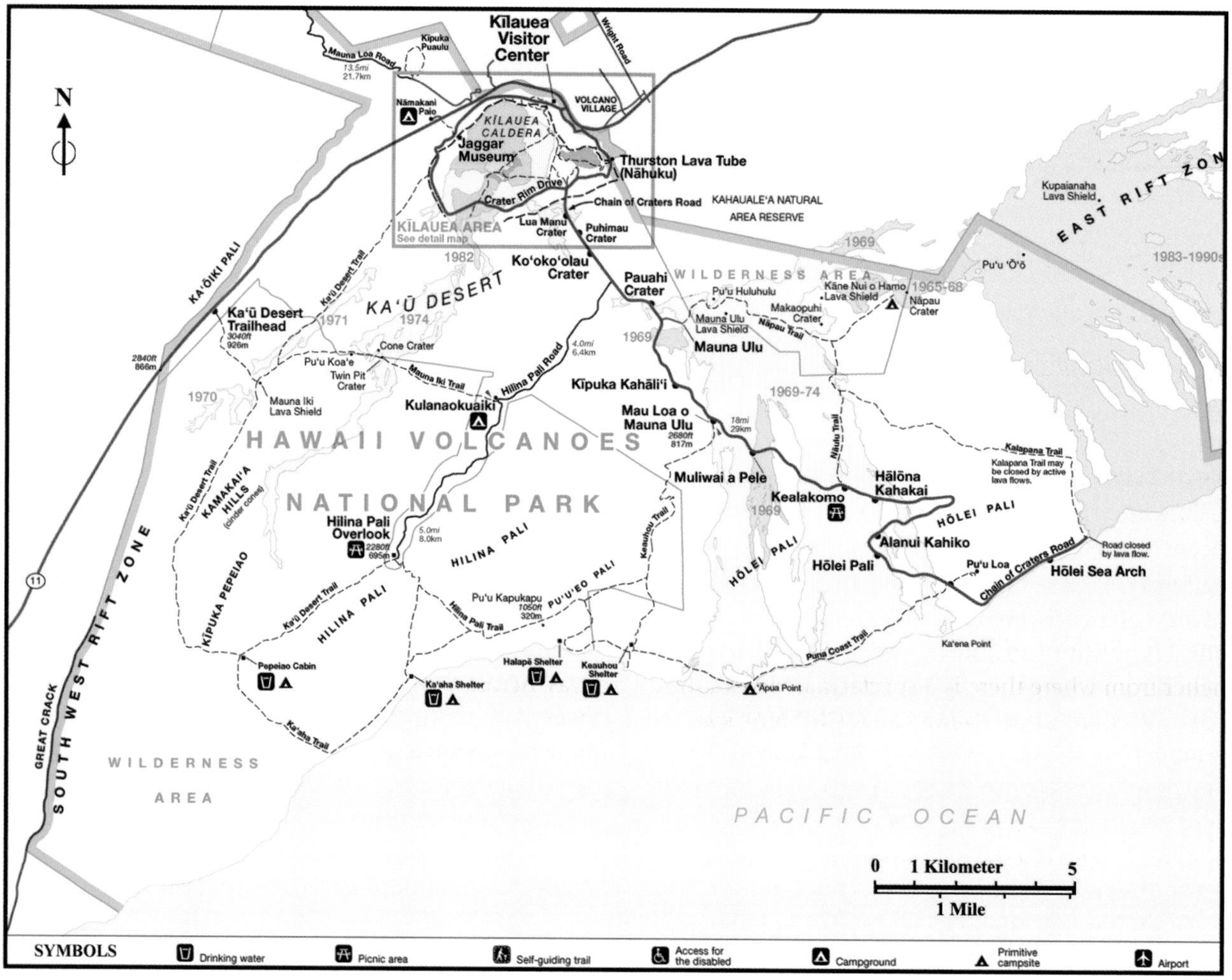

Map of Kilauea showing major roads and trails. Note that lava flows continue to pour down the slopes of Kilauea and may affect the Chain of Craters Road and trails. Check the current road and trail situation with the National Park Service. (Modified from National Park Service figure.)

This lava flow from the Pu'u O'o eruption in Hawaii shows how guidebooks and maps of active volcanoes can easily become outdated. Lava flows do not respect stop signs, though this one is keeping under the speed limit. Roads in the Kilauea National Park have been buried by lava flows numerous times. (Photograph courtesy of Scott Rowland.)

coast. It now dead-ends where the Pu'u O'o lavas cut the road. Allow at least a full day to explore the major attractions along this road, and more if access to the active lava flows is still possible.

Just south of the junction with Crater Rim Drive, the Chain of Craters road goes through the East Rift Zone, where the landscape is marked by spatter cones, fractures, and the pit craters for which the road is named. Several of the pit craters are marked stops. Lua Manu ("Bird Pit") is about 90 m (300 feet) wide and 15 m (50 feet) deep. The young pahoehoe lava that spilled into the crater is from the 1974 eruption. Puhimau ("Ever-Smoking") crater is over 200 m (700 feet) wide and 150 m (500 feet) deep. Its floor is covered with rubble from collapses and landslides. Ko'oko'olau crater is smaller and older than the others, and is now covered by vegetation. It is named after a plant used for making medicinal tea.

Next comes the intersection with the 13 km (8 miles) long Hilina Pali Road, which leads to a picnic shelter from where there is a spectacular view of the coast and of the steep Hilina Pali ("Cliff Struck by Wind"). The Pali, about 518 m (1,700 feet) high, was formed by a faulting and slumping process that is still going on. Kilauea's southern flank is breaking up along faults and gradually sliding into the ocean. The reason is simple: magma is being injected along Kilauea's rift zones, but much of the magma is not erupted to the surface – hence it builds up stress. Every once in a while, the flank gives way and shifts towards the ocean, relieving the stress. Not surprisingly, the area is subject to earthquakes. On November 29, 1975, a magnitude 7.2 earthquake caused the coastline to drop by as much as 8 m (26 feet) in some places. The earthquake generated a 15 m (50 feet) high tsunami that killed a member of a Boy Scout troop and a fisherman who were camping on the coast.

Two trails meet at the Hilina Pali overlook: the Kau Desert Trail and the steep Hilina Pali trail, which leads to the coast. If you are planning to hike these trials, you will probably want to use the shelters that are available for overnight hikers (a permit from the Park Headquarters is required).

Beyond the Hilina Pali intersection, the Chain of Craters Road passes by two major craters, Hi'iaka and Pauahi. *Hi'iaka*, named after Pele's younger sister, is an impressive crater over 300 m (1,000 feet) wide and 75 m (250 feet) deep. The last eruption in this area was in 1973, when fissures opened on both sides of the road and on the crater's southeast wall, pouring lava onto the road and into the crater. Visitors soon get the idea that the Chain of Craters Road is frequently overrun by lavas. Across the road from the Hi'iaka overlook is a short path across the 1973 flow. At the end of the path there is a good view of a fault scarp cliff, part of the Koa'e fault zone that cuts Kilauea, linking the East and Southwest Rift Zones. The part of the volcano south of the Koa'e fault zone is slowly sliding into the ocean.

Pauahi ("Destroyed by Fire") crater comes next and deserves to be seen mostly because of its size and shape: a 520 m (1700 feet) long, 100 m (350 feet) deep figure-of-eight, formed by partial merger of three craters. A lava lake from 1973 left behind a distinct "bathtub ring." The 1973 eruptive fissure cuts across the Pauahi's northeastern wall. You may see Pele's hairs near the short trail from the parking lot to the overlook. The lava trees are easy to spot. These weird formations are the result of lava cooling quickly against a tree, forming a solid lava "skin." When the rest of the lava flows away, this skin is left standing as a lava "tree."

The next highlight along the Road – and one of the best – is *Mauna Ulu*, a small shield about 100 m (330 feet) high. It grew as successive lava flows piled over one another from 1969 to 1974, during the second-longest known eruption from Kilauea's East Rift Zone. The extensive flows radiated from several vents that eventually merged to form a long-lived lava lake at the summit of the shield. Several of the flows reached the ocean and the Big Island became bigger by some 0.8 km^2 (200 acres).

Napau Crater Trail

Starting at the Mauna Ulu parking lot, this trail (16 km, 10 miles one-way) provides close views of Mauna Ulu and plenty more. Try to do at least the first part of the trail (0.6 km, 1 mile) that takes you on the northern side of Mauna Ulu and up Pu'u Huluhulu ("Shaggy Hill"), a 60 m (200 feet) high spatter cone about 400 years old. Near the foot of Pu'u Huluhulu, the trail crosses pahoehoe lava from Mauna Ulu, which has numerous tree molds. The hike to the top of Pu'u Huluhulu rewards you with panoramic views of the East Rift Zone, Mauna Loa and, to the east, the cone of Pu'u O'o. About halfway between Mauna Ulu and Pu'u O'o is a lava shield about 600 years old and now covered by vegetation, called Kane Nui o Hamo (an old name, whose meaning is not known). Its collapsed western flank is the site of the pit crater Makaopuhi ("Eye of the Eel"), the East Rift Zone's largest. Prior to 1965, this it was a double crater – its eastern pit was about 300 m (1,000 feet) deep but it has largely been filled in by lavas from 1965 and 1972–3, which barely invaded the western pit. The trail continues by the eastern slope of Kane Nui o Hamo and through a fern forest. It crosses more lavas – one of the many flows from the huge 1840 eruption is exposed on the southern side of the trail and those of 1965, 1968, and 1969 on the northern side. The trail comes to a stop at Napau ("The Endings") crater, a wide pit about 1 km (3,200 feet) in diameter. It has been largely filled in by lava flows, most recently by those from early episodes of the Pu'u O'o eruption. The eruption started in 1983 at this crater and, in January 1997, activity returned to the same site, marking the start of the eruption's 54th episode.

Back on the Chain of Craters Road, there are overlooks worth stopping at for some fine views of Mauna Ulu. The first is called *Muliwai a Pele* ("River of Pele") from where you can see a prominent channel in a 1974 flow from Mauna Ulu. Kealakomo ("The Entrance Path") and *Halona Kahakai* ("Seashore Peering Place") have panoramic views of Mauna Ulu and its long lava flows, as well as of Kilauea's collapsing southern flank. Kealakomo has a picnic area and is a good place to stop for a while. Across the road is the start of the Naulu Trail, which goes up to Makaopuhi Crater, which you can also reach from the Napau Crater Trail described above. Looking down towards the coast, notice the number of "steps" in the topography. Before the next overlook, the road crosses extensive pahoehoe and aa flows from Mauna Ulu before descending down the Holei Pali ("Cliff of the Holei Tree"), a 460 m (1,500 feet) high slump scarp. From the next overlook, *Alanui Kahiko* ("Old Road"), you can see part of the old Chain of Craters Road between fingers of 1972 Mauna Ulu lavas. Nearly 20 km (12 miles) of the old road were buried by the eruption, in some places to a depth of nearly 100 m (330 feet). The *Holei Pali* overlook offers a good view of where Mauna Ulu lavas poured over the steep cliff. The lavas cover the face of the cliff and the changing texture of the pahoehoe lava as it goes down the steep slope and becomes "jumbled" is quite noticeable. Prominent green kipukas stand out among the black lava.

Pu'u Loa Petroglyphs Trail

This short trail (1.6 km, 1 mile one-way) starting by a parking area leads to one of the best petroglyphs sites in Hawaii. The petroglyphs are geometric designs carved on lava boulders, depicting a variety of subjects including human figures. The meaning of many of them is not known, but it is likely that they had religious significance. You can also see hundreds of little holes made into the rocks, which served as receptacles for umbilical cords – this was supposed to ensure a long life for a child. The Pu'u Loa ("Long Hill") trail is part of the ancient trail linking Puna to Kau, which can still be followed to a junction with the Kau Desert Trail.

Holei Sea Arch

A short walk towards the sea leads to an overlook of this fine lava arch, carved by the action of waves.

The end of the road

The Chain of Craters Road is at present blocked off near where it was covered by lavas from Pu'u O'o. If the eruption is still going on, you may be able to hike to a designated viewpoint to see the active lavas. If the activity has ended, at least you will be able to stand on the USA's newest land.

Visiting during activity

The opportunity to see volcanic activity is a major draw for visitors to Hawaii, and the Park service does their best not to disappoint anyone. Trails are set up to allow volcano watching at a safe distance and the Visitor Center is always ready to give information about the best places from which to view the eruption. True adventurers may find this "managed volcano watching" restricting, but beware of trying to get to off-limits areas – you could be arrested. I heard of two volcanology students who were arrested while doing their field work without an official permit. They were soon released, but learnt that the local police had become quite sensitive to looters going into the evacuated town of Kalapana. Soon after this incident, I visited the active part of the flow in the company of staff from the Hawaiian Volcano Observatory. Two policemen became suspicious of my rental car, parked close to the active area. To my great surprise, the policemen came walking over the hot flow to check if the car's driver did indeed have permission to be there. Local police and Park rangers may at times appear to be overly zealous, but one must sympathize with them. No other active volcano in the world attracts the number of visitors that Kilauea does. Most visitors have no concept of the dangers of a volcano and might get seriously hurt if restrictions were not imposed. The island's police and Park rangers really try to do the best job possible to help volcano tourism and education. During the Pu'u O'o eruption, schoolchildren have been bused in to watch lava flows and, when lava moved close to the town of Kapoho, police escorted car convoys in and out of the town every 15 minutes. During the Kilauea Iki eruption, a lookout was arranged by the Park at a safe distance so that spectators could have an unforgettable view of the fire-fountains. That eruption alone drew about 175,000 visitors.

At the time of writing, the Pu'u O'o eruption is still going strong and lavas continue to flow towards the southern coast. Visitors are usually allowed to hike into the lava field from the end of the Chain of Craters Road. Follow directions in signs posted by the Park, wear boots and long pants, and take plenty of water. Carry a flashlight and spare batteries if you are planning to hike back after dark. Remember that the new coastline is often unstable, and signs alert you to the danger of collapse. For the less adventurous, helicopters and light plane tours are options in Hawaii.

If the area of activity is outside the National Park, it is still worth checking with the Visitor Center on the state of the activity, though they are unlikely to encourage you to go near an active area outside Park boundaries. Local residents can be excellent sources of information in these circumstances. If you can afford it, hire a helicopter to fly over the active area and take a map with you for orientation. You'll gain a good sense of what type of activity is taking place and what the routes in and out of the area are – you may even see hikers. An added bonus is that helicopter pilots tend to be excellent sources of information.

The decision to get close to an active area is a personal one and depends on the circumstances at the time. Read the safety guidelines in Chapter 4 and use common sense. Kilauea is a relatively safe volcano to watch erupt, as demonstrated by the very few known fatalities in spite of intense tourism. The worse tragedy happened back in 1790, when an eruption killed some 80 Hawaiian warriors. It is said that the warriors were overcome by fumes and gases while crossing the Kau desert, but it is more likely that they suffocated because of heavy ash fall. An interesting stop along the Hawaii Belt Road, which heads southwest from Kilauea caldera, is the "Footprints." The preserved footprints in the hardened 1790 ash are thought by some to be those of the fleeing warriors. However, since the footprints do not appear to have been made by anybody in a state of running, they are more likely to date from after the eruption.

In 1992, a 24-year-old woman was killed on Kilauea, but the tragedy did not involve erupting magma. The woman and a friend were sitting at the edge of a steam vent when a particularly hot gush made them lose balance and fall into a hole less than a meter (2.5 feet) wide. Her companion managed to climb out, but the woman fell about 6 m (20 feet) down and was scalded by the heat. Deaths resulting from falls into natural steam vents are not uncommon in volcanic areas and show that some of the greatest dangers are not the obvious ones. An unfortunate tourist died during the Pu'u O'o eruption when a lava bench near the ocean collapsed.

Volcano photography

One of the joys of visiting Kilauea while it is active is to be able to capture the eruptions on film. The general guidelines are no different than for other volcanoes, but here you are often able to get closer to the active vent or flow than would be sensible on many other volcanoes. There are numerous books and postcards showing spectacular photos of Kilauea's eruptions to serve as inspiration. You may be able to photograph lava entering the sea, partly crusted lava lakes and, if you are really lucky, gushing fire-fountains. A telephoto lens is essential to capture details such as the wonderful patterns created by lava as it splatters on the ocean.

Remember to be careful about your equipment – you may be walking over lavas that are still warm at the surface and setting your camera down on the ground can be a very bad idea. Even a tripod may get burnt feet if left in a single place for too long. As for yourself, move if you start to smell burnt rubber from your boots. When near the ocean, remember that there are acidic droplets from the plumes. Ash and Pele's hair can get blown everywhere, so be sure to protect the equipment as much as possible and to clean everything at the end of the day.

Hawaiian-type eruptions are the most spectacular and photogenic, as illustrated by this nighttime view of a fire-fountain during the Pu'u O'o eruption. Fire-fountains can easily reach tens of meters above the ground. Best of all, they can often be viewed safely – just keep your distance from falling fragments and watch for other potential dangers such as lava flows. (Photograph courtesy of Scott Rowland.)

Other local attractions

Diamond Head, Oahu

If you are traveling to Kilauea from outside the Hawaiian islands, you will most likely fly via Honolulu on the island of Oahu. Known as "the gathering place," Oahu gets plenty of visitors but has not seen any volcanic activity for at least 6,000 years. One of its most prominent volcanic features is Le'ahi, better known as Diamond Head. Another famous volcanic landmark is Punchbowl, the site of a famous military cemetery. Both Diamond Head and Punchbowl are tuff cones formed during a late stage in the activity of the Ko'olau volcano, which makes up the eastern part of Oahu. These tuff cones were formed by violent phreatomagmatic eruptions, most likely caused by hot magma rising into shallow groundwater. Tuff cones have a characteristic broad, saucer-like shape, and Diamond Head is a great example. Most of the cone's glassy basaltic ash has been altered to palagonite. The tuff also has many fragments of coral, torn up from the underlying reef, and calcite crystals, which British sailors thought were diamonds – the reason for the far-fetched name. It is easy to visit Diamond Head, as a road leads right into the cone. After reaching the parking area, take the trail that climbs up the western flank, to about 170 m (560 feet) high. The trail goes through a long, dark tunnel, so it is best to take a flashlight. The view from the top is spectacular and well worth the short climb.

Mauna Loa

The volcano

Mauna Loa is the world's largest active volcano and although few people realize it, it is also the largest mountain on Earth. It stands over 9 km (30,000 feet) above the ocean floor and 4,169 m (13,677 feet) above sea level. The mountain's enormous volume makes other volcanoes look like dwarfs – we could fit

Aerial view of Moku'aweoweo, Mauna Loa's summit caldera. The snow-capped Mauna Kea volcano looms in the background. (Photograph by D.W. Peterson, courtesy of US Geological Survey.)

125 of California's imposing, snow-capped Mt. Shasta inside Mauna Loa's bulk.

The volcano's huge mass has been formed slowly, by thousands and thousands of lava flows, many no more than 3 m (10 feet) thick, piling up on top of each other. Flow by flow, the volcano grew to be a mighty Hawaiian shield, typical in its form, topped by an impressive caldera and crossed by rift zones. The caldera was given the peculiar Hawaiian name Moku'aweoweo, meaning "Section of the Aweoweo Fish", probably because the red of the fish is similar to that of glowing lava. At the end of an arduous climb, the caldera is a truly rewarding sight – about 5 km (3 miles) long and 2.4 km (1½ miles) wide, its elongated shape outlined by sheer walls as high as 180 m (600 feet). However, most of the volcano's lavas erupted during the last 150 years did not come from here, but rather from vents along the two main rift zones that radiate down the flanks. One rift extends from the caldera down to the southwest, towards the southern tip of the island, and the other to the northeast, towards the city of Hilo. A much less pronounced rift zone extends northward to the Humuula Saddle, located between Mauna Loa and its giant neighbor Mauna Kea.

Mauna Loa is one of the world's most active volcanoes – it has erupted 39 times in the last 150 years. The eruptions are often spectacular and can be very voluminous, usually producing more lava per eruption than its more frequently active neighbor Kilauea. It is thought that the magma chamber feeding Mauna Loa is much larger than the one underneath Kilauea. Mauna Loa has both summit and flank eruptions and there is a general pattern that flank eruptions follow within 3 years of summit eruptions, but the interval between a flank eruption and the next summit eruption is longer. At the time of writing, Mauna Loa's last activity was a flank eruption in 1984. It only lasted 21 days, but seriously worried the residents of Hilo when they realized that the lava was advancing towards their city. This was the seventh time since Hawaii was settled that Pele threatened to destroy Hilo and we are sure it will not be the last.

Mauna Loa is potentially the most hazardous

Shuttle Imaging Radar (SIR-C) image of Mauna Loa shows clearly the summit crater, rift zones (shown in orange), and many lava flows spreading out from the rift zones. The different colors of the lava flows are caused by varying roughness of their surfaces. Smoother flows (pahoehoe) appear red, while rougher flows (aa) are shown in yellow and white. The blue area seen in the lower left of the image represents the South Kona District, known for cultivation of macadamia nuts and coffee. The area shown is about 41.5 × 75 kilometers (26 × 47 miles). (Image courtesy of JPL/NASA.)

volcano in Hawaii because it can send out very long lava flows, some of which have reached over 40 km (25 miles). Unfortunately for its residents, Hilo lies directly on the path of flows from the northeastern rift. The city barely escaped destruction during the great eruption of 1880–1. In 1935, when another eruption sent a flow towards Hilo, the government of Hawaii decided to try a daring and unprecedented procedure to divert the course of the lava: to literally go to war against the volcano. This was not done at the instigation of the military, but rather that of Thomas A. Jaggar, director of the Hawaiian Volcano Observatory. A fleet of US Navy bombers dropped 2,750 kg (6,000 pounds) of explosives onto the flow, blowing up a lava tube and opening new paths for the lava to flow through. The results, however, are debatable, because soon after the bombing the eruption stopped of its own accord. Seven years later, the maneuver was repeated as another lava flow threatened the city. This time everything, including the eruption itself, had to be kept secret because of World War II. Hawaii, a potential target for the Japanese, was under a black-out and a Hawaiian eruption can be more effective than a lighthouse. The secrecy didn't work, as the famous "Tokyo Rose" congratulated Hawaii for its fine eruption the second day after it started. The Navy went to war against the lava, dropping bombs onto the flow's hardened margin in an attempt to form a lateral breach. Although the attack on the lava was successful, the new flow paralleled the old, thus defeating the military exercise. However, Hilo was lucky once again, as the flow never reached the city.

The 1984 eruption also threatened Hilo, but this time no diversion was considered. It is clear that lava diverted from its course goes somewhere else – maybe towards another town, village, or farm. These days, the legal consequences of a lava diversion in the USA are not even worth thinking about and it is unlikely that Jaggar's visionary if somewhat questionable tactics will ever be attempted again in Hawaii. Hilo, like everywhere else on the island, is now entirely at the mercy of Pele.

The great eruption of 1880–1

The night the eruption started, residents of Hilo noticed a "red glow" from Mauna Loa and lava "leaping like a fiery fountain into the sky." It was November 5, 1880, the start of a long period of agony for residents of Hilo. Lava erupted from a fissure vent at an altitude of about 3,380 m (11,100 feet) and the first flow came down towards Mauna Kea, running by the north side of the prominent cone known as Red Hill. Shortly thereafter, a second flow broke out from lower down the same fissure and traveled to the southeast. It was the third flow, erupted from still lower down the same fissure, that became truly ominous, advancing towards Hilo.

Nervous residents appealed to authorities in Honolulu for help but they had to face another big problem. Oahu had been stricken with a smallpox epidemic and was under a strict quarantine. After some deliberation, the Surveyor General of the Kingdom of Hawaii, Professor W. D. Alexander, was allowed to leave Oahu and come to Hilo, "to determine what Madame Pele was doing on this side of the island". The Professor arrived at Hilo near the end of November and by then the massive aa flow was 32 km (20 miles) away from Hilo. He went to see the flow from a close range, guided by Judge D. H. Hitchcock, a keen mountaineer and volcano watcher whose son eventually became famous for his paintings of Hawaii's volcanoes. The Professor's visit to the Big Island was brief, because the flow had apparently stopped. Thinking the eruption was over, he returned to Oahu.

However, Madame Pele was far from finished with her work. It wasn't long before pahoehoe lava broke out from the same vent and flowed over the earlier lava, heading directly towards Hilo. For months the advance of the pahoehoe was slow, averaging less than 1 km (about half a mile) a month, but the lava did not stop. By June 30, the flow was only about 8 km (5 miles) from Hilo and its front was about 1.5 km (1 mile) wide. Soon the situation got even worse, as the flow front narrowed down to about 400 m (¼ mile) and its speed increased greatly. Around that time, the quarantine on Oahu was lifted and one of its most frustrated residents was able to come to the Big Island.

E. D. Baldwin was a young surveyor and civil engineer working for the Hawaiian government. We know him mostly through his journal, in which he wrote a vivid account of the last stages of the eruption. Baldwin and several colleagues arrived in Hilo on July 12, when the lava was only 5 km (3 miles) away from town. They camped near the active flow and Baldwin methodically documented the eruption's progress as well as their everyday life during this exciting time. "We were new to a live lava flow and did not dare go near it," he wrote. "This amused us soon afterward, as we practically lived on the flow for 2 weeks." The entry for July 15 reads ". . . the flow had crossed the Alenaio stream and formed a natural dam and reservoir in which one could enjoy any kind of bath, from cold to scalding hot. We cooked every meal on the lava flow. We found a place where the lava was oozing out in a small way and on this set our pots and fried our pancakes. Occasionally our frying pan floated down with the lava." On one occasion, when Baldwin and his friends explored the region, they found their way back to camp cut by newly emplaced lava. "After testing the tongue of new lava by throwing heavy stones on it," he wrote, "we sprang over it with as few steps as possible, only slightly scorching our shoes."

On July 29, Baldwin and his companions had to return to their jobs in Honolulu. By then, the flow was only 2.5 km (1½ miles) from Hilo and closer still to the Waiakea sugar mill and plantation. As they walked towards Hilo, they caught sight of the Hawaiian Princess Ruth and her retainers, all decked out in red bandannas, chanting by the edge of the lava flow.

Princess Ruth (Luka) Ke'elikolani, one of the last descendants of Kamehameha the Great, is one of the most colorful characters in Hawaiian history. She was a rather tall, imposing woman, vehemently anti-Christian and fiercely devoted to Hawaiian ways. She was rumored to have been a beauty in her youth, but some necessary surgery and, more unfortunately, a fight with her second husband caused her nose to become permanently disfigured. As she grew older, she became heavier, and by 1881, at the age of 55, she weighed more than 180 kg (400 pounds). The Princess decided to return a plea for help from her people and make the journey from Oahu to Hilo, with the purpose of appealing to Pele to stop the lava. Her journey to Hilo was a rather difficult one. When her ship arrived at Kailua, it is reported that she had to be unloaded with the aid of a cattle sling. When she climbed into a waiting carriage, the axle broke. Eventually, with her retainers helping the horse, the Princess arrived at Hilo and went to her house. For the next several days, lava inched towards the city. When the flow was only about a kilometer (⅝ miles) from Hilo and things looked rather desperate, the Princess ordered her bookkeeper, Oliver Kawailahaole Stillman, to go out and buy as many red silk handkerchiefs as he could, plus a bottle of Pele's favorite drink, brandy (it seems that she changed her preference to gin at a later date). The Princess and her retainers then went to the edge of the creeping flow, where she chanted and offered Pele the red handkerchiefs. Then, in a dramatic gesture, she threw the bottle onto the hot lava. When the brandy went up in flames, the Princess was satisfied that her job was done. She ordered everyone to go to the tents pitched nearby for a sumptuous dinner of roast pig. The next morning, the flow had apparently ceased to move.

The Princess returned to Honolulu to be greeted as a heroine by her people. It is interesting that there were no newspaper accounts of her feat. At the time, the local press was controlled by Christians who preferred to ignore this potentially embarrassing event. The story survived only in Hawaiian oral history and in a couple

The Mauna Loa sulfur flow is a rare example of this type of flow on Earth. The sulfur flow is only accessible by helicopter due to the inhospitable conditions on the upper slopes of Mauna Loa. The author (with hat on) and colleagues came here in 1993; the author returned in 2000. (Photograph courtesy of T.N. Gautier.)

of contemporary books. Baldwin mentioned in his journal that the natives believed their Princess had stopped the lava.

In reality, the flow had not stopped, though it had slowed down considerably. The lava crept along slowly for a few days after the Princess departed. The people of Waiakea Plantation started to build a stone wall on the flow's path, hoping to protect their mill. A civil engineer from Hilo, W.R. Lawrence, sent by the government to help save Hilo, came up with a plan to divert the lava. The flow was by then following the Alenaio stream's bed, so Lawrence's advice was to build an embankment on the Hilo side of the stream's gulch. Baldwin, back in Honolulu, started to place orders for 1,000 picks, shovels, and other necessary tools. However, the plan was never emplaced because the lava stopped on August 9 – close enough to Princess Ruth's visit for folk history to brand her as the savior of Hilo.

A Personal view: Expedition to the sulfur flow

Sometime around 1950, a peculiar event volcanic event happened high up on the Southwest Rift Zone. Unseen and unknown to all at the time, yellow sulfur deposits that had accumulated on the slopes of a cinder and spatter cone were melted by a passing lava flow and the sulfur flowed a short distance down the steep cone. The result was a bright yellow, fan-shaped deposit about 27 m (88 feet) long. This minor volcanic event could be considered insignificant had it not been for two factors. First, sulfur flows are extremely rare on Earth. Second, sulfur flows may exist on the surface of Jupiter's moon Io. Planetary geologists like myself try to find formations on Earth that are similar to those we see on other planets. We call these "terrestrial counterparts" and the idea is that if we understand how they formed on Earth, we can extrapolate the mechanisms to other planets, taking into account gravity and other differences in environment. Planetary geologists became particularly interested in sulfur flows after 1979, when images from the Voyager spacecraft showed that sulfur flows might be happening on Io.

In 1983, three planetary geologists from the Arizona State University took a helicopter up to the remote Sulfur Cone. Before their visit, Mauna Loa's sulfur flow had only been described once, in 1970, in a two-page article in an obscure scientific journal. The author, a B.J. Skinner from Yale University, had visited the site in July of 1967. He wrote that the Sulfur Cone had been mapped and named in 1921, by E.G. Wingate, a US Geological Survey topographic engineer. Skinner managed to contact Wingate, who informed him that there was no sign of a sulfur flow in 1921. Skinner thus concluded that the flow had been formed in 1950, when an eruption caused a local heating of the Sulfur Cone area which made the sulfur melt and flow down the cone's steep slope. The Arizona group agreed with this interpretation and thought that sulfur flows on Io might be formed in the same way.

In 1993, I decided to see the sulfur flow for myself. I contacted colleagues at the University of Hawaii to arrange a small expedition. At first, we considered hiking to Sulfur Cone. The problem was to find a feasible route. We knew that the Arizona group had rented a helicopter, because they did not know if it was possible to hike up there. Three other colleagues had attempted the hike, but they were forced to turn back after trying to cross a field of dangerous, shelly lava

Fire-fountains rise along fissures during Mauna Loa's 1984 eruption. This is a typical and magnificent start of the volcano's eruptions. (Photograph by J. D. Griggs, US Geological Survey.)

that broke under their feet. I soon found out that nobody we talked to knew exactly how to get to Sulfur Cone on foot – not even Jack Lockwood, a famous volcanologist also known as "Mr. Mauna Loa." Jack's knowledge of the volcano is truly remarkable, but it turned out that he had never been to Sulfur Cone. We had no doubt that there had been large changes on Mauna Loa's upper slopes since the time Wingate and Skinner had visited the site and their hiking route might have been cut off by more recent lavas. I was truly disappointed at giving up the hike, but for the sake of being sure to get to Sulfur Cone, we decided to take the easy way up.

My colleague Scott Rowland, a volcanologist at the University of Hawaii, headed our six-person expedition and took care of the local arrangements, which included renting a helicopter. We had to travel two at a time in the helicopter, so that it could reach altitudes over 3,000 m (10,000 feet). Much to my surprise, Scott also had to obtain a permit from the local landowners – it turns out that Sulfur Cone is on private land. Trespassing is not tolerated in Hawaii and we know of at least one lawsuit brought against a volcanologist who wandered into private land. Even if you are going to a remote location over 3,000 m (10,000 feet), trespassing is not worth the risk.

We met our pilot at the Mauna Loa Weather Observatory early in the morning and were taken two at the time up above the barren slopes of the volcano. It was a truly spectacular ride. From above, the brown lava fields looked deceptively easy to walk on. After what seemed to be only a few minutes, we were landing on a flat patch of whitish ground about 1 km (5/8 mile) away from Sulfur Cone. The landscape was remarkably beautiful – white and yellow sulfur deposits covered the ground, steam rose from fumaroles, and ominous fissures crossed our path, reminding us to be careful. The deposits were friable and we walked slowly, stopping here and there to photograph exquisite sulfur crystal formations. Sulfur Cone rose in the distance, its white and yellow blanket contrasting against a deep blue sky. We were alone for miles, in a place where few had walked before.

The most striking thing about Sulfur Flow is its bright, uniform, pale yellow color. It is composed of almost pure sulfur, in fact, analysis of a sample showed that it to be 99% pure sulfur. Most sulfur samples from volcanoes contain impurities that affect their color, making them anything from a dull yellow to green to muddy brown. Mauna Loa had managed to produce an exceptionally "clean" specimen. The trip turned out to be only a reconnaissance, as the flow margins I had hoped to make measurements of were too eroded to produce reliable data.

I returned to the sulfur flow in 2000, while filming *Planet Storm*, a television documentary. The helicopter again carried us there two at a time, ferrying us from the observatory. One of the most pleasantly eerie feelings I ever had was being up on the flow with my colleague Bill Smythe, waiting for the helicopter to return to pluck us out of the mountain. We were the last people to go back at the end of the shoot and we wondered what we would do if somehow the chopper failed to come back. We were truly in the middle of nowhere, without a known route to return. Bill joked that I take him to the nicest places.

I still have a lingering desire to find a hiking route to Sulfur Cone. At the same time, I'm glad that there is no easy route. The sulfur formations are extremely fragile and most likely would be destroyed if the place became popular. Above all, it good to know that Mauna Loa still has some remote, untamed places from where the volcano's solitary beauty can be fully appreciated.

Visiting during repose

Climbing to the summit

On an island where so much of nature is accessible by car, including Kilauea's drive-to crater, Mauna Loa's remoteness is truly refreshing. Relatively few people make the arduous journey to the summit but most who do wonder who was the first person to get there. Only the history of Mauna Loa's ascent by Westerners is known, leaving us to ponder if native Hawaiians climbed the volcano before Captain Cook's arrival. It is possible that they didn't, simply because they might have considered such a journey unnecessary and foolhardy, something that most of today's visitors to Hawaii agree with. Those who disagree are immensely rewarded when they reach the top – here you can truly feel on top of the world.

A little history

The first known attempt to climb Mauna Loa took place during Captain Cook's last voyage, in which he dropped anchor at Kealakekua Bay. John Ledyard, a corporal of marines aboard Cook's ship *Resolution*, wrote a book in which he described his attempt to climb the mountain. He says the party consisted of himself and three others from the ship, including David Nelson, the expedition's botanist who "met with great success" in his quest to collect plant specimens. A number of Hawaiians went with them but it seems that they refused to carry anything and became fatigued before the Westerners. Impenetrable vegetation and rough lava blocked the routes they attempted and, after 2 days, Ledyard and his party were forced to give up.

Sixteen years later, the naturalist, botanist, and surgeon Archibald Menzies, a member of the George Vancouver expedition to the Hawaiian islands, led three expeditions to try to reach Mauna Loa's summit. The first two attempts, one in 1793 and the other in January 1794, tried to ascend from the western flank as Ledyard had done before, and met with equal failure. Menzies, who was described as a clever, middle-aged, and rather persistent Scot, was not ready to give up. In a wise move, he sought advice from Kamehameha, the Hawaiian king. Kamehameha assured Menzies that the most likely way to succeed was to ascend the mountain from the south. Furthermore, the king provided canoes to take the Westerners to the southern coast and entrusted them to the care of Rookea, a local chief who, according to the monarch, was well acquainted with the proper route. It is doubtful that the chief himself had ever gone up as far as the summit, as he and most of his men gave up the journey before getting to the snowline. What really mattered was that Rookea knew how to reach the snowline.

Menzies and his men traveled by sea and land, being met along the way by local people who provided them food, hospitality, and all the help they could. The day before Menzies reached the summit, when the party was at an altitude of about 3,000 m (10,000 feet), Rookea and most of the Hawaiians were forced to abandon the journey. They were barefoot and seriously feeling the effects of the cold. Rookea strongly advised Menzies to give up the attempt, fearing they would all die. The Scot reassured him that they would go no higher than the edge of the snow. Rookea and most of his followers went back to a camp and waited. Given the conditions, it is amazing that some of the Hawaiians got to the summit at all: Menzies' account tells us of two of them waiting for him as he and three of his men hiked around the caldera, and of an unspecified number of others who returned to camp earlier with a Mr. Howell, who was much fatigued and whose shoes were "torn by the lava."

Mauna Loa's caldera was not active at the time, but Menzies wrote of "smoke . . . which we conceived to issue from hot springs." He described perpendicular walls about 400 yards in height and the bottom of the caldera as being quite flat and covered with cooled lava. He correctly ascertained that Mauna Kea was higher than Mauna Loa and used a mercury barometer to calculate Mauna Loa's height to be 13,634 feet (4,156 m) – only 43 feet short of the actual height. Menzies and his companions returned safely from their adventure, which he described as "the most persevering and hazardous struggle that can be perceived."

The next known ascent of Mauna Loa did not occur until 40 years later. David Douglas, the botanist for whom the Douglas fir is named, reached the summit on January 29, 1834. He wrote that even after this 40-year interval, the Hawaiians still remembered Menzies and his role as a doctor and botanist, calling him "the red-faced man who cut off the limbs of men and gathered grass."

Douglas himself was an impressive character. In the 3 weeks before his successful ascent of Mauna Loa, he had already reached the summits of Mauna Kea and Kilauea. He was enormously impressed by Mauna Loa, calling it "the Great Terminal Volcano." Some even say he became mentally unhinged after his journey to the summit, as his writings became inaccurate and confusing. Months later, on the way back to Europe, his ship paused on the northern side of the Big Island. Douglas went ashore and decided to walk to Hilo, meeting with an untimely death on the way. His body was found in a pit into which a wild bull had also fallen. It is not known if he was murdered or fell in the pit and died from injuries inflicted by the bull. "I must go back to the volcano," he had written, and it is possible that he died en route to see Mauna Loa's summit one more time.

Present-day hike to the summit

The highlight of a visit to Mauna Loa is undoubtedly its summit, but this hike is not suited to everyone. It is a strenuous journey best taken as a 2-day or even a 4-day trip, depending on the chosen route. It is possible to reach the summit and come back in a single day via the Observatory Trail – if you are very hardy and start early. Although the distance is less than 21 km (13 miles), and the elevation gain only 770 m (2,527 feet), the hiking is all above the 3,000 m (10,000 feet) level, which means slow. A fit person can do the round trip in 7 to 9 hours, which leaves little time to enjoy the summit. Whatever your chosen route, remember that Mauna Loa is not to be taken lightly. Altitude sickness affects many hikers, sometimes seriously impairing their judgment. Weather conditions can deteriorate fast and it is often freezing at the top. Snow, fog, rain, high winds, and icy conditions are possible anytime of the year. At other times, the air is extremely dry and dehydration can come swiftly, worsening the effects of altitude sickness. If fog comes in, visibility can be very poor so, even if you think you can hike the mountain in one day, bring plenty of food, water, and warm clothing. If you plan to spend the night at one of the cabins provided by the National Park, you must obtain a permit beforehand, in person, from Kilauea Visitor Center and Park Headquarters. The (unpurified) water supply as well as the space in the cabins is limited, so the Park must restrict the number of visitors. Even if you plan only a day hike, it is still a good precaution to let the Park know of your plans. A word of caution: never rely on finding water at any of the marked waterholes – even the water supply at the cabins cannot be guaranteed.

There are two main trails up Mauna Loa, the Observatory Trail and the Mauna Loa Trail. Both join up with the Summit Trail at about 4,000 m (13,000 feet). A third trail, the Ainapo, climbs along the Southwest Rift Zone, but it has rarely been used since the nineteenth century and should not be attempted without an experienced guide.

Observatory Trail (up to junction with Summit Trail)

Round trip: distance 12.5 km (7¾ miles), elevation gain 594 m (1,950 feet), hiking time 7 to 9 hours. Combined with the Summit Trail, this is the most direct route to the top and also the most strenuous. It can be done as either a day hike or a 2-day trip, stopping at the Mauna Loa Cabin overnight. The Weather Observatory, at 1,036 m (11,150 feet) elevation, is where important studies of the amount of carbon dioxide and ozone in the atmosphere are made. The Observatory is not open to the public, but has a public parking lot. To get there, you have to first drive along the Saddle Road that links the Kona coast to Hilo. The side road leading to the Observatory starts just east of Pu'u Huluhulu, an old cinder cone covered with trees and surrounded by lavas from 1935. The side road is unmarked, but it is 12.4 km (7¾ miles) past Mauna Kea State Park (a camping and recreation area) if you are coming from Kona. If you are planning to do the 1-day hike to the summit, camp there the night before and be sure to arrive at the trailhead by the Observatory no later than 10 a.m.

The drive from the Mauna Kea State Park to the Observatory takes about 1 hour – about 29 km (18 miles) on a one-lane road. Note that although both the Saddle and the Observatory roads are paved, ordinary rental cars are technically not allowed on either. If you want to obey the rules, be sure to rent a four-wheel-drive vehicle. The Observatory road starts out over lavas from 1935 and 1936, but after only 300 m (1,000 feet), it goes over prehistoric lavas that are paler in color and look distinctively smoother and more eroded than the recent lavas. At 870 m (½ mile) away from the junction, you reach a darker historic flow, dating from 1899. At 11.4 km (7 miles), the road rises to go over flow from 1855 and 1856, then at 15.6 km (10 miles), you are back onto 1899 lavas. While you drive over the historic flows, note the kipukas (islands) of older, mostly prehistoric lavas. It is worth making a brief stop at 22 km (14 miles) to look uphill towards one of the many fissure vents on Mauna Loa, this one marked by a spatter rampart where spatter fell and accumulated along the sides of the fissure. The road reaches the Observatory after crossing lavas from the

1942 eruption. Go to the public parking lot and look for the trailhead sign. On a clear day, the view from here can be spectacular: to the north, you can see Mauna Kea's summit dotted with white domes housing astronomical telescopes, to the northwest is the eroded Kohala volcano and further beyond is Haleakala on Maui.

The trail begins as an unpaved road but in less than 800 m (½ mile), a sign directs hikers off the road, to the south-southeast, past a cairn and then into lava flows about 1,000 years old. Even if you have a four-wheel-drive vehicle, driving further up than the Observatory is not a good idea – rough roads crisscross and it is easy to take the wrong one. Besides, driving will save you less than 3 km (about 2 miles) on the trail, as there is a locked gate farther up.

About 2.4 km (1½ miles) from the trailhead, the trail passes west of an old brown cinder cone and then between a pair of cairns, placed there so as to mark off a collapsed lava tube (you can carefully look inside it). Soon the trail meets the four-wheel-drive road again, and follows it for about 500 m (⅓ mile), passing a cone covered with ribbons of congealed lava. Going off the road again, the trail goes by an old, red and black colored vent, then meets an old pahoehoe lava flow. Crossing the four-wheel-drive road one more time, you are soon at the junction between the Observatory Trail, the Mauna Loa Cabin Trail, and the Summit Trail. The junction is at about 4,000 m (13,000 feet) altitude.

If you are on the summit day-hike, follow the Summit Trail straight up (see below), skirting the western side of North Pit. Otherwise, take the Cabin Trail, which crosses over young lavas from 1942. Very soon you will see Jaggar's Cave, a large, cinder-floored pit where Jaggar and his colleagues used to bivouac on their trips to watch summit eruptions. There is a "waterhole" here, meaning that (dirty) snow or ice can usually (but not always) be found. Just a few steps further is the rim of the caldera's North Pit (sometimes called North Bay). The edge of the pit is covered by lavas from 1942, which also extend along the north-western edge of the caldera. You can see the impressive walls of Moku'aweoweo from here. The Mauna Loa Cabin is about 3 km (about 2 miles) away at 4,039 m (13,250 feet), on the edge of Moku'aweoweo's east wall. The route, well marked by cairns, goes over lavas between 200 and 1,500 years old, skirting the southern side of Lua Poholo. This pit crater formed after 1840

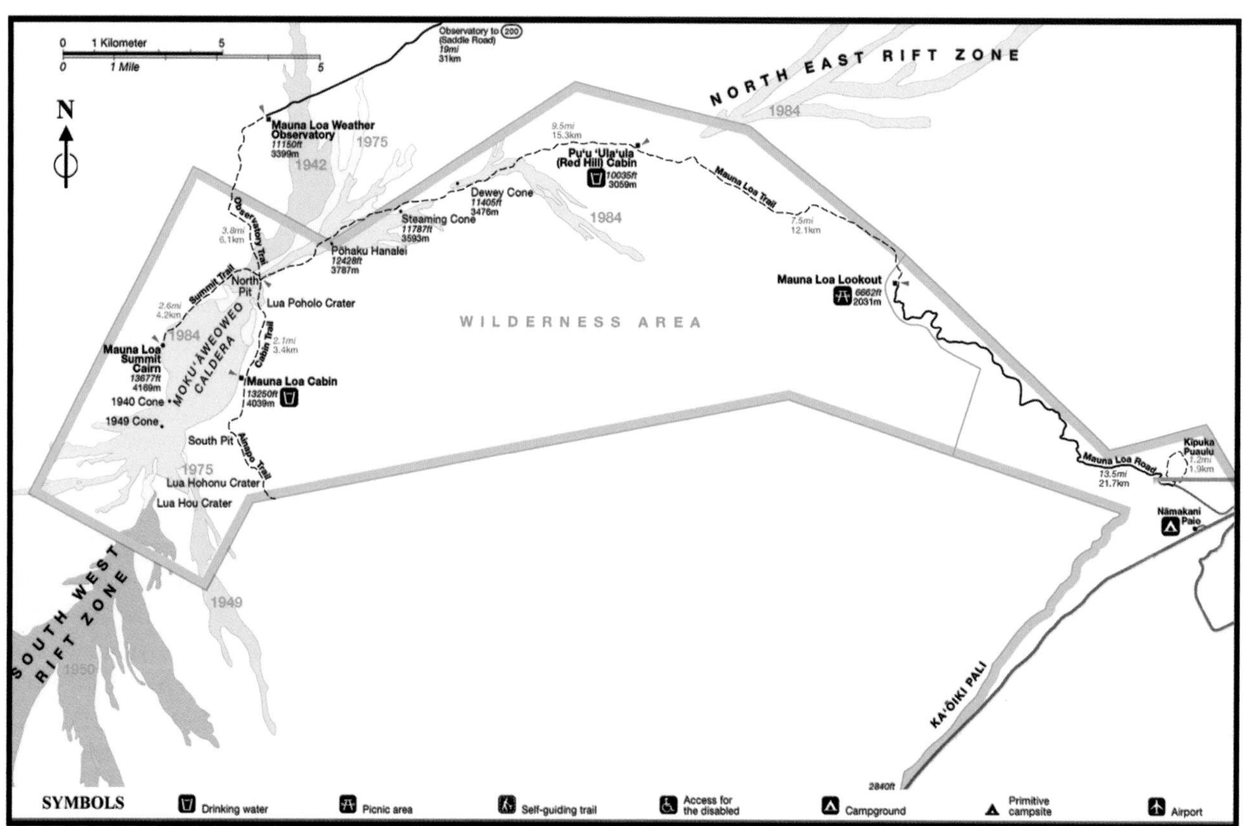

Map of part of Mauna Loa showing the National Park boundary (green) and the two trails to the summit. (Modified from National Park Service map.)

but probably before 1880, because the flow cascading into it is thought to be from 1880.

You can stay overnight at the cabin and follow the Summit Trail the next day. Note, however, that the cabin is at nearly 4,000 m (13,000 feet) altitude. If you started out from sea level, you may suffer from altitude sickness the next day, making your journey to the summit anything from unpleasant to downright impossible.

Mauna Loa Trail

Return trip to Mauna Loa Cabin: 62 km (38 miles), elevation gain 2,008 m (6,588 feet), 14 to 16 hours. Return trip to summit: 77 km (48 miles), elevation gain 2,291 m (7,515 feet), 20 to 22 hours. This is the most popular trail up Mauna Loa because it minimizes the possibility of altitude sickness. It was built in 1915 by the US Army, at the instigation of Jaggar, who wanted access to the volcano's summit eruptions. It is an easier trail than the Observatory's, but it is not for those in a hurry: you will need to allow at least 2 days to go up to the summit and at least one full day to hike down. Most people prefer to allow 3 days up and 2 days down. The Red Hill and Mauna Loa Cabins are overnight stops – remember that you will need a permit. It is best to think of this trip in three parts (or days), the first to Red Hill Cabin, the second to Mauna Loa Cabin, and the third to the summit itself.

From Mauna Loa trailhead to Red Hill Cabin (12 km, 7½ miles): drive up the Mauna Loa road to Kipuka Puaulu and continue on the one-lane, windy road that ends at the Mauna Loa Trail parking lot. The trail starts out crossing ancient lavas that are now vegetated with shrubs of ohelo (*Vaccinium reticulatum*), sacred to Pele, and kukae nene (*Coprosma ernodeoides*), a type of berry named for its unfortunate resemblance to the nene goose's excrement. Soon the vegetation becomes sparse and the terrain rough. Here and there you will find stretches of old, smooth pahoehoe, which are a welcome relief from the rubbly aa lavas. Be careful to follow the cairns and don't wander off the path – collapsed lava tubes are a good reminder that the terrain is dangerous. There are spectacular views of Kilauea from the trail – a good reason to stop frequently and look around. The Red Hill Cabin, the first overnight stop, is located inside the Pu'u Ula'ula ("Red Hill") cone, at 4,039 m (10,035 feet), a product of an eruption several thousand years ago. A short, steep trail leads to a marker on top of the cone, which points to various sights including Mauna Kea and Haleakala. East of the cone is a dark, young-looking lava flow from the 1984 eruption. West of Red Hill is the second flow of the 1880–1 eruption and, beyond it, more lavas from 1984. It is best to go up Red Hill early in the morning, when the weather is most likely to be clear.

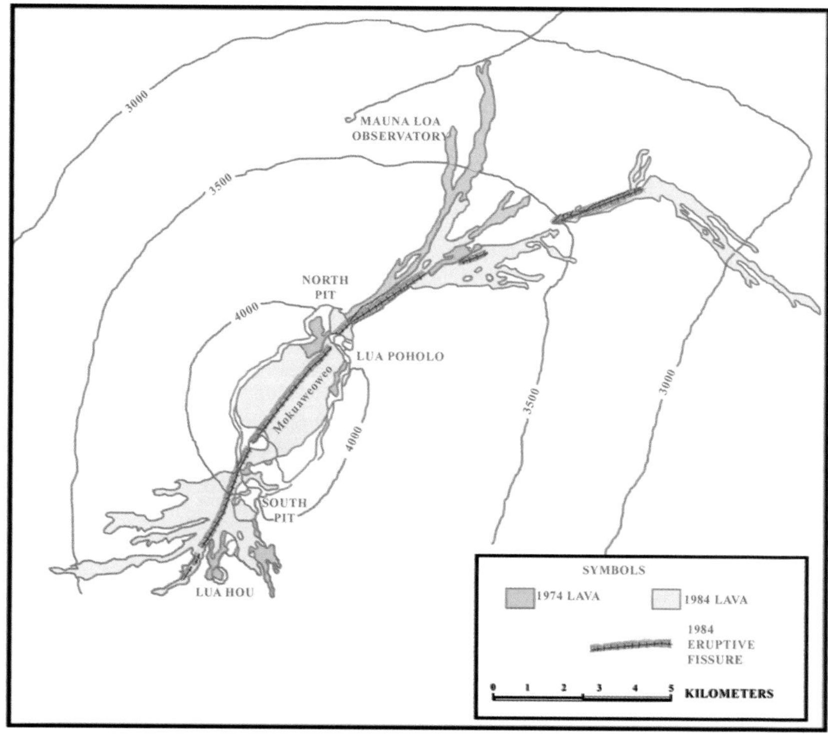

Map of the summit of Mauna Loa showing lava flows from 1974 and 1984. (Modified from Lockwood *et al.*, 1987.)

From Red Hill to Mauna Loa Cabin (18.7 km, 11½ miles): follow the trail from the west side of Red Hill. You will cross the 1880 and the 1984 lava flows and pass near several spatter cones and collapsed lava tubes. Keep an eye on the cairns, as it is not safe to wander off the trail. The landscape is particularly colorful along this trail, as the lava flows are stained with red and golden yellow hues. There are several interesting sights along the way, such as the old Pukauahi Cone at 3,374 m (11,071 feet), to the north of the trail, which is covered with glassy lava. Further along and marked by a sign is another old cone, Dewey, at 3,593 m (11,405 feet). At 3,593 m (11,787 feet) is the photogenic, red-capped Steaming Cone from 1899, which is surrounded by lavas from 1984. If you are lucky, the cone will live up to its name and puff out a little steam. About halfway to the Mauna Loa Cabin, the trail reaches a marked waterhole which may or may not contain some ice – do not rely on it. The trail gets rougher after this point, going over young lavas from 1975 and 1935, until it reaches the Pohaku Hanalei Cone at 3,787 m (12,428 feet). Beyond it is the crater rim and the 13,000 feet marker near Jagger's Cave and the junction of the Observatory and Summit Trails (described above). You have two options here. One is to make your way towards the cabin for a well-deserved night's sleep. The other, if you still have enough energy and daylight left, is to follow the Summit Trail to the top.

Summit Trail

Return trip from Mauna Loa Cabin: distance 15.3 km (9½ miles), elevation gain 283 m (927 feet), about 6 hours. Return trip from the junction with the Observatory Trail: 8.9 km (5½ miles), elevation gain 176 m (577 feet), about 3 to 4 hours. The true summit of Mauna Loa is located on top of the caldera's west wall. To reach it, you must start from the junction with the Observatory Trail described above. All the hiking is above 4,000 m (13,000 feet), so allow plenty of time. The trail goes over flows and the fissure vent of the 1942 eruption, skirts the caldera's North Pit, which is flooded with 1940 lavas, and crosses over numerous fissures. Needless to say, never attempt to negotiate this trail after sunset! Even in daylight, the trail is rough and hard to follow in places. Use the cairns as markers – they should take you to the seismograph station labeled "K Summit Seismograph." As you approach the summit, the cairns direct you southeastward towards the caldera. The view from the rim is truly amazing – the steep walls of Moku'aweoweo seem to disappear into the immense pit. On the caldera floor, there are billows of congealed lava, mostly from 1940, 1949, and 1984. The sight makes you wonder what the caldera must be like when it is filled with a glowing lake, as was the case in 1873, when Victorian globe-trotter Isabella Bird arrived here on muleback. The volcano was in the midst of a spectacular summit eruption and, according to Miss Bird, there were "great fountains of fire below." At night, she could not sleep and crawled back to the edge of the crater to watch the spectacle alone – a scene of "aweful sublimity," she wrote, "light at once of beauty and terror, unwatched by any mortal eyes but my own."

Even in its bleak, quiet state, Moku'aweoweo is hard to draw your eyes away from. However, the true summit, if you feel you must get there, lies a little farther along the path, marked by a cairn, benchmark, and a "register box" where you can record your name. This is a good place to stop and admire the landscape and most of all to feel good about being on top of the largest mountain on the planet, as long as strong winds, which are common here, do not make standing up unpleasant.

The summit

Moku'aweoweo stretches about 5 km (3 miles) in a north–south direction and coalesces at both its northern and southern ends with two nearly circular pit craters. These are called, unimaginatively, North Pit and South Pit. Southwest of South Pit are two other large pit craters, named Lua Hou ("New Pit") and Lua Hohonu ("Deep Pit"), which were formed sometime after 1840. At the northeastern edge we find two more craters, East Pit and Lua Poholo ("Sinking Pit"). If you look south from the summit benchmark, you can see a gap in the caldera wall, beyond which is the South Pit. The prominent cinder cone to the south is from the 1940 eruption. Looking to the north, you may be able to see the summit of Mauna Kea rising above the northern wall of Moku'aweoweo. Looking east across the caldera, you can see Mauna Loa Cabin. It is possible to return to the cabin by hiking on the southern side of the caldera, but the trail is not well defined. Whichever way you choose to hike back, make sure to start well before sunset.

Other (and easier) options to visit Mauna Loa

Summit aerial tour

This is not one of the usual advertised tours for tourists, but it is possible (though expensive) to organize. You will need to make arrangements well in advance to book a helicopter or light plane capable of flying to that altitude and to obtain a permit from the Park. Good weather is essential for the trip, so you will need to be flexible. If you can get a permit from the Park to land a helicopter and find a pilot willing to do it, you can ask to be dropped off on the summit and picked up several hours later. A less expensive option, if you don't mind having only a bird's eye view, is to rent a light plane to fly around the summit area.

Kaumana Cave

This is a rather large lava tube in the 1880–1 pahoehoe flow, the lava that threatened Hilo. The entrance is located just outside Hilo on Kaumana Drive, which becomes the Saddle Road – follow signs to Kaumana Caves Country Park. You reach the tube entrance by means of a staircase down a skylight. There are no trails and no lights, so you will need a good flashlight (plus a spare) to venture into the cool, damp tube. You can get nearly a kilometer into the tube if you are willing to crawl as well as walk. The walls of the tube expose the insides of the lava flow, including layering formed by multiple spillovers of lava from the channel that later became the roofed-over tube. The roof is about 6 to 7.5 m (20 to 25 feet) thick and you need not worry about it collapsing – the skylights you see were formed during or shortly after the eruption and the roof has been stable since. Many lava stalactites drip from the upper walls and roof, formed as lava dripped from the roof as the level of the tube lowered. If you get as far as about 450 m (1,500 feet) into the tube, you will see a smaller tube that was formed when a roof arched over a small channel. Further along still, the pahoehoe floor changes to rubbly aa. Turn back when you start feeling uncomfortable and remember that not many people visit Kaumana, so be particularly careful if you go into it alone.

South Point (Ka Lae) and the Green Sand Beach

South Point is a popular spot in the Big Island, perhaps only because it is the USA's southernmost tip of land. There is, frankly, not much here, and many car rental companies forbid you to drive on the narrow South Point road. The Point itself is part of a strip of prehistoric Mauna Loa lava covered by Pahala ash, an important geologic unit on Hawaii because it serves as a time-marker between lava layers older and younger than that eruption. Looking west from South Point, you can see Pali o Kulani, a slump scarp that was formed when a large submarine slide broke across the submerged southwestern slopes of Mauna Loa. You may be able to see many thin units of Mauna Loa lava flows forming the face of the scarp. Further west is Pu'u Hou, a large cone located about 8 km (5 miles) away at the edge of the 1868 lava flow. Only half of the cone survives, as it has been eroded by the rough waves along this coast.

Much more interesting and unusual than South Point is the Green Sand Beach. It was named for the olivine crystals that weathered out of a tuff cone formed about 7,750 years ago by lava from the Awawa Kahao eruption. Wave action has eroded away about half the cone, leaving the remaining open to the sea. The beach is about 3 km (2 miles) east of South Point and there is a four-wheel-drive road that links the two sites. You may find patches of green sand along the seafront road, in fact, I thought the sand on these small patches looked a brighter green than that on the actual beach. The road ends at an overlook of the beach. It is best to stop here, not only because it is hard to climb down onto the narrow beach but also because of concerns about preserving this rare site. If you choose to climb down, remember that swimming here is not recommended, as the sea is extremely rough and treacherous.

Visiting during activity

Eruptions from Mauna Loa are similar to those from Kilauea, except that, in general, they are less accessible and less frequent. The most recent eruptions of Mauna Loa have been predicted, as the volcano gave out warning signs in the form of ground inflation and seismic activity. The longest known interval between eruptions on Mauna Loa was between the 1950 and the 1975 eruptions. At the time of writing, Mauna Loa's last eruption was in 1984 and the volcano may be due for another soon. If Mauna Loa follows its usual pattern, the next eruption is likely to be at the summit and it will be followed by another flank eruption, with the interval between them being shorter than that between the 1984 and the next summit eruption.

A flank eruption of Mauna Loa typically has three phases. It often starts with activity at the summit and, a few hours later, the true "opening phase" begins. This is the most spectacular phase, with a curtain of lava fountains erupting along a fissure or a series of fissures that can extend for miles. During the second phase, called "cone building," the length of the erupting fissure decreases and the fountains become concentrated in a segment usually less than 500 m (1,650 feet) long. The fountains may reach their maximum height during this time and the spatter around them starts to accumulate, forming ramparts and cones. One or more lava flows start pouring out from the fissure vents, though flows can start as early as the first phase. The final stage of the eruption, called the declining phase, happens when the fluid pressure at the vents decreases and the fountains start to die out. Lava flows may continue to pour out for weeks or even months after the fountains shut off. Mauna Loa eruptions can be long-lived, often lasting weeks to months.

Watching Mauna Loa's eruptions can be as easy as watching Kilauea's, or considerably harder. Mauna Loa's summit and a small part of its flanks are within the National Park, so the restrictions discussed in the Kilauea chapter also apply here. Mauna Loa's eruptions are similar to Kilauea's, so the general advice for eruption viewing is the same, with one important difference: if the eruption is at the summit and you are allowed to climb up, extra precautions are in order. While the sight of a fiery lake inside Moku'aweoweo is worth a few risks, do not forget to use common-sense precautions. Climb to the summit only if you are physically fit but remember that even athletes can suffer from altitude sickness. Having impaired judgment on an active volcano is definitely a bad idea, so take time to adjust to the altitude. Hike with a friend who can help if you get into trouble, or try to join a group of hikers. It is quite possible that the Park Services would only allow hikers to go up during an eruption if they are part of an organized group led by a ranger or guide.

If the eruption is down on the flanks, viewing conditions will be more similar to those on Kilauea, but a lot will depend on how accessible the location is. Eruption fans who do not live in Hilo might hope for a flank eruption that threatens the town – a flow coming near Hilo is likely to be quite easy to get to. In fact, a repeat of the 1880–1 eruption would attract many modern-day Baldwins to come to the Big Island.

Other local attractions

Hualalai and the Kaupulehu lava flow

Perched between Mauna Loa and Kohala is Hualalai, a small volcano that last erupted in 1801. This is the first volcano that many visitors to the Big Island see, as Keahole airport in Kona is built on Hualalai's Hu'ehu'e flow, dating from the last eruption. Hualalai is only 2,521 m (8,271 feet) high, but you cannot climb to the summit, as it is on private land. However, you can visit Hualalai's most interesting location, the Kaupulehu lava flow, which is conveniently located just off Highway 190 (the Hawaii Belt Road). This flow's date is uncertain – it may also be from the 1800–1 eruption – but it is a truly special lava. The erupted magma brought up nodules of green olivine to the surface and today you can find them scattered along the flow's channel. Finding the channel is not easy without a map, so I recommend getting the 1:24,000 map of Kiholo, available from the US Geological Survey and from good bookstores in Hawaii. From Kailua-Kona, take Highway 190 to the east, going past Kalaoa. The road follows the topographic contour of Hualalai at about 600 m (2,000 feet). Note the Hu'ehu'e Ranch, which is located along the rift zone of Hualalai. There is a convenient overlook right on the Kaupulehu lava flow, about 5 km (3 miles) from the Ranch. Park here and walk 500 m (⅓ mile) farther along the road and look for the lava channel on the northern side of the road. Note that this is private land, so stay on the road.

Olivine nodule (to the left of coin) from Hualalai volcano's lava flow known as Kaupulehu. The erupted magma brought up nodules of olivine (xenoliths) to the surface. (Photograph by the author.)

The flow's composition is alkali basalt and it had very low viscosity when it erupted. It flowed very rapidly down the volcano's slopes, with an estimated speed of 20 km/hour (30 miles/hour). There are believable though unverified accounts of people in coastal villages being overwhelmed by the lava. You may notice splash and overflow features on the margins of the channel, attesting to the runny state of the lava as it erupted.

Look for the brownish-green nodules, also called xenoliths (literally meaning foreign rock). The low viscosity lava drained away, leaving the heavier xenoliths exposed. It is not hard to find nodules 3 to 8 cm (1 to 3 inches) in diameter, but their size ranges up to half a meter (20 inches) across. Please remember Pele's curse and resist the temptation to take samples – this is a rare geologic location that should be preserved for future generations to learn from.

Mauna Kea

This volcano, the highest in Hawaii, is now famous for its international astronomical observatory, one of the best in the world – 90% of all stars visible from Earth can be seen from here. Mauna Kea's summit is noteworthy to geologists because it is the only place in the Hawaiian islands known to have had glaciers. The volcano had a permanent ice-cap during the last Ice Age, estimated to have extended for some 72 km^2 (28 square miles). Glaciers leave characteristic markings on the landscape, mostly because the ice moves slowly downhill under the influence of gravity, carrying along large amounts of rock and debris. When glaciers melt and retreat, the rock and debris are left behind as ridges, which are known as moraines. Some moraine deposits are well preserved on Mauna Kea. Another type of deposit, called outwash, can be seen in deep canyons high on the mountain. They were probably formed by floods, caused by eruptions melting the old ice-cap. Keen visitors can also look for small areas of polish and striation on rocks near the summit, formed by the moving glacier abrading the underlying rock. Sometimes the rocks are eroded into rounded knobs which have been given the fanciful name "roches moutonnées", because of their vague resemblance to sheep. If you are interested in glacial geology, make time to hike the upper slopes of Mauna Kea.

It is possible to drive up to the volcano's summit; however, the road requires a four-wheel-drive vehicle and the observatory is not open to the public. Summit guided tours that visit parts of the Observatory are available; call the Visitor Center in advance (808-935-7606). Hardy types can drive an ordinary car up to the Visitor Center at 2,835 m (9,300 feet) altitude and walk up to the summit from there. To get to the Visitor Center, follow the Saddle Road from Kona towards Hilo. Just east of Pu'u Huluhulu (see Mauna Loa Observatory road description), an unmarked paved road heads north up the slopes of Mauna Kea. You will see a number of cinder cones before reaching the Ellison Onizuka Visitor Center, named after the late *Challenger* astronaut who was born in Hawaii. Do not be tempted to drive your car further unless it is a four-wheel-drive vehicle.

Whether you drive or walk, follow the road to the summit past numerous cinder cones, some of which are a striking red color. The road is 14 km (8¾ miles) long and the altitude gain is 1,407 m (4,616 feet) – allow at least 8 hours for a round trip on foot. The road is paved only near the summit, in order to keep the ash content in the air down so as not to interfere with astronomical "seeing." The sight of telescopes perched amongst the cinder cones is quite stunning, the white and silver domes contrasting with the stark volcanic landscape. Remember that drop-in visitors are not welcome here and do not knock on doors. There are no public facilities of any kind at the summit, so you will have to carry your own water and food. The true summit, 4,205 m (13,796 feet) high, can be reached by a footpath near the University of Hawaii 88-inch telescope. On a clear day, you can see the summit of Mauna Loa and its caldera from here, as well as Kohala, Hualalai, and, beyond the ocean, Haleakala. On your way back, watch out for the parking lot on the east side of the road at about 4,000 m (13,000 feet). There is a short trail (800 m, ½ mile) starting from here that goes to Lake Waiau – a tiny, high-altitude lake that the ancient Hawaiians believed to be bottomless. The lake is atop Pu'u Waiau, a cone heavily altered by steam and water which percolated through its walls near the end of the cone's eruption. The alteration produced impermeable clay, which allowed the lake to form. It also increased the runoff from rain and melting snow, resulting in deep gulleys down the cone. On the north side of the lake you will find part of a flow that erupted from the cone of Pu'u Hau Kea, located to the northeast. The flow must have stopped against ice, as it has characteristic features such as lava pillows and mosaic fractures. West of Lake Waiau is Pohakuloa Gulch, where drift from the earliest glaciation is exposed. If you have walked rather than driven to the summit, you may choose to come back down to the Visitor Center via the Mauna Kea Trail, which intersects the Lake Waiau Trail.

Aerial view of Haleakala Crater on the East Maui volcano. (Photograph courtesy of Scott Rowland.)

Haleakala

The volcano

East Maui Volcano, better known as Haleakala ("House of the Sun"), is one of the world's most spectacular volcanoes and the major attraction in the island of Maui. Haleakala's last eruption was around 1790 and, since the island of Maui has drifted past the Hawaiian hot spot, it may not erupt again. Haleakala's size is an impressive 53 km (33 miles) across and 3,055 m (10,023 feet) above sea level, but this only represents 7% of the mountain's total mass. Like all Hawaiian volcanoes, Haleakala is a giant rising from the ocean floor.

Seekers of active volcanoes should not be discouraged from visiting Haleakala as, even in its sleepy state, it can offer some of the most breathtaking views that most of us are likely to see. The summit crater is vast – 49 km^2 (19 square miles), meaning that the whole of Manhattan could fit inside it. Many visitors compare the sight to that of the Grand Canyon – an immense geologic wonder, with walls that plunge down nearly a kilometer (3,000 feet) in places and enclose a barren but colorful landscape. The crater floor is covered with ash deposits in hues of gray, brown, red, and russet, which are broken up here and there by steep cinder cones and dark lava flows.

There are some similarities between Haleakala crater and the Grand Canyon. Both were enlarged by erosion and both are easy to see from the rim – there is a paved road all the way to the summit of Haleakala. Getting down to the bottom is a different matter – the crater is crisscrossed by trails but, unlike the Grand Canyon, there is not a drop of water down there. Haleakala does offer one distinct advantage: the possibility of walking to the coast instead of hiking back up the steep walls. There is no doubt that hiking inside the caldera is the best way to understand its volcanic history.

Haleakala is the youngest volcano in a complex that forms the islands of Maui, Lanai, Molokai, and Kohoolawe. The present island of Maui consists of two major volcanoes, East Maui (with its crater Haleakala) and the older West Maui volcano. Haleakala emerged from the sea as a small island about 1,250,000 years ago, at time when West Maui was near the end of its

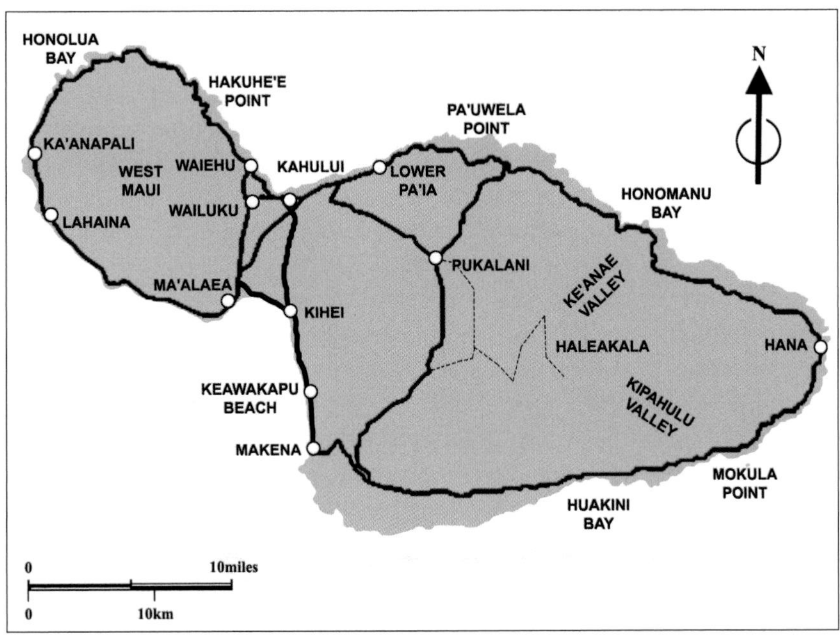

(top) Map of Maui showing major roads and population centers. The winding road up Haleakala is sketched out in dashed lines. (bottom) The island of Maui as imaged by Shuttle Imaging Radar (SIR-C). The cloud-penetrating abilities of radar expose the usually cloud-covered higher-elevation terrains. The multi-wavelength capability of the instrument also allows differences in the vegetation to be identified. The light blue and yellow areas near the isthmus (lowlands) are sugar-cane fields. The middle flanks of the East Maui volcano have rainforests (yellow) and grasslands (dark green, pink, and blue). The major towns appear as small yellow, white, or purple mottled areas. (Image courtesy of JPL/NASA.)

active life. Haleakala grew over the Hawaiian hot spot and became a large shield volcano, its lavas banking against West Maui and forming the broad plain of the Maui isthmus. Maui's major town, Kahului, and airport are located on the north end of the isthmus.

The oldest part of Haleakala is a shield volcano formed by pahoehoe and aa flows of tholeiitic basalt. These lavas are now known as the Honomanu Volcanic Series and they are only exposed in a few places. The next stage of Haleakala's life is known as the Kula Volcanic Series, consisting of hawaiite basaltic lavas and of a greater quantity of ash and tephra from explosive eruptions. Many large cinder cones were formed during the Kula stage. Most of the flows from this period are aa; they are well exposed in the walls of the crater and on the surface elsewhere.

Haleakala's activity started to decline about 750,000 years ago and the next stage in the volcano's life was marked by erosion. By about 400,000 years before present, stream erosion had begun to enlarge the summit crater and to carve giant valleys down the flanks. Erosion was most rapid on the rainy northern and eastern sides. Some streams cut faster than others, became major pathways for drainage, and carved giant valleys, such as Ke'anae on the north, Waihoi and Kipahulu on the east, and Kaupo on the southeast.

Ke'anae and Kaupo valleys had the shortest and steepest courses and eventually the heads of the two valleys expanded and merged, forming a single huge depression that cut across the top of the volcano. Erosion was therefore responsible for forming the most spectacular feature on Haleakala – its giant summit crater. Erosion also destroyed the former summit, which is thought to have been about 900 m (3,000 feet) higher than the present crater rim. Before the formation of the crater, Haleakala probably looked like Mauna Kea does today.

About 40,000 years ago, volcanic activity started again. The main rift zones, the Southwest and the East Rift Zones, reopened, and the great valleys formed by erosion were flooded with lava flows and dotted with cinder cones. Lava and ash from this last period of activity are known as the Hana Volcanic Series. The rock types are essentially the same as those in the Kula series. Both the Hana and Kula eruptions were often separated by long periods of time; in fact, the intervals were long enough for the tops of lava flows to become weathered and eroded before they were covered up by the next flow. One marked difference between the lavas from the two periods is that, in general, the Hana lavas were thinner and more fluid than the Kula lavas. In some places, such as the walls of Ke'anae Valley, the Hana lavas were so fluid that when they poured down steep scarps and valley walls, most of the flow drained away and only a thin coat was left against the wall, sometimes only a few inches thick.

More explosive eruptions also took place during the Hana period, forming some large cinder cones. A major event was the formation of the palagonite tuff known today as Molokini, located about 5 km (3 miles) off the southwestern coast. Erosion has taken away half of Molokini, and now it is a picturesque crescent-shaped islet. After this, Haleakala became less and less active, and only one historical eruption is known.

The 1790 eruption

We have no written accounts of what happened during Haleakala's last eruption. The events have been pieced together from the local geology, Hawaiian oral history, and a couple of contemporary maps. The geology is quite straightforward. We know that aa lava erupted from two vents in the Southwest Rift Zone and flowed into the sea at the present day Cape Kina'u, on Maui's southwestern coast. The first flow came out of a vent 470 m (1,550 feet) above sea level, near the prehistoric cinder cone Pu'u Naio and, a short time later, the second flow erupted from a vent only 175 m (575 feet) above sea level. We know that the lower flow is the youngest because it covers part of the flow from the higher vent. It is actually common for eruptions to progress downslope in this way. Lava from both flows covered a total area of about 5.7 km^2 (2.2 square miles) on land and entered the sea, forming a fan on the north side of the present La Perouse Bay. This eruption was unremarkable, except for the fact that it was Haleakala's last. If the volcano does not erupt again, we can say that it ended its eventful life with a little whimper.

Although the eruption is always referred to as being from 1790, the actual date is uncertain by a few years. The first attempt to date the eruption was made by Lorrin A. Thurston, editor of the *Honolulu Advertiser* and a keen volcano enthusiast, after whom the famous lava tube on the Big Island is named. Thurston visited Maui in 1879 and was told about the eruption by Father Bailey. The missionary had noticed the flow as being rather fresh-looking when he had first come to Maui in 1841. He questioned some of the local people, who told him that their grandparents had seen the eruption. Thurston got another clue to the date of the eruption in 1906, when he talked with a Chinese–Hawaiian cowboy named Charlie Ako. The cowboy's father-in-law, who died in 1905 at the age of 92, had told him that his grandfather saw the eruption when he was a boy old enough to carry two coconuts from the beach to the upper road (a distance of about 8 km (5 miles) and a climb of some 600 m (2,000 feet)). This was Thurston's basis for his proposed date of 1750, a calculation based on the assumption that a Hawaiian generation at the time spanned 33 years. It is a wonderful example of Hawaiian oral history being used to help science, although great accuracy cannot be expected.

The currently accepted date was proposed by B.L. Oostdam in 1965, after he compared charts of the coastline of Maui made by a French expedition in 1786, led by Count de La Perouse, and by the Vancouver expedition in 1793. There is a marked difference in the charts that is consistent with the flow being emplaced sometime between the two expeditions. Thus 1790, give or take a few years, is our best estimate for Haleakala's last eruption, as well as the only date we know for an eruption on Maui.

A personal view: House of the Sun

Every visitor should make sure to get to Haleakala's summit at least twice. The first time should be during the day, to grasp the scale of the crater and become acquainted with its general geology. The second trip should be at night or, more specifically, before dawn. There is a very special reason why Haleakala got its "House of the Sun" name, but you must be at the summit before sunrise to fully understand it.

Hawaiian mythology tells us that in the distant past, the island of Maui received only a few hours of sunlight each day, because La, the sun, would race across the sky so it could go back to sleep. Maui, half man and half god, lived in Hana with his mother Hina, who made tapa, a cloth from tree bark. The short hours of daylight made it hard for Hina to dry her tapa, so Maui thought of a clever way to give his island more hours of sunlight. He knew that La rose over Haleakala by thrusting one sunbeam after another over the crater, like a spider climbing over a rock. Maui carried some strong 'ie'ie vines from Hana up to the summit and waited for La to appear. He then used the vines as lassos and tied up the sun's rays one by one. Trapped, La had to agree to Maui's demand that he always walk slowly and steadily across the sky.

Visitors today can make sure that La is keeping his word – all that is required is to wake up before he does. It is far easier to do this is you can stay overnight at one of the cabins on Haleakala, rather than drive up the windy, slow mountain road in the early hours of the morning. There are three cabins in the crater: Paliku, Kapalaoa, and Holua – permits to stay overnight can be obtained at Park Headquarters. Not surprisingly, the cabins are very popular and are assigned by the Park months in advance, using a lottery system. Camping outside the cabins is not allowed, but there are no rules against sleeping inside your parked car. Whatever way you choose, you can be sure the sight will be worth missing some sleep for. Mark Twain's words say it best: "the sublimest spectacle I ever witnessed, and I think the memory of it will remain with me always."

Visiting during repose

Haleakala crater

The magnificent crater is the highlight of a visit to Maui. The steep, windy Haleakala Crater Road (Hawaii 378) climbs up the western flank and is the only road leading to the summit. Watch out for cyclists – a

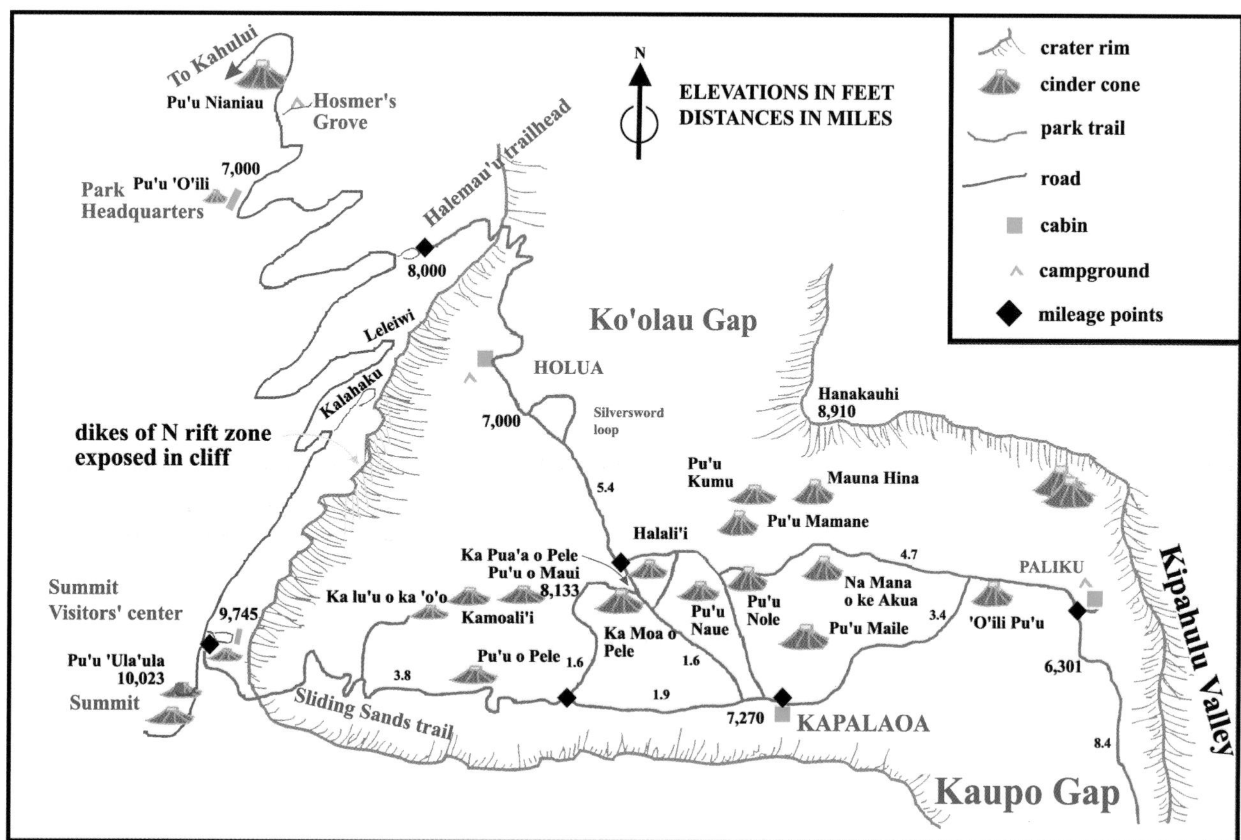

Map of the Haleakala National Park showing the windy road to the summit, the local topography, and major trails and attractions. (Modified from http://www.haleakala.national-park.com, courtesy of Scott Rowland.)

popular tour in Maui takes visitors and bicycles up to the summit and allows them to cycle down the road, which takes several hours. The road is rather scenic and has several overlooks with signs giving information on what you see. All the rocks exposed along the road are basalts from the Kula period and most are alkali basalts rich in green olivine and black pyroxene crystals.

Stop at *Park Headquarters*, at about 2,100 m (7,000 feet), to see exhibits and to collect general information and trail maps. Further along the road is the *Leleiwi Overlook*, from where there is a great view of Haleakala's summit. To the west, you can see West Maui, Molokini, and as far out as Molokai and Lanai. The next stopping place along the road is *Kalahaku Overlook*, from where you can see the largest concentration of silversword (*Argyroxiphium*) plants on Haleakala. Silverswords used to be much more common, but they have been endangered by people and goats. Each plant flowers only once, between June and October, and then dies.

The *Crater Rim Visitor Center* at 2,970 m (9,744 feet) provides a wonderful view of the crater as well as displays on the geologic history of Haleakala. You can buy a variety of books and maps here and, if the weather is inclement, marvel at the crater view from behind a panoramic window. From the Visitor Center, it is only a short walk to the Pu'u 'Ula'ula (Red Hill) overlook. This is the highest point in Maui, 3,055 m (10,023 feet) high. The astronomical domes nearby are part of an observatory from where scientists search for asteroids that may come close to the Earth.

Haleakala crater is 12 km (7½ miles) long by 4 km (2½ miles) wide and nearly 1 km (3,000 feet) deep. Trails crisscross the crater floor, the main ones being the Sliding Sands Trail and the Halemau'u Trail. Exploring all the trails in the crater can take several days, but you can see the highlights in one full day by going down the Sliding Sands Trail on the southern side, crossing the crater, and coming out at the trailhead for the Halemau'u Trail on the northwestern side. You will either need to arrange to be picked up at the end of the trip or else take the trip in 2 days. The 2-day trip can be done either by starting at each end and going down to a halfway point or by staying overnight in the Holua Cabin.

Sliding Sands Trail to Halemau'u Trailhead

Distance 21 km (13 miles); elevation change 512 m, (1,680 feet) downhill; hiking time 7 to 8 hours. The famous Sliding Sands Trail starts by the Visitor Center parking lot. As you get to the crater rim, you can see the trail stretching along the crater wall and the immensity of Haleakala really becomes clear. The beginning of the trail goes over cinders from the Red Hill cinder cone, dating from the Kula period. It is worth stopping before the trail descends into the crater to get your bearings. To your left is the Koolau Gap, in the center is a line of cinder cones from the Hana period and, to the right and far away, is the Kaupo Gap. On a clear day, you may be able to see the summits of Mauna Loa and Mauna Kea beyond the Kaupo Gap.

Notice, as you go down, the native plants such as shrubby dubatias and the famed silverswords, both relatives of the California tarweeds. After about 3 km (2 miles), the Sliding Sands Trail intercepts a subsidiary trail to Ka Lua o Ka Oo ("Digging Stick Pit"), a cinder cone from the Hana period. Follow the side trail up to the cone, going over a photogenic landscape of red-brown cinders dotted with grey silversword and green dubautia plants. The trail leads to the edge of the cone and then up to the rather colorful vent area. Geologists will be interested in the flow this cone produced, basaltic in composition with very large olivine crystals.

Back on the Sliding Sands Trail, continue east through a colorful landscape of brown, pink, golden, and even purple tones that is worth many photo stops. The next highlight is the Pu'u o Pele (Pele's Hill) cinder cone. The trail goes over aa lavas, then bears east-southeast and zigzags down to the junction with the trail to Holua Cabin. Take this trail and head further into the crater, going over Hana period flows. You will notice the highest cone in the crater, Pu'u o Maui (Maui's Hill), on the western side. The trails ascends the reddish Ka Moa o Pele ("The Chicken of Pele"), a cone from the Hana period, and you can see its gaping vent. Soon "Pele's Pig Pen" comes into view – this is a small cinder and spatter cone with a lava flow that has prominent levees, resembling a "pig run." From here you can take a 1.2 km (¾ mile) long loop trail to Kawilinau, a 20 m (65 feet) deep fissure. Follow the loop around the Halalii cone, walking through an area so colorful that it is known as "Pele's paint pot." At the end of the loop around Halalii, you join the Halemau'u Trail. Note the bombs and the accretionary lapilli scattered along the trail and on the flanks of the cone. Accretionary lapilli are small spheres of fine ash, typically formed in eruptions involving some water, possibly by ash accreting around water droplets. The trail also crosses a flow rich in

Inside Haleakala Crater. (left) View of the Pu'u o Maui cinder cone from Ka Moa o Pele. You can get here by following the Sliding Sands Trail. The colorful landscape is worth many photo stops. (Photograph courtesy of Haleakala National Park, US National Park Service.) (right) Interior of the crater viewed from the Sliding Sands Trail. (Photograph courtesy of Scott Rowland.)

olivine and augite crystals that erupted from the base of Halalii.

On the way to Holua Cabin, you may want to take a short detour onto the Silversword Loop Trail to see this rare plant. The vegetation increases as you get nearer the cabin, contrasting with the barren landscape of the crater floor. Once at Holua Cabin, either retrace your steps (not recommended, as it is a lot of effort to go up "sliding sands") or go further onto the Halemau'u Trail to where it meets the Haleakala Highway.

Skyline Trail

This little-known trail is also known as Skyline Drive, though you cannot actually drive on it. Skyline is a trail following the Southwest Rift Zone and it offers spectacular views of Haleakala's lower flanks, the oceans, and the islands beyond. To get to the trailhead, drive from the Visitor Center towards the summit building at Pu'u Ulaula. Take the left-hand turn about 400 m (¼ mile) from the top and bear left. You will soon come across a gate marked "Skyline Drive, Hikers Welcome."

Pu'u Ola'i and the 1790 flow

These are easily reached by driving south from Kihei, along popular beaches and condominium-land. Follow the Wailea-Alanui road past Wailea Beach Park and, farther south, on the inland side of Pu'u Ola'i towards Makena State Park. Pu'u Ola'i is a prominent, 110 m (360 feet) high cinder cone on the coast, dating from the Hana period of rejuvenated activity on Haleakala. The cone was built by mild explosive activity, but near the end of the eruption, a lava flow was erupted. You can follow a trail around and to the top of the cone. On the south side the vent is exposed by erosion, a black basalt dike low on the flank. This is a good example of a dike feeding magma into a lava flow and it is possible to see the point where the dike and the flow connect. From the top of the cone, there is an excellent view of Molokini. There are two lovely beaches adjacent to Pu'u Ola'i: Big Beach to the south and Little Beach to the west, the latter being a well known though illegal nudist beach.

Back on the road, continue towards the south, into the 'Ahini-Kina'u Natural Area Reserve. The road cuts the 1790 flow, a distinct black, chunky aa. Just over 1.6 km (1 mile) inside the reserve there is an excellent view of Haleakala's Southwest Rift Zone. If you look uphill towards the rift zone, you can see the split cone that is the vent of one of the 1790 flows. The road ends on the flow at La Perouse Bay.

Hana Highway

Named the "Highway to Heaven" by tourism promoters, this famous road takes you halfway around the volcano along the northern and eastern coastline. It starts at the town of Kahului and ends at Kipahulu, at the mouth of the valley of the same name. Many visitors to Maui drive the Hana Highway and few forget the experience – 93 km (58 miles) of narrow, windy, and sometimes single-lane road through tropical forest. It is best to take the drive at a leisurely pace and to arrange overnight lodgings at Hana. There is a first-class hotel here that is a favorite with the rich and famous; however, less expensive alternatives can be found, including the very basic Ohe'o Campground near Kipahulu.

The first part of the highway is Hawaii 36, heading

east from Kahului and crossing the alluvial plain of the isthmus between Haleakala and West Maui. A few kilometers east of the town of Lower Pai'a the road goes across Haleakala's Northwest Rift Zone, the least active of the three rift zones. Just before milepost 9, the road crosses a small patch of alkalic basalt flows from the Kula period. Look out for views of the rift zone – the cinder cones inland from Ho'okipa Beach were also formed during the Kula stage of rejuvenated (alkalic cap) volcanism. Soon the road becomes Hawaii 365 and then Hawaii 360, dipping in and out of valleys and eroded channels. Most of the exposed rocks are basalts from the Kula period. Near milepost 11 there is a short trail to Puohokamoa Falls. Water from the river pours over the core of an aa flow and into a pool that many like to take a dip into. The next highlight is Ke'anae Valley, a major erosion valley coming down from the crater. The valley cuts deep and its floor is made up of old lavas from the Honomanu flows; however, they are mostly buried by sediments and Hana lavas. You can go down into the valley and visit Ke'anae Arboretum, where a patch of taro (the plant from which the infamous poi is made) is cultivated in traditional style. Near milepost 19 there is an overlook of the valley that is worth stopping at.

You may also want to stop at some of the other waterfalls along the way to Hana. In Pua'aka'a State Park there are two particularly beautiful waterfalls that drop across successive lava flows and into natural swimming pools. Geologically speaking, the road gets more interesting after milepost 24, when it crosses the East Rift Zone and the landscape is dominated by cinder cones and flows from the Hana and Kula periods. Just beyond the turnoff for Hana Airport, there is a side road to Wai'anapanapa State Park and its black sand beach, formed by waves eroding away at the basaltic lava cliffs. A sea arch on a lava lobe that entered the sea is a clear indication of the eroding power of the waves. There is a large lava tube here that is popular with visitors. Groundwater has flooded the tube and the open skylights serve as swimming holes. An ancient trail leads to Popolana Beach, just north of Hana.

The next attraction is Heavenly Hana itself and its equally heavenly beach of mixed lava and coral sand. The prominent cinder cone by Hana Bay is Kauiki Hill. Hawaiian legends say that "the sky comes close to Hana" and tell us of a deity who stood atop Kauiki Hill and threw his spear right through the sky. The cinder cone also features in Hawaiian history as the site of a fortress, and of battles fought between the Maui islanders and invaders from the Big Island.

The attractions of Hana include a small cultural museum and the venerable Hasegawa General Store, which gave its name to a song. Just west of the town is Pu'u o Kahaula, an old cinder cone from the Kula period which sticks out from the younger flows and tephra that cover its lower flanks. There is a side road from Hana that takes you to a lookout, from where you can see the cone and Haleakala's East Rift Zone.

Kipahulu section of the Park and the Pi'ilani Highway

If you survived the Hana Highway, you can take the next challenge and drive its continuation, the Pi'ilani Highway, from Hana to Kipahulu. Note that the mileposts count down from 52 and that Hana is your last chance to get gasoline and other necessities. The road becomes even narrower and more windy than the

The famous 'Ohe'o Pools (also known as the Seven Sacred Pools) are among Maui's major attractions but it takes effort to get there. They are located near the Pi'ilani Highway, the continuation of the windy, narrow Hana Highway. (Photograph courtesy of Scott Rowland.)

Hana Highway, but the sights are worth the journey. The first highlight is Wailua Falls, which drop 29 m (95 feet) over vertical basalt columns. Next comes the site responsible for most of the traffic on this road: the 'Ohe'o Pools, sometimes wrongly called Seven Sacred Pools. They are wonderful for swimming, amazingly beautiful, and often alarmingly crowded.

The road enters Haleakala National Park just beyond milepost 43 and soon crosses the mouth of Kipahulu Valley, a gigantic, remote valley that cuts the volcano's flanks from the crater down to the sea. The valley is a Scientific Research Reserve and is closed to the public. Lavas poured down the valley during the Hana period and you can see some of the flows exposed in the Palikea Stream Gulch, just beyond the town of Kipahulu. A 3 km (2 mile) trail goes from the Ohe'o campground up the gulch to Waimoku Falls, where the stream spills over ledges of basalt. There are other trails and dirt roads in the Kipahulu area that are worth exploring if you have the time. Driving on the highway beyond Kipahulu is not recommended unless you have a four-wheel-drive vehicle. The unpaved section of the road is dry, rutted, and squeezed between the ocean and the steep flanks of the volcano. You can see the spectacular Kaupo Valley either by hiking about 8 km (5 miles) from the end of the paved section of the road or by driving in from the western side, on the other segment of the Pi'ilani highway. If you have a four-wheel-drive vehicle, you can drive all around the southern side of the volcano and up on the western side back towards Kahului.

Visiting during activity

It is unlikely, but not impossible, that Haleakala will erupt again. The volcano is monitored and, in fact, small earthquakes do occur from time to time. However, these are not the result of magma moving up but rather of subsidence of the crust from the weight of the volcano. Maui is settling on the ocean floor at the rate of a couple of centimeters every decade.

It is hard to predict what an eruption on Maui would be like, or what effect it would have on the island. If we assume that a future eruption might be like that of 1790, we can expect a mild effusive event. However, an eruption here is a big unknown, and local officials would most certainly be extremely cautious. It is unlikely that visitors would have the relatively easy access to the eruption that has been possible on Kilauea. Haleakala is not an old friend, and it might turn out to be like the demigod Maui – a cunning trickster who will strike when least expected.

Other local attractions

West Maui volcano

This shield volcano rises 1,764 m (5,788 feet) above sea level, is 29 km (18 miles) long and 24 km (15 miles) wide. Its youngest lavas are about half a million years old and now the volcano is thought to be extinct. West Maui followed the usual pattern of Hawaiian volcanoes, with mild eruptions of tholeiitic basalts during its early life and more explosive eruptions of more evolved lavas in its later years. The older lavas are called Wailuku basalts and they are generally darker than the younger lavas, which are called Honolua Formation. Rejuvenated activity after these two periods built four cinder cones of highly alkalic basalt, just upslope from the town of Lahaina. These youngest volcanic rocks are called the Lahaina Formation.

It is easy to drive all around the volcano on Highway 30, which follows the coast. Many visitors come to West Maui to see Lahaina, an old port town connected to local whaling history. The most interesting place to go to see the geology of this old volcano is inside its eroded caldera, in the head of the 'Iao Valley. You can drive right into it by following the 'Iao Valley Road (Highway 32), which intersects Highway 30 on the west of Kahului. Along the first 3 km (2 miles) after the intersection, you will be able to see Wailuku basalts exposed in the valley walls. Just over 4.3 km (2¾ miles) from the junction, the road passes through Black Gorge, a tributary of the main 'Iao Valley stream. Many visitors make a stop here because of the mass of dark rocks on one side of the gorge are thought to resemble the profile of John F. Kennedy. The rocks, which date from the Honolua period, are gabbro – intrusive rock that is like basalt in composition but has larger crystals.

The road ends at the 'Iao Valley State Park, inside the old caldera, which has now been carved out by erosion. The natural beauty of this area has made it very popular with visitors and it is often crowded. The first striking sight is the 'Iao Needle, standing about 370 m (1,200 feet) above the valley floor. Although it looks like an isolated rock, it is in fact a knob on the crest of a ridge that divides two stream valleys. The Needle is made up of thin Wailuku basalt flows and it has resisted the erosion that carved out the valley. You can see dikes cutting through the old basalt flows, formed by lava from later eruptions filling in the cracks in the Needle.

You can walk around the Park and see outcrops of lava flows and rubble that partly filled the crater, as well as flowering plants and cascades in the 'Iao stream. It is not easy to see the edge of the old caldera from anywhere in the Park, but if the weather is clear you may be able to trace it from the 'Iao Needle overlook.

You get to the caldera's western rim by taking a road inland from Honokowai on the western coast. The paved road continues as a dirt road, then as a trail about 9 km (5½ miles) long. The trail ends at Pu'u Kukui, the present summit of the volcano, perched on the edge of the caldera at 1,764 m (5,788 feet). Pu'u Kukui, named after a native plant, should perhaps be known as La's Revenge, as it is one of the wettest places on Earth, with annual rainfall of more than 10,000 mm (400 inches). It is not surprising that this is not a popular hike with most visitors to Maui.

Molokini

This crescent-shaped island, very popular with snorkelers and divers, is a half-submerged tuff cone, formed when hot lava encountered shallow seawater. It is easy to get there by boat; in fact, it is almost difficult for a visitor to avoid going snorkeling or diving there. Coral grows in the inside part of the tuff, where snorkelers are taken to. Divers are taken to the outer part of the tuff ring, which has the advantage of being less crowded. Molokini is a Marine Life Conservation District Seabird Sanctuary, so access to the islet is restricted. Frankly, there is not much interesting geology to be seen from the water, but you may still find the experience worthwhile. It is not often that one can swim inside a tuff ring, particularly one that is full of brightly colored, over-friendly, over-fed fish. If you don't like swimming, you can take a sightseeing boat trip or get a good view of the island from McGregor Point, on the south shore of West Maui volcano. McGregor Point is also a good place from which to see Haleakala, Lanai, and Kahoolawe.

References

Lockwood, J. P., *et al.* (1986) *Mauna Loa 1974–1984: A Decade of Intrusive and Extrusive Activity*. US Government Printing Office.

Tilling, R., C. Heliker, and T. L. Wright (1990) *Eruptions of Hawaiian Volcanoes: Past, Present and Future*. US Geological Survey.

8 Volcanoes in the continental USA

The western USA

The United States are both blessed and cursed with some of the most beautiful, fascinating, and dangerous volcanoes in the world. The USA ranks third in the world as the country with the most historically active volcanoes, being topped only by Indonesia and Japan. The USA, however, can be considered tops in terms of variety. The almost-constant, relatively benign activity of the Hawaiian volcanoes is contrasted by the imposing mountains in the Cascades Range and in Alaska, potentially lethal volcanoes that are a source of fear but also pride for the local population. Who does not remember the May 1980 eruption of Mt. St. Helens and the remarkable change it brought to the peaceful landscape? It could erupt again, as could many of the other equally impressive volcanoes along the western edge of the Ring of Fire. Unknown to many people, there are volcanic areas further inland in the USA that still have the potential to unleash violent activity, such as Yellowstone, once the site of a gigantic caldera-forming eruption.

There are whole books dedicated to volcanoes in the continental USA. Here I have chosen to highlight four that are rather diverse, easily accessible, and representative of the variety of volcanic landforms in the continental USA. Lassen Peak is a dome made up of viscous dacitic lava. Mount St. Helens is fascinating both as a composite volcano and as a site of a recent, devastating eruption. Yellowstone, a large caldera, is one of the world's prime geothermal areas, full of attractions beyond the famous Old Faithful. Sunset Crater in Arizona represents a smaller, milder type of volcano, the cinder cone, and is probably the world's most picturesque example of one.

Visiting these volcanoes comes with special bonuses, as most are located in areas that are geologically and scenically stunning. The "other nearby attractions" recommendations could fill many pages. It would be very ambitious to try to visit all the four volcanoes highlighted here, plus others described as nearby attractions that should not be missed, in a single trip. Yellowstone deserves at least 1 week, and at least another week would be well spent at Sunset Crater and the Grand Canyon, Meteor Crater, and a myriad of other geological wonders in the area. It is possible to visit Lassen Peak and Mt. St. Helens in one trip, but remember to allow enough time for at least a quick visit to Mt. Shasta, Crater Lake, and Mt. Rainier. Keen mountaineers may want to spend extra time in the area so they can climb Mt. Shasta, Mt. Rainier, or some of the other nearby volcanoes, such as Mt. Hood.

The proximity of the volcanoes in the western USA to densely populated cities and cultivated farmland means that a constant threat is present and the volcanoes must be watched around the clock for any signs of increasing activity. Monitoring of all the Cascades volcanoes, plus Long Valley and Mono Lake in California, is carried out by the US Geological Survey's Cascades Volcano Observatory (CVO). The observatory is not open to the public, but it provides information on the volcanoes by phone and via a homepage.

It is extremely unlikely that any of these volcanoes could erupt without warning. However, they are capable of extremely violent eruptions, as the placid-looking Mt. St. Helens demonstrated so well in 1980. Even the most up-to-date monitoring techniques cannot yet predict the date and magnitude of an eruption from these volcanoes, which are significantly more complex than their Hawaiian cousins. If CVO issues a warning to stay away from an area, be sure to do so. There are plenty of volcanoes to see in the western USA and, with the exception of Yellowstone's geysers, it is best to visit when all is quiet.

Tectonic setting

The American West is a complex volcanic region that spans not only a large geographical area but also a

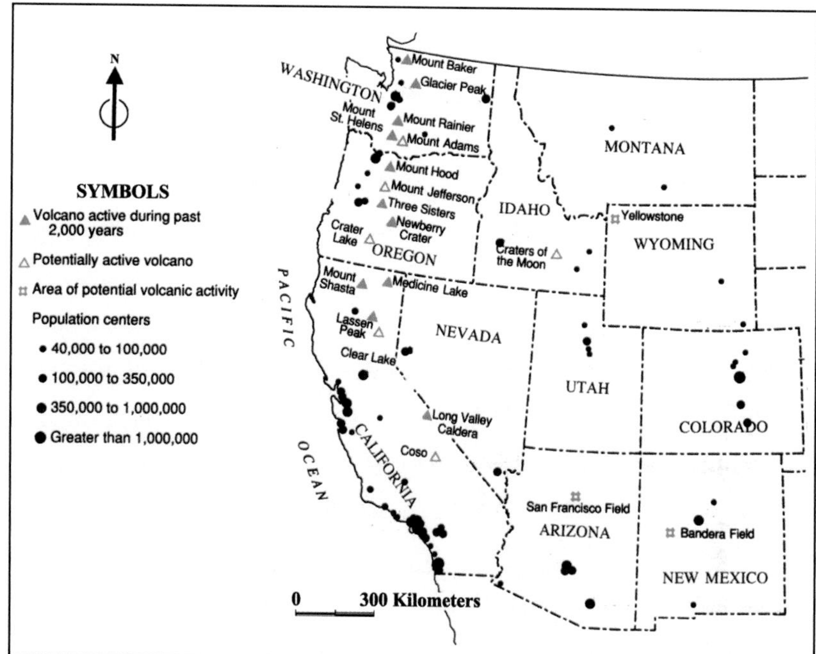

Active and potentially active volcanoes in the USA. Note that several of the Cascades volcanoes are located close to major population centers. (Modified from Brantley, 1995.)

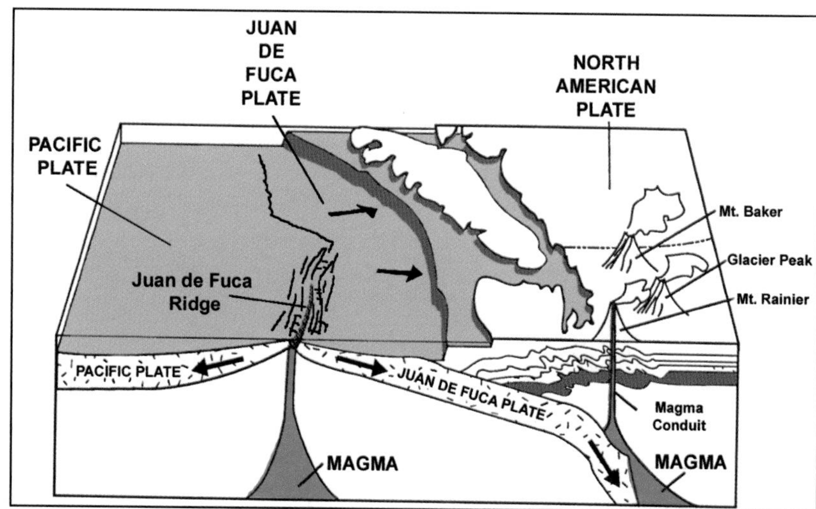

Tectonic setting of North America. The Juan de Fuca plate plunges underneath the North American plate. The boundary between the two plates is marked by a broad submarine mountain chain known as the Juan de Fuca Ridge. The ocean floor is spreading along the Ridge. (Modified from Brantley, 1995.)

wide range of volcanic styles and rock types. The tectonic setting involves subduction along the Cascades Range and hot spots, rifts, and faults elsewhere. The activity of the western USA during the last 5 million years or so has been very mild in comparison with that during the mid-Tertiary period, when violent caldera-forming eruptions took place. Why these violent eruptions occurred when and where they did is still not well understood. More is known about the relationship between the volcanism going on today and the tectonic setting, but many details still need to be worked out.

The tectonics of the Cascades region is dominated by the subduction of the Juan de Fuca plate beneath the North American Plate. The Juan de Fuca plate is separated from the Pacific plate by the Juan de Fuca Ridge, a zone of spreading in the Pacific Ocean. The present rate of convergence between the plates is relatively slow at 3 to 4 cm (1¼ to 1½ inches) per year and is considered to be about half of the rate of about 7 million years ago. This may be why volcanism has declined since that time, and why there is relatively little seismic activity along the subduction zone.

The Cascades volcanoes have the usual characteristics of subduction-zone volcanoes: explosive eruptions as well as lava flows have formed steep-sided volcanoes, magma types range from basalt to andesite and dacite, and pyroclastic flows and mudflows have occurred many times. There are 11 major volcanoes in the Cascades that have erupted within the last 2,000 years and several others that are potentially active.

Away from the Cascades Range which spans northern California, Oregon, and Washington State, there are other large volcanic areas that are considered potentially active. These include the Craters of the Moon and Snake River Plains lava fields in Idaho, Yellowstone in Wyoming, the San Francisco volcanic field in Arizona, the Bandera field in New Mexico, and Long Valley Caldera and Mono Lake in Eastern California.

The Snake River Plains–Craters of the Moon and Yellowstone regions are part of an 80 km (50 miles) wide, 450 km (280 miles) long swath of basaltic and rhyolitic volcanism that can be considered the most dynamic volcanic area in North America. This may seem surprising, since there have been no historical eruptions in these areas. However, the whole region is moving to the northeast at a rate of 3.5 cm (1½ inches) per year. In about 20 million years, it will be at the Canadian border, having played some havoc with the Montana landscape. The Snake River Plains–Yellowstone region is very complex, as it involves a hot spot, major faults, and great subsidence of the crust between the faults. The volcanism has included the violent formation of great calderas, such as Yellowstone, but also the effusion of enormous quantities of lavas, such as the flood basalts of the Snake River Plains.

Another major volcanic region is the Colorado Plateau. Volcanism has not occurred within the plateau itself, but rather near its boundaries in the states of Utah, Arizona, New Mexico, and Colorado. It is speculated that perhaps the repeated vertical movement uplifting the plateau over millions of years opened up paths for magma to come to the surface. The volcanoes on the edges of the plateau are mostly small, isolated fields of monogenetic cinder cones such as Sunset Crater, created during a single phase of activity and not expected to erupt again.

There are two other major volcanic provinces in the USA, the Rio Grande Rift–Jemez Zone volcanic field in New Mexico and the Eastern California region which includes Mono Lake and Long Valley caldera. Volcanism in New Mexico is a result of two major tectonic features, the Jemez Zone and the Rio Grande Rift. The two major volcanic fields of this region, called Jemez and Taos Plateau, are found at the intersection of these two great tectonic systems. Eastern California has scattered volcanoes, mostly basaltic fields of monogenetic cinder cones. Regional stresses control volcanism in this region, but the formation of the rhyolitic Long Valley caldera is considered something of an anomaly. The eruption that formed the caldera about 700,000 years ago was much more violent and voluminous than any other in California. From time to time there is still unrest in the caldera, in the form of swarms of earthquakes, changes in hot springs, and ground uplift. Eruptions from the Mono–Inyo Craters Volcanic Chain, located to the northwest of Long Valley, have occurred as recently as 550 years ago. Future eruptions are possible here, and very likely at many of the other volcanoes in the western USA.

Practical information for the visitor

The National Parks and Monuments

Most of the volcanoes highlighted here are part of the US National Parks system, with Sunset Crater and Mt. St. Helens being National Monuments, Mt. Shasta being part of a Wilderness area and the others being within National Parks. The Monuments and Parks do a remarkable job of offering the best to visitors: plenty of wilderness areas for the adventurous types, and guided walks, exhibits, and lectures for those who want to learn about volcanoes but would rather not explore remote locations. The postal and World Wide Web addresses of the National Parks are given in Appendices I and II.

Transportation

The USA is the land of the private car. Rental cars are widely available and are generally cheaper than in most other countries. Visitors under the age of 21 may not be able to rent a car and should check with several companies well in advance. Public transportation is generally scarce. The National Park and National Monument Headquarters are good sources of information on currently available public transportation to and within the Parks and Monuments.

Tours

All National Parks and Monuments offer naturalist and ranger-led tours. The Yellowstone Institute, an educational institution, offers guided tours to Yellowstone National Park and a variety of classes and short courses, including one on the geology of Yellowstone.

Tours from the air, using helicopters or light planes, are available (and widely advertised) at Mt. St. Helens. For other locations, inquire at the local airport (Flagstaff for Sunset Crater, Redding for Lassen Peak, West Yellowstone for Yellowstone (summer months only), Seattle/Tacoma for Mt. Rainier, Klamath Falls and Medford for Crater Lake, and Montague/Yreka for Mt. Shasta). The Mount Shasta Balloon Company (916-841-1011) and Dream Chaser Balloon Adventures (916-938-2315) operate hot-air balloon flights that come close to Mt. Shasta for some different volcano sightseeing.

Lodging

A wide range of accommodation is available within easy driving distance of each volcano described here. Camp grounds, youth hostels, and inexpensive motels are generally easy to find. A great source of information on lodging is the American Automobile Association (AAA). Their guidebooks for each state are available for only a few dollars from AAA offices. The National Park and National Monument offices should be contacted for current information on campgrounds and accommodations within the parks. All the Parks and Monuments require that visitors register at headquarters or obtain a wilderness permit for overnight stays in the backcountry.

Sunset Crater National Monument

The town of Flagstaff, on the historical Route 66 highway, is the obvious choice for lodging near Sunset Crater, offering a wide variety of mostly budget to moderately priced accommodations. Camping on Sunset Crater is allowed on the Bonito Campground, operated by the US Forest Service, and located across from the Visitor Center.

Yellowstone National Park

The National Park itself provides lodges, cabins, and camping facilities within the boundaries of four out of five of its "countries." The Park is so large that, if you can spend the time to see a significant part of it, it is worth moving lodgings during a single visit. Geyser Country offers the Old Faithful Inn, a logs-and-stone National Historic Landmark building dating from 1904. The inn is not cheap but it is full of charm, with an enormous stone fireplace in the lobby and an elegant dining room that serves sumptuous meals. If your budget is tight, stay nearby at the Old Faithful Lodge Cabins. I recommend booking one of the cabins that has a private bathroom. The extra money is worth spending to avoid a possible night-time encounter with one of the bison that seem to like wandering around the cabins. If you visit the Park during the winter, Old Faithful Snow Lodge, also in Geyser Country, is one of only two choices. The other is Mammoth Hot Springs Hotel and Cabins in Mammoth Country, open year-round. The Canyon Lodge and Cabins, located near the spectacular Lower Falls, are convenient for those visiting Canyon Country. Another alternative is the beautiful Dunraven Lodge, which opened in 1999. The Roosevelt Lodge Cabins in Roosevelt Country consists of a 1916 rustic log lodge and cabin facility near the Tower Fall area, which was a favorite of President Teddy Roosevelt's. In total, there are some 600 hotel rooms and 1,600 cabins inside the Park, but booking early is essential due to the large number of visitors.

Those wishing to camp have a choice of 11 campgrounds operated by the Park. Seven of these are available on a first-come, first-served basis. Three other campgrounds can be booked well in advance, while the sites at the Bridge Bay Campground can be reserved up to 8 weeks in advance. For those traveling with recreational vehicles (RVs), there is Fishing Bridge RV Park, with more than 300 full hook-up sites. For all camping and RV Park reservations, call (307) 344–7311.

Mount St. Helens National Monument

There are many public and private campgrounds in the area, though not within the Monument, and a variety of motels and inns are available in the towns surrounding the mountain. Because of the layout of roads, Mt. St. Helens is best visited in three stages: the western side, the northeastern side, and the southern side. To save driving time, you may want to change lodgings during your visit.

Campgrounds include Seaquest State Park (1-800-452-5687) on Highway 504 on the western side, Iron Creek (1-800-280-2267) on Forest Road 25 on the northeastern side, and Kalama Horse Camp (1-800-280-2267) on Country Road 81 on the southern side, all well-situated public campgrounds. Additional campgrounds are operated on the southern side by Pacificorp (503-464-5035).

For motels and inns, try the towns of Castle Rock (western side), Cougar (southern side), and Randle (northeastern side). There are plenty of standard-fare motels and bed-and-breakfast places available. Consult a AAA guidebook for up-to-date listings and prices.

Lassen Volcanic National Park

The only lodging inside the Park is the Drakesbad Guest Ranch, located in the Warner Valley area, 29 km (18 miles) from the town of Chester. The ranch offers meals, a geothermally heated pool, and horseback tours to local attractions. However, it is only open during the summer and its location is not particularly convenient for the western part of the Park, where most of the attractions are. There are numerous places to stay outside the Park – ask the Park services for a list of lodgings or use the AAA guide. Lodgings in the town of Mineral, 16 km (10 miles) from the southwest entrance to the Park, are very conveniently located for Lassen Peak and various other attractions of the Park. Those willing to camp have a wide choice. The Park offers several campgrounds, including those at Manzanita Lake, Juniper Lake, Butte Lake, and Warner Valley. Check with the Park Headquarters for availability. There are also several campgrounds outside the Park that are privately operated.

Mount Shasta

Motels, hotels, and bed-and-breakfast places are available at Mt. Shasta city, Weed, Dunsmuir, McCloud, and other nearby towns. The Siskiyou County Visitor's Bureau has a toll-free number (800-446-7475) for information on lodgings. Camping is allowed inside the Mt. Shasta Wilderness area.

Crater Lake

The famous Crater Lake Lodge is the best choice though not the cheapest. Built in 1909, the lodge narrowly escaped demolition in the late 1980s but public protest forced its restoration and it reopened in 1995. It is only open during the summer and has only 71 rooms, so booking ahead is essential. The lodge's most impressive feature is its view of the magnificent lake. If you cannot stay at the lodge, consider coming to dinner in the original ponderosa pine and flagstone dinning room. A cheaper alternative for lodgings is the Mazama Village Motor Inn (call the lodge for reservations). Camping is also available: the Mazama and Lost Creek Campgrounds, both located within the Park, operate on a first-come, first-served basis and are open during the summer only. Winter lodgings are not available inside the Park. Motels and bed-and-breakfast places can be found in the nearby towns of Klamath Falls and Medford.

Mount Rainier

There are two imposing, rustic lodges inside the National Park: Paradise Inn, open only during the summer months, and the National Park Inn at Longmire, open year-round. Numerous motels and bed-and-breakfast places are available in and around Ashford, located just outside the Park's Nisqually Entrance. There are five campgrounds inside the Park: Cougar Rock, Ipsut Creek, Ohanapecosh, Sunshine Point, and White River. Several of these are open all year and they all operate on a first-come, first-served basis.

Safety and emergency services

The USA probably has the best emergency rescue services and medical facilities in the world, but they come at a price. Visitors from abroad should obtain travel medical insurance before leaving their home countries. The US volcanoes described here are located in areas where violent crime is not a considerable problem, but remember to use caution. Hitch-hiking (or picking up strangers) is not considered safe in any part of the USA and neither is sleeping in your car in public "rest areas" on roads.

Maps

The US Geological Survey publishes topographic and geologic maps of all the USA – you can buy them by mail from the US Geological Survey (see Appendix I for their addresses). The 1:24,000 topographic map series is particularly useful, and geological maps are available for all of the volcanoes described here. Less detailed but useful maps will be given to you at the entrance to the National Monuments and Parks. A variety of maps are available for sale at the various Visitor Centers and local bookstores. Don't forget to check the World Wide Web as maps are often available (see Appendix I for addresses).

Lassen Peak

The volcano

Lassen Peak is the world's largest plug-dome – a volcano formed by viscous lava that erupts as an almost rigid piston, eventually plugging its own vent. Lassen Peak is the southernmost volcano in the Cascades Range, as well as its tallest non-stratocone, rising 610 m (2,000 feet) above its surroundings and reaching an elevation of 3,187 m (10,457 feet). Before the eruption of Mt. St. Helens in 1980, Lassen held the honor of being the most recently active volcano in the continental USA. Although its eruptions are infrequent, Lassen more than earns its place amongst the world's best volcanoes to visit because it is an

Lassen Peak seen from Reflection Lake. (Photograph courtesy of Nick Gautier.)

outstanding example of a lava dome. There is still plenty to remind visitors that this volcano is far from dead: steaming active fumaroles on the summit area, hot ground on the northern flank, and extensive geothermal areas in the vicinity.

Lassen is easy and enjoyable to visit. Its location is outstanding, not only because of magnificent scenery but also because of the variety of volcanic landforms nearby. The Lassen Volcanic National Park was created in 1916, during the last major eruption, although Lassen Peak itself and its neighbor Cinder Cone had been declared National Monuments as early as 1907. Lassen is a rather special volcanic National Park, not only because of Lassen Peak itself but also because of the most extensive collection of active geothermal features in the Cascades Range. Among the major attractions is Bumpass Hell, a spectacular geothermal area complete with boiling mudpots, hissing fumaroles and colorful sulfur deposits. Another highlight is Cinder Cone, a very fine example of its kind. The Park even has some small, Icelandic-type shield volcanoes of andesitic lavas, plus Chaos Crags, a group of dacitic domes covering an area of about 5 km^2 (2 square miles). Chaos Jumbles, formed by collapse of one of the domes, is a textbook-type example of a rockfall avalanche. Visitors wanting wider geologic variety can seek out glacial features and tectonic faults. Lassen is not, however, interesting only to the geologically inclined. The spectacular scenery and range of facilities offered make the Park an ideal place for families to come and enjoy nature. There are campgrounds, educational exhibits, extensive hiking trails, horse-riding, boating and fishing in azure lakes, and countless opportunities for watching wildlife.

The varied volcanic landscape of the Park was produced by a combination of different types of eruption that begun with the formation of an andesitic stratovolcano, Mt. Tehama, about 600,000 years ago. Mount Tehama (sometimes called Brokeoff volcano) is thought to have been a stratocone about 3,350 m high (11,000 feet) and 24 km (15 miles) in diameter, built mostly by andesitic flows (specifically, olivine-augite and hypersthene-augite andesites) interbedded with pyroclastics. Its evolution may have been similar to that of Mt. Mazama in Oregon, now site of Crater Lake. Dating of lavas from Mt. Tehama show that the volcano was active for about 200,000 years. Mount Tehama was cut by a series of faults and it eventually collapsed, forming a caldera some 4 km (2½ miles) wide. The main vent of Tehama is thought to have been where Sulphur Works is now located. Tehama was deeply eroded by glacial action and by hydrothermal activity, making the rocks friable, but we can still see the remnants of the caldera in the southwest corner of the Park: Brokeoff Mountain, Mt. Diller, Pilot Pinnacle, Diamond Peak, and Mt. Conard form a broken circle that marks the caldera.

After the extinction of Tehama, there was a major change in the character of volcanism in the area. Activity shifted to the north flank of Tehama and the erupted lavas were more silicic. Three sequences of silicic lavas and a group of hybrid (mixed) lavas were extruded in the last 400,000 years, starting with the eruption of small rhyodacite lava flows and a rhyolite dome. Some of the lavas are still preserved on Raker Peak and Mt. Conard. This activity was quickly followed by an explosive eruption of more than 50 km^3 (12 cubic miles) of rhyolitic magma, forming ash flows and producing a small caldera on the northern flank of the old Tehama. The caldera was later filled by lavas from the next sequence of activity. These lavas formed 12 lava domes and various thick flows associated with them. The second sequence lasted from about 250,000 to 200,000 years ago, extruding some 15 to 25 km^3 (3.5 to 6 cubic miles) of lava. The main vents were concentrated along the northern edge of the Tehama caldera and the most prominent domes are Bumpass Mountain, Mt. Helen, Ski Heil Peak, and Reading Peak. The third sequence of activity also involved the formation of lava domes, which now form the northern and western portions of the dacite dome field. The domes were emplaced in about 12 separate episodes, starting about 100,000 years ago and involved not only the extrusion of pasty lava as short, thick flows, but also some pyroclastic flows. The most prominent lava domes from this period are Eagle Peak (about 55,000 years old), the Sunflower Flat domes (35,000 years), Lassen Peak (25,000 years) and Chaos Crags, erupted only about 1,050 years ago. The vents of these domes are concentrated along linear chains near where the

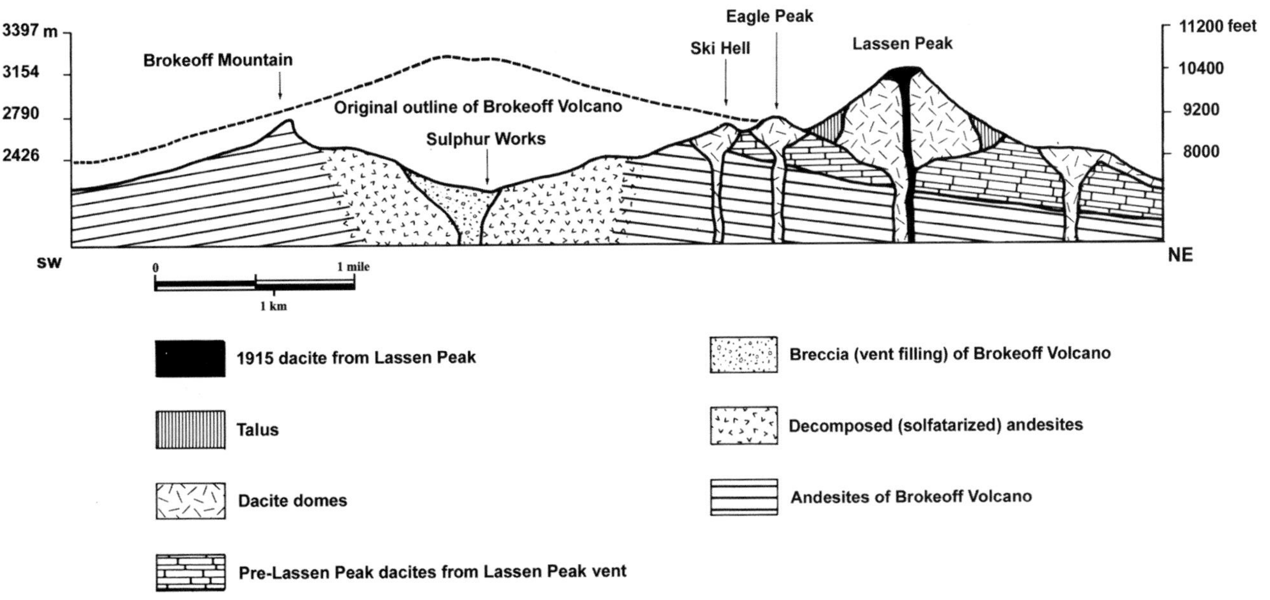

Schematic cross-section through Brokeoff Volcano and Lassen Peak. (Modified from Kane, 1990.)

western edge of the Tehama caldera is thought to have been.

Lassen Peak was born as a volcanic vent on the northern flank of Mt. Tehama. Before the Lassen dome was built of pasty dacitic lava, the same vent extruded large quantities of fluid lavas, called the "pre-Lassen dacites." The largest remaining example of these lavas is now known as Loomis Peak. After the fluid dacites, at least one pyroclastic flow was erupted from the vent. Its deposits are found in Kings Creek Meadows area and were dated as 11,000 years old. The lava dome itself was erupted immediately after the pyroclastic flow. At that time, glaciers still mantled the remains of Mt. Tehama.

The formation of Lassen Peak would have been a terrifying event to witness. As the massive body of lava moved up, parting the overlying rock, strong earthquakes no doubt rocked the land. The pasty mass emerged, slowly building the great dome. It is thought that the slow extrusion lasted about 5 years and, after that, the volcano's enormous mass plugged the vent and no more magma came out. As the lava cooled and fractured, it slid down and formed a talus slope, on which glaciers left behind their signature in the form of small moraines. It is possible to find some places where the original dacitic lava can still be seen, giving us an idea of what the steep original dome must have looked like.

Following its birth, Lassen Peak suffered the effects of glacial erosion. During the Pleistocene, the Lassen area was covered by large ice sheets. The eruption of Lassen occurred close to the end of the Pleistocene, when a large ice sheet had melted away. Later on, during a return of the cold climate, valley glaciers formed and eroded shallow cirques on the north and northeast sides of the dome. South of Lassen Peak, Lake Helen and Emerald Lake formed in cirque basins. Evidence of glacial erosion on Lassen is apparent in many areas. Glacial debris was added to the volcanic debris and both types of material are found mixed together in deposits from mudflows and avalanches.

The 1914–21 eruption is the only one to be recorded at Lassen Peak, but there are clues to earlier eruptions. It is known that before 1914 the dome was topped by a large, bowl-shaped, tephra-floored summit crater about 300 m (1000 feet) in diameter and up to 110 m (360 feet) deep. This crater was formed by a large explosive eruption, probably at least as violent as any in the 1914–21 episode. Other evidence for activity before 1914 comes from mudflow deposits that had the Lassen dome as their source. These deposits are found along Hat Creek, Lost Creek, and east of the Devastated Area. Some of them may be less than 500 years old.

The last sequence of lavas in the Lassen area were a group of hybrid (mixed) andesite lavas that formed lava flow complexes with cones marking their vents. About 10 km^3 (2.4 cubic miles) of these lavas were erupted around the margins of the dacite dome field. The eruptions started about 300,000 years ago and were grouped in ten major episodes. The products from the 1915 eruption of Lassen fall into this group, as do Hat Mountain and Cinder Cone. Some of the most recent activity in the area dates from a few thousand years ago, when eruptions took place from vents along a line running from south of Old Station and through Prospect Peak. The Hat Creek lava flow, thought to be less than 2,000 years old, was erupted from the northwest end of this line, about 10 km (6 miles) north of Badger Mountain. Another flow was erupted from the tephra cone located in the depression between West Prospect and Prospect Peaks. The blocky basalt flow reached 6 km (4 miles) to the north and still appears fresh.

About 1,200 years ago, the Chaos Crags complex begun to be formed, initially as low cinder and tephra cones built along fissures a few kilometers north of Lassen Peak, at the head of the Manzanita Creek. Several pyroclastic flows from these cones run down Lost and Manzanita Creeks. Soon after came the extrusion of dacitic lavas, forming the plug-domes. About 300 years ago, part of one of the domes collapsed creating a rockfall avalanche and the deposit now known as Chaos Jumbles. The tallest dome in the group was reportedly still emitting large quantities of steam as recently as 1857.

One of the youngest features in the Park is the 225 m (750 feet) high, beautifully symmetrical Cinder Cone. Its summit crater has two distinct rims, plus segments of two others, indicating that several eruptions took place from here. Four basaltic andesite lava flows erupted from this vent. The oldest of the eruptions probably took place around 1567. The eruptions deposited an ash blanket that covered about 300 km^2 (116 square miles) and two of the flows are younger than this ash. The most recent flow was documented by eye witness accounts in 1851.

Although historical records of the Lassen area go back less than two centuries, the volcanic forces must have been known to Native Americans. One of the Indian names for Lassen Peak is "Amblu Kai," meaning "Mountain Ripped Apart" or "Fire Mountain." The

Lassen area was a meeting place for four tribes in the warmer months – the Atsugewi, the Mountain Maidu, the Yahi, and the Yana – but it is not conducive to year-round living. The tribes were not pottery-makers, so they left behind little evidence of their presence and no archeological sites that could help us date deposits. The first white man to come to Lassen was Jedediah Smith, on his way to the west coast in 1821. The California gold rush brought settlers to the area and Lassen Peak was eventually named after one of them – Danish blacksmith Peter Lassen, who came in 1843 or, according to different sources, during the 1830s. He obtained a large land grant from the Mexican governor of California, started a ranch, and begun guiding other pioneers west through the Lassen Road, located north of the present Park. It is not clear that Peter Lassen ever climbed the mountain that came to bear his name. There is an unconfirmed story that he often got lost while guiding his hapless pioneers, because he couldn't tell the difference between "his" mountain and Mt. Shasta. The story says that, on one occasion, a pioneer forced Lassen at gunpoint to climb the Peak so that he could get his bearings. However, the earliest recorded ascent of Lassen Peak is credited to the party of Grover K. Godfrey in 1851. The second known ascent, in 1864, included a lady, Mrs. Helen Tanner Brodt, whose purpose was to sketch the beautiful landscape. It is fitting that she gave her name to Lake Helen, one of the most scenic areas of the Park.

The 1914–21 eruption

The eruption was long and varied, with many explosive episodes and the extrusion of lava flows, mudflows, and a devastating pyroclastic flow. It was truly notable for its many moments of awesome beauty and for showing us that the majestic Cascades volcanoes are far from dead. Lassen's eruption attracted eager spectators and must have been an ideal eruption for visitors to watch from a distance, provided they could stay for long enough, as the volcano took long rests between events.

The first sign of activity that we know of was a brief eruption of ash and steam from the summit crater on May 30, 1914. Intermittent explosions continued for a year, with observers listing 170 separate events. It is likely that there were many more and that they went unobserved because of winter storm clouds hiding the Peak from view or because they happened at night – Lassen was merely clearing its throat and the material thrown out was non-juvenile, that is, just bits of old dacite. At this stage, glowing magma was not yet coming out. Ash from these initial explosions reached the towns of Manton and Mineral, respectively 32 km (20 miles) to the west and 18 km (11 miles) to the southwest of the volcano. Magma was slowly pushing its way upwards and Lassen soon put on a better show. On the night of 14–15 May, 1915, observers reported a red glow around the summit. On May 18, the postal mistress at Manton saw, through a gap in the summit crater rim, a "black mass" slowly pushing its way upwards. Magma had finally arrived at the surface.

In the early hours of May 19, lava filled the summit crater and spilled over the rim. Lava flowed down the west flank, much to the delight of those watching. A Mr. G.R. Milford was one of the spectators observing from the town of Volta, 31 km (19 miles) west of the volcano. He later wrote about the "wonderful spectacle" in a letter to a geologist: "The glow commenced to increase in brilliancy, now brighter, now still brighter, until behold, the whole rim of the crater facing us was marked by a bright-red fiery line which wavered for an instant and then, in a deep-red sheet, broke over the lowest part of the lip and was lost to sight for a moment, only to reappear again in the form of countless red globules of fire about 500 feet below the crater's lip." He went on to describe the "globules" as they "rolled down the mountainside." Mr. Milford probably did not know that such behavior is expected of a blocky lava flow emplaced over steep slopes – blocks break away from the edge of the flow and roll downhill. There was another, unreported lava flow down the northeast side that caused large quantities of snow to melt, creating mudflows down the Lost Creek and Hat Creek valleys that reached more than 20 km (12 miles). The mudflows carried large blocks of lava

Eruption of Lassen seen from Anderson, California, on March 22, 1915. (Photograph by R.I. Meyers, courtesy of US National Geophysical Data Center, National Oceanic and Atmospheric Administration.)

downhill, washing away trees and the buildings in four ranches. One of the blocks carried down is now known as "Hot Rock" – because, like others of its size, it required a week to become cool to the touch. The mudflow debris caused permanent damming of Hat Creek, forming Hat Lake.

A few days later, on May 22, 1915, the eruption reached its climax with a violent explosive event. At about 4.30 p.m., a huge column rose some 9,000 m (30,000 feet) above the volcano and, about the same time, a pyroclastic flow ran down the northeastern side, following the path of the earlier mudflow and blowing down trees, creating the Devastated Area. No one saw the pyroclastic flow, but the eruption column was seen from most of northern California. Ash begun to fall, continuing late into the night, and reached some 500 km (300 miles) away in Elko, Nevada. Bombs fell around the summit, some reaching as far as Prospect Peak, about 15 km (8 miles) away.

Following this violent event, Lassen settled down somewhat, erupting steam and ash until early 1917, when the explosions became more intense during May and June. As with the previous year, there was probably a correlation between the snow melt in the spring and more violent explosions, because of the water seeping down and reaching the hot magma. After June, the volcano settled down, with only occasional puffs of steam coming out until 1921. By the mid-1930s, about 30 fumaroles were still active in the summit area and, even now, a few fumaroles still puff away at the north and northwest base of the Peak. We do not know when Lassen will erupt again but, when it does, it will put on a show worth watching.

A personal view: A volcano for photographers

Lassen Volcanic National Park is perhaps the most pleasant of all the volcanic National Parks in the USA. It does not feel crowded or over-commercialized, even though it has numerous attractions. The Park is ideal for those who wish to hike, camp, and enjoy nature in an unspoiled setting, as well as learn about volcanoes. I find Lassen comparable to Crater Lake in beauty but far more varied in its volcanic attractions – and far more peaceful. There is no way of predicting how long the crowds will stay away, but I am sure that a new eruption would make Lassen as popular as Mt. St. Helens and Yellowstone are today. My advice on visiting Lassen is to go before it blows. A new eruption could be spectacular and exciting to watch, but the opportunity to enjoy the tranquil scenery away from crowds, shops, and IMAX theaters might be gone forever.

Keen photographers should consider Lassen one of the top choices of volcanoes to visit. Lassen Peak is probably one of the most photographed volcanoes in the world – its postcard-perfect views are often seen in nature calendars. It is hard to be anywhere on the National Park not worthy of a picture, but there are some particularly good places that photographers should not miss. The vivid colors of Bumpass Hell and Sulphur Works are ideal for color pictures. The placid beauty of Lassen Peak is best (and most often) photographed from Manzanita and Reflection Lakes, both located at the western end of the Park. The hike around Manzanita Lake offers excellent views of the mountain, particularly from the western and northern shores. The lake reflects Lassen Peak in all its glory, as does the nearby Reflection Lake, as its name implies. Early evening is the best time for pictures, as the sun is directly on Lassen Peak and the wind dies down, so the reflection is unrippled. Early fall is probably the best season, as there is snow on the Peak and the leaves of the trees by the lake's shore have started to turn golden. Although spectacular in color pictures, these views of Lassen Peak are also ideal for black-and-white photography in the style of Ansel Adams.

Visiting during repose

Lassen Volcanic National Park is open year-round but, because of heavy snows, it is best to visit in the summer. The trans-Park road is closed from the end of October to early June, considerably limiting your options during that period. If you ski, you can take advantage of the winter sports area that operates near the south entrance to the Park. However, if you wish to climb Lassen Peak and enjoy the volcanic scenery at its best, come in late August or early September, after the school holidays are over. Note that some of the trails (including Bumpass Hell) are not opened until July 1. Snowbanks at higher altitudes persist even during the summer, adding to the beauty of the landscape. The weather is usually very good during the summer; however, afternoon thunder showers are common. Do not go up to the summit of Lassen Peak, Brokeoff Mountain, or other peaks during these showers, as lightening strikes present a real danger.

Lassen offers much to explore. The Park is vast – 400 km^2 (160 square miles) in area crossed by over 240 km (150 miles) of foot trails. The Lassen Park Road crosses the western half of the Park, winding around three sides of Lassen Peak, but to reach the major attractions of the eastern half of the Park you have to drive in from outside using other Park entrances. It is therefore

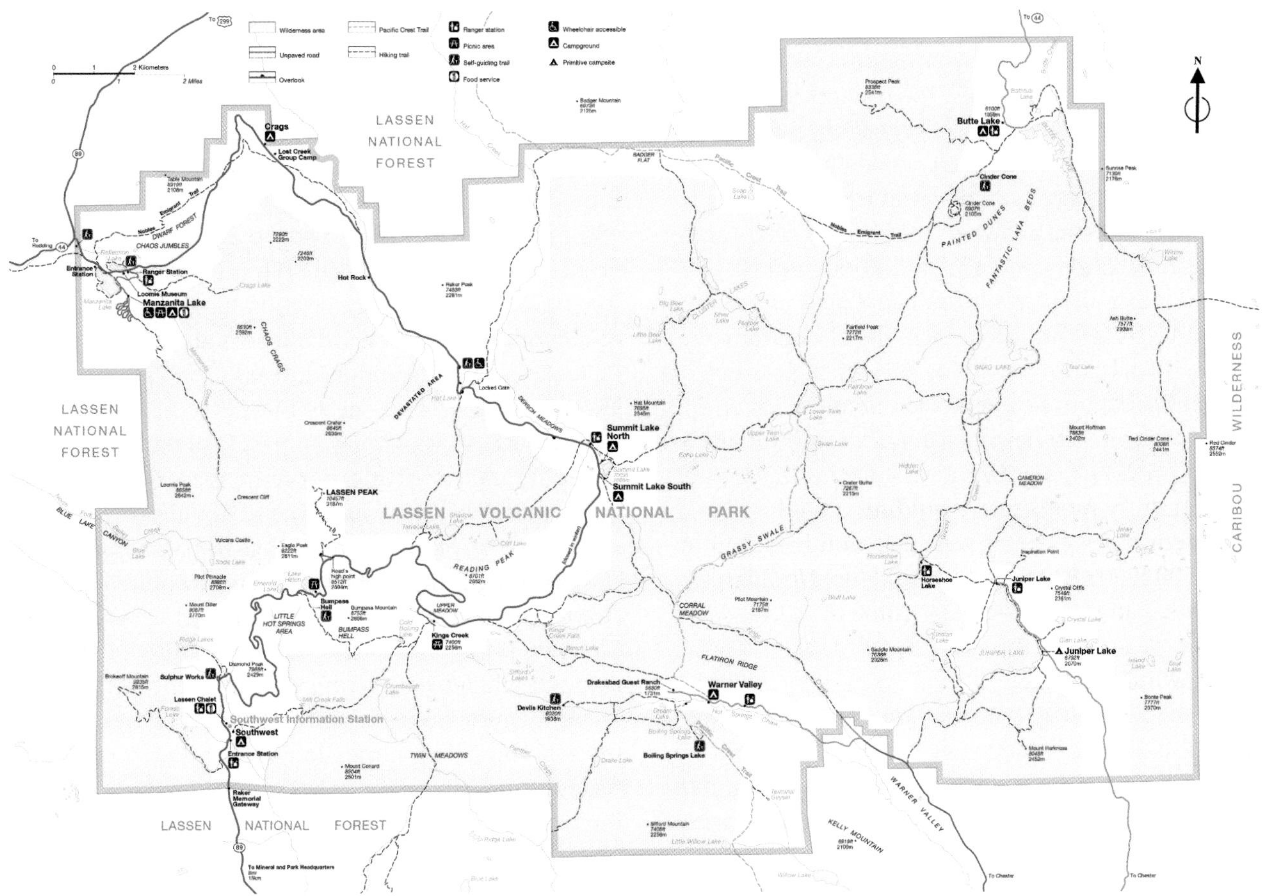

Map of Lassen Volcanic National Park, showing main attractions. (Modified from US National Park Service, Lassen website.)

logical to plan your visit in three parts: the western side, the Butte Lake area, and the Warner Valley area. Most of the attractions described below are concentrated in the western half, but I highly recommend making the effort to visit at least Cinder Cone in the Butte Lake area.

The Park's two main entrances are located to the southwest (Sulphur Works) and to the northwest (Manzanita Lake). Both have information centers and exhibits that are open from early June to late September. Publications about Lassen, maps, and trail information are available at both locations, as well as a schedule of naturalist-led trail hikes. The Park Headquarters in Mineral offers information and sells publications year-round.

There are several books entirely devoted to Lassen that contain far more detail than can be given here (see Bibliography for recommendations). The major trails all have signs explaining the geology and natural history of the area, as well as self-guiding trail leaflets. The itinerary below covers the volcanic highlights of the park.

The western side

Highway 89 turns into the Lassen Park Road as it crosses the Park. It is somewhat preferable to start your visit at the southwestern end, which has the most attractions nearby. The order below follows the Park Road, and has the added benefit of starting your visit at the remnants of the old Mt. Tehama. The trailhead for the Brokeoff Mountain trail is located about 800 m (½ mile) before the southwestern entrance. Just past the entrance booth are the southwest campground, the Southwest Information Center, and the Lassen Chalet, where winter sports equipment can be rented.

Brokeoff Mountain Trail

This is a moderately steep trail (round trip 11 km, 7 miles, 4 hours) which climbs the remnants of Mt. Tehama, an andesitic stratovolcano. The trailhead is located by road guide marker number 2, next to a stream. The trail takes you south at first, away from Brokeoff, but soon turns west, passing through a forest of Western pine and red fir. After the first 500 m (⅓ mile), the trail turns northwest and there are good views of Brokeoff Mountain. Notice the talus slopes, marking where rockslides have come down. The trail now begins to climb steeply. The unmarked trail that branches off to the right is short (160 m, 500 feet) and leads to Forest Lake. You can see the lake from further up the Brokeoff trail, and decide if it is worth the detour. As you climb towards the summit, there are views of Mt. Harkness to the east. About 400 m (⅓ mile) from the start, the trail goes along the west side of Brokeoff and soon gets in the alpine zone above the timberline. Wildflowers such as Lassen paint brush and sulphur flowers are abundant in the spring. At the summit, the views of Mt. Lassen and Lake Helen (to the northeast) are fantastic. To the north, you can see Chaos Crags and, to the northeast (on the foreground from Lassen), the remnants of the rim of the old Mt. Tehama between Brokeoff and Mt. Diller. On a clear day, you can easily see Mt. Shasta.

Sulphur Works Trail

This short trail (300 m, ¼ mile round trip, 25 minutes) starts just north of the parking lot on Lassen Park Road, by guide post number 5. It is the easiest trail in the Park and the only one accessible to those in wheelchairs. Sulphur Works is a hydrothermal area that never fails to delight visitors because of its bubbling mudpots, hot springs, steaming fumaroles, and colorful deposits. The Sulphur Works name comes from an old mining operation that started here in 1865 but that was largely unsuccessful. The temperature of bubbling water and mud mixture is around 76 °C (195 °F) and, around 1940, tourists used to bathe in it. Since the water contains sulfuric acid, it probably did not do too much good to the tourists' skin and, maybe for this reason, the bath business also failed.

Sulphur Works is an important site in the volcanic history of the Lassen region because it is thought to be where the main vent of Mt. Tehama was once located. It is fun to stand here and imagine Mt. Tehama before its collapse – the mountain was over 3,000 m (some 11,000 feet) high. The characteristic "volcano smell" of hydrogen sulfide reminds you that the mountain may be gone but the active volcanic area remains. The beautiful hues of yellow, orange, green, and red make this a great site to photograph. Minerals found here include hematite, pyrite, kaolinite, alunite, and sulfur. Unlike many areas in the Park, Sulphur Works can easily be visited during the winter, as it is just a short cross-country ski trip away from the Ski Area. At all times, stay on the trails. The ground in this whole area is friable and could easily collapse under a person's weight.

Bumpass Hell Trail

This trail (5 km, 3 miles round trip, 3 hours) goes around the largest hydrothermal area in the Park, Bumpass Hell, which is also one of its most popular attractions. The area was named after Kendall V. Bumpass, who discovered it in 1864. It is said that, while guiding a newspaper editor around the area a year later, Bumpass stepped on friable ground, breaking through the thin surface and plunging his leg into the boiling mud below. He was so badly burned that he lost his leg and supposedly said "The descent to Hell is easy."

The trailhead is by the Bumpass Hell parking lot, between markers 16 and 17 on the road. Before going on the trail, walk to the south end of the parking lot where you can see numerous boulders with glacial striations. There is a prominent boulder about 2 m (6½ feet) in diameter, which is a pyroxene dacite from Bumpass Mountain. Before setting off on the trail, pick up a self-guided leaflet from the box at the trailhead. The leaflet describes the many points of interest along the way and tells you that you are standing on remains from the old Mt. Tehama. To be more precise, Bumpass Hell is located on the outer slopes of the old Mt. Tehama caldera and, unlike Sulphur Works, was not the site of one of its main vents. However, the area is the focus and the major thermal outflow from the Lassen geothermal system, which consists of a central vapor reservoir at a temperature of about 235 °C (455 °F). The vapor reservoir is underlain by a reservoir of hot water. Bumpass Hell is cut by fissures that penetrate down deeply enough to allow volcanic gases and heat to escape to the surface.

The trail gives a good view of the western wall of the Tehama caldera, marked by Brokeoff Mountain, Mt. Diller, and Pilot Pinnacle. The trails then enters into the bowl of Bumpass Hell, an area of about 64,700 m^2 (16 acres), carved out from andesite lava by the action of the hot sulfurous gases eroding the rocks. The action is still here, in the many pits of boiling mud and superheated fumaroles. During the summer of 1988, one of the fumaroles reached 161.4 °C (322 °F). The colorful, friable ground is largely covered by orange and yellow sulfates. The gray and black mudpots contain pyrite, which is also seen as scum floating on the surface of pools. For safety reasons, the Park has built a series of wooden platforms to allow people to approach the vents without suffering the same fate as Kendall Bumpass.

Most people return to the parking lot from here, but there is an alternative. Another trail leads from

Bumpass Hell geothermal area, one of the most colorful and interesting places to visit in the Lassen Volcanic National Park. (Photograph courtesy of Nick Gautier.)

Bumpass Hell to Cold Boiling Lake (3 km, 1¾ miles away) and to Kings Creek picnic area by the Park road (a further 1 km, ⅝ mile). Along the trail to Cold Boiling Lake there are good views of Mt. Conard, marking the eastern remnant of the Tehama caldera, and of the scenic Lake Crumbaugh. Cold Boiling Lake is named for the gas bubbles that continuously come up through the cold water. Going all the way to Kings Creek is an ideal thing to do if you can arrange to be picked up at the parking lot there, otherwise it is a 5-hour round trip from the Bumpass Hell trailhead.

Lassen Peak Trail

This climb (8 km, 5 miles round trip, 4.5 hours) is the highlight of the Park and, although somewhat steep, it is easy for most hikers. Note that it is cold and windy at the top, and bring appropriate clothing. The trailhead and parking lot are located by road marker number 22. Pick up a trail leaflet from the parking lot or visitor center. There are several signs along the way explaining what you see and mile posts indicate how much further you have to get to the top. Stay on the zigzag trail even if it is tempting to go straight up – for reasons of conservation and safety, the Park takes a dim view of visitors who shortcut.

As the trail winds up the dome, there are wonderful views of the western wall of the Tehama caldera, with Brokeoff Mountain and Mt. Diller clearly visible looking west from the trail. Most of what you see of Lassen Peak along the trail are the talus slopes, as the dome mass itself is only exposed in a few outcrops. Glaciers have left their signatures in the form of small moraines on the talus slopes. Plant and animal life are scarce along the trail, but hawks and eagles often nest

Hiking trail up Lassen Peak. (Photograph courtesy of Nick Gautier.)

on inaccessible crags. I met a couple of rather friendly (and hungry) squirrels on my first visit to the mountain. More rarely, one sees a pika, a shy relative of the common rabbit.

Lassen Peak is unusual among plug-domes not only because of its size but also because of the presence of summit craters from which pyroclastic material has been ejected. For example, Chaos Crags is made up of several dacite plugs but does not have a summit crater. The crater bowl on Lassen's summit contains a large dacitic lava flow from 1915 which covered over the 1914–15 vent and a pre-1914 vent. The flow poured down the western side of the mountain and was observed on May 19, 1915. A third vent area in the summit also dates from 1915 and the most recent vent is a crater blasted out of the northwest corner by a series of steam and ash explosions in 1917.

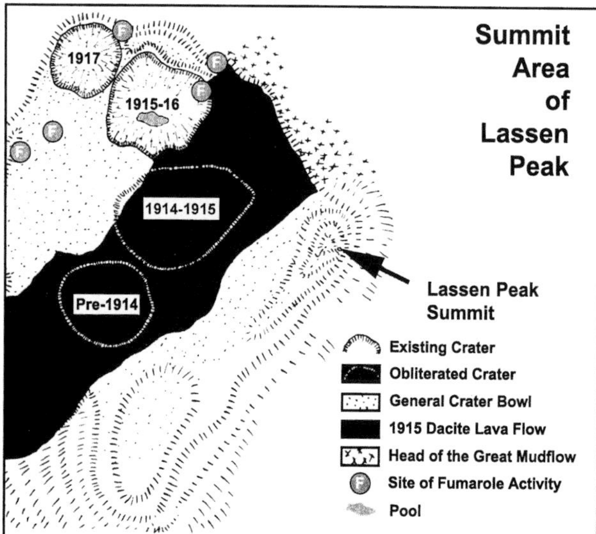

Map of the summit area of Lassen Peak. (Modified from US National Park Service *Lassen Peak* trail leaflet.)

When the trail reaches the summit area, it stops its long climb and begins to descend into the 300 m (1000 feet) wide depression, blasted out during the first year of the 1914–15 eruption. The trail follows the southeast side of the depression and up to the actual summit at 3,187 m (10,457 feet) elevation. Notice the black dacite flow filling the bottom of the crater and spilling out on the southwestern side. Look to the northeast and you will notice that the lava is no longer seen there, as it is covered by a mudflow deposit from May 19, 1915. The mudflow sweeps down into Hat and Lost Creeks.

The trail continues across the black dacite flow up to two small explosion pits on the northwest side. The one closest to the lava flow was formed on May 22, 1915, when a large eruption cloud was observed. This crater cuts into the lava (along the eastern side) and is a great place from where to study the deposits from the last eruption. The second, smaller explosion pit is located to the northwest. It was formed during the first half of 1917.

The views of the rest of the Park from the summit of Lassen are truly spectacular. Take a map and look for the other features of geologic interest: Mt. Tehama remnants to the southwest, Chaos Crags to the northwest, Hat Mountain to the east and, beyond it to the northeast, Prospect Peak and Cinder Cone.

Devastated Area, Hat Lake, and Hot Rock

The devastation refers to the effects from Lassen's last eruption. The Forest Service at the time estimated that between 4.5 and 5 million feet of timber were destroyed. Like other areas of its type, this became a natural laboratory for studying the recovery of vegetation. Park naturalists point out that both the Devastated Area and Chaos Jumbles are recovering directly to conifers without passing through the preparation stage of herbaceous plants, a type of recovery observed for the first time at Lassen.

The mudflow from May 19, 1915, battered the trees in the area but a more striking effect was that of the blast and pyroclastic flow from May 22, which shot down across the heads of Lost and Hat Creeks. An account by Benjamin Franklin Loomis, who visited the area in June of 1915, tells of all trees in a space about a mile and a quarter wide (2 km) being blown down by the blast. Examining the trees, which were lying down on the ground, he and his party found that the upper sides of the trunks (which would have faced uphill towards the volcano when the trees were standing) showed where the May 19 mudflow had stripped off the bark and scarred the wood, but left most of the trees still standing. They correctly concluded that the May 19 mudflows had not been accompanied by a horizontal blast.

Hat Lake is located by marker 42 of the trans-Park road, just south of the Devastated Area. Hat Lake was formed by the May 19 mudflow, as it dammed the west fork of Hat Creek. There is a trail from Terrace Lake (by trans-Park road marker 27) to Hat Lake that takes you to Paradise Meadows, a particularly beautiful area in midsummer because of its many wildflowers. The trail is 4.5 km (2¾ miles) long and will take you 2.5 hours one-way downhill from Terrace Lake. It is worth doing this hike if time allows, particularly if you can arrange to be picked up at the parking lot by Hat Lake. Otherwise, just walk a little way from Hat Lake and the Devastated Area to see the remains of the 1915 mudflows.

Further north along the trans-Park road, marked by a sign, is Hot Rock, one of many impressive-looking blocks of hot lava carried down by the May 19 mudflow.

Chaos Crags and Chaos Jumbles

The Chaos Crags and Crags Lake Trail (5.8 km, 3½ miles, 3 hours round trip) starts by the Manzanita Camp road (near the creek crossing) and follows the edge of the Jumbles, through a pine forest. Those short on time can view the Jumbles and the Crags from the road between Crags campground and the west entrance to the Park. The pyroclastic deposits from Chaos Crags can best be seen along the drainages of the Manzanita and Lost Creeks.

The formation of the Chaos Crags volcanic domes about 1,100 years ago started out with explosive activity building a tephra cone, followed by two pyroclastic flows, and by the extrusion of lava that plugged the vent. About 70 years later, the dome complex was destroyed by a violent eruption that produced another pyroclastic flow. Chaos Jumbles is a huge rockfall avalanche formed by the partial collapse of the most northwesterly dome in Chaos Crags. It is one of the most spectacular avalanche deposits in the USA and a favorite stop for geologic field trips. The deposit, made up of dacitic blocks and pulverized dacite, covers an area of about 8 km^2 (3 square miles), is about 1.5 km (1 mile) wide and has a thickness of up to 40 m (130 feet). The total volume is estimated to be some 150 million m^3 (200 million cubic yards). The avalanche moved rapidly in the northwestward direction away from the dome, across the present trans-Park road to Table Mountain and, deflected by the mountain, westward past the Reflection and Manzanita Lakes area. Reflection Lake is a watertable pond in one of the many surface depressions of the tongue-like avalanche deposit. Manzanita Lake was also created by the avalanche, as the debris dammed Manzanita Creek.

Detailed studies show that the avalanche actually consisted of three units, with the first being the largest and the others getting progressively smaller and shorter, but also thicker. Radiocarbon dating of wood from trees submerged in the lake indicates that the three avalanches occurred in quick succession around the year 1690.

The surface of the avalanche is a real jumbles – lots of angular blocks of debris and a wavy topography of ridges and furrows. The topography is typical of high-velocity avalanches whose movement was probably lubricated by a cushion of air trapped underneath the material, lowering the friction. As the avalanche moved, the air gradually seeped upwards and eventually the edges of the debris came to an abrupt rest. The sudden deceleration at the margins produced "shock waves" that traveled through the deposits, creating ridges and furrows.

Chaos Jumbles must have moved quite fast, because it managed to travel about 120 m (390 feet) up the steep slope of Table Mountain before it was deflected westward. Calculations show that the avalanche must have had a speed of over 160 km/hour (100 miles/hour). There have been various suggestions as to what exactly caused the dome collapse and the avalanche. An earthquake could have triggered the event, causing the already unstable, stiff cliffs on that side of the Crags to collapse. The trigger event could also have been the intrusion of a new dome, causing over-steepening of slopes, or else steam explosions at the base of the dome. It is sobering to think that any of these scenarios could happen again anytime. A study of the hazards from avalanches and pyroclastic flows in the Manzanita Lake area of the Park caused the closure of several facilities and living quarters in the area in 1974.

Loomis Museum

Located near Manzanita Lake, the museum was named after Benjamin Franklin Loomis, who came to the area in 1874. His hobby was photography and, when Lassen begun erupting, he was among the first to photograph the eruptions and make inspections of the areas affected, providing valuable scientific information for posterity. His eruption photographs became so famous that, after the death of his only child Louisa Mae, Benjamin and his wife built the Mae Loomis Memorial Museum as a place to display his unique collection and honor their daughter's memory. The museum opened in 1927 and, 2 years later, the couple deeded the building and surrounding land to the National Park. Today, the museum is open during the summer, offering information about the Park and selling publications.

Reflection Lake and Manzanita Lake

Both these scenic lakes are located near the Manzanita Lake (western) entrance to the Park and are must-see attractions for nature-lovers. Reflection is located north of the road and can be reached by walking along a short trail, while Manzanita Lake is on the south side and can be reached by driving along a paved road. There are camping, restaurant, and picnic facilities at Manzanita. Easy trails go around each of these lakes and they are wonderful places for birdwatching.

The lower end of Reflection Lake reflects Lassen Peak, hence its name. It is also one of the best places from where to photograph Lassen Peak, and to see Chaos Crags and the May 19, 1915 lava flow described by Mr. Milford. The flow is visible as a black tongue that extends down the mountain for about 300 m (1000 feet). The black color is unusual because dacitic lavas are normally gray or pink. The dark color of this flow is due to the very rapid cooling of the lava, which formed glass, so that the black iron and magnesium compounds were left as clouds of particles in the rock mass and did not form individual black crystals.

Manzanita Lake also provides magnificent views of Lassen Peak and the short hike around the lake (about 1 km, ⅝ mile) is a must for keen photographers. The lake was formed by the first Chaos Crags avalanche and, in 1912, its water level was raised by the construction of a small earth dam at the southwestern end. During the fall, the lake is a stopping place for migrating waterfowl, such as Canada geese and wood ducks.

The northeast (Butte Lake) area

This area can be reached via a short side road from Highway 44 that runs north of the Park. The side road enters the Park at the Butte Lake Ranger Station and ends at Butte Lake, a highly scenic, out-of-the-way area where one of the Park's campgrounds is located. Two main trails start from here, one going around the eastern shore of Butte Lake and crossing the eastern end of the Park and another heading southwest towards Prospect Peak and Cinder Cone. There is a trail up to the top of Prospect Peak. The Peak is one of four shield-type volcanoes formed in the Lassen area during the late stages of the development of Mt. Tehama. The others are Raker Peak, Red Mountain, and Mt. Harkness. All four are formed mainly of andesitic lava flows. Although Prospect Peak is an interesting geologic feature, the views from its summit are not sufficiently striking to make this 4-hour hike truly worthwhile, unless you have plenty of time to spare. The main attraction of this part of the Park is Cinder Cone. Those camping in the Butte Lake area may wish to visit Cinder Cone by taking a long trail from the Butte Lake trailhead to Lower Twin Lake via the Nobles Emigrant Trail, then east to Rainbow Lake and Cinder Cone. This whole circuit is 19.8 km (12¼ miles) long and will require about 6.5 hours. Those short on time should just go to Cinder Cone, one of the youngest and most striking volcanic features in the Park.

Cinder Cone

This trail is 2.2 km (1½ miles) one-way to the base of the cone, another 0.8 km (½ mile) to the top. Trails around the rims add another 1.5 km (1 mile) to the journey. Allow 3.5 hours for the round trip. The trail starts by the Butte Lake Ranger Station and leaflets are available at the trailhead. Cinder Cone is a textbook example of a cinder and scoria (pieces of broken volcanic rock) cone. It last erupted in 1851, producing a series of lava flows from the base of the cone that poured into Butte Lake. The composition of the flows is unusual – basaltic andesite but containing quartz xenocrysts (meaning "foreign crystals"). Because quartz does not occur naturally in this type of magma, it is

Lassen's Cinder Cone, one of the top attractions in the Lassen Volcanic National Park. (Photograph courtesy of Nick Gautier.)

thought that these flows came from a magma chamber that was contaminated with a more felsic magma before it erupted. This type of hybrid lava is sometimes called a quartz basalt.

As soon as you start on the trail you will notice that the ground is peppered with black cinders. In fact, the whole eastern end of the Park is covered by cinders from Cinder Cone and other Strombolian-type vents, such as the nearby Prospect Peak and Fairfield Peak. The trail to Cinder Cone has some historical interest, because it is part of the Nobles Emigrant Trail, which was established in 1852 by William Nobles as a shorter alternative to Peter Lassen's trail. The first sight of major geologic interest along the trail are the dark, rubbly and fresh-looking flows from Cinder Cone. The flows are marked on maps as the Fantastic Lava Beds, though they don't live up to the name. The quartz crystals are the only somewhat unusual aspect of these flows. The crystals are white and can be easily seen at close range (the flow's white freckles that you can see from a distance are actually lichens). The most recent flows, from 1851, are not visible from the trail as they flowed between and over older lavas. Near the trail and the black lava are charred remains of the pine trees that are abundant in the area. The fires that damaged the trees may have been ignited by the lava flow or, possibly, by lightning strikes.

Cinder Cone is a beautiful, symmetrical cone about 225 m (750 feet) tall, with a maximum diameter across the base of 800 m (½ mile). Rounded, glassy bombs can be seen around the base of the cone. They were ejected from the top and rolled down the slopes while still hot. It is easy to follow the trail to the top of Cinder Cone, though the loose cinders and steepness make most people stop now and then to catch their breath. The slope of the cone is about 35 degrees, which is the angle of repose, meaning the steepest angle for a structure of this type. This angle is typical for uneroded cinder cones, as the material accumulates around the vent at the angle of repose. As you climb, you may notice some white pumice fragments among the dark cone material – the pumice is from the 1914–21 eruption of Lassen Peak.

The climb is worthwhile, as there are spectacular panoramic views from the top. If you look down the cone towards Snag Lake, you can see the black basaltic flow from 1851 that emerged from vents near the base of the cone. Snag Lake was, in fact, created by an earlier flow from Cinder Cone that blocked and dammed Butte Creek and then flowed into Butte Lake.

After admiring the panorama, turn your attention to the inside of the cone. You can walk around the three crater rims that represent different eruptions. The main crater is 72 m (240 feet) deep and 300 m (1000 feet) in diameter. It is, in fact, a double crater, evidence of at least two eruptive episodes. One or possibly two additional rim remnants can be seen on the northwest side of the cone. It is very instructive to walk around the top, and the views and opportunities for photography are outstanding. One trail completely encircles the double crater and a spur trail goes down inside it. Another trail encircles the south side of the cone. The trails have markers corresponding to descriptions in the trail leaflet that are self-explanatory. A particularly beautiful view from the trail is that towards the Painted Dunes, located to the southeast of the cone. These colorful features are thought to have been "painted" by the 1666 eruption of the cone. It is thought that the lava flow from that eruption (which later became known as the "dunes") was bombarded with tephra from the cone while the block lava was still hot. As cinders and ash fell on the hot lava, heat and steam oxidized the iron in them, creating the beautiful colors.

The southeast (Juniper Lake and Warner Valley) area

The Warner Valley area can be reached via trails from the Lassen Park road, but visitors short on time may prefer to drive here from the town of Chester, located southeast of the Park. There are two short roads from Chester into the Park: one goes to Juniper Lake campground and the other to the Drakesbad Guest Ranch (the only lodgings inside the Park) and the Warner Valley campground. The summit of Mt. Harkness can be reached by trails starting from either road. A word of caution: although bears are not common in the park, they have been seen near Juniper Lake (and also Summit Lake). It is a good idea to make some noise while hiking in this area to give them a chance to get away from you. Take precautions while camping, and most of all, never keep food in your tent.

Drakesbad Guest Ranch

The ranch is named after one of the past owners of the land, Edward R. Drake. The "bad" part of the name comes from the German word for bath or spa and reflects the presence of hot springs suitable for bathing in this area. The guest ranch was established in 1900 and is now run by the Park. It provides lodgings, meals, and horse-riding, but reservations are needed for everything. Hikers can drop by to use the telephone and to buy drinks, snacks, and maps. There are several interesting hikes that start from the ranch. The most interesting, from a volcanological point of view, are the hikes to Devil's Kitchen, Boiling Springs Lake, and Terminal Geyser. The names alone make you want to see these geothermal areas. Devil's Kitchen is an active hot spring and solfataric area similar to Bumpass Hell, with bubbling mudpots and fumaroles. Boiling Springs Lake is 200 m (630 feet) in diameter and quite striking because of its very brown, and sometimes red-colored water. Submerged hot springs are responsible for the lake's name, though the color of the water and steam vents around the shore give it a frightening appearance that probably led to the alternate name Lake Tartarus, after the infernal lake of Homer's Iliad. Terminal Geyser is not a true geyser, nor it is likely to be terminal. The reason for the name remains a mystery, though the misnomer geyser probably refers to the irregular fountaining of this active hot spring.

Devil's Kitchen from Drakesbad

This trail is 3 km (1¾ miles) one way; allow 2.5 hours for the round trip. You can go to Devil's Kitchen on horseback from Drakesbad Ranch, but this is a pleasant trail for hiking. The trail starts by the parking lot west of Warner Valley campground, crosses a meadow near Drakesbad, which is pretty when the wildflowers are out, then enters a forest. After 2.25 km (1½ miles) from the start, the trail forks: Devil's Kitchen is to the right, the other fork goes to Drake Lake, 1.6 km (1 mile) away. A little further, the trail descends and crosses Hot Springs Creek, then passes through a beautiful grove of pine and cedar trees, before descending steeply into Devil's Kitchen. Hot Springs Creek runs through the area and the rather stinky mudpots are boiling away on a shelf just above the creek. It is easy to imagine the Devil cooking something unappetizing in this setting. The Kitchen is not as large as Bumpass Hell, but is an attractive geothermal area and convenient for those who choose to stay at this end of the park.

Terminal Geyser Loop via Boiling Springs Lake

The trail is 1.4 km (just under 1 mile) to Boiling Springs Lake, 4.3 km (2¾ miles) to Terminal Geyser; allow 3 hours for the round trip. This is one of the best hikes in the Park – and judging from the few hikers one sees, it must be one of the Park's best-kept secrets. The trail follows part of the Pacific Crest Trail which crosses the Park roughly north–south. It is well signposted and self-guiding leaflets are available explaining what you see along the way. The first 600 m (⅓ mile) are the same as the Devil's Kitchen trail, starting from the parking lot. Just after the fork to Devil's Kitchen is another fork, at which the right fork goes to Drake Lake and the left to Boiling Springs Lake. Another fork comes at marker 16, with the left going to Terminal Geyser and the right to Boiling Springs Lake. The right fork will take you around the lake and then towards Terminal Geyser.

Boiling Springs Lake is another multi-hued geothermal area worth spending a lot of film on. The lake's water is maintained hot (about 52 °C, 125 °F) by steam rising through underground vents. On cool days, water vapor can be seen rising from the lake surface. The water's color is a brownish-yellow, due to clay, opal, and iron oxide particles suspended in the water. The shallower parts, near the shore, look green, thanks to algae that have adapted to this environment. The mudpots along the lake's southeastern shore are some of the best in the Park.

You can hear Terminal Geyser before you see it: it sounds like cars on a freeway. At one time, it had the more appropriate name of Steamboat Springs, since it is not a geyser but a violently hissing fumarole, emitting lots of steam. There are boiling mudpots in this area too, by trail marker number 37. Use caution when walking here and stay on the trail. The crumbly surface near the mudpots is friable and it is easy to sink through – this could well be the reason why this area is called terminal.

Mount Harkness from Juniper Lake

This trail is 3.1 km (nearly 2 miles) one way to the summit of Mt. Harkness; allow 2.5 hours for the return trip. Loop trail is 9 km (5½ miles), allow 3.5 hours. Mount Harkness is a small andesitic shield volcano about 382 m (1,256 feet) high, with a cinder cone at its summit. This trail is not a "must-do" from a volcanological point of view, but I recommend it because of its scenic views and the opportunity of seeing wildlife, including deer, mountain quail, and grouse. It is also a very pleasant hike. The trail starts Juniper Lake campground and ascends through a pine forest. There are beautiful views of the lake and of some of the lava flows that make up this shield and you can see a few lava outcrops along the trail. The view from the Mt. Harkness lookout is fabulous: to the west is Lassen Peak and many other features of the Park, to the north is Cinder Cone and, beyond, the white-capped Mt. Shasta. You can go back to Juniper Lake the same way (down the northeast flank of Mt. Harkness) or go down the northwestern side and then across the base of Mt. Harkness by the shore of the lake. Looking up towards the summit of Mt. Harkness you can see rock slides down the flanks. The flanks of this shield are steeper than is typical for basaltic shields, because the andesitic lava flows are generally thicker than basaltic flows.

Visiting during activity

Lassen Peak, Mt. St. Helens, and Mt. Shasta are considered the Cascade volcanoes most likely to erupt again in the near future. In recent years, some of Lassen's geothermal areas have shown temperature increases and some of the waters in hot springs have shown temperatures above the boiling point. It could be that the next eruption is not too far off in the future.

What will Lassen's next eruption be like? The 1914–21 eruption is the only recorded period of activity at Lassen, but it is likely that it was fairly typical of the type one can expect from the volcano. Since Lassen is within a National Park, visitors will be advised by Park personnel about restrictions and safety procedures. The danger of pyroclastic flows and mudflows will be a serious concern. Lassen's eruptions are best watched from afar, just as Mr. Milford did during the last event. Luckily the mountain is easily seen from many miles around, so the challenge for visitors will be to find the perfect spot. Park rangers will no doubt know about these spots and advise people on the best and safest places. I expect these places will be very crowded.

Cinder Cone and Prospect Peak may be ideal locations, as long as they are accessible to the public at the time of an eruption from Lassen Peak. Bombs from the violent event of May 22, 1915 reached that far, so the danger will be a concern for Park personnel. Access will no doubt depend on the level of activity. Mount Harkness in the southeastern end of the Park is another good bet. Vantage points that require some hiking up will be less crowded than the more easily accessible places.

I can imagine that photos from Reflection and Manzanita Lakes would be spectacular, but unfortunately the area is a hazardous one, as pyroclastic flows and mudflows from Lassen Peak could easily reach that far. The lakes might be great places for remote cameras, but they better be of the disposable type.

Other local attractions

Subway Cave

This lava tube is located about 24 km (15 miles) north of the Manzanita Lake entrance to the Park. Follow Highway 89 and look for the signposted turnoff to Subway Cave. The tube formed in the Hat Creek basaltic flow and is probably less than 2,000 years old. It is not particularly memorable as a lava tube, but visitors who have not seen better examples may want to stop by. Access to the inside is via a skylight. The tube is quite regular in cross-section, with a nearly flat floor and an almost semicircular roof. It does remind one of a subway tunnel, but the origin of the name is not known.

Mount Shasta

This magnificent snow-capped mountain, 4,316 m (14,161 feet) high, is located only 120 km (75 miles) to the northwest of Lassen. It is the largest volcano in the Cascades Range, with a volume of about 350 km^3 (84 cubic miles), and is second only to Mt. Rainier as the highest. Like most of the Cascades Range volcanoes, Shasta is a stratovolcano. Its bulk is made up of four overlapping cones of different ages, produced by magmas that are andesitic to dacitic in composition. The activity during the last 10,000 years built the two youngest cones: Hotlum, which forms the present summit, and Shastina, which, if it stood alone, would be the third tallest peak in the Cascades.

Shasta is one of the most potentially dangerous volcanoes in the USA and is considered very likely to erupt again. The volcano has been restless during the last 1,000 years, having erupted at least three times. This recent activity has been concentrated on Hotlum and Shastina, though some activity also occurred at a group of overlapping domes known as Black Butte. Shasta's most recent eruption, from Hotlum, is thought to have happened in 1786. La Perouse

The majestic, snow-capped Mt. Shasta. (Photograph courtesy of Nick Gautier.)

observed a volcanic eruption in the direction of Shasta while sailing off the northern California coast. While it is not certain that the eruption was indeed from Shasta, the account is supported by radiocarbon dating of the most recent deposits. This last eruption sent pyroclastic flows about 12 km (7½ miles) down the flanks of the mountain.

Visiting Shasta can be done in various ways. One can simply drive around the mountain, photographing it from different viewpoints. You can drive all around Mt. Shasta using Highway 97 and a dirt road called the Military Pass. The best towns to use as a base are Weed and Mt. Shasta, located by the Interstate 5. There are two campgrounds located by the Everitt Memorial Highway, a road that brings you onto Bunny Flat on the volcano's southern flank. There is a popular trail from Bunny Flat to the Sierra Club cabin at Horse Camp, located on the timberline. The trail is easy (5.5 km, 3½ miles) and is the first step towards climbing to the summit. Horse Camp (so called because climbers used to tether their horses here before going to the summit) is a popular spot for overnight camping, though the inside of the small cabin is for emergencies only. In summer, there is usually a custodian to help climbers. The cabin's first custodian, Mac Olberman, built Olberman's Causeway using large flat stones as a path that goes up nearly a mile towards the summit. Mac worked nearly 9 years alone building the Causeway, and he was nearly 60 years old when he started. On his 70th birthday, he climbed to the summit in 5 hours flat, never stopping along the way. The summit is only 6.6 km (4⅛ miles) away from the cabin, but more than 1,800 m (6,000 feet) up.

For most people, the climb to the summit is arduous and requires the use of ice-axes. The Shasta Mountain Guides, based on Mt. Shasta town, offer guided climbs at various levels of ability. The most popular route is the John Muir/Avalanche Gulch, used by Captain E. D. Pearce in 1854 to make the first recorded ascent of the mountain. The climb is considered only a "1" in terms of difficulty and can be done in a day. However, many climbers choose to stay overnight at Horse Camp or at Helen Lake, a flat area at 3,120 m (10,400 feet). Like Lake Helen in Lassen, this was named after a lady climber – Helen Wheeler made the ascent of Mt. Shasta in 1924. This lake, however, is usually under snow and rarely seen. Above Helen Lake is the most strenuous part of the climb, a 750-meter (2,500-foot) snowdrift that steepens to 35 degrees near the top. As you climb, aim toward the right side of Red Banks, a prominent exposure of orange welded pumice that is one of the volcano's most recent products. The Red Banks are also the major source of rockfalls in Avalanche Gulch and this is the main danger of this climb. To increase safety, stay at Helen Lake overnight and climb early in the morning, before the sun's heat can cause rocks to become loose. There is a small saddle between Red Banks and Thumb Rock at 3,840 m (12,800 feet) where climbers often rest. From here, walk behind Red Banks on the edge of the Konwakiton Glacier, avoiding the crevasse, then regain Red Banks and follow the ridge to its top. The crevasse is at times too large for safe passage and the alternative is to climb up from one of several shallow gullies (known as chimneys) in the Red Banks. The last part of the climb is up Misery Hill to the flat summit snowfield that is often sculpted by the wind into beautiful shapes. Cross the plateau and head for a col between the summit pinnacle (to the east) and a smaller pinnacle (to the west). The final ascent is a scramble on the summit pinnacle's northwestern side. Because of snow, not many rocks are exposed on Shasta's upper flanks, but there are steaming fumaroles to remind you that the volcano is far from dead.

Shasta could erupt again with a vengeance. The permanent snow at the summit and the steep slopes mean that catastrophic mudflows and avalanches could occur. It is hard to envisage how one could visit Shasta during an eruption – most likely a large area around the volcano would be evacuated and, sadly, the mountain's serene beauty might be gone for many years. For now, it remains as poet Joaquim Miller described it at the end of the nineteenth century: "Lonely as God and white as a winter's moon."

Mount St. Helens

The volcano

Mount St. Helens became one of the world's best-known volcanoes when it erupted violently in May 1980, killing 57 people and dramatically changing the landscape all around it. The eruption still lingers in people's minds, perhaps not for the deaths or the devastation, which were by no means the worst a Plinian eruption can do, but rather because of the surprise factor. The volcano sent out plenty of warning signs. Volcanologists correctly predicted that an eruption would occur but, up to the last moment, there was an element of disbelief that the beautiful, serene, snow-capped mountain – known then as the Fuji of America – could, in a matter of minutes, lose over 400 m (1,300 feet) of its height and become an asymmetrical cone surrounded by barren wasteland. The change in the landscape was truly stunning. Forests, the glorious Spirit Lake, the beautiful setting that had attracted campers and settlers for so many years – all disappeared in a few cataclysmic moments. It is impossible to clean up or rebuild. Only time and nature can make the landscape green and the mountain serene again, but it will take centuries. A visit to Mt. St. Helens is both educational and sobering. We can still see ample evidence of the 1980 devastation and gage the volcano's destructive powers while, at the same time, witness signs of nature's amazing ability to recover. The speedy resurgence of life has been one of the eruption's most valuable lessons.

Before 1980, Mt. St. Helens had been dormant for over a century and its eruptions in historic times were not nearly as violent. Native Americans, who must have witnessed some eruptions, called the mountain "Louwala-Clough", meaning "Smoking Mountain" – a more appropriate name than its present one, which was given by Commander George Vancouver in honor of his friend Baron St. Helens, who never actually saw the volcano. However, the Native Americans seem to have been more concerned about other volcanoes in the area. Their legends tell of battles between the brothers Wy'east (Mt. Hood) and Pahto (Mt. Adams) for the fair maiden Loowit (Mt. St. Helens). The bad behavior of the two brothers included hurtling hot rocks at one another and sending forth streams of liquid fire.

The true nature of St. Helens becomes apparent from its geologic record, which shows it to have been the most active volcano in the Cascades Range during the last two millennia. Evidence of violent past activity is given by the many deposits from pyroclastic flows, debris avalanches, and explosive eruptions. The beautiful, symmetrical stratocone that stood 2,950 m (9,677 feet) high before 1980 had been almost entirely built in the last 2,200 years. Volcanic activity in the area prior to the formation of the St. Helens cone goes back some 40,000 years. This ancient period involved

(a) Mount St. Helens viewed from Spirit Lake before the 1980 eruption. The old Mt. St. Helens stood 2,950 m (9,677 feet) high; the new volcano is only 2,549 m (8,364 feet) high. Spirit Lake was a popular recreational area before being devastated by the eruption. (Photograph by Jim Nieland, US Forest Service, courtesy of US Geological Survey.)

(b) Mount St. Helens viewed from Spirit Lake after the 1980 eruption. (Photograph by Lyn Topinka, courtesy of US Geological Survey.)

explosive and effusive eruptions of dacites and silicic andesites, with a few rare rhyodacites, forming domes, pyroclastic flows, mudflows, and depositing tephra. The eruptive stage known as Spirit Lake started 3,900 years ago and extends to the present day. Geologists have divided it into seven periods. The first period, called Smith Creek (3,900 to 3,300 years before present), ended a 5,000-year slumber. Studies of deposits from this period show that the largest yet known eruption from St. Helens happened about 3,500 years ago. This massive eruption generated about 4 km^3 (1 cubic mile) of tephra, that is, more than 13 times the amount produced by the 1980 eruption. The area covered by this tephra, known to geologists as "Yn," stretches nearly 900 km (560 miles) to the northeast, reaching Canada. Studies of the deposits show that the eruption must have been similar to that of Vesuvius in AD 79, showing us what St. Helens is capable of unleashing.

The second eruptive period, known as Pine Creek, lasted from about 2,900 to 2,500 years before present. Intermittent eruptions produced a thick fan of pyroclastic flows on what is now the south side of the mountain. Silver Lake was formed during this period, when Outlet Creek was dammed by a series of enormous mudflows, formed by breakouts of lakes upslope. Mudflow deposits from this period were a warning sign to geologists after the 1980 eruption, as they realized the danger of water breaking out from the clogged Spirit Lake and rushing downstream as a destructive mudflow. Preventive measures were taken and now a tunnel drains water from the lake.

The onset of the eruptive period known as Castle Creek about 2,200 years ago marked a major change in the volcanic activity at St. Helens. Numerous lava flows were erupted from all sides of the volcano, some of which were basaltic in composition, ranging from olivine basalt to basaltic andesite to trachybasalts, though andesites and dacites were also erupted during this period. One of the basaltic eruptions formed the Cave Basalt flow about 1,700 years ago. This flow is notable because it contains the Ape Cave lava tube, a now a major attraction in the area. It is not known exactly why St. Helens had these unusual fluctuations in composition of the lavas erupted. It might have been because of stratification in the magma chamber, with the denser lavas such as basalts accumulating at the bottom of the chamber, maybe over thousands of years. The eruptions from this period may have tapped the deeper parts of the chamber. The Castle Creek period ended about 1,600 years ago, by which time St. Helens had grown nearly as high as it was before the 1980 eruption.

The fourth eruptive period, Sugar Bowl, is named after its major product, the Sugar Bowl dome, which can still be seen on the north flank of the volcano, just east of the present crater's mouth. There are not enough carbon-14 dates to bracket the extent of this period, but it is thought that the dome was formed about 1,150 years ago. In the light of 1980 events, it is interesting that the growth of this old dome seems to have been accompanied by two lateral blasts. The largest of the two deposited lithic (meaning no new magma) material up to 10 km (6 miles) from the vent. Around the same time, another dome was formed, now known simply as East Dome because of its location on the eastern base of the volcano. Both domes are made up of rhyodacitic lavas, which are some of the most silicic yet found on St. Helens.

The fifth eruptive period, Kalama, begun in 1480. The year was deduced from dendrochronology (tree-ring dating), and we even know that the huge tephra fall that started this period happened in the winter or the early spring. The dacitic tephra (known to geologists as the "Wn" tephra) was the product of the largest tephra fall from the volcano during the last 4,000 years, with a volume of about six times that produced by the 1980 eruption. This tephra fall was followed a year later by another, not as large, but also widespread. Episodic activity continued after that, including eruptions of voluminous flows of andesite that piled up with the earlier flows from Castle Creek and intermingled with tephra, pyroclastic flows, and mudflow deposits to produce the landscape and symmetrical cone so admired before 1980.

The next period of activity, Goat Rocks, lasted only from 1800 to 1857. It extruded the Goat Rocks dome on the northwest flank of the mountain, about 700 m (2,300 feet) below what was then the summit. The dome can no longer be seen, as it was destroyed by the 1980 eruption. Events during this period included the eruption of the Floating Island lava flow, a high-silica andesite flow that is now covered by products from the 1980 eruption. A major explosion in 1842 deposited ash 100 km (60 miles) downwind. The event was witnessed by the Revd Josiah Parrish and other missionaries, who recorded the date as November 22. There are other contemporary accounts of activity during the Goat Rocks period, but most are not specific and some are contradictory. A particularly interesting – and visually stunning – historical record is a fine painting made by artist Paul Kane in 1847. The painting, now housed in the Royal Ontario Museum in Toronto, shows the dome growing on the side of St. Helens and a fiery eruption at the summit. The vent was apparently the Goat Rocks dome. The last major activity of this period happened in 1857 and reports tell of "dense smoke and fire." There may have been minor eruptions in 1898, 1903, and 1921, but the reports are unconfirmed and the deposits have not been identified. Mount St. Helens then settled down to a slumber that would last until 1980, the start of the eruptive period known simply as Modern, which has not yet ended.

The 1980 eruption

Mount St. Helens begun to stir from its 123-year slumber around March 16, 1980, with tiny earthquakes that were felt only by seismographs. On March 20, at 3.47 p.m. local time, a magnitude 4.2 earthquake gave scientists the first sign that something might be happening, but nobody in the region reported feeling the tremor. Earthquakes became more frequent during the following week and, between March 25 and 27, the seismic activity was at a frenzy, with 174 shocks of magnitude 2.6 or greater recorded. People close to the volcano only felt the largest events but scientists knew that the volcano might be waking up. Still, they were reassuring to the local press. There was no cause for alarm – yet. As one geologist pointed out, volcanoes burp from time to time.

On March 27, at 12.36 p.m. local time, St. Helens woke up with a particularly bad case of indigestion, in the form of a thunderous phreatic explosion spewing out ash and steam. This was the first eruption in the Cascades since the Lassen Peak event of 1914–17. The plume rose some 2 km (6,000 feet) into the air and a new, 75 m (250 feet) wide crater was born within the larger, snow-filled summit crater. The new crater looked like a black blemish on the pristine mountain. David Johnson, a 30-year-old volcanologist working for the US Geological Survey, talked with reporters about the event. Although the explosion had not been large and no new magma had yet arrived at the surface, Johnson knew that it was just the opening salvo. He compared the mountain to a powder keg with the fuse lit. The problem was that nobody knew how long the fuse was.

In the following weeks, phreatic explosions continued, and the volcano ejected ash and steam in bursts lasting from a few seconds to tens of minutes. The ash blown out was all from the old summit dome, blasted out by steam explosions. The first new crater was joined on the west by another, slightly larger. As the activity continued, both craters became larger and eventually merged. Several avalanches of snow and ice, darkened by ash, streaked down the mountainside. The white, pristine look had gone. Earthquakes continued to shake the area, but were not strong enough to cause people to leave. The US Geological Survey moved in a team of scientists for what they thought would be a long period of volcano watching. Local officials prepared evacuation plans and set up roadblocks on highways close to the volcano. Journalists moved in and St. Helens became daily news. Volcano watchers and tourists started arriving from all over the world. A 21-year-old Washington man disregarded the barricades and climbed the mountain on April 3. He said he could see the ground moving "like waves on the ocean" and was showered with ash from the crater. The air space around the mountain became congested with planes carrying scientists, journalists, and visitors. Merchants begun to sell volcano souvenirs, including T-shirts with the rather premature slogan "Survivor, Mount St. Helens Eruption, 1980."

While a carnival atmosphere prevailed on the surface and harmless steam explosions delighted visitors, what was happening inside the volcano was far from reassuring. Magma was moving up inside the mountain, causing large and rapid deformation of the ground. The hot magma heated up groundwater, which flashed explosively to create the steam-blast eruptions at the summit. The volcano was literally being wedged apart, becoming highly unstable. Pictures of the mountain taken before March 27 were compared with others taken a month later and showed that an ominous bulge was growing at the head of Forsyth Glacier, about 2,500 m (8,000 feet) up the north flank of the volcano. The area had expanded about 90 m (300 feet) and was continuing to grow at the amazing rate of 1.5 m (5 feet) per day. The bulge was caused by the pressure of magma moving inside the mountain and the danger was clear: even if there was no eruption, the glacier could race down the north slope into Spirit Lake. One expert described the glacier as a "time bomb sitting on marbles."

Evacuation of the Spirit Lake area was clearly necessary and Washington State governor Dixy Lee Ray declared a "Red Zone" of 5 miles around the peak, ordering everybody except scientists and law-enforcement personnel to get out. The decision did not meet with the approval of some residents. Harry Truman, an 83-year-old Spirit Lake resort owner, refused to leave. Other Spirit Lake property owners followed the evacuation orders, but were far from pleased with the governor's decision. To avoid a confrontation, law-enforcement officials escorted a caravan of 35 Spirit Lake cabin owners into the Red Zone to retrieve their possessions. The date was Saturday, May 17. Local Sheriff Bill Closner described the exercise as "playing Russian roulette with the mountain."

The morning of Sunday, May 18, dawned with no apparent signs of what would soon happen. David Johnson was on duty at the observation post about 10 km (6 miles) north of the volcano. The observation post had been set up on a ridge and was considered to be a safe location. At 7 a.m. local time, he radioed fellow scientists in Vancouver, Washington, about 60 km (40 miles) away, to report the results of some laser-beam measurements he had just taken. These measurements, like many others used in monitoring the volcano, showed no unusual changes. A little after 8.32 a.m., a magnitude 5.1 earthquake shook the ground. Suddenly, the unstable, bulging north flank of the volcano collapsed, triggering a rapid series of events that would soon lead to the deaths of 57 people. David Johnson radioed his colleagues again, but they never heard him. An amateur radio operator was the only one who heard Johnson's final call and was later able to say that he sounded excited rather than frightened when he shouted "Vancouver! Vancouver! This is it . . ."

High above the volcano, geologists Keith and Dorothy Stoffel were flying in a small plane. They first

On May 18, 1980 at 8.22 a.m. local time, Mt. St. Helens erupted spectacularly and violently. Some 400 m (1,300 feet) of the peak collapsed or blew outwards, devastating a large area as a result of the blast, debris avalanche, and mudflows. Fifty-seven people were presumably killed; some bodies were never recovered. Five more explosive eruptions occurred during 1980, including this spectacular event of July 22. (Photograph by Michael P. Doukas of the US Geological Survey/CVO.)

noticed landsliding of rock and debris inward into the summit crater. Then, within about 15 seconds, the whole north side of the crater began to move. They reported that, at this moment, "the entire mass began to ripple and churn up, without moving laterally. Then the whole north side of the summit begun to sliding to the north along a deep-seated slide plane . . . We took pictures of this slide sequence occurring, but before we could snap off more than a few pictures, a huge explosion blasted out of the detachment plane. We neither felt nor heard a thing, even though we were just east of the summit at this time."

The pilot of their plane was able to outrun the eruption cloud that threatened to engulf the small plane. The Stoffels' account was invaluable for understanding the sequence of events. They had witnessed the collapse of the north flank, which produced the largest debris avalanche recorded in living history. It is estimated that the avalanche started about 7 to 20 seconds after the triggering earthquake. The debris avalanche moved northwards at speeds of nearly 300 km/hour (180 miles/hour). Part of the avalanche surged into Spirit Lake, but most of it flowed towards the west into the North Fork of the Toutle River. The momentum of the avalanche was enormous – at one location about 6 km (4 miles) from the summit, the mass of volcanic debris, glacial ice, and, possibly, water from Spirit Lake, flowed over a ridge more than 350 m (1,150 feet) high. The enormous avalanche covered an area of 60 km^2 (24 square miles), changing it into a moon-like landscape.

The worst part of the eruption happened a few seconds after the triggering of the avalanche. As the Stoffels had described, a huge explosion blasted out of the detachment plane, which they neither heard nor felt. The collapse of the north flank caused an almost instantaneous expansion of high-temperature, high-pressure steam which until then had been dissolved in the magma that formed the bulge. The avalanche "uncorked" the volcano, unleashing an enormous lateral blast of rock, ash, and hot gases that sped downhill as a deadly fan which, in moments, totally devastated an area of about 600 km^2 (230 square miles), mostly to the north of the volcano. The lateral blast began a few seconds after the avalanche, but the blast's speed was much greater and it soon overtook the avalanche. The blast's initial velocity was about 350 km/hour (220 miles/hour), but it quickly increased to near-supersonic velocities of about 1,100 km/hour (670 miles/hour). The volcanic debris in the blast included the first new magma, released from what had been the bulging magma dome. The magnitude of the blast was estimated to have been about 500 times greater than that of the Hiroshima atomic bomb. David Johnson was directly in its path. A fellow geologist who got to the ridge by helicopter a few days later found that the area had been wiped clean – there was no sign of David Johnson, his jeep, or his camper. Everything, including all the trees, had been blown away.

Keith Ronnholm and Gary Rosenquist were at the Bear Meadows camping area, about 18 km (11 miles) northeast of the volcano when the earthquake struck. Both grabbed their cameras and recorded the sequence of events showing the collapse of the mountain. They were able to get away and their photos were invaluable to help scientists understand the sequence of events. Some others on the fringes of the blast also managed to survive. Bruce Nelson and his girlfriend saw the blast cloud coming at their camping site, which was located well outside the Red Zone. They clung to each other as trees crashed all around and, even though they were nearly buried by ash, they managed to stay alive. Their two friends, only a short distance away, died as their tent was crushed.

Very shortly after the lateral blast, St. Helens unleashed the next demon: a strong explosion created a column of ash that rose more than 19 km (12 miles) in less than 10 minutes, and expanded into a mushroom-shaped cloud. For the survivors beyond the devastation zone, as well as for residents of nearby towns, this was the most frightening event. Near the volcano, the swirling ash particles in the atmosphere triggered lightning strikes, which in turn ignited forest fires. Prevailing winds sent much of the cloud drifting towards the east and northeast at speeds of about 100 km/hour (60 miles/hour). When the cloud reached the town of Yakima at 9.45 a.m. and Spokane at 11.45 a.m. it was dense enough to turn day into night.

The possibility of mudflows and pyroclastic flows had worried scientists and Mt. St. Helens lived up to their fears. Destructive mudflows begun a few minutes after the blast, as the hot material in the debris avalanche, the lateral blast, and ashfall from the erupting column met with ice and snow on the flanks of the volcano. Mudflows in the upper reaches of the South Fork of Toutle River were seen around 8.50 a.m. The largest and most destructive mudflows developed several hours later in the North Fork of Toutle River when the water-saturated deposits of the debris avalanche begun to slump and flow. On the upper slopes of the mountain, the mudflows travelled as fast as 150 km/hour (90 miles/hour), slowing to about 5 km/hour (3 miles/hour) in the flatter and wider parts of the Toutle River drainage. The mud was hot – even after traveling tens of kilometers the temperature remained warm, in the range of 29 to 33 °C (84 to 91 °F). Like wet cement, it flowed down both forks of the Toutle River down to the Cowlitz and Columbia Rivers below. On the way, the mud overran two logging camps 19 km (12 miles) away from the volcano, killing at least three people. A couple who

Reid Blackburn's car destroyed by the eruption, about 16 km (10 miles) north of Mt. St. Helens. Blackburn, a photographer for *National Geographic* magazine, was killed by the eruption. (Photograph taken on May 31, 1980 by Daniel Dzurisin, courtesy of US Geological Survey/CVO.)

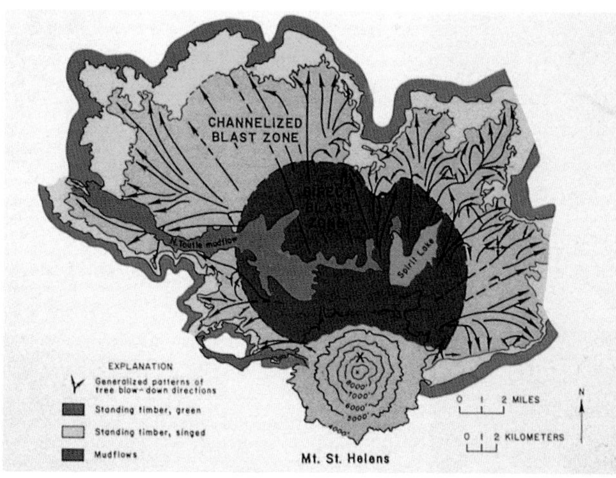

Geologic map of eruption products, showing range of mudflows and other deposits. (From Kieffer, 1981; courtesy of S. Kieffer)

were camping 37 km (23 miles) away on the banks of the Toutle River barely escaped death as the river of mud carried them for over a kilometer (3/4 mile) before they were able to wade to the shore. The mudflow reached the town of Toutle, about 40 km (25 miles) away from the volcano, destroying all that was close to the river. More than 1,000 people were evacuated from the valley, 150 houses were destroyed and many more were flooded, while the high temperature of the river killed all the fish. In some places, the mudflow reached 110 m (360 feet) deep, leaving "bathtub rings" on the valley walls. The debris eventually poured into the Columbia River, decreasing the depth of its navigational channel from 12 m (39 feet) to 4 m (13 feet). Freighters upriver in Portland were stranded and river traffic came to a halt. Other forms of transportation were doing no better: ash forced the closure of roads, air traffic stopped, and everywhere in the region people were stranded as their cars, trains, and buses could go no further. It is estimated that about 10,000 travelers were stranded.

Pyroclastic flows had been the other major worry and St. Helens made sure there were some. The first pyroclastic flow was seen shortly after noon, but it is likely that flows started to form just after the blast. At least 17 pyroclastic flows rushed down the mountain during the next 5 hours, covering layers of debris deposited earlier. Eyewitnesses reported that the largest flows started with the upwelling of material in the crater, then the mass flowed northwards, escaping through the breach in the crater – as one scientist described it, like a pot of oatmeal boiling over. The material forming the pyroclastic flows was mostly juvenile pumice, that is, new magmatic material that came out of the volcano. The flows formed a fan-shaped deposit of overlapping sheets and lobes. Two weeks after the eruption, some of the deposits still had temperatures of up to 420 °C (785 °F). Pyroclastic flows had been a major concern of scientists but the damage they caused was relatively small compared to that of the blast and of the mudflows.

As the fateful day of May 18 wore on, the eruption continued unrelenting. The most vigorous phase lasted for about 9 hours, constantly feeding the drifting ash column. About 540 million tons of ash fell over an area of more than 57,000 km^2 (22,000 square miles). By the end of the day, the eruption finally began to subside. It finally stopped the next day. By that time, the ash cloud had spread to the central regions of the country. Ash from the eruption fell as far away as Oklahoma.

The change in the mountain was a shock to many, even to those who had realized how violent the eruption had been. The stately, 2,950 meters (9,677 feet) peak was torn apart. One geologist said that it looked like some huge hand had scooped out an entire side of the mountain. A giant amphitheater now dominated the whole northern side. The summit crater was 1.5 km (1 mile) wide and 3 km (2 miles) long. The crater rim stood some 900 m (3,000 feet) lower than it had been before the eruption. Even more shocking was the change in the landscape around the volcano. Gray humps of congealed mud and debris had replaced the lush forests. The fallen trees, looking like broken matchsticks, covered some 400 km^2 (150 square miles). President Jimmy Carter, flying over the area a few days later, said the scene was "literally indescribable in its devastation."

Next came the hard task of finding the survivors, cleaning up the ash, and housing the homeless. Rescue crews had to contend with numerous difficulties, including the fact that helicopters could not land in many areas because of the loose ash on the ground. Despite the problems, nearly 200 survivors were rescued within a few days of the eruption. Ash removal then became the major problem to be solved. In the town of Yakima alone, clearing the ash took 10 weeks at a cost of over 2 million dollars. The economic loss from the eruption was estimated at 1 to 3 billion dollars, primarily due to timber, civil works, and agricultural losses. An inevitable effect of the eruption was loss of revenue from tourism, but this turned out to be only temporary. Mount St. Helens has more than regained its appeal for visitors. They come to see the

The extraordinary devastation caused by the eruption is clear in this view of overturned trucks and caterpillars on the log camp on the South Fork Toutle River. The photograph was taken on May 19, 1980, one day after the eruption. (Photograph by Phil Carpenter, courtesy of US Geological Survey/CVO.)

devastation first-hand, to learn about what volcanoes can do and how nature recovers from such a catastrophic event. In 1982, the Mount St. Helens National Volcanic Monument was created, comprising 485 km^2 (120,000 acres) around the volcano. Since then, St. Helens has often been referred to as "America's favorite volcano."

A personal view: Lessons from the eruption

Where we you when the mountain blew? Even so many years later, people living within sight of Mt. St. Helens still ask one another this question. The eruption made headlines around the world and is still remembered by many, while more deadly events – such as El Chichón in 1982 and Nevado del Ruíz in 1985 – seem to have been largely forgotten by the public and media. The St. Helens eruption killed 60 people, while the toll was between 2,000 and 3,000 for El Chichón and 25,000 for Nevado del Ruíz. Why, we might ask, did St. Helens get all the press?

I have no doubt that the main reason for the eruption's impact was simply because it happened in the USA, where the most modern volcano-monitoring technology was available, as were many of the world's best-trained geologists. The volcano was being monitored day and night, an eruption was deemed likely, and a large area was evacuated – but not large enough. Many of the people who perished were well outside the Red Zone, in areas considered safe. The lesson the world learnt was that neither science nor technology could reliably predict when the eruption would happen or what its magnitude would be. People living near other, apparently placid, volcanoes realized that the St. Helens tragedy could happen to them.

I remember where I was when the mountain finally blew – on a rather quiet Kilauea volcano. It was my first trip to Kilauea and I sincerely hoped it would blow, but evidently Pele was taking a vacation in the mainland. My colleagues and I gathered around a speakerphone at the Hawaiian Volcano Observatory, hearing the latest news from one of Observatory's scientists who, like many others, had gone to Mt. St. Helens when the signs of unrest became clear. I thought of the volcano as Humpty Dumpty falling off the wall, while the scientists and their instruments – like all the king's horses and all the king's men – were powerless to prevent the catastrophe and unable to repair its damage.

What we were able to do was learn. Mount St. Helens has justifiably earned its place in history not because of the magnitude of its eruption or the number of lives lost, but because of how much it taught us both scientifically and socially. The hardest lesson was how we can be surprised by even a closely watched volcano. Mount St. Helens was a well-monitored volcano before the eruption but it still managed to do the unexpected. In particular, the phenomenon of a giant dome growth on the side of the volcano had never before been studied in detail and the resulting collapse and sideways blast were completely unpredicted. It was not an unknown event – in fact, according to scientists who worked on the Mt. St. Helens crisis, they discussed the 1956 eruption of Bezimianny volcano in Russia, in which a bulge on the side of the volcano eventually led to a blast that felled trees up to 25 km (15 miles) away. The Bezimianny blast was even mentioned in an article in the local paper, the *Tacoma News Tribune*, published on May 6, 1980. Jack Hyde, a local geologist, is quoted in the article speculating about the potential instability of the volcano's northern slope, which could lead to massive landslides. With hindsight, it may seem odd that the catastrophe was not predicted more accurately. The problem was the lack of knowledge of this type of eruption, which meant that the possibilities could not be properly assessed. The Bezimianny scenario remained just one of the many possibilities.

Prior to the Mt. St. Helens crisis, knowledge about stratovolcanoes was glaringly lacking in one particular aspect: what kind of signs did they give before an eruption? The most intensely monitored volcano in the world up to then had been Kilauea, a shield volcano. The responses of Kilauea had become well known and volcanologists working in Hawaii were generally able to predict when and where an eruption would happen. However, the behavior of stratovolcanoes such as Mt. St. Helens is very different. The monitoring before, during, and after the 1980 eruption has been invaluable to extend our knowledge of how stratovolcanoes work. Years later, much of the knowledge was put to good use at Mt. Pinatubo in the Philippines, averting what could have been a major tragedy.

Visiting during repose

Mount St. Helens is located near two state capitals: Seattle, Washington, and Portland, Oregon, and within the Mount St. Helens National Volcanic Monument. Access to the Monument is easy and the facilities are very good, making Mt. St. Helens a popular destination – in fact, a major tourist attraction. This has some advantages, such as excellent museums and trails, and the availability of maps, books, films, and air tours, not to mention campgrounds and lodgings. Mount St. Helens is an excellent destination for families and schoolchildren who want to learn about volcanoes. On the down side, it is easy to feel that the mountain has become over-exploited. Luckily, it is still possible to get away from the beaten path and to find incredible scenery and even some degree of isolation.

The best times to visit are from mid June to late September, when the days are long, sunny, and clear. However, the weather can change quickly at any time, so be prepared for rain and sudden drops in temperature. Nights can be very cold, even in summer, and snowstorms above the timberline can occur at almost any time. Many of the Monument's roads are closed in winter and climbing the mountain at this time can be dangerous because of snow avalanches. If you must visit during winter, call the Monument Headquarters and go to the visitor centers (open year-round) to get information on roads and climbing conditions. Winter facilities include two Sno-park areas for cross-country skiing and snowmobiling.

The most impressive sights on Mt. St. Helens are the devastated areas and the summit crater and dome. To understand the lay of the land, it helps to review the different devastation zones. The major units are the blast zone, the debris avalanche deposit, the pyroclastic flow deposits, and the mudflow deposits. The lateral blast zone is usually described as three separate parts. The closest to the volcano is the direct blast or "tree-removal" zone, about 13 km (8 miles) in radius. Almost everything within this area was totally obliterated and carried away by the blast and the flow of material was not at all diverted by topography. The second zone is the channelized blast zone or "tree-down" zone and extends out to about 30 km (19 miles) from the volcano. Everything was flattened in this area and the force and direction of the blast is obvious from the parallel alignment of large trees, which were broken off at the base and toppled like broken matchsticks. The topography in this zone channeled the flow to some extent, hence its name. The third and outermost zone, on the fringe of the area impacted by the blast, is called seared or "standing dead" zone. The trees in this area remained standing, but were singed brown by the hot gases in the blast. Mudflow deposits extend beyond the blast zone in the valleys of the north and south forks of the Toutle River, and around the summit cone to the south, east, and west.

One of the best ways of seeing the devastated zone is from the air. Thanks to the popularity of Mt. St. Helens as a tourist attraction, tours by helicopter and light planes are offered routinely and are fairly inexpensive. Many of the helicopter tours circle the lava dome growing inside the crater, offering the best and closest view at present, as ground access to the crater is forbidden.

On the ground, the best place to start your visit is the Visitor Center at Silver Lake. The Center stocks maps as well as books about the volcano which provide more details than can be given here (see Bibliography for recommendations). The suggestions below are what I consider the highlights of a visit to Mt. St. Helens. Because of the layout of local roads, it

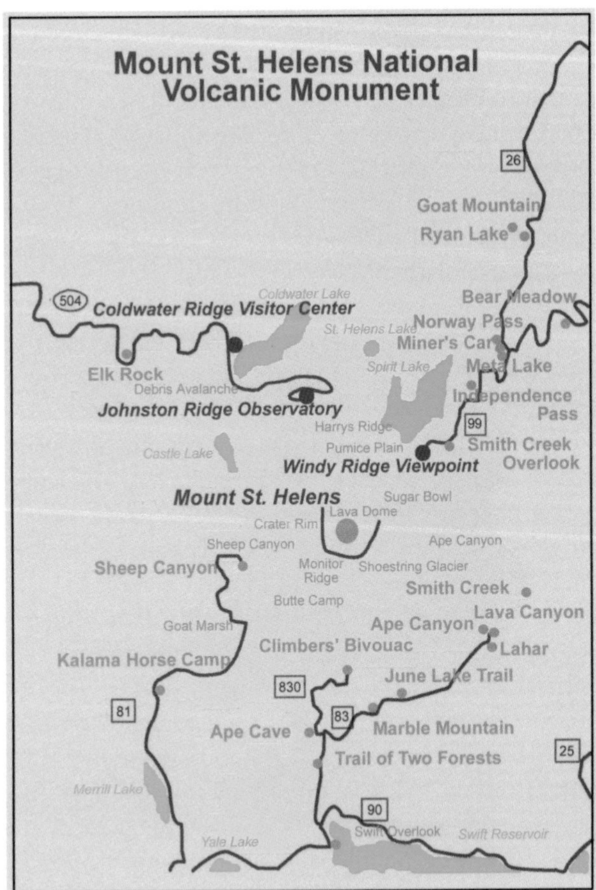

Map of Mount St. Helens National Monument, showing major roads and points of interest. (Modified from US National Park Service.)

is best to plan your visit in three stages, allowing 2 days for each: the western side, the northeastern side, and the southern side. I recommend starting on the western side, stopping at the visitor centers for an overview of the volcano and of the 1980 eruption. It is not a good idea to stop at all the visitor centers in the same day: you will get an overdose of exhibits and multimedia.

The western side

Interstate 5 runs about 50 km (30 miles) west of the volcano, linking Portland to Seattle. The nearest towns to St. Helens along Interstate 5 are Castle Rock and Woodland. The Spirit Lake Memorial Highway (504) starts by Castle Rock, at exit 49 from Interstate 5. A Cinedome is located by this exit, showing an impressive, giant-screen movie of St. Helens. Once on the Memorial highway, head towards the Visitor Center at Silver Lake. The highway cuts high above the valley of the North Fork of the Toutle River, ending at Johnson Ridge Observatory where David Johnson perished in 1980. It crosses the Toutle River in several places, and overlooks allow you to see some of the dramatic effects of the 1980 eruption, as well as deposits from ancient events. Over the years, the river has cut into layers of mudflows exposing their deposits, which can be recognized by rounded boulders that are loosely cemented in a mass of dried mud typically 1 to 3 m (3 to 10 feet) thick. The many layers of mudflows attest to their frequency in the eruptive history of Mt. St. Helens. Mudflows can happen even if there is no eruption taking place, as ash and friable rocks can be loosened by heavy rains or snowmelt. The deposits from the 1980 eruption make the river valleys and drainage areas particularly susceptible to new mudflows.

Mount St. Helens Visitor Center at Silver Lake

Located about 50 km (30 miles) to the northwest of the volcano and 8 km (5 miles) east of the town of Castle Rock, this visitor center is the best first stop on a visit to St. Helens. Among the attractions is a theater showing an award-winning film about the volcano and the 1980 eruption, a large walk-through model of the volcano, and displays about the geology and history of the area. On a clear day, you can see the crater from the terrace in front of the building, with Silver Lake in the foreground. Silver Lake was formed when a prehistoric mudflow from St. Helens dammed one of the Toutle River's forks. Across the highway is Seaquest State Park, which offers recreational facilities and a campground. This is a particularly convenient place to camp when visiting the western side.

Continue east on the highway for about 35 km (22 miles) up to *Hoffstadt Viewpoint and Hoffstadt Bluffs Visitor Center*. Stop here for a beautiful view of the volcano. In the foreground are the remnants of the first sediment dam built after the 1980 eruption across the Toutle River valley. The purpose of the dam was to stop the runoff of loose ash and avalanche debris during winter storms, but it did not last long, as it was filled and overtopped by mudflows during 1981 and 1982. The Visitor Center, operated by Cowlitz County, offers information, a bookstore, and a restaurant. This is also the place from where to get helicopter rides to the crater. The site is named after the Hoffstadt Bluffs, which can be seen from here and from the highway. These cliffs are eroded remnants of rocks from volcanic eruptions that occurred millions of years ago, before the cone of Mt. St. Helens was formed.

Continue along the highway, which becomes steeper after this point, for another 10 km (6½ miles) up to the

Charles W. Bingham Forest Learning Center. As the name implies, the exhibits focus on the forest destruction and recovery. The entry hall contains a re-creation of the pre-1980 forest. A path leads into the "eruption chamber" where there is a re-creation of the blast zone and a multi-media program entitled *You Are There*. The next exhibits tell of the salvage efforts of downed timber and of the natural recovery of the landscape. You can take a "forest video tour" that simulates flying over the volcano in a helicopter and let the kids play in the volcano-theme playground.

If you stop here, make sure to walk to the *Elk Viewing Area*. Elk can often be seen from here but the highlight is the great view of the North Fork of the Toutle River valley. This is where the deadly avalanche swept down during the early stages of the 1980 eruption. Floods and mudflows have eroded and covered up much of the avalanche's deposits, but if you look upstream you will see a distinct hummocky texture that is characteristic of the avalanche deposit. The trees in this area were all blown down, but much of the timber was salvaged after the eruption and, since then, new trees have been planted. On the morning of the eruption, four loggers were working near here, behind North Fork Ridge, with the volcano hidden from their view. Only one of them survived.

Continue on the road for about 6 km (4 miles) up to the *Elk Rock Viewpoint* to see a panorama of the volcano with the Toutle River Valley in the foreground. To the west of Mt. St. Helens is Spud Mountain, so called because of its potato-shaped summit, which now divides the "tree-down" zone (on its east side) from the "standing dead" zone (on its west side). On the eastern side of Mt. St. Helens, you can see Johnston Ridge and beyond it Mt. Adams, 50 km (30 miles) away.

As the road continues through Elk Pass, notice the pink and green hues of the old strongly altered rocks. Make sure to stop at the *Castle Rock Viewpoint*, about 9 km (5½ miles) from Elk Rock. This is one of the best places from where to see the 1980 avalanche debris deposits. The avalanche dammed Castle Creek and created the small Castle Lake across the valley to the south. The lake level is kept stable by means of an outlet. Note also the erosional terraces along the North Fork Toutle River exposing layers of old mudflows.

Continue driving for another 8 km (5 miles) up to *Coldwater Ridge Visitor Center*, one of the major highlights of a visit to Mt. St. Helens. The focus of the exhibits is on nature's recovery of the devastated area. It is worth photographing the Mt. St. Helens crater and lava dome from here and to stop by the Center's information desk, bookstore, and restaurant. The Winds of Change Interpretative Trail is a short loop (0.5 km, ⅓ mile) that illustrates the recolonization of the environment. Many visitors like to stroll down to Coldwater Lake where there is now a fishing and boating area. The 1980 debris avalanche deposit is easily recognized near the lake by its hummocky surface. Some of the hummocks poke out of the lake's south end as small islands. The Birth of a Lake interpretive trail (0.4 km, ¼ mile) explains how the lake formed during the 1980 eruption, when the avalanche debris dammed Coldwater Creek.

To view the avalanche deposits close-up, drive another 3 km (1¾ miles) on the highway up to *Crater Rocks Trail*. This short, barrier-free loop (400 m, 1,300 feet) goes through the hummocks of the avalanche deposits. An extended trail goes down to the North Fork Toutle River (3.5 km, 2¼ miles round trip). Notice the large chunks of pre-1980 material that were carried by the avalanche. Their different colors and textures are due to the different ages and compositions. Light gray chunks are dacites from the Goat Rocks period, while those with a bluish tinge are dacites from the older Kalama period. Dark, reddish brown chunks are basaltic andesites also from the Kalama, while the black chunks are Castle Creek basalts and basaltic andesites. Some "breadcrust" bombs from the 1980 eruption can be seen on top of the avalanche deposits. The bombs are pieces from the dacitic mass that intruded below the volcano in 1980. They were ejected hot, cooling quickly on the outside as they fell.

Johnson Ridge Observatory

This is the absolute must-see for any visitor to Mt. St. Helens. The view from the Observatory is stunning – by far the best view of the crater from the northern side. This is also the best place on the northern side from which to photograph the lava dome with the 1980 avalanche deposits and the pumice plain between the volcano and Spirit Lake in the foreground. The crater is only 8 km (5 miles) to the south of here, but the view makes you think you are right inside it. It is chilling to imagine standing here, as David Johnson did, when the whole northern face of the mountain collapsed.

Johnson Ridge is a working volcano observatory, housing seismometers and other monitoring equipment. A video camera is pointed at the volcano and the images are put on the World Wide Web every 20 minutes throughout the day. Technophiles can sit at terminals and access the Web and video disks with information about the volcano. A variety of other exhibits tell the story of the 1980 eruption and show visitors how to interpret the landscape they see. Talks are given from time to time in the 280-seat theater.

The building itself is worth seeing – partly dug into the rock, it blends with the landscape. The feeling of being in a volcano museum inside a volcano makes this by far the most interesting locale in the Monument accessible by road. The Observatory is the final stop along the Spirit Lake Memorial Highway, but you can continue exploring on foot. The short Eruption Trail (0.8 km, ½ mile) starts by the Observatory and features a memorial to the 57 people who died during the 1980 eruption.

Norway Pass Trail

This trail is part of Boundary Trail number 1. Distances one-way from Observatory: to Harry's Ridge, 5.2 km (3¼ miles); to Norway Pass, 19 km (12 miles); to trail-head on Road 26, 23 km (14 miles). This trail, which also starts by the Observatory, is considered one of the best hiking and backpacking trails in the Monument. The many ups and downs in topography make it either a seriously strenuous day hike or a backpacking trip. Be aware that snow can persist on north-facing slopes late in the season, making the trail tricky to find. If you are short on time or not in good physical shape, hike as far as Harry's Ridge. There is an alternative, shorter route to Norway Pass from Road 26 which is described later.

Starting from the Observatory, the trail takes you through the debris avalanche deposits up to Harry's Ridge, named after Harry Truman (killed by the avalanche). *Harry's Ridge Viewpoint* has the premium view in the Monument of the crater, lava dome, and Spirit Lake. After this point, either return to the Observatory or backpack all the way to the trailhead on Road 26 (be sure to arrange a pick-up). The views from the trail are spectacular – such as the classic views of the crater reflected on Spirit Lake – but you can see the best of them by hiking from the trailhead on Road 26 up to Norway Pass, which has the advantage of being a considerably shorter route. What you get by hiking all the way from Harry's Ridge is the chance to see the beautiful, uncrowded backcountry. Small green lakes shine amid the gray ash and pumice and, during the summer, a surprising number of wildflowers add color to the landscape.

The summit dome growing inside Mt. St. Helens. The dome is about 1,100 m (3,600 feet) in diameter. (Photograph courtesy of Nick Gautier.)

The northeast side

To get to this side of the mountain, drive north from Castle Rock up to Highway 12, which cuts east–west about 30 km (20 miles) north of the volcano to the town of Randle. The road follows the Cowlitz River and gets close to its two reservoirs, Mayfield and Riffe. The Randle Ranger Station is located 5 km (3 miles) east of Randle and operates during the summer months, providing information on road conditions and on the volcano in general. Highway 12 continues to Mt. Rainier. To get down to the northeastern side of Mt. St. Helens, take Road 25 south of Randle. The Woods Creek Information Station is located 10 km (6 miles) south of Randle where you can ask questions at the booth without leaving your car. From here on the drive gets more interesting. Take Road 26, a two-way but one-lane country road with no services. There are turnouts for passing traffic but the pace is obviously very slow. The reward is the scenery, including water-falls and majestic old conifers. Seeing these trees live and contrasting the scenery with that of the devastated zone really puts the magnitude of the blast into per-spective. Similar trees to these, some up to 60 m (200 feet) high were destroyed in seconds, blown down like matchsticks.

Ryan Lake Trail

About 20 km (13 miles) from the start of Road 26 is the Ryan Lake trailhead parking lot. This is an easy loop trail (1km, ⅝ mile) that climbs up to the Green River valley overlook and lets you see the effects of the blast. Ryan Lake is located just inside the boundary of the "tree-down" area and signs along the trail tell visi-tors about the destruction as well as the salvage efforts and the natural recovery of the landscape since the eruption. The ash on the trail, forming a layer about 8 cm (3 inches) thick, is from the blast cloud. Look for places where the trail cuts into this ash and notice the different layers. On top of the ash you can find small pieces of pumice. The small Ryan Lake, located just north of the trail, is surrounded by downed trees that have been left in place since 1980, though other timber in the area was salvaged after the eruption. One person died here during the blast, a camper who was on the north shore of the lake.

Back on Road 26, drive for another 2.5 km (1½ miles) up to a small turnout from where there is a great view of the "tree-down" area. The trees in this area have been left just as they were after the blast. Unlike the trees blown down nearer to the crater, which point neatly away from it, these trees show more variation in their alignment due to the effect of the topography and gravity on the motion on the blast cloud. The best way to view the "tree-down" area close up is to take the Norway Pass to Independence Pass trail. The trailhead is located 4.4 km (2¾ miles) further down Road 26.

Norway Pass to Independence Pass Trail

The distances are 10 km (6 miles) one-way to Road 99, 3.5 km, 2¼ miles) to the top of Norway Pass. This is one of the most highly recommended hikes in the National Monument. Arrange for a pick-up at the other end of the trail on Road 99, otherwise you may consider returning to the trailhead on Road 26 after reaching the top of Norway Pass. Starting at the trailhead on Road 26, the trail climbs about 330 m (1,100 feet) for about 3 km (2 miles) through the blown-down forest up to Norway Pass at 1,400 m (4,500 feet) altitude, from where there is a stunning view of Spirit Lake, the Mt. St. Helens crater, and the lava dome. Charred and blown-down trees along the trail contrast with new vegetation and wildflowers, making the scenery truly spectacular and one that keen photographers should not miss. If you continue down to Independence Pass, you will walk right through the blast zone, seeing first-hand the destructive power of a volcano.

Meta Lake Trail and Miner's Car Site

From Road 26, take the turnoff to Road 99 towards Spirit Lake and the Windy Ridge viewpoint up to Meta Lake. The small emerald-green lake, surrounded by downed trees, is accessible by a short, barrier-free trail (0.8 km, ½ mile). Meta Lake provides an interesting lesson in survival and recovery. The small, healthy Pacific silver firs around the lake were very young trees in 1980 – so small that they were able to survive the eruption beneath the cover of ice and snow, while the older trees were killed. Ice and snow covering the lake also enabled some fish, amphibians, and insect larvae to survive the blast. Sadly, a family perished near here during the eruption. Their car, which was thrown some 18 m (60 feet) away by the blast, can still be seen near the road. The family had camped overnight in a miner's cabin, which was totally destroyed by the blast.

Harmony Viewpoint

This viewpoint over the north end of Spirit Lake is located about 8 km (5 miles) further down Road 99. The lake was formed over 3,000 years ago when debris from a violent eruption dammed the Toutle River. Since then, other eruptive episodes have caused the water level in the lake to rise and fall. The lake level rose some 70 m (240 feet) following the 1980 eruption because of the blockage of its natural outlet into the North Fork Toutle River and also because of debris falling in and reducing its depth. Shortly after the eruption, the lake's surface was covered by a mass of logs. Some still float on the surface, though many more have sunk since then. When the 1980 avalanche hit the lake, the water displaced surged upslope. Notice that tree logs are sparse on the ground upslope from the lake's shore, because the displaced water washed them down into the lake. The logs there point downslope, while those felled by the blast point uniformly away from the volcano. The lake's level is still higher than it was before the 1980 eruption, but it has now been stabilized by means of an outflow channel connecting the lake to South Coldwater Creek. In 1982, the lake came dangerously close to overflowing and you can still see the "bathtub ring" of logs that were left behind.

Harmony Trail to Spirit Lake

The distance is 5 km (3 miles) round trip; allow 1.5 hours. This short though steep trail is the only legal access to Spirit Lake. The trail descends the wall of a cirque and goes down 180 m (600 feet) to the lake's east shore. Most of the rock exposed along the trail are old breccias (broken pyroclastic rocks and tuffs, but halfway down the trail there is an exposed lava flow showing columnar jointing. As the trail flattens out, look towards the northern side of the cirque – a scar is visible, left by water displaced from the lake as the 1980 debris avalanche flowed into it.

The name Spirit Lake was given by Native Americans

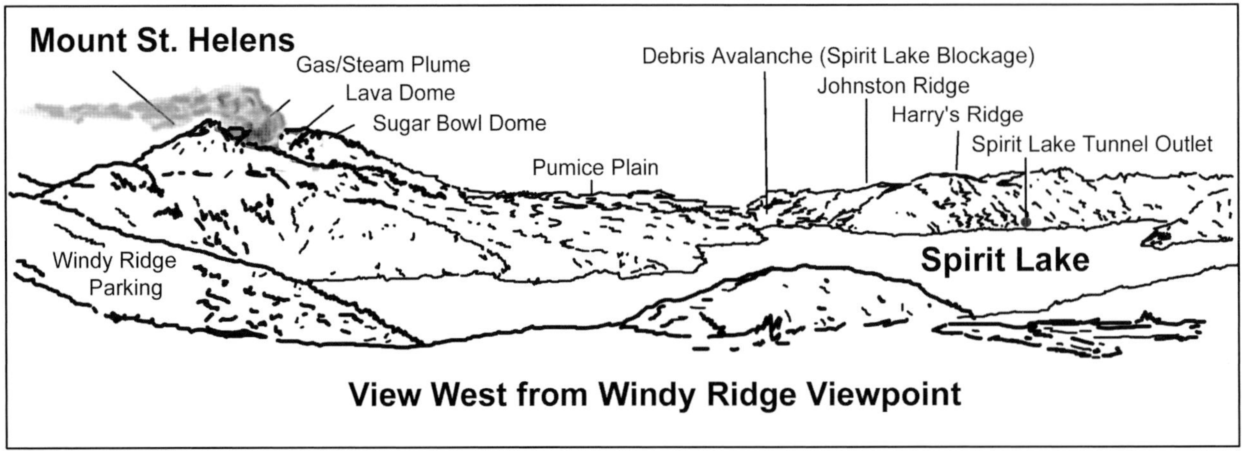

Panorama from Windy Ridge, one of the Mount St. Helens National Monument's most visited sites. (Modified from drawing by US Geological Survey/Cascades Volcano Observatory.)

who believed it was inhabited by evil spirits. The legend fits the place, as the devastation of the area would make even the most evil of spirits feel at home. It is hard to believe that before 1980 this was one of the most scenic spots in Washington State. Although it will be many years before the landscape loses its moon-like aspect, nature is slowly recovering. The chemical changes in the lake following the eruption were so drastic that scientists were unsure if the lake could ever support life again. In an amazing lesson in recovery, the lake's waters were almost back to normal within 5 years, thanks to fresh water added by rain and snowmelt and the stirring action of wind and waves. Spirit Lake now supports life again, including frogs and algae.

Windy Ridge Interpretive Site

This viewpoint, at an elevation of about 1,200 m (4,000 feet), has one of the best panoramas in the National Monument: the crater, Spirit Lake, Johnson Ridge, Harry's Ridge, and the blast zone. There is an outdoor amphitheater where talks are given during the summer months (ask for a schedule at any of the visitor centers or information stations). The view is particularly good from the top of the 361 steps leading from the parking lot up a small hill – it is worth the climb. This is a great place from which to photograph the devastated landscape and the crater, which is only 6 km (4 miles) away. The hilly deposits you see between Johnson Ridge and Harry's Ridge are from the avalanche deposit. The avalanche ran over Johnson Ridge where it killed the young volcanologist. Harry Truman died when the avalanche buried his Spirit Lake resort.

Truman Trail to Harry's Ridge

Named after Harry Truman, this trail (11 km (7 miles) one-way, allow 4 hours for the round trip) winds its way from the Windy Ridge viewpoint down to the pumice plain and leads to the closest legally accessible point to the lava dome. Harry and his lodge are buried here, covered by some 100 m (300 feet) of pyroclastic flow deposits and pumice fall. The devastation here was complete, with no surviving plants or animals. Note that this is a restricted research area, where natural recolonization has been studied since the eruption. The trail passes by several research plots – it is very important to follow posted directions and stay strictly on the trail.

To end your visit to the northeast side of the mountain, drive back to Randle by taking Road 99 and then turning north on Road 25 which takes you to the Bear Meadow viewpoint. It was from here that Gary Rosenquist and Keith Ronholm shot the world-famous sequences of photos of the initial phase of the eruption. They were very lucky to survive, as Bear Meadow is a notch right between the northern and eastern zones devastated by the blast.

The south side

The main reason to visit the south side is to climb the volcano, though Ape Cave is worth a detour. Since the 1980 eruption did not significantly affect the southern side, it can give you an idea of what Mt. St. Helens used to be like. To explore the various attractions, drive to Cougar on Highway 503, which follows the Lewis River valley, and continue on Road 90. If you need information, drive to the Pine Creek Information Station, located 27 km (17 miles) east of Cougar on Road 90. The information station is open daily during the summer and offers help and directions, as well as books and video sales and a short movie about the volcano. To get to Ape Cave, head north along Road 83 and follow signs to Ape Cave, located on side Road 8303. To reach Lava Canyon, continue north and east up to the end of Road 83. To climb the mountain starting at Climbers' Bivouac, turn off from Road 83 at the intersection with Road 81, then head east on Road 830 up to where it ends.

Ape Cave

Discovered in 1946, Ape Cave is 3.9 km (12,810 feet) long. It is famous for being the longest intact lava tube known in North America as well as one of the longest in the world. The tube was named after a local group of young outdoorsmen, the St. Helens Apes, who explored it during the 1950s. The tube was formed about 1,700 years ago by the Cave Basalt pahoehoe lava flow during an eruption very different in style from that of 1980. You can go into the tube down a skylight and walk the lower (downhill) part easily using a flashlight. Lantern rentals and an informative trail leaflet are available at Apes' Headquarters, located by the tube entrance and open during the summer. They will also provide the schedule of cave tours, guided by volunteer speleologists, which are given on summer weekends.

The lower part of the tube is about 1,200 m (4,000 feet) long, while the upslope part extends for some 2,100 m (7,000 feet). The upper part is much harder to explore as it involves scrambling over piles of broken rocks. When exploring the tube, notice the lava stalactites hanging from the ceiling, formed by lava drips, and the lava stalagmites on the floor, formed by the accumulation of drips. A sandy mudflow flowed into the cave sometime after its formation, possibly around AD 1480 or 1482, leaving behind deposits that include pumice fragments.

The nearby *Trail of Two Forests* is a short (0.4 km, ¼ mile) boardwalk trail that allows you to see lava tree molds formed by the Cave Basalt lava as it flowed over a forest. The trailhead is located near the lava tube entrance on side road 8303. The trail's name refers to the present forest and to the old forest destroyed by the flow. The tree molds were formed when lava cooled and hardened against tree trunks. The trees were burnt, but the lava "molds" remain. Bring a flashlight if you want to crawl through one of the molds.

Lava Canyon Trail and Lahar Viewpoint

Road 83 crosses a lahar (mudflow) that swept down Pine Creek on May 18, 1980, creating yet another stretch of barren landscape. You can see the mudflow deposit from Lahar Viewpoint about 1.5 km (1 mile) from the end of the road. Lava Canyon, located at the very end of Road 83, is an old valley that was filled by a lava flow about 3,500 years ago. Deep channels and potholes were cut into the lava by stream erosion but these were later filled by mudflows and pyroclastic deposits. The 1980 mudflow ripped through the valley with enough speed to erode some of the older deposits and to expose the channels and potholes. The first kilometer (⅝ mile) of the Lava Canyon Trail is paved for easy access and leads to a viewpoint of a waterfall where the Muddy River plummets into the canyon. The middle section of the trail is a 1.5 km (1 mile) loop that crosses over the turbulent Muddy River. The lower part of the trail (4 km, 2½ miles) is quite steep and rugged as it descends into the canyon, but provides beautiful views of several waterfalls. Beware of slippery rocks and steep drops.

Climbing to the summit

The highlight of a visit to Mt. St. Helens is no doubt the scenic ascent to the summit. The view from the crater rim is breathtaking: the lava dome nestled inside the crater and, beyond it, the devastation of the 1980 eruption. It is only from the top that the scale of events becomes truly apparent. The most popular climb is the Ptarmigan Trail–Monitor Ridge route. Although this route is fairly easy and no special equipment is needed during the summer months, it is not for everyone. It requires from 7 to 12 hours for a round trip, the high altitude causes problems for some people, and the terrain is rugged, requiring some scrambling. During the winter, ice-axes, crampons, and ropes may be needed.

Permits are required for climbing above 1,500 m (4,800 feet). During the summer months there is a small fee. Each permit is valid for 24 hours (annual passes are also available). To get a permit, go to Jack's Restaurant and Store on Road 503, about 37 km (23 miles) east of Woodland. Climbing is limited to 100 people per day during the summer months. Advance reservations can be made by mail by contacting the Monument Headquarters. There are always 40 permits available on a daily basis at Jack's and, if the demand exceeds this number, a lottery is held. Be sure to get at Jack's early in the morning during the

The author on the summit of Mt. St. Helens, September 1995. (Photograph courtesy of Nick Gautier.)

peak months (July and August) if you don't have an advance reservation. For current information, call the Climbing Information Line at 360-247-3961.

The Monitor Ridge climb starts at Climbers' Bivouac (road directions above). This route ascends 1,400 m (4,500 feet) in 8 km (5 miles) to reach the crater rim at 2,550 m (8,365 feet) elevation. Follow the Ptarmigan Trail (216A on maps) from Climbers' Bivouac. The trail gently climbs 330 m (1,100 feet) up to the timberline about 3.7 km (2¼ miles) away. From then on, the trail follows Monitor Ridge, getting rougher and steeper, going over blocky lava flows on the lower slopes and over loose pumice and ash on the upper slopes. The trail is unmarked for the last 400 m (1,300 feet), but by then you can easily see the crater rim. Take care on the descent, as a minor detour can put you off the marked route.

The view from the top is magnificent. Be sure to start the climb early so that you will have plenty of time to walk part-way around the crater rim and photograph the dome, the steep crater walls where fumaroles steam away, and the other Cascade volcanoes: Mt. Adams is 55 km (34 miles) to the east, Mt. Hood is 100 km (60 miles) to the southeast, and Mt. Rainier is 80 km (50 miles) to the north.

For safety reasons, stand back from the crater rim, as it is crumbly and unstable in places. It is a long way down to the bottom – some 600 m (2,000 feet). In winter, watch out for snow cornices that develop over the rim; they sometimes persist during the summer. They can be very icy and slippery and sometimes have large glide cracks or bergschrunds.

Lava dome

Visitors are not allowed to enter the crater, so viewing the lava dome must be done from a distance – the crater rim is just about the best place, unless you take one of the helicopter trips. The dome is 270 m (880 feet) high and 1,100 m (3,600 feet) in diameter. It is made up of dacitic lava and is the result of a series of events, mostly involving the pasty lava being pushed upwards, each time cracking and breaking away some of the old, congealed material. A tube of toothpaste being periodically squeezed is a good analogy, though the dome growth also involved some small explosions that blasted out whole pieces.

The dome began to grow inside the crater on October 18, 1980. The growth was partly from the accummulation of lava and partly endogenic (from within), meaning that the pasty magma rose up inside the conduit and into the dome, swelling it and causing cracking of the walls in a radial pattern and thrust faulting of the crater floor. Typically, the intrusions happened 1 to 3 weeks before magma actually came out as short, thick flow lobes. The cracking and other deformation of the dome became quite useful for predicting when and where a new flow lobe would come out. In general, the intrusion affected only part of the dome, often the oldest exposed part at the time. This may have been because the older rocks had suffered more alteration and became weaker, thus making it easier for the magma to push away.

One of the most conspicuous features of the dome is the talus apron mantling its flanks. Most of the talus is the result of hot rockfalls that happened either during extrusion of the lobes or during rapid deformation of the dome as material pushed its way from inside. You may also be able to distinguish some of the flow lobes. Most of the lobes were fed from the dome's summit region. They were typically only 200 to 400 meters (650 to 1,300 feet) long, and 20 to 40 m (65 to 130 feet) thick. The lobes piled up on top of each other in an overlapping, somewhat messy form.

A total of 17 episodes of dome growth happened between October 18, 1980, and October 22, 1986; 14 of these episodes produced one flow lobe each and three produced two lobes each. Between episodes, the dome slowly subsided because of the effect of gravity on the hot, pasty core.

Visiting during activity

Mount St. Helens is now considered to be in a quiescent state, but this does not mean that new activity could not occur. The dome growth episodes ceased in 1986, but the volcano did not settle down immediately for a long rest. Small explosions producing ash continued to take place occasionally in the crater until 1991, but the ash consisted of pulverized old dome fragments rather than new magma. The events were most likely triggered by rain or melting snow percolating down into the still-hot dome. Small but deep earthquakes were recorded in 1994 and increased in 1995, maybe suggesting a pressure build-up of gas that is still coming out of the cooling mass of magma. There is no indication that fresh magma is rising, but monitoring efforts will continue to be intense for the foreseeable future.

Entry into the crater is strictly forbidden. Small explosions could still occur but another good reason is to preserve the delicate natural recovery that is being intensively studied. It is unlikely that visitors will be allowed to climb the dome anytime in the next few years.

St. Helens has taught us a valuable lesson about visiting volcanoes during activity. We often tend to view civil authorities as being overprotective of our safety during a volcanic crisis, which translates to closing off an area much larger than strictly necessary. In 1980, the authorities were not cautious enough and the unfortunate consequence was that some of people who died were outside the Red Zone. Residents of the Spirit Lake area, in their attempts to retrieve possessions, missed being killed by only a few hours. Others entered the blocked-off areas without permission and some paid with their lives for it. Topographical maps, hastily marked with roadblocks, enabled many people to make use of little-known logging roads. Police and other civil authorities were unable to monitor all roads and the consequences were tragic.

The explosion itself was classed as "very large," with a VEI of 5. However, the damage caused by mudflows, including the lives lost, is a good reminder that the consequences of an explosion can also be devastating. If Mt. St. Helens, or a similar type of volcano, threatens to explode again, stay well away from river valleys and topographic lows, not to mention from any areas blocked off by authorities. The dangers from this type of volcano – including mudflows and pyroclastic flows that can reach a long way – are among the worst that volcanoes can unleash.

Other local attractions

Mount Rainier

Considered the most dangerous volcano in the USA, Mt. Rainier rises to 4,392 m (14,410 feet) and reigns as the largest mountain in Washington State. The danger comes not from the frequency of eruptions but from the proximity of urban centers. Mount Rainier is located just 35 km (22 miles) from the Seattle–Tacoma metropolitan area which has a population of over 2.5 million – and more than 100,000 people live on top of the volcano's old mudflows. We know that Mt. Rainier is not only capable of very large explosive eruptions, but that these eruptions can trigger floods from its 26 glaciers as well as long-reaching mudflows. Even in the absence of an eruption, hazards such as glacier outburst floods, mudflows, and avalanches can occur, as the volcano's structure is essentially unstable: structurally weak rock deposited by explosive eruptions, capped by snow and ice. Earthquakes and ground deformation caused by magma moving up into the volcano could cause the collapse of a large section of the mountain, with catastrophic results. The volcano is being closely monitored and recent studies suggest that it is more frequently active than previously thought. Evacuation plans have been laid out and schoolchildren in the threatened areas have practice drills. Visitors to the Mount Rainier National Park are warned that if they are near a river and notice a rapid rise in water level or hear a roaring sound, they should move quickly to higher ground.

Records of Mt. Rainier's past activity are poor. Accounts by early pioneers in the area indicate that several small eruptions took place in the 1800s, the last one in 1894, but active fumaroles in the summit attest to its active state. Native American legends

Mount Rainier, considered the most dangerous of the Cascades volcanoes. (Photograph courtesy of Kathy Cashman.)

mention a cataclysmic eruption, but the account has not been verified. Mount Rainier is a stratovolcano similar to Mt. St. Helens and many of the lessons learned from the 1980 eruption are valuable to assess the hazards from Mt. Rainier. Perhaps the most valuable lesson is that the volcano could erupt again violently after a long period of repose. It is worth seeing the majestic volcano and its old forests before the landscape is changed forever.

The Mount Rainier National Park is situated about 160 km (100 miles) northeast of Portland, Oregon, and about 100 km (60 miles) southeast of Seattle, Washington. The Nisqually entrance to the park, located in the southwest corner, is open all year, but many park roads are open only in summer. Most visitors head for the Henry M. Jackson Memorial Visitor Center and to the nearby Paradise, a subalpine meadow from where many hiking trails start. Hiking is a popular activity, but climbing Mt. Rainier is no easy feat. A permit is necessary and although it is not obligatory to climb with a guide, it is wise to do so. The climb is dangerous and, since the first documented ascent in 1870, at least 68 people have died attempting it. Most have been victims of avalanches, falls, crevasses, or simply getting lost and succumbing to exposure. About 5,000 people succeed to climb the mountain every year. If you want to be one of them, consider going with Rainier Mountaineering, Inc. (RMI), a company that offers climbing instruction and 2-day guided climbs. The most popular route starts from Paradise and goes up to Camp Muir at 3,100 m (10,200 feet) for overnight bivouac. Climbers start the summit ascent around 2 a.m. in the morning in order to get to the top and come down before the mid-afternoon sun begins to melt the snow, opening crevasses and triggering avalanches. Mount Rainier poses many dangers even when dormant. If there is a threat of eruption, remember the lessons learned from Mt. St. Helens – a new eruption could turn out to be even more deadly than predicted.

Crater Lake

This magnificent volcanic caldera is somewhat closer to Lassen Peak than to Mt. St. Helens in distance; however, it is more convenient for visitors to Mt. St. Helens to drive down to Crater Lake, since the major

Crater Lake is an ancient caldera famous for its beauty. Wizard Island, in the center, is a steep-sided cinder cone. (Photograph courtesy of Nick Gautier.)

airport closest to both these volcanoes is Portland. Crater Lake is a water-filled, 10 km (6 miles) wide depression of the collapsed stratovolcano known as Mt. Mazama. The dark blue lake, about 600 m (2,000 feet) deep, is nearly perfectly circular, surrounded by steep cliffs, and beautiful beyond description. Breaking through the lake's surface is a steep-sided cinder cone about 240 m (800 feet) tall named Wizard Island due to its resemblance to a sorcerer's hat. Crater Lake is the result of the catastrophic transformation of Mt. Mazama, some 3,600 m (12,000 feet) high, into a caldera where the tallest walls reach only 2,400 m (8,000 feet). The eruption happened about 6,900 years ago and was one of the most violent of post-glacial times; about 10 times greater than that of Krakatau in 1883. So much magma was ejected by Mazama's Plinian-type eruption that the top of the volcano could no longer be supported and collapsed. Deposits from pyroclastic flows are 75 m (250 feet) thick in places and ashfall covered an area of over 800,000 km^2 (300,000 square miles). During the following millennium, renewed activity occurred on the crater floor but much smaller in scale, producing small domes and Wizard Island. The volcano has been dormant for about 5,000 years but, given its past history, the possibility of violent eruptions is still a cause for concern.

Most visitors to Crater Lake are drawn by the scenery rather than by its volcanic history. Little organic material enters the lake, because it has no inlets or outlets and its shores are mostly barren. Because of this, the lake's clarity is superb. The depth – it is the deepest lake in the USA – is the reason for its amazing cobalt-blue color. Crater Lake National Park offers much in terms of hiking, sightseeing, and boating activities. You can drive all the way around the crater using the Rim Drive. It is best to start your visit at the Rim Village Visitor Center and to hike from there to Discovery Point (4.2 km, 2½ miles round trip) along the Pacific Crest Trail. It is also possible to hike on Wizard Island. Boats to the island leave from Cleetwood Cove on the northern side of the lake. There is a trail from the boat dock on Wizard Island that climbs for a few hundred meters, then forks. The left fork trail goes over blocks of andesitic lava up to Fumarole Bay, a popular swim spot, even though the water is very cold. The trail's right fork leads to the summit of the island from where the views of the lake are breathtaking. Another highlight of the Park is Devil's Backbone on the northwestern side of the lake. The Backbone is an andesite dike that is exposed from the lake's edge to the top of the crater rim. The dike stands out from the cliff because it is more resistant to weathering and erosion than the more friable volcanic materials around it.

Sunset Crater

The volcano

Sunset Crater is the youngest of the 400 or so cinder cones in Arizona's San Francisco Volcanic Field. It is remarkable both because of its beauty and its near-pristine condition, despite being nearly a millennium old. The crater's poetic name was given by John Wesley Powell in 1885 when, as head of the US Geological Survey, he explored the San Francisco Volcanic Field.

Sunset Crater is a large cinder cone in the San Francisco Volcanic Field near Flagstaff, Arizona. (Photograph courtesy of Nick Gautier.)

The rough aa texture of the Bonito flow viewed from the Lava Flow Nature Trail. In the background is Sunset Crater. (Photograph courtesy of Nick Gautier.)

He wrote in his journal that "the contrast of colors is so great that on viewing the mountain from a distance the red cinders seem to be on fire. From this circumstance the cone has been named Sunset Peak." In reality, the red and yellow hues are the result of steaming gases that oxidized iron in the lava fragments and deposited sulfur, gypsum, and limonite near the crater rim.

Sunset Crater is the only monogenetic volcano included in this book. Monogenetic means that the volcano was built over a short period of time by a single eruption or series of related eruptions, and is not expected to become active again. Even without the prospect of a possible eruption, Sunset Crater is certainly worth visiting. It is one of the world's prime examples of a monogenetic volcano. The 300 m (1,000 feet) high cone is remarkably photogenic, intriguing, and scenic. It is located in an area of geologic wonders: within easy driving distance are the Grand Canyon, Oak Creek Canyon, and Meteor Crater. The formation of Sunset Crater had great impact on the indigenous people of the time, and visitors interested in history will be fascinated by the wealth of archeological findings, such as the Wupatki dwellings.

Sunset Crater has been well preserved thanks to the region's dry climate. Ironically, what nature protected for so long was endangered in recent years. In 1928, a Hollywood film company planned to dynamite part of the crater to simulate an eruption. Angry protests from local groups put a stop to this threat, and eventually led to President Hoover declaring Sunset Crater a National Monument in 1930. The key person lobbying for the protection of the crater was Harold Colton, whose work in archeology was fundamental in piecing together the history of the indigenous people.

A different threat to Sunset Crater became evident years later, when its flanks were being seriously eroded by the large number of visitors climbing up the cone. In 1973, officials responsible for the preservation of the crater closed access to its summit. Although this is certain to disappoint many visitors, there are still ways to make a visit here more than worthwhile. The Bonito ("beautiful" in Spanish) lava flow at the base of the cone is cut by a trail that shows visitors a wide variety of fresh-looking flow patterns. There are a number of interesting volcanic and archeological sites in the area and it is still possible to hike up neighboring cones.

The geologic history of Sunset Crater is a small part of that of the San Francisco Volcanic Field, a collection of mostly basaltic volcanoes near the south margin of the Colorado Plateau. Almost all the volcanoes are cinder cones and there are more than 600 of them within an area of about 4,700 km^2 (1,800 square miles). The volcanic activity started out about 6 million years ago, near the present town of Wiliams. With time, the activity migrated east and northeast, up to Sunset Crater which is about 80 km (50 miles) away from the starting point. Sunset Crater erupted in very much the same way as many of the other cones in the San Francisco field – basaltic magma came out in the form of Strombolian explosions and lava flows. The volume of magma erupted at Sunset Crater was actually quite large for a typical Strombolian eruption – about 3 km^3 (0.7 cubic miles), divided between cinders building the cone, two lava flows, and finer materials that were widely scattered, reaching as far away as 1,300 km (800 miles). Because there are no written records, what we know about the history of the eruption comes from a combination of geologic and archeological studies. Sunset Crater created its own Pompeii in the USA, with the fortunate difference that no lives seem to have been lost.

The eruption of Sunset Crater

Sunset Crater shares many similarities with Parícutin in Mexico and it is likely that their eruptions started in a similar way. Many people have heard the story of Parícutin, the volcano that started in a cornfield in 1943, giving a local farmer a truly unexpected surprise. The eruption of Sunset Crater must surely have been a shock to the Sinagua, the people who farmed this area around 1,000 years ago.

The first eruption occurred sometime between the growing seasons of AD 1064 and 1065. It most likely started like Parícutin, as a small crack in the earth. Soon explosions were sending spatter and ash high in the air. The larger fragments of spatter fell to the ground and begun to build the cone, while ash blanketed the countryside.

There is no historical record of the eruption, but the burial of native dwellings by the ash and other archeological evidence have allowed dating of the eruption. Accurate dates were obtained thanks to the ponderosa pines that still give the region an alpine feel. Beams of these pines were used in the construction of the Sinagua's pit-houses. Experts in dendrochronology, the study of the annual growth of rings within trees, have been able to date the eruption accurately by examining the beams found in Sinagua houses. The youngest ring from houses buried by the eruption dates from the growing season of AD 1064, while the oldest ring in beams of post-eruption houses date from AD 1071. This narrows down when the eruption occurred, but dendrochronology can do better. Growth rings also reflect growing conditions and scientists studying pines damaged but not killed by the eruption found that up to and including 1064 the rings appear normal. However, rings from 1065 were thinner, indicating some upsetting event for the trees. Rings for the next few years show that the trees slowly recovered from the trauma, that is, from the eruption that must have taken place between 1064 and 1065.

Paleomagnetic dating was also important in dating the eruption. The directional alignment of iron particles in the solidified lava flows were measured and compared to today's magnetic north. The deviation gives the number of years since the particles were deposited. These and other geologic studies reconstructed the various events in the eruption, which did not take place only from Sunset Crater. The first activity was the opening of a fissure some 15 km (9 miles) long, with fire-fountains erupting all along. One can only imagine what the people thought of this spectacle. Soon the eruption became localized at the northwest end of the fissure, slowly building Sunset Crater. The heavier cinders fell near the vent to build the cone, and ash and small fragments went up as an eruption column that reached several hundred meters. Two lava flows came out before the end of the explosive phase. First was the Kana-a flow, which erupted from the east side of the cone and traveled down a wash for about 11 km (7 miles). While explosive activity continued building the cone, the Bonito flow erupted from the base of the cone on its western and northwestern sides, eventually covering an area of 4.6 km^2 (1.8 square miles). The flow took away portions of the cone, carrying them like rafts on the lava surface.

The total duration of the eruption is not precisely known, but it is thought that Sunset Crater's last burst of activity came in 1250, when magma containing relatively high amounts of iron and sulfur erupted from the top of the cone. This magma eventually oxidized and produced the red and yellow colors that have given the crater its current name and, before Powell, the descriptive designations of "Yellow-Topped Mountain" by the Navajo and "Red Hill" by the Hopi. The total duration of the eruption, over two centuries, is unusually long for monogenetic volcanoes – perhaps even unique. Parícutin, for example, was done in just over 9 years.

In the absence of any recorded history, it is impossible to tell how the Sinagua felt about Sunset Crater erupting in their backyards. We know that the Sinagua were a hardy lot. Their name means "without water" in Spanish and reflects the fact that they were desert people, able to farm and thrive in the dry conditions of the region. The eruption must have given the Sinagua time to flee, as no bodies have been found. We know, however, that many of them came back. New houses were built even before Sunset Crater gave its last eruptive gasp. Ash from the eruption eventually turned the soil even more fertile than it used to be, and soon new pueblos were built around the cinder and lava fields. Sunset Crater has actually been credited with being the reason for one of the first population explosions in the USA.

A personal view: A geological wonderland

Northern Arizona is a remarkable part of the world. If you are visiting the area for the first time, try to spend at least a full week here. Nature put so many geologic wonders here that you may feel overwhelmed. I remember little of my first visit to Sunset Crater, probably because it happened at the end of an exhausting 3 days of driving around, trying to take in the Grand Canyon, Oak Creek Canyon, Meteor Crater, the Painted Desert, and more. Seeing all these places in a single trip makes sense if you are coming from far away, but the order is important – it is best to save the Grand Canyon for last. It is probably the most impressive geologic feature most of us will ever see, but its magnitude can make anything else you see look puny.

My personal recommendation for a visit to the area is to use Flagstaff as a base, unless you want to camp or backpack. Flagstaff is located on the historical US Route 66 and seems to have become a curious mix of motel-land for overnight stops, major railway stop, university town, far-west remnant, and New Age center. There is no shortage of places to stay or eat, though I have not found any in the outstanding category. Convenience is the key word here, as you can be out of town and on your way to the local attractions in no more than a few minutes' drive. Sunset Crater and Wupatki are good places to visit first. They will give you an easy way to grasp the local geology, the indigenous history, and how the eruption of Sunset Crater influenced the people living in the area. The theme continues at Walnut Canyon, where pueblos and cliff-dwellings are so well preserved that it is hard to believe that nobody has lived there for over seven centuries.

One of the most beautiful Arizona landscapes is the Painted Desert. I still vividly remember seeing it for the first time, from the loop road between Sunset Crater and Wupatki. The view was breathtaking, even after my senses had been somewhat anesthetized by the Grand Canyon.

After visiting volcanoes, the desert, and archeological sites, head for something really out of this world: Meteor Crater, the best preserved impact feature on Earth. Don't be put off by the theme-park feel of the side road leading to the crater, or by the steep admission fee. This is a unique place, and the museum is actually very good, focussing not only on meteorites and impact craters but also on the Apollo program (astronauts going to the Moon were trained on field work at Meteor Crater). Weather permitting, take the crater rim tour (bring walking shoes and a windbreaker).

Before heading for the Grand Canyon, take a trip to the Oak Creek Canyon, the setting for the spectacular Red Rock Country of Sedona. The road from Flagstaff to Sedona (Highway 89A) winds through the bottom of the canyon, which was carved into the red sandstone by Oak Creek. The red rock formations are as stunning as those in the Grand Canyon, but on a more manageable scale. You can stop at Slide Rock State Park and hike over one of the several trails. Better still, go for a swim in Oak Creek, taking advantage of a natural rock chute. The road continues to the town of Sedona, which is worth seeing for its breathtaking setting amongst the red rock pinnacles. Unfortunately, Sedona has evolved into a major resort and something of a tourist trap. It has also become a New Age mecca, thanks to the belief that spiritual energies concentrate around the town in vortices.

I recommend saving the Grand Canyon for last, but it should be seen in good weather, so keep an eye of the forecast. The canyon is spectacular any time of the year, including in the winter when snow hugs the rock pinnacles and the light is almost ethereal. There are numerous books written about the Grand Canyon's geology, history, and how to plan a visit. Options include hiking, going down on muleback, traveling down the river on rafts, and taking sightseeing flights. Remember that this is one of the most popular spots on Earth and that organized trips such as those on muleback or rafts need to be booked months or even a year in advance. If you only have a short time, drive around the South Rim stopping at the numerous lookout points – and plan to come back another time.

If you are coming from far away to visit the wonders of Arizona, consider extending your trip to include some of the geologic wonders of the neighboring state of Utah, particularly those located in the southern part of the state which are not far from the Grand Canyon. My personal favorite is Zion National Park, which has fantastically colorful canyons and mesas. Avoid the crowds of summertime if you can. Also worth a trip is Bryce Canyon National Park, where high-altitude overlooks and clear air provide spectacular view of the pinnacles and spires.

Visiting Sunset Crater

Sunset Crater is not expected to erupt again, so we can only visit it during repose. It is possible that new eruptions will occur in the San Francisco Volcanic Fields, but maybe not for many thousands of years. Although experts differ in whether to classify Sunset Crater as active or dormant, visitors should consider the crater as a learning ground for monogenetic volcanoes and their products, rather than as an active volcano. A visit to Sunset Crater itself need not take long – most visitors only spend a day here, calling in at the Visitor Center and hiking on the Lava Flow and the Lenox Crater trails. However, there is much more to explore in the San Francisco region, both in terms of geology and archeology, and at least another day is needed to visit Wupatki National Monument.

To get to Sunset Crater from Flagstaff, take Highway 89 north for 19 km (12 miles) and turn onto the Sunset Crater–Wupatki loop road. The best way to visit the area is to drive all around the loop road (58 km, 36 miles one way), stopping at the locations described below. Be sure to get food and gas before leaving Flagstaff.

Visitor Center

Located about 3 km (2 miles) down the loop road, the Visitor Center should be your first stop. The Center sells books and maps, gives out leaflets to guide you over the Nature Trail, and has an informative exhibit that includes an eruption video. Inquire here about special programs such as the Lava Walk, and about permits for backpacking and camping. From the Center, carry on driving for about 2.2 km (1½ miles) until a turnoff to the Lava Flow Nature Trail parking lot.

Lava Flow Nature Trail

This easy, self-guided trail (1.6 km, 1 mile loop) is mostly level and paved at the start, with a short portion accessible to wheelchair users. The trail goes onto the Bonito flow and allows you to see many interesting features common on basaltic lava flows, such as squeeze-ups, lava blisters, hornitos, and even a small tube. Signs point out various interesting features on the flow, all described in a helpful leaflet available at the Visitor Center. The black, rubbly lava looks like it was erupted only a short while ago. The flow is still so barren and lunar-like that it was used as a training ground for Apollo astronauts in the 1960s. NASA engineers brought a mock lunar rover here to tour around, and the unfortunate astronauts had to walk over the lava wearing their space suits.

The Bonito is also a good training ground for geologists. It is unusually thick for a basaltic flow, reaching 30 m (100 feet) in places, because the lava ponded at the foot of the cone. Most of the flow is of the aa type, but parts are pahoehoe, with characteristic ropy surface texture. Among the features pointed out in the trail are squeeze-ups, formed when molten lava from the flow's interior oozed out of the cracked crust. Some areas of the flow contain lava blisters or bubbles, some are intact and some are burst. These are formed by the escape of gas from inside the lava when the crust had almost, but not quite congealed. At one stop in the trail, xenoliths are pointed out. These xenoliths ("foreign rock") are pieces of yellowish-white Kaibab limestone, which were broken off and fused with the molten lava. Unfortunately, one of the most interesting features of the flow, a 68 m (225 feet) long lava tube – called the Ice Cave – is at this time no longer open to the public, though you can still see its entrance. A portion of the tube collapsed in September 1984 (luckily when nobody was inside) and it is now considered dangerous. The Ice Cave name comes from the fact that ice can be found there for most of the year and, during the 1880s, was used to supply the saloons in Flagstaff.

The dry climate has preserved the Bonito flow's surface, and only a few scattered plants and trees, such as the ponderosa pine, struggle to grow on the black lava. Notice that lichens are widespread on the lava surface – they are extremely important for breaking the hard lava into soil, which they do by secreting a weak acid. Lichens live for a long time, and those you see here today may have begun forming while the lava was still warm. Mosses usually follow lichens, growing in tiny pockets of soil, and eventually more conventional plants and trees grow. If you come here in June or July, look out for the rare, lovely pink tubular flower *Penstemon clutei*, an endangered species restricted to this area.

O'Leary Peak

The peak of this volcanic center is worth climbing if the weather is clear, as you can get spectacular views of Sunset Crater and the Painted Desert from its summit. Named after Dan O'Leary, a guide for General George Crook during the Indian Wars, this volcanic center is made up of two dacitic lava domes, flows of andesite and dacite lavas, and several small rhyolitic domes. The volcanic activity here happened between 170,000 and 220,000 years ago. O'Leary Peak and the Sugarloaf rhyolite dome (on the northeast flank of San Francisco Mountain) are the youngest of the major silicic eruptive centers in the San Francisco Volcanic Field.

To get to the Peak, head west from the Sunset Crater Visitor Center (back towards Highway 89) for about

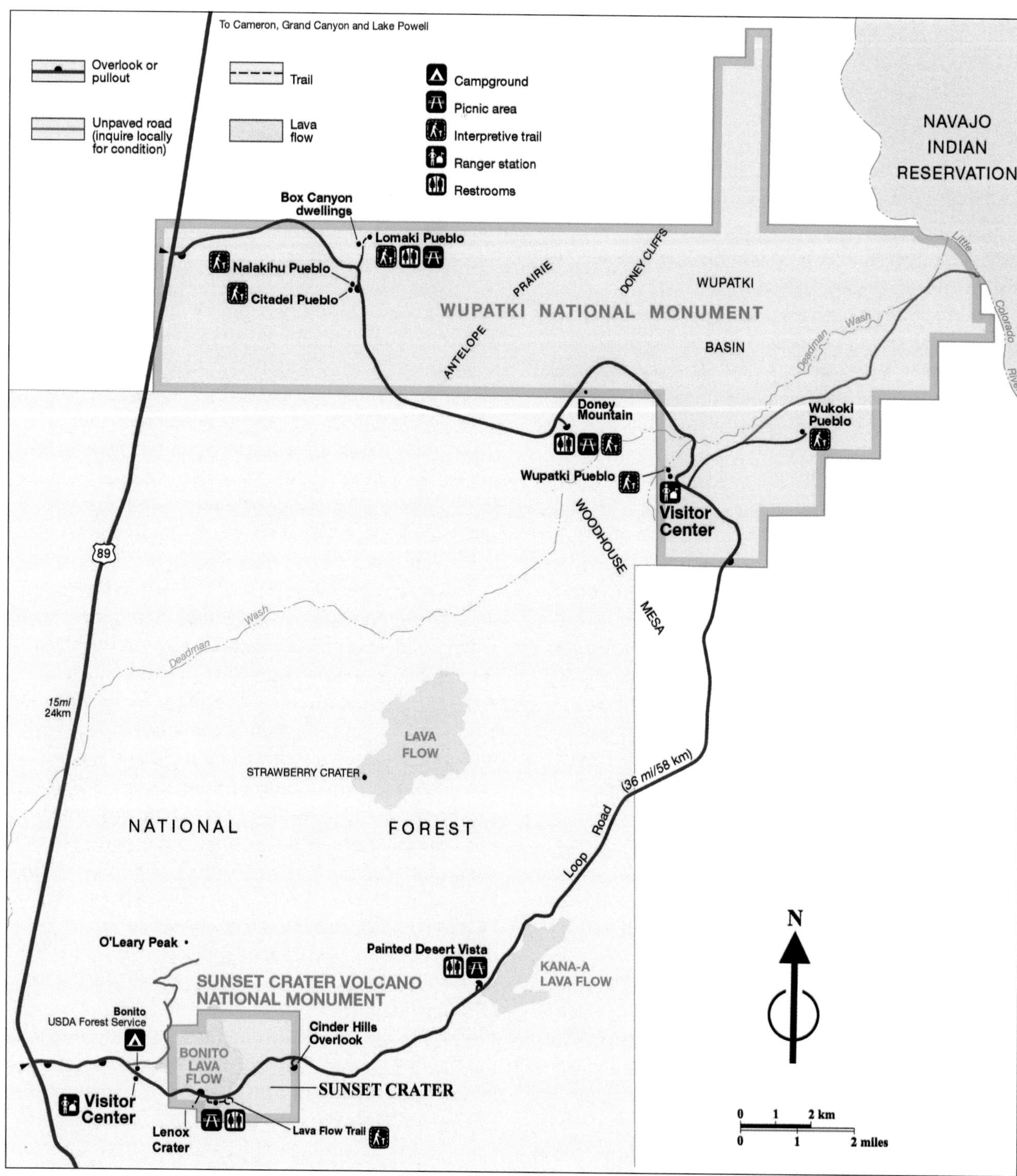

Local map showing the Sunset Crater and Wupatki National Monuments. (Modified from US National Park Service map.)

400 m (¼ mile), then take Forest route 545A. This steep dirt road climbs for about 8 km (5 miles) up to a fire lookout tower at the summit, but sometimes a gate blocks access to the last mile. If this is the case, just park and walk. If possible, come here in the later afternoon when the colors of Sunset Crater and the Painted Desert are at their most photogenic.

Lenox Crater Trail

This steep trail (1.6 km, 1 mile round trip; elevation gain 85 m, 280 feet) gives you an opportunity to climb a cinder cone near Sunset Crater. Allow 30 minutes to climb up and about 15 minutes to come down. The trailhead is located about 1.6 km (1 mile) east of the Visitor Center.

Cinder Hills overlook

Back on the loop road, continue towards Wupatki, crossing lavas from Sunset Crater. Stop at the Cinder Hills overlook just before the eastern entrance to the Sunset Crater National Monument. You will see the steep north flank of Sunset Crater and also the south flank of a smaller volcano that was blanketed with cinders from Sunset Crater eruption. Looking towards the east and northeast, you can see a line of reddish cinder and spatter cones, the "cinder hills," formed by spectacular fire-fountains that were a part of Sunset Crater's eruption. At the edge of the "cinder hills" is Merriam Crater, the probable source of a copious lava flow that caused the Little Colorado River to be diverted onto the canyon wall that is now known as the Grand Falls.

On leaving the Sunset Crater National Monument, watch out for cinder deposits cut by the road; you can see deposits as thick as 3 m (10 feet), with several distinctive layers indicating different phases of Sunset Crater's eruption. About 1.8 km (1⅛ mile) after leaving the Monument, the road crosses and then follows the Kana-a lava flow. A large part of the flow has been covered by cinders and ash from the later phases of Sunset's eruption. It is best to stop at the *Painted Desert Vista* for a magnificent panorama. Make sure to stop here in the early morning or late afternoon, when the colors are at their most vivid. The multiple shades of red, orange, gray, green-blue, and purple show up in tall escarpments of rock, creating a photographer's delight. The exposed rocks are multi-colored sedimentary rocks of Mesozoic age, called the Chinle Formation. The Painted Desert is also famous for its petrified wood, best seen in the Petrified Forest National Park, and for its fossils, including dinosaur and early mammal bones.

Wupatki National Monument

The eruption of Sunset Crater forced the Sinagua to evacuate the farmlands they had cultivated for 400 years. They moved to Wupatki basin and were joined there by the Kayenta Anasazi from northeastern Arizona and the Cohonina from the west. Wupatki (Hopi for "Tall House") became a thriving cosmopolitan center, with three diverse groups sharing technology and advancing as a result. The location helped the community grow. A thin ash layer deposited by Sunset Crater absorbed moisture, helped prevent evaporation, and conserved heat in the soil, extending the growing season. The youngest rock layer, the Moenkopi red sandstone, made an ideal building material and large, multi-storied houses replaced the rough pit-houses the Sinagua had left behind. Wupatki was also a beautiful place to be, as pueblos were built on rock outcrops isolated by erosion, providing spectacular views of the landscape.

For unknown reasons, people began to leave the area and Wupatki had been abandoned by about 1225. A drought beginning in 1150 has been suggested as the explanation, but violence amongst the various tribes is also a possibility. However, there is no evidence of large-scale conflict, as only one excavated site, now known as House of Tragedy, shows remains of humans who suffered violent deaths. Archeological research points to the tribes migrating south to Verde Valley and north to the Hopi mesas. Hopi legends trace some of their ancestry back to Wupatki.

Wupatki Ruin was once a pueblo occupied by the Sinagua Indians, who farmed in the area when Sunset Crater came to life. (Photograph by the author.)

The Wupatki monument is the best place in which to learn about the effects of Sunset Crater on the local population. The 145 km^2 (56 square miles) of dry, rugged land contains an estimated 2,700 archeological sites. Only a few of these are open to the public, but they are also the most impressive. Make your first stop at the Visitor Center, where there are exhibits of Indian artifacts and a reconstruction to show how the interior of a Wupatki room might have looked. The most impressive of the ruins is Wupatki itself, which you can visit by following a trail from the Visitor Center. One of the most intriguing features is a ball court that resembles some found in Mexico. Several other ball courts have been discovered in northern Arizona and it is thought that they had religious significance, like their Mexican counterparts, though not to the same grisly degree. The Aztecs used the ball courts for games between two men, one of whom (not always the loser) was sacrificed to the gods at the end. Another curiosity of the Wupatki ruin is a blowhole just east of the ball court, an opening in the Kaibab limestone layer that is connected to other blowholes in the area by a system of underground cracks, probably created by earthquakes. Air sometimes blows out or rushes into the blowhole, a fact that may have had religious significance for the tribes.

Other ruins in the Monument that are worth visiting are the Wukoki ("Big and Wide House"), the Citadel – a fortress-like pueblo, the Nalakihu ("House Standing outside the Village"), and the Lomaki ("Beautiful House"). If you plan to be in the area for a longer visit during April or October, book in advance the ranger-led overnight backpack hike to Crack-in-the-Rock ruin. The round trip is about 22 km (14 miles) and climbs to the ruins atop a mesa, from where the views of the Little Colorado River are magnificent.

Don't miss the easy, 800 m (½ mile) hike up *Doney Mountain* – which is not a mountain but a cinder cone, named after a pioneer who prospected around Wupatki looking for a lost mine. The views from the top of the cone are spectacular, and the trail has plenty of signs explaining the local geology and ecology. Of particular importance is the Doney Fault, marked by a group of reddish cones, of which Doney Mountain is the tallest. The movement across the fault has been dramatic, with rocks on the east side being dropped about 60 m (200 feet) in relation to those on the west side. Doney Mountain itself was formed after the major movements across the fault had ceased. We know this because Doney's cinders and ash layers were not dislodged or cut by the fault. However, the magma seems to have taken advantage of the fault cutting into the Earth's crust and used it as a pathway, erupting along the fault to form Doney and a number of other cinder cones.

While in the Doney picnic area, or from the top of the trail, look towards the east and you will be able to see the Hopi Buttes (or mesas) on a clear day. These tower-like forms are the eroded remains of a group of ancient volcanoes that erupted about 7 to 8 million years ago. The eruptions that formed the Buttes were initially phreatomagmatic, that is, magma erupted under shallow water or mud. The powerful steam explosions created large craters that were eventually filled with lava. The lava was much harder, and more resistant to erosion, than the ash and fragments forming the surrounding crater, so now only the lava "core" remains. You can get another good view of the Hopi Buttes from Interstate 40 between Flagstaff and Winslow (from Flagstaff, look to the northeast after milepost 217).

Other local attractions

The following sites are not as famous or spectacular as the Grand Canyon and Meteor Crater, but are more directly relevant to the Sunset Crater region's volcanology and archeology.

Museum of Northern Arizona

Located in Flagstaff, the museum has extensive exhibits and information on the natural and cultural history of the region. This is a good place to learn about the geology of the Colorado Plateau.

Walnut Canyon National Monument

This beautiful canyon located east of Flagstaff has the ruins of about 300 Sinagua dwellings along its cliffs. The Sinagua took advantage of the ledges eroded out of the canyon's Kaibab limestone to build cliff-dwellings. The Sinagua lived here from about AD 1120 to 1250, while Sunset Crater was still active. It is likely that the people living near Sunset Crater relocated here because of the eruption. Walnut Canyon must have been a very good place for them to live – Walnut Creek at the bottom of the canyon brought water, the farmland was good, and there were game-filled forests nearby. Why the Sinagua left these dwellings is not known.

You can take a couple of trails and get a feel for how the people lived, and stop at the Visitor Center to see artifact displays and an interesting exhibit about the Sinagua. Other Sinagua ruins are located at Tuzigoot and Montezuma Castle National Monuments, both south of Flagstaff.

SP Crater and flow

This crater and the black tongue of lava at its base are well-known to volcanologists. Photographs of the cinder cone, crater, and of the blocky, andesitic lava flow have been reproduced in many geology textbooks, mostly because of the nearly symmetrical shape of the cone and the well-preserved lava ridges on the flow. SP Crater looks very much like Sunset Crater; it even has reddish cinders on its rim. It is located on private ranch land and was named by the original owner, C. J. Babbit, in the 1880s. He was not, alas, as poetic as John Wesley Powell. The bowl-shaped crater and the black spatter on the rim reminded this earthly person of a pot of excrement, and the name stuck. Map-makers couldn't bring themselves to spell out the name, so it became "SP" – probably the only volcano in the world to be called after a rude acronym.

Getting to the crater is not easy, as you have to take dirt roads. From Flagstaff, drive north on Highway 89 to Hank's Trading Post (milepost 446). Turn left (west) on the unmarked road just south of the trading post. Note that this road is impassable when wet. You can recognize SP because of its symmetry and height (about 300 m, 1000 feet). Keep left where the road forks and drive for about 800 m (½ mile). The road skirts the edge of the SP flow and you can see its blocky surface. The crater will be on your right, before the road gets to another fork. Take the right fork (the left will take you to Colton Crater) for about 100 m (100 yards), then the dirt track on your right. You can drive on the track for about 800 meters (½ mile), depending on weather conditions and your vehicle. Park and take the foot track to the edge of the cone. There is no trail up the steep cone, so you will have to scramble up the loose cinders. The view from the top is well worth the effort: a panoramic view of the San Francisco Volcanic Field and of the black tongue of lava below. The SP flow extends about 7 km (4¼ miles) to the north of the cone and is about 9 to 15 m (30 to 50 feet) thick. It is a fine example of a blocky lava flow, with such a remarkably well-preserved surface that it is hard to believe that it erupted some 70,000 years ago. The crater rim is littered with bombs, many are up to about 1 m (3 feet) long. You can walk around the rim and down onto the flow, but scrambling down into the 110 m (360 feet) deep crater is not really safe. While on SP Crater, remember that you are on private land. The owners of Babbit Ranch have been hospitable and have not insisted on permits for visiting the crater, but visitors should not abuse their goodwill.

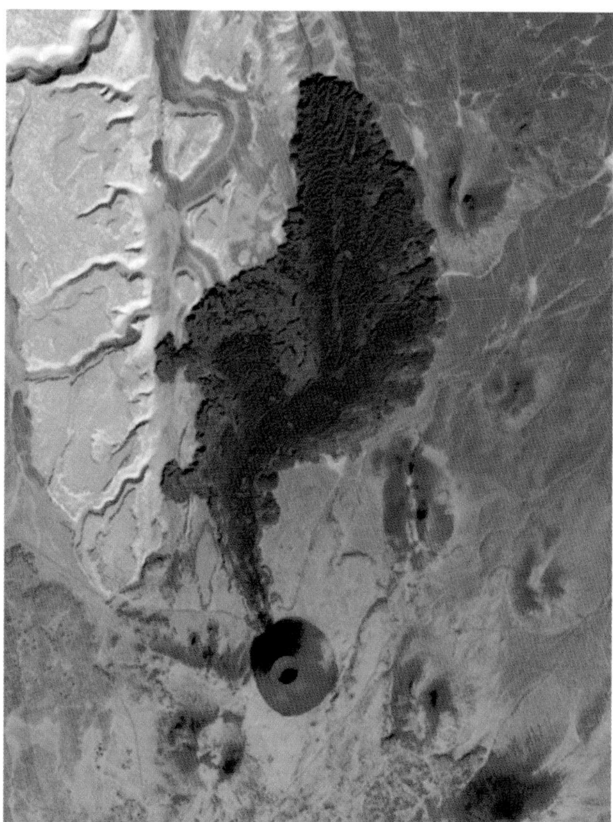

The SP Crater and flow imaged by Advanced Spaceborne Thermal Emission and Reflection Radiometer (ASTER), on NASA's Terra satellite in July 1986. (Image courtesy of Elsa Abbott, JPL.)

If you want to see another crater, go back onto the road and take the fork to Colton Crater. It is an easier hike up to the top, as the crater is only about 100 m (300 feet) high. This interesting crater is the result of a phreatomagmatic eruption. You can easily climb to the bottom and see the innards of a cinder cone. At the bottom of the crater, there is a surprise: a second, reddish cinder cone, only about 150 m (500 feet) across – probably one the prettiest mini-volcanoes anywhere.

Yellowstone

The volcano

Yellowstone Plateau is the most famous and visited volcanic area in the USA and holds the honor of being the world's first National Park. The US Congress established Yellowstone National Park in 1872, a year after an expedition led by geologist F.V. Hayden had explored the area and reported on its incredible

Lower Geyser Basin, Yellowstone National Park. (Photograph by the author.)

features. An important member of the Hayden expedition was artist Thomas Moran who, despite describing the scenery as beyond the reach of human art, produced numerous drawings and paintings, including a magnificent oil canvas entitled *The Grand Canyon of Yellowstone*. Together with the comprehensive report by Hayden and the excellent photographs by W. H. Jackson, Moran's work helped persuade Congress that Yellowstone's uniqueness required its preservation "for the benefit and enjoyment of the people." The first National Park was thus created and Yellowstone became one of the USA major tourist attractions. Moran's painting, bought by Congress, still hangs in the US Capitol.

The members of the Hayden expedition were the first to document in detail the wonders of Yellowstone, but were by no means the first to see them. Native Americans had lived in the region for thousands of years, drawn by the hot springs and the plentiful supply of obsidian for making weapons and tools. The Wyoming Shoshones, also known as the "Sheepeaters," even had their own name for geysers which can be translated as "water that keeps coming out." As far as we know, the first white man to visit Yellowstone was John Colter, who had been a guide on the Lewis and Clark expedition. In 1807, he traveled alone and on foot to Yellowstone, traded with the Indian peoples, and was hurt in a battle between tribes. After managing to make his way to a trading post in Montana, he described an incredible place of "fire and brimstone," but his stories were attributed to delirium and dismissed. Nobody seems to have believed the famous mountain man Jim Bridger either, when in 1857 he began spreading tales of spouting water, boiling springs, and a mountain of glass. Eventually, however, enough reports of this type begun to raise curiosity and even some credence. In 1870, Henry Washburn, the Montana Territory's Surveyor General, led an expedition that traveled all around Yellowstone Lake. Reports from this venture were the trigger for Hayden's expedition.

Visitors coming to Yellowstone today still marvel at what they see, but not all realize that this volcano is much more than a collection of geysers and hot

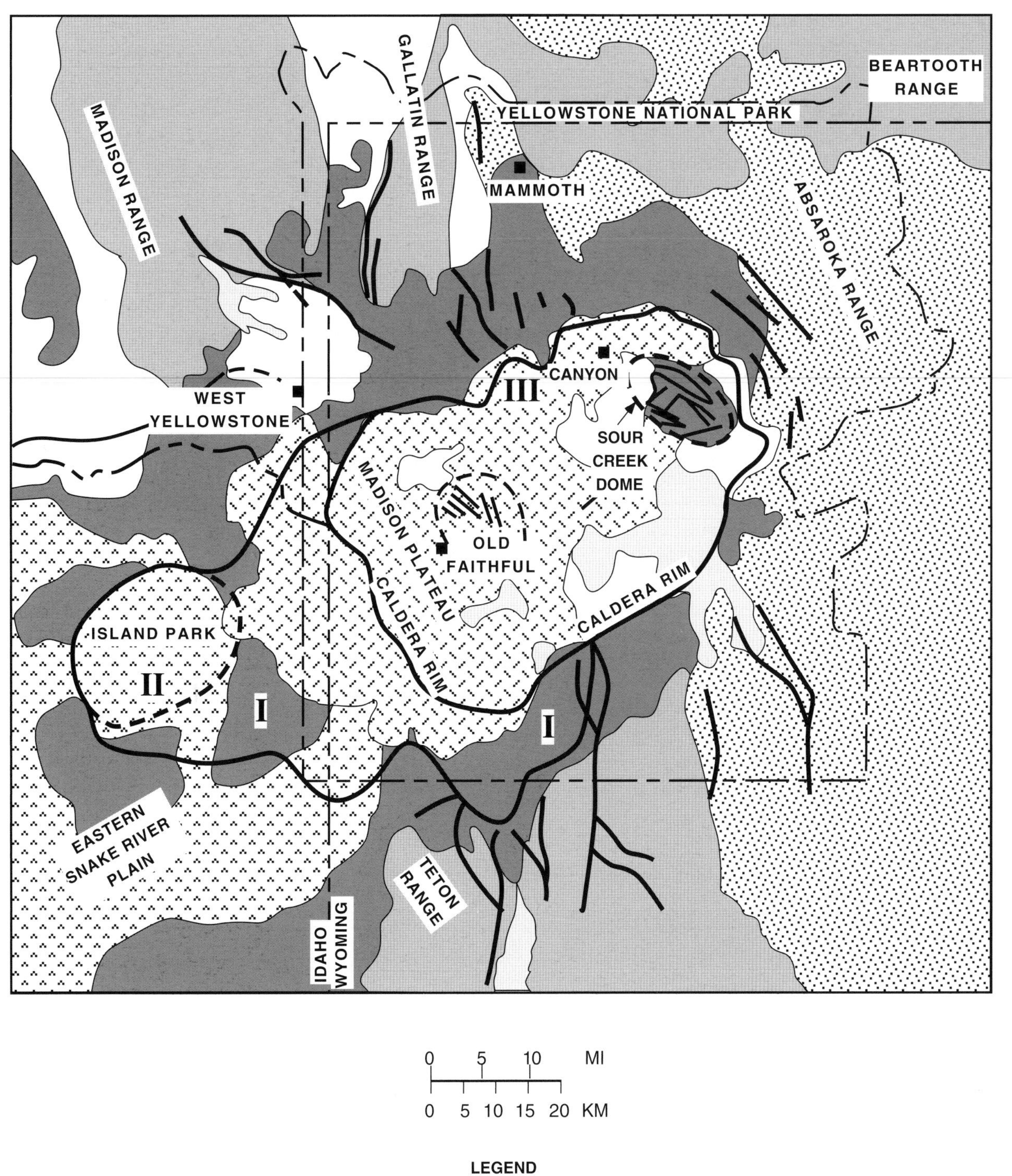

Simplified geological map of Yellowstone caldera and surrounding area. (Modified from Smith and Braille, 1994.)

springs. The Yellowstone Plateau lies at the center of one of the world's largest volcanic fields and has been the site of eruptions of truly cataclysmic sizes. Yellowstone is also a relatively young volcano: all the lavas in the area have been erupted during the last 2.5 million years. The eruptions during this time have been enormous – Yellowstone has erupted some 6,000 km^3 (1,400 cubic miles) of magma in total – though this is only a small fraction of the phenomenal volume of magma that has been intruded under the crust. On the surface, the eruptions have produced gigantic ash flows and three nested, collapsed calderas, the youngest of which is the Yellowstone caldera. The two older calderas form part of a conspicuous circular basin at the west edge of the volcanic field, called Island Park. The Yellowstone caldera is rather subtle in topography, because it is filled by younger rhyolitic lavas.

Each of the three calderas represents the climax of a cycle of activity, in which two types of magma were involved: basalt (olivine-tholeite) and rhyolite, which commingled at a shallow depth. This is an unusual combination, because the silica content of the two types of magma is so diverse. At the start of each cycle, both types of magma were erupted. As the cycle continued, large volumes of rhyolitic lavas came out of ring fractures formed by pressure from the huge amount of magma intruded from below. Each cycle ended with a big bang, when even larger volumes of rhyolitic magma were erupted from the ring fractures, releasing magma from the high-level parts of the chamber. These final cycle eruptions involved hundreds to thousands of cubic kilometers of magma ejected in a matter of few hours or days, forming ash flow sheets that blanketed vast areas. Such rapid withdrawal of magma caused the collapse of the magma chamber, forming a caldera each time. Between cycles, post-collapse eruptions of rhyolitic lavas inside the calderas filled them in and smoothed the topographic contours, so that the calderas are not so easy to recognize now.

The three cataclysmic eruptions at the end of each cycle formed ash flow deposits (tuffs) that define the stratigraphy of the region. The first of the three eruptions, about 2 million years ago, emplaced the Huckleberry Ridge tuff, which is by far the largest, having a volume of about 2,500 km^3 (600 cubic miles). This eruption also produced the largest caldera, which was about 80 km (50 miles) in diameter but can no longer be seen. The second end-of-cycle eruption created the Mesa Falls tuff about 1.3 million years ago. This is the smallest tuff in volume (280 km^3, 67 cubic miles) and corresponds to the smallest of the three calderas, Henrys Fork, about 20 km (12 miles) in diameter. Both the caldera and the outcrops of this tuff are now only exposed in the Island Park area.

The most recent cycle of activity produced the Lava Creek tuff and the Yellowstone caldera about 600,000 years ago. The tuff has a volume of about 1,000 km^3 (240 cubic miles). The caldera, measuring about 70×40 km (40×25 miles), is located in the center of the Yellowstone Plateau and is of the resurgent type, meaning that upwelling or resurgence of the caldera floor happened after it was formed. In the case of the Yellowstone caldera, two domes grew on the caldera floor, followed by eruptions of yet more rhyolitic lavas along the ring fractures. After a long period of repose, activity started up again some 150,000 years ago in the western part of the ring fracture zone, followed by more dome uplift. Some researchers think that this restart of activity, which was followed by other eruptions about 110,000 and 70,000 years ago, may indicate a fourth volcanic cycle, but the matter is still being debated. During this recent phase of activity renewal, voluminous flows of rhyolitic lavas poured inside the caldera from vents at both the western and the eastern fracture zones. The lavas filled the central part of the caldera and overflowed its western rim, forming two plateaus: Madison (110,000 years ago) and Central (70,000 years ago).

Apart from volcanic eruptions, the Yellowstone area has also been tectonically active. It is cut by numerous extentional normal faults, formed by forces that expanded the crust by pushing it upwards from below. To the south of the Yellowstone Plateau, the faults trend northwards, while north of the plateau they trend northwest–westward. The vents for the Madison and Central plateau eruptions were ultimately controlled by this fault system that extended into the caldera from its margins.

Although there have been no magmatic eruptions from Yellowstone since the Central Plateau was formed, ground deformation has continued to happen, indicating movement of magma underneath the caldera. The caldera floor was uplifted by about 1 m (3 feet) between 1923 and 1984 and since the mid-1980s the floor has subsided at rates as high as 1.5 cm/year (⅝ inch/year). There are many other signs that magma still underlies the caldera at a shallow depth, such as swarms of small, shallow earthquakes. The most obvious of these signs is Yellowstone's hydrothermal system, which is the world's largest. There are, in fact, more geysers and hot springs here than in the rest of the world combined. The hydrothermal system produces so much heat that the heat flow from the Yellowstone caldera is 40 times greater than the Earth's average.

Yellowstone is, in many aspects, one of the world's most unique environments. It was designated a Biosphere Reserve in 1976 and a World Heritage Site in 1978. In terms of geology, Yellowstone is an outstanding example of a long-lived, large-volume volcanic system caused by a hot spot underneath a continental plate. The current manifestations of volcanism – geysers and hot springs – make this area one of the most spectacular volcanic settings in the world.

It is possible that magmatic eruptions will happen again at Yellowstone, perhaps small eruptions of rhyolite and basalt at the margins of the plateau, or medium-size eruptions of rhyolite within the caldera. There is even still a possibility that a major eruption will occur, producing ash flows and another caldera collapse. We can be confident, however, that Yellowstone will stir and give us plenty of warning before waking up again.

Old Faithful geyser viewed from the east. Old Faithful erupts more frequently than any of the other big geysers, although it is not the largest or most regular in the park. (Photograph courtesy of Susan Kieffer.)

The daily geyser show

There are more than 400 geysers within the Park – that is about two-thirds of all the known geysers in the world. Yellowstone is one place where you can be sure to see this volcanic phenomenon numerous times every day. Old Faithful behaves almost like clockwork, but there are many other active geysers in the Park that erupt frequently, making this an ideal place for understanding how they work.

The system that creates geysers and hot springs works like a natural, giant boiler, in which hot magma supplies the heat to turn rainwater into pressurized hot water. Rainwater trickles down from the surface through fractures and seeps down through porous rock, which collects the water like a sponge. Heat from the magma chamber below reaches the permeable rock layer by conduction and heats the water. Because of pressure from the water and from the rock above it, the water trapped in the permeable rock layer may become superheated, reaching temperatures above 260 °C (500 °F) without boiling. This superheated water is lighter than the cool water that is coming down into the system, and it rises towards the surface. Exactly how it rises up determines whether it reaches the surface as a fumarole, as a hot spring, or as a geyser.

If the path to the surface is unobstructed and a relatively small amount of water rises through it, the water will boil as the pressure drops and emerge at the surface as a fumarole, also known as a steam vent. If superheated water mixes with cool water on the way up and does not boil, it surfaces as a hot spring (or pool) or a mudpot, depending on whether mud from surrounding rocks has mixed in with the rising water. The surrounding rocks lining a hot spring crater can easily disintegrate and turn into mud because of the corrosive action of sulfuric acid. The acid is formed from hydrogen sulfide in the hot spring water, which becomes oxidized both by chemical reactions and by primitive types of bacteria, such as *Sulfolobus*. The appearance of mudpots can vary during the course of the year, depending on the amount of rainfall and snowmelt. They range from "paint pots" of thin consistency to rather pasty "mud volcanoes." Sometimes a mudpot dries up completely and reverts to a fumarole.

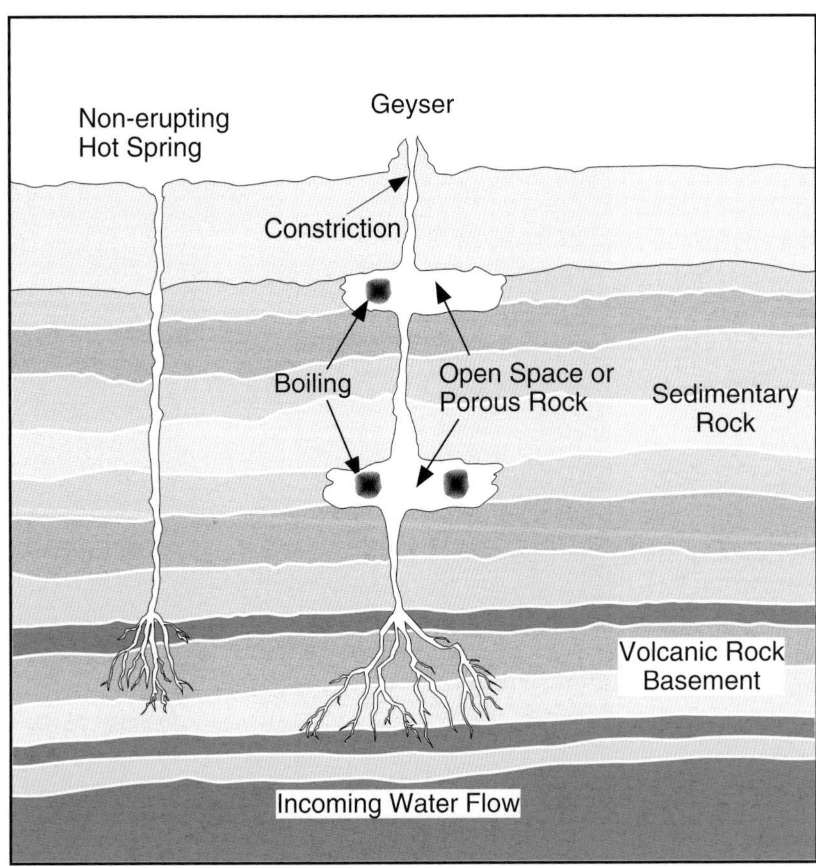

Sketch of a cross-section showing the plumbing systems of a hot spring and a geyser. (Modified from Bryan, 1990.)

The fumarole is, in fact, always there underneath the mud and the rising gases is what makes the pots "bubble."

If the hot spring has enough subsurface water to prevent mud and debris from filling up the crater, and the water from boiling away, a thermal pool is formed. The temperature of the water is usually close to boiling point, so it does not become stagnant. The water can be remarkably clear and vividly colored, though it only looks colored. The water is, in fact, very clear and clean and absorbs all colors of the spectrum except the blues and greens, much in the same way as oceans and clear lakes do. The shade of the pool – usually turquoise, azure, or green – depends on the depth and volume of its water. Bacteria and algae can thrive in the cooler edges and runoff channels from hot springs and geyser craters, making the water vividly colored and quite spectacular. Different temperatures of water allow different types of algae and bacteria to grow. In the case of a typical hot spring, the runoff channel is clear near the source, because only a few single-cell bacteria can live in water that is about 93 °C (199 °F). Away from the source, as the water gets slightly cooler, thermophilic (hot-water-loving) bacteria develop into long, hair-like strands. Further away, as the water cools down to 75 °C (167 °F), colorful bacteria, such as cyanobacteria, can thrive, forming microbial mats.

Fumaroles, ordinary hot springs, and thermal pools can be found in many places throughout the world. Geysers, however, are much rarer events. Geysers are a special type of hot spring, characterized by intermittent discharge of water ejected turbulently and powered by pressure from the vapor phase of gases (mostly steam) rising from below. The way a geyser works is fairly simple. Superheated water rises into pockets of groundwater that are under sufficiently great pressure so that boiling does not occur. The temperature of the mixing waters rises until a small amount of water boils despite the pressure. The steam, which has a volume much greater than that of the water that formed it, pushes up and out of the underground pocket, carrying some of the water with it. When the steam and water exit the system, more superheated water turns into steam and ejects the remaining groundwater out to the surface. The geyser gushes out. As it subsides, the process begins again, with groundwater from above accumulating in the pocket and superheated water from below entering it.

What makes geysers rare is that very special circumstances are needed to produce them. First, they require very large volumes of water. Some geysers eject tens of thousands of gallons of water in a single eruption; therefore geyser basins have need to have an abundant supply of water from rain and snowmelt. Second, a strong heat source underground is required to heat up all the water. Third, the system needs to be pressure-tight. If the rocks beneath the geyser basin can leak pressure, a geyser cannot form. The seal is provided by silica dissolved from rocks. Enough silica must be present to seal in the geysers, so the local volcanic rocks must be very silica-rich, such as rhyolite. Some of the silica contained in these rocks is dissolved in water and, little by little, some of it is deposited, forming a type of opal called geyserite or siliceous sinter. Geyserite (sinter) deposits can be easily seen around geyser basins, often forming ornate patterns, though the critical geyserite needed to form a geyser is, of course, deep below ground.

The fourth and last requirement for the formation of a geyser is a special plumbing system. A geyser requires a narrow spot or constriction, usually close to the surface. Water above the constriction acts as a lid, helping to keep the boiling water below under pressure. When the geyser erupts, it blows off the lid. Activity from different geysers varies in terms of appearance, duration, and interval between eruptions.

The appearance of eruptions depends on the geyser's surface structure. There are two major types of eruption: cone-type and fountain-type. Most of the famous geysers, such as Old Faithful, are cone-type. The characteristic of this type is a very narrow opening just below the surface and, often, a geyserite cone above the surface. The narrow opening acts like a nozzle during an eruption, resulting in a forceful jet of water that can reach great heights. Fountain-type geysers have an open crater at the surface that fills with water before or during an eruption. Since the jet has to rise through the pool of water, the eruption appears less forceful than those of cone-type geysers and its appearance is that of a spray or fountain. This type of geyser is far more common than the cone-type. Most of the small geysers in Yellowstone are fountain-type; some of the larger examples are Imperial Geyser (Lower Geyser Basin) and Grand Geyser (Upper Geyser Basin). Iceland's Geysir is also a fountain-type.

The duration of a geyser's eruption depends mostly on how large the plumbing system is. Most geysers erupt for minutes, but a few can last for days. Eruptions stop because they run out of water and go into a steam phase or, more commonly, because they run out of heat, in which case the water drains back and the eruption shuts off quickly.

Since a geyser's eruption uses up all of the heat and/or water, the geyser has to rest and recover between eruptions. The time of repose is called the interval. As the geyser refills, reheats, and gets ready to erupt again, it may give some signs, such as bubbling and small splashes, which are technically called pre-play. This is a good indication of how close a geyser is to its next eruption.

The basics of geysers are easy to understand but the details of specific geysers are far more complex. Old Faithful is the best-studied geyser in the world, as well as the most famous. It blows its top every 79 minutes, on average, but the interval times range from 45 to 105 minutes, depending on the amount of water left in the system after it has run out of steam. In 1984, a team of scientists started a detailed investigation of Old Faithful. They measured temperature and pressure every 5 seconds at eight different depths along the top 21.7 m (70 feet) of the geyser throat, its only accessible portion. The team included the brilliant volcanologist Susan W. Kieffer, but even she could not quite make sense of the data, which did not match what would be expected from the current theories of geyser behavior. Susan told her colleague Jim Westphal, a geologist at the California Institute of Technology, that she really would like to see what the inside of Old Faithful looked like. This seemed impossible at the time but, by 1992, miniature video cameras had become available and they were able to lower an insulated 5-cm (2-inch) video camera into the vent. It turned out that the vent had a much more complicated shape than they had expected. Susan Kieffer then realized why the data had not made sense. She had assumed a simple geometry for the fissure and had expected a complicated physical explanation for the geyser's behavior. The video camera showed that the reverse was true.

The vent is not a smooth pipe but rather a crack running in an east–west direction that reached at least 14.3 meters (47 feet) in depth. Some portions of the crack are wide enough so that the camera (which had a 1.8 m (6 feet) field of view) could not see the walls. Other parts of the crack were extremely narrow – in one place the walls were only 15 cm (6 inches) apart. The camera also showed that the walls are riddled with cracks and that water enters the system continuously at several different depths.

When the researchers put together the map with measurements of temperature and pressure, it became

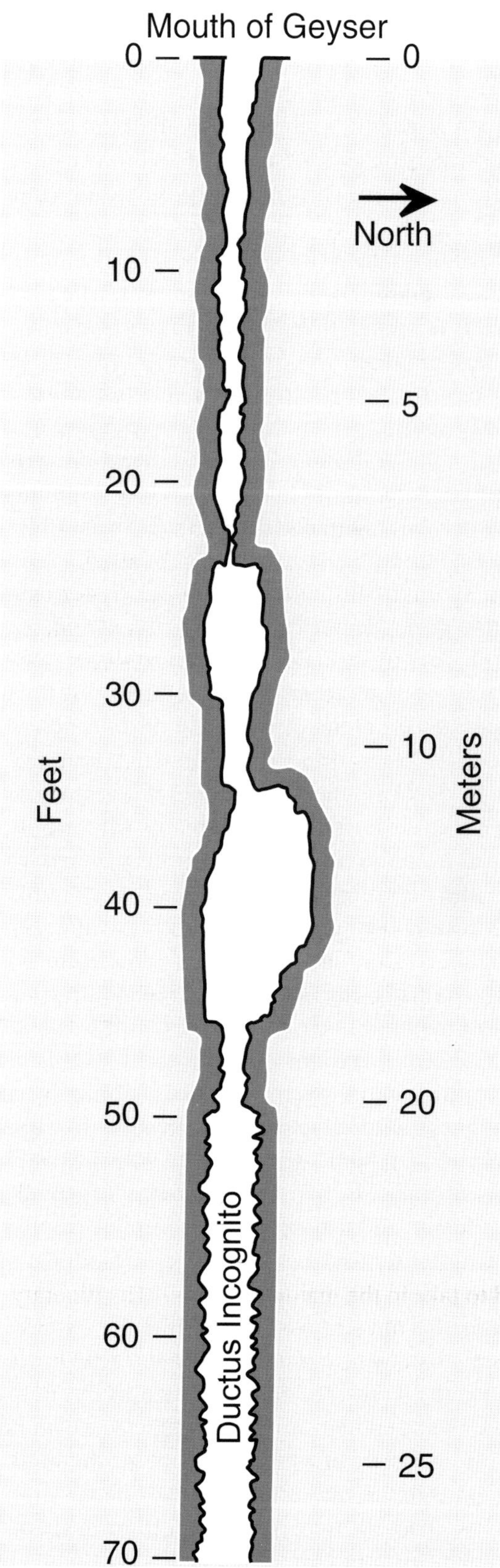

Map of the upper 14.3 m of the interior of Old Faithful. J. Westphal and S. Kieffer used a video camera to probe the geyser's internal plumbing. (Modified from a drawing by J. Westphal, Caltech.)

clear that the narrow 15-cm opening plays a critical role. For the first 20 to 30 seconds of each 3 to 5 minute eruption, steam and boiling water shoot through the narrow gap at the speed of sound. The narrow opening is critical because it limits the geyser's discharge rate. Once the pressure driving the eruption falls below a certain value, the eruption slows down and the geyser begins to shrink.

Although these results were groundbreaking, the detailed behavior of Old Faithful is still not completely understood. The researchers observed that, at times, the water in the fissure receded for several minutes to deeper levels than their equipment could observe. This happens despite the fact that groundwater constantly replaces the water in the fissure. What is really going on at the deeper levels of Old Faithful is still unexplained.

Like humans, geysers get old and die, though some geysers can be active for thousands of years. The silica that is essential to the formation of geysers is ultimately responsible for their demise, as more and more accumulated silica eventually clogs the fissure. Visitors to Yellowstone are lucky because Old Faithful, only some 300 years old, will still delight them for many years to come.

Dangers in Yellowstone

Yellowstone is a true wilderness, offering both wonders and dangers. The Park has become such a mecca for family vacations that it is common for visitors to think only of its wonders. Because Yellowstone is not currently exploding or disgorging streams of lava, it gives visitors a false sense of security. It is, however, a wild and treacherous place. A whole book has been written about deaths in Yellowstone and it is not pleasant reading.

Some of the dangers can be quite surprising and unexpected. During my first visit to Yellowstone, I had a rather close encounter with one of the bison that wandered freely around the Old Faithful cabins, where I was lodged at the time. Bison are hard to see in the dark and I had neglected to carry a flashlight. Park rules say it is illegal for visitors to wander within 25 yards (23 m) of wildlife, 100 yards (91 m) in the case of bears. Unfortunately, the beasts are not subject to the reciprocal rule. I was lucky as the bison I met decided to ignore me, but some of them are not so mild-mannered. There is an animated feature in the Park's web site showing a tourist being gored by one.

On the whole, it is easy to stay safe, but be aware of the many dangers and act sensibly. Be particularly careful not to start a fire, as fires in Yellowstone can easily get out of control. The 1988 fires destroyed many thousands of trees and caused enormous devastation in the Park. There are still thousands of dead trees, known as snags, left standing – be careful around them, as they can fall with very little warning. Yellowstone's wildlife is one of the Park's major attractions but also one of the major dangers visitors might encounter. Black bears (*Ursus americanus*) and, in particular, grizzly bears (appropriately named *Ursus horribilis*) are a major concern for hikers and campers. Aside from all the sensible precautions, such as not leaving food on the ground near your tent, it is a good idea to make plenty of noise while hiking to give bears a chance to get away. Banging saucepan lids is a successful tactic but singing loudly and badly is said to work best. If the worst happens and you encounter a bear, don't run. Bears will give chase and they can outrun even Olympic sprinters. Black bears and young grizzlies can climb trees, so the old advice doesn't often work. The best course of action depends on what the bear does. If it has not seen you, detour slowly and quietly. If it sees you but shows no interest, back away slowly. If it charges, and you have nerves of steel, stay standing and don't move. It turns out that bears like to bluff their way out of a threatening encounter, so often they will charge but veer off at the last minute. If you are attacked, play dead. Drop to the ground, curl up with your knees to your chest and your hands clasped over your head. Many people have survived bear attacks this way.

Other wildlife in the Park includes elk, wolves, coyotes, moose, lynx, bighorn sheep, and bobcats. These are not usually a danger to adults, but if you bring young children keep a very close eye on them. Children, in fact, make up a significant number of those killed in unfortunate accidents in the Park. Seven children, ranging in age from 3 to 15, have become victims of the hot pools, some by falling off boardwalks. Three adults are also known to have died in the thermal pools. One of the deaths was far from accidental – a young man donned scuba gear and was scalded to death while attempting to dive into one of the hot pools.

Use great caution when near geothermal areas: stay on trails and hold the hands of young children at all times. The ground near mudpools, geysers, and steam vents is usually soft and friable, and can easily break under a person's (even a child's) weight. Winter can make the situation even more dangerous, as trails become covered by snow. A young man was killed some years ago while skiing, when he accidentally fell into a thermal pool near Shoshone Lake. Even more tragic was the death in 1997 of Rick Hutchinson, the Park geologist, in a snow avalanche. These disasters show that although Yellowstone does not pose the usual dangers of an active volcano, it has other, less obvious but still potentially fatal hazards.

Visiting Yellowstone National Park

Yellowstone National Park is big – nearly 9,000 km^2 (3,400 square miles) of steaming geysers, lakes, waterfalls, wildlife, and breathtaking views. It is the largest National Park in the continental USA and, during summer, it tends to be quite crowded, with as many as 30,000 visitors coming in during a single day. The Park is located in the northwest corner of the state of Wyoming, but it is accessible from the states of Montana and Idaho. There are five entrances to the Park: the north entrance (Highway 89 from Interstate 90 at Livingston, Montana), the northeast entrance (Highway 212 from Interstate 90 at Billings, Montana or Highway 296 from Cody, Wyoming), the west entrance (Highway 191 from Bozeman, Montana, or Highway 20 from Idaho Falls, Idaho), the east entrance (Highway 16 from Cody, Wyoming), and the south entrance (Highway 89 from Jackson, Wyoming). Once inside, visitors have a choice of more than 1,900 km (1,200 miles) of hiking trails, 600 km (370 miles) of paved roads, some 10,000 geothermal features (including about 250 active geysers), five major visitor centers, and two museums. It can take you a whole day just to drive the 228 km (142 miles) Grand Loop Road through the Park. In summer, roads can get clogged with traffic and in winter many are impassable and closed. To find out current information on road conditions and closures, check the Park's website or call their information line (see Appendix I).

Most visitors drive the figure of eight Grand Loop Road to take in the major attractions. The itinerary below is one of volcanological highlights. There are many books devoted entirely to Yellowstone, or to single aspects of the Park, such as its geology, history, and wildlife. It is possible to spend a month visiting the Park and still not see much of it. I recommend spending at least 5 days, avoiding the peak school vacation period (July and August). Late June and early September are particularly pleasant times in Yellowstone. Most of the Park's facilities and roads are closed from mid October to mid May, though in some years roads can open as early as mid April. Lodgings start to open in early May and close on staggered dates

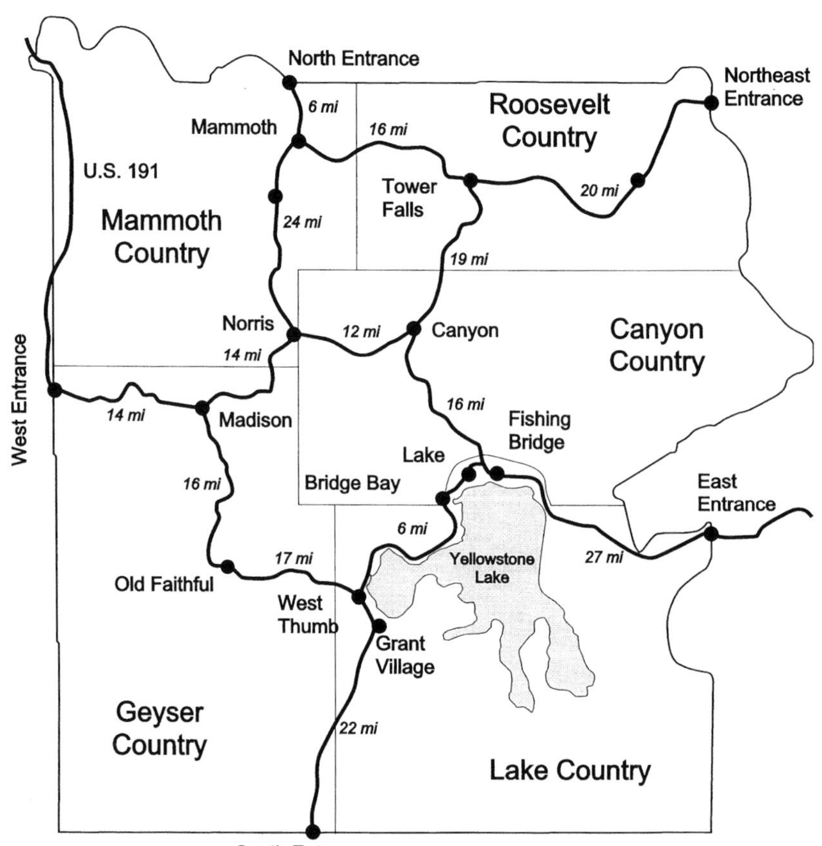

Yellowstone National Park map showing the five countries and main roads and attractions. (Modified from US National Park Service map.)

from late August to mid October, with only two being open in winter. Limited access is available during the winter, and only through the north entrance, but people still come to ski and it is possible to rent snowmobiles and to tour some of the sights by snowcoach.

Before starting a visit, it helps to know the Park's geography. There are five "countries": Geyser, Mammoth, Canyon, Roosevelt, and Lake. Each country has a main village with facilities that include visitor centers. The Grand Loop Road interconnects these villages and side roads lead to the five entrances to the Park. The itinerary below starts from the south entrance, convenient for visitors who arrive via Jackson Hole airport. The five legs can be done comfortably in a day each, though you can do two legs in a day if you rush and skip some stops. Leg 5 involves little driving, as it is entirely in the Old Faithful area (Upper Geyser Basin) and the nearby Shoshone Lake. Be sure to pick up an information leaflet and map (free with admission) at any of the Park's entrances. The Park's leaflet contains detail on the major attractions, as well as maps of each of the five villages. For approximate interval times, durations, and heights of the main currently active geysers, see Table 8.1.

Leg 1: South entrance to Grant Village

The drive from Jackson Hole to the Park's south entrance passes through the Grand Teton National Park (see "Other nearby attractions" section) and crosses the Snake River by Flagg Ranch. To the west, the north end of the Teton Range plunges beneath the Yellowstone Plateau. Ahead on the skyline is Pitchstone Plateau, a large rhyolitic lava flow erupted from the Yellowstone caldera. Just beyond the Park's entrance, on the right, is the trailhead and parking lot for Moose Falls.

Moose Falls and rhyolite lava flow

Take the short trail to Moose Falls (40 m, 130 feet). The waters from Crawfish Creek fall over a rhyolitic lava flow about 0.9 million years old, emplaced before the formation of Yellowstone caldera and its associated Lava Creek tuff.

View of Lewis Canyon and Lava Creek tuff

Back on the road, look for a roadside turnout about 7.4 km (4½ miles) from the south entrance. Lewis Canyon rhyolites, emplaced after the Lava Creek tuff and after the Moose Falls rhyolite described above, are

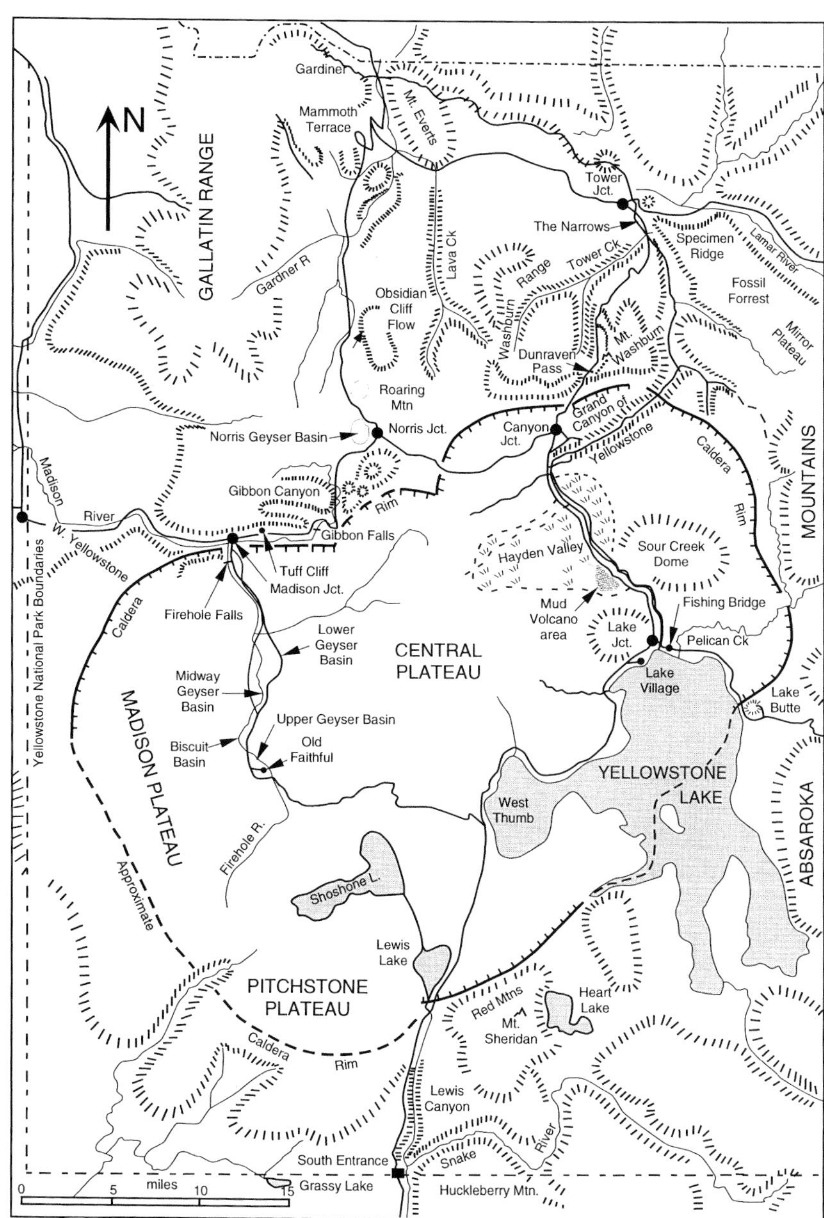

Map of Yellowstone National Park, showing the Yellowstone caldera and its main thermal features. (Modified from Parsons, 1978.)

well exposed inside the canyon. Just across the road, the contact between the Lava Creek tuff and the rhyolite can be seen – look for the glassy, fragmented part at the top of the rhyolite and, above it, bedded fallout ash from the tuff.

Lewis Falls and the Yellowstone caldera

Continue for another 5.3 km (3⅓ miles), following the rim of Lewis Canyon. There are good exposures of the Lava Creek tuff along the road, the tuff gets progressively thicker as you approach the Yellowstone caldera. Park just before the road crosses the Lewis River bridge (there are turnouts on both sides of the road). The road crosses the caldera margin here and this is a good place to orient yourself with a map. To the west is the Pitchstone Plateau, a large rhyolite flow and, to the east, the Red Mountains terminate abruptly at the southern caldera margin. It is hard to see the topography of the caldera, but you may notice the margin just north of the Red Mountains: the topography becomes progressively lower as it approaches the Lewis River. West of the river, there is no clear topographic indication of the margin, because it is buried by the Pitchstone Plateau rhyolites. Inside the caldera, to the east, is the Aster Creek flow, an older rhyolite.

Lewis Falls are located to the west side of the road as it crosses the Lewis River bridge. Lewis Falls and River

Table 8.1. *Yellowstone Park's major active geysers*

Geyser	Location	Interval	Duration	Height (m, feet)
Occasional	West Thumb	20–40 min	4–5 min	1–4 m, 3–12 feet
Steamboat	Norris Basin	Days–years	Minutes–hours	30–115 m, 100–380 feet
Vixen	Norris Basin	Minutes–hours	Seconds–50 min	1.5–9 m, 5–30 feet
Echinus	Norris Basin	30–120 min	3–15 min	15–30 m, 50–100 feet
Monument	Gibbon Basin	None	Steady	0.3–1 m, 1–3 feet
Spray	Lower Basin	1–5 min	3–10 min	3–8 m, 10–25 feet
Fountain	Lower Basin	Hours–years	30–60 min	15–21 m, 50–70 feet
Steady	Lower Basin	None	Constant	0.6–5 m, 2–15 feet
Narcissus	Lower Basin	2–8 h	5–10 min	3–5 m, 10–15 feet
Bead	Lower Basin	25–33 min	2.5 min	5–8 m, 15–25 feet
Pink Cone	Lower Basin	6–16 +h	30 min–3 h	6–11 m, 20–35 feet
White Dome	Lower Basin	12–24 min	2 min	3–9 m, 10–30 feet
Great Fountain	Lower Basin	8–12 h	45–60 min	25–45 m, 75–150 feet
Rusty	Biscuit Basin	2–3 min	20–45 secs	1–2 m, 4–6 feet
Jewel	Biscuit Basin	5–10 min	60–90 secs	3–9 m, 10–30 feet
Shell	Biscuit Basin	1.5 +h	20–90 secs	1.5–2.5 m, 5–8 feet
Avoca Spring	Biscuit Basin	1–18 min	10–30 secs	3–6 m, 10–20 feet
Mustard Spring	Biscuit Basin	5–10 min	5 min	1–2 m, 4–6 feet
Spouter	Black Sand Basin	1–2 h	10–11 h	1.5–2 m, 5–7 feet
Artemisia	Upper Basin	6–16 h	10–30 min	3–8 m, 10–25 feet
Riverside	Upper Basin	7 h	20 min	25 m, 75 feet
Rocket	Upper Basin	1h–2 days	Variable	1–3 m, 3–10 feet
Grotto	Upper Basin	1h–2 days	3–13 h	6–9 m, 20–30 feet
Daisy	Upper Basin	78–144 min	2.5–4.5 min	25–45 m, 75–150 feet
Grand	Upper Basin	6–15 h	9–16 min	45–60 m, 140–200 feet
Turban	Upper Basin	15–25 min	4–5 min	1.5–6 m, 5–20 feet
Sawmill	Upper Basin	1–3 h	15–90 min	1.5–12 m, 5–40 feet
Castle	Upper	9–11 h	1 h	18–27 m, 60–90 feet
Lion	Upper Basin	2–5 +hours	2–4 min	9–18 m, 30–60 feet
Aurum	Upper Basin	3–4 h	1 min	3–9 m, 10–30 feet
Pump	Upper Basin	Nearly constant	Nearly constant	0.5–1 m, 2–3 feet
Sponge	Upper Basin	1 min	Seconds–1 min	0.3–0.6 m, 1–2 feet
Anemone	Upper Basin	3–8 min	Seconds–2 min	0.5–3 m, 3–10 feet
Plume	Upper Basin	25–36 min	1–2 min	3–9 m, 10–30 feet
Beehive	Upper Basin	7 +hours	4–5 min	25–60 m, 150–200 feet
Depression	Upper Basin	3.5–5.5 h	2–3 min	2.5–3m, 8–10 feet
Old Faithful	Upper Basin	45–105 min	1.5–5 min	30–55 m, 110–185 feet
Lone Star	Upper Basin	3 h	30 min	11–12 m, 35–40 feet
Minute Man	Shoshone Basin	1–3 +min	2–10 secs	3–12 m, 10–40 feet
Soap Kettle	Shoshone Basin	9–21 min	1–3 min	1–2 m, 4–6 feet

were named after Captain Lewis of the famed Lewis and Clark expedition. The falls plunge 11 m (37 feet) down and are quite scenic and you may see some moose in the swampland next to the river, which is one of their natural habitats.

Lewis Lake

Just beyond the bridge is Lewis Lake campground and a turnout on the left side of the road from where there is a good view of the lake. The lake is contained by rhyolitic lavas and the effect is very picturesque. Beyond the western edge of the lake is the Pitchstone Plateau.

Back on the road, you will soon pass the Continental Divide, marked on the Park's map and also by a sign. Waters south of the Divide drain into the Pacific Ocean via the Snake River and Columbia River, while waters to the north drain into the Atlantic via the Mississippi and Missouri rivers. When you come to a Y-junction, follow signs to Grant Village and West Thumb Geyser Basin (the west fork leads to Old Faithful).

Grant Village

Facilities in the village include a general store, motel, restaurant, campground, ranger station, and shops. The Grant Village Visitor Center has exhibits that provide a good introduction to the Park and its auditorium exhibits a slide show at frequent intervals.

West Thumb Geyser Basin

This small geyser basin, located only about 3 km (2 miles) away from Grant Village, is the most scenic in the Park because of its location by the shore of Yellowstone Lake. Its strange name was given by members of the 1870 Washburn expedition, who thought that Yellowstone Lake looked like a hand and this part looked like the thumb. The geysers and hot springs in West Thumb tend to be more variable in temperature and discharge rate than those in the Park's other major thermal areas. Sometimes the activity is vigorous, but at other times it is disappointing and you may even see abundant algae, a sign of lower temperatures than usual (algae cannot grow in the truly hot springs and pools). The variations in activity are thought to reflect the effect of the lake's waters in recharging the plumbing underneath the basin. Silica may at times form a seal that prevents water from coming in, causing the activity to decrease. When fractures break the seal, cold water from the lake can come in again and the basin is recharged.

West Thumb is thought to be a relatively small caldera within the much larger Yellowstone caldera. It was formed by an explosive eruption during the period of volcanism after the collapse of the Yellowstone caldera. The West Thumb caldera is not at all obvious as its edges are now nearly completely buried by younger rhyolite flows.

To visit the basin, stop at the parking area and take the short trail (0.8 km, ½ mile) at the north end of the parking lot. You will soon come up to *Occasional Geyser*, a rather old feature as indicated by its pre-glacial geyserite formation, nearly 6 m (18 feet) thick. On the way back to the parking lot, take the Loop Trail to the edge of the lake, passing the rarely active *Twin Geysers* and the dark-green *Abyss Pool* which, at 16 m (53 feet), is the deepest known in Yellowstone. *Black Pool* owes its name to the combination of the natural, blue-looking water and the orange algae that lines its bottom. The pool's temperature is only about 55 °C (132 °F), allowing algae to thrive. Next comes *Fishing Cone*, a dormant geyser at the lake's edge that got its name from early explores who found it convenient to fish from here, as they could cook their fish over the steaming vent without even taking them off the hooks. *Lake Shore Geyser* has long periods of dormancy but when it erupts the water reaches 6 to 9 m (20 to 30 feet) high. There are two major hot springs in this basin: *Blue Funnel*, named for the optical illusion that makes you think that the vent moves as you walk around it and *Surging Spring*, named after its occasional surges of water. The *West Thumb Paint Pots* are bubbly mudpots that were one of Yellowstone's highlights for early tourists.

Leg 2: Grant Village to Canyon Village

Continue north on the Grand Loop Road towards Lake Village. About 16 km (10 miles) north of Grant Village is the *Pumice Point* parking area, from where there is a good view across West Thumb basin down to the Red Mountains at the south wall of the caldera. Back on the road, park at a turnout on the right 10 km (6¼ miles) from Pumice Point and walk to the viewpoint and picnic area by the edge of Yellowstone Lake.

Yellowstone Lake Main Basin

To the northeast is the Sour Creek dome, formed after the collapse of the caldera. South Creek has a prominent area on its south side, called Sulphur Hills, where the rocks have been altered by hydrothermal action. The Yellowstone caldera margin continues west, along the front of the Absaroka Mountain Range, and across the lake. Just north of the lake is an area of present-day uplift, which happens at rates of as much as 250 mm (1 inch) per year. The uplift may be a sign of future volcanic activity.

Bridge Bay, Lake Village, and Fishing Bridge

This scenic area offers lodgings, food, and other facilities. There are ranger stations at both Bridge Bay and Lake Village and a visitor center at Fishing Bridge that feature the Park's birds. Yellowstone Lake is North America's largest mountain lake. It is 32 km (20 miles) long, 23 km (14 miles) wide and reaches 98 m (320 feet) at its deepest level. Boating is allowed in the lake, but permits are required – inquire at one of the ranger stations. Swimming is discouraged because the lake's waters are so cold that they can cause hypothermia in minutes.

Mud Volcano hydrothermal area

This small, very gas-rich area contains several mudpots and shows surface deposits that have been heavily altered by the hydrothermal activity. Iron sulfide is responsible for the dark gray, black, or brown water, which is very acidic. The acidity of the water is due to gases that come up with the steam and dissolve in the condensate, particularly hydrogen sulfide gas, which gives out the rotten-egg smell so characteristic of volcanic areas. Take the loop trail (about 1.2 km, ¾ mile) from the parking lot to see the various geothermal features, including Black Dragon's Caldron, Mud Volcano, and Dragon's Mouth. Mud Volcano used to be a lot more active in 1870, when it was named by the Hayden expedition. Both the Hayden and the Washburn expeditions described a sound coming that could be heard from several miles away, resembling distant artillery. Another small geothermal area, *Sulfur Cauldron*, is located just about 300 m (1000 feet) north of the parking lot for Mud Volcano. It is an unusual hot spring that has a high acidity (pH 1.2) and harbors bacteria that are responsible for its yellow color.

View of the northeast part of the Yellowstone caldera

Continue north for about 7.6 km (4¾ miles) from the Mud Volcano parking lot up to a viewpoint where there area large turnouts on both sides of the road. To the east, across the river, is the north end of the Sour Creek dome. Mountains to the north mark the wall of the caldera. Northeast of the road is the edge of the Hayden Valley flow, a rhyolitic lava flow from the Central Plateau, located to the west of the viewpoint.

Artist Point and the Grand Canyon of Yellowstone

Continue north on the road and turn right at the sign to Artist Point. There is a parking lot at the end of the road and a short trail (100 m, 110 yards) to an overlook over the canyon, which plunges some 300 m (1,000 feet). This is considered by many, myself included, to be the most beautiful view in Yellowstone.

The canyon was carved by the Lamar River, a tributary of the Yellowstone River, mainly into rhyolitic lavas and tuff. The deepest part of the canyon is carved into the Canyon flow. Hot ground water and hot springs caused hydrothermal alteration of the rocks, resulting in a dazzling array of colors: reds, browns, and most remarkable, the "yellow rocks" which gave Yellowstone its name. The "yellow rocks" are actually altered rhyolite. Near the bottom of the canyon, hot springs still bubble and fumaroles puff, continuing the process of alteration of rocks.

At the head of the deepest part of the canyon is Lower Falls, which cascade down 94 m (310 feet) over the edge of an unaltered rhyolitic flow that is more resistant than the tuff and the softened rhyolitic rocks below. The smaller Upper Falls are located above the Lower Falls. A third waterfall, Tower Falls, is near the canyon's end and can be seen from a lookout near Tower–Roosevelt (Leg 3). On the way back to the main road, stop at the Uncle Tom's Trail parking lot to explore the short trails that offer beautiful close-up views of both the Upper and Lower Falls.

Canyon Village and Canyon lookout points

Canyon Village offers lodgings, a campground, food, shops, and a visitor center. Take the one-way, 4 km (2½ miles) loop road that leads from the village to Inspiration Point. There are spectacular views of the canyon from here, as well as from Grandview Point. Stop at Lookout Point for a beautiful view of the Lower Falls. A short trail takes you down for an even closer look at the falls.

Leg 3: Canyon Village to Mammoth Hot Springs

From Canyon Village follow the road north towards Tower Junction and park at the Washburn Hot Springs overlook on the right-hand side of the road, about 5.3 km (3⅓ miles) from Canyon Junction. Park at the overlook for a view of the caldera.

View of Yellowstone caldera

The northern caldera wall is visible from this overlook, though it is not easy to see. The easiest way to orient yourself is to look north to the Washburn Range and Mt. Washburn, which are located to the north of the caldera wall. In the foreground inside the caldera is the tree-covered surface of the Canyon flow, reaching the caldera wall. The flow is cut by the Grand Canyon of Yellowstone. To the southeast is the gentle rise of the Sour Creek dome, broken by a northwest-trending graben. On a clear day, you can see the southern wall of the caldera from here, located just in front of the Red Mountains. Next to the parking lot is a roadcut exposing an andesitic debris flow cut by an andesitic dike (about 4 m (13 feet) across) from the Mt. Washburn volcanic center, which dates from the Eocene.

Back on the road, you soon leave Yellowstone caldera and cross Dunraven Pass, the highest point on the Grand Loop Road (elevation 2,700 m, 8,859 feet). The wildflowers in the summer can be quite spectacular in this area. There is a parking lot and the trailhead for the Mt. Washburn summit trail (about 5.8 km (3½ miles) long). An alternative trail is located further down the road, by the Chrittenden parking area. The view from the summit of Mt. Washburn is fantastic: you can see much of the Park and, on clear days, down to the Grand Tetons. Be careful when hiking in this area, as it is a favorite of bears.

Tower Fall and The Narrows of the Grand Canyon of Yellowstone

Stop at the Tower Fall parking area and walk down the short paved trail to the Tower Fall overlook. The beautiful Tower Fall drops 40 m (132 feet) over andesitic breccias from the Eocene, some of which have been eroded to form the "towers." Note the large boulder near the top of the fall that looks quite precariously located. In 1871, members of the Hayden expedition placed bets on the hour when this very boulder would fall.

From the overlook, there is a short foot trail (0.5 km, ⅓ mile) that goes down to the base of the fall. This part of the Yellowstone River canyon tells a complex history of erosion. Before about 2.5 million years ago, the river drained along a broad open valley in which basaltic lavas had ponded and the Huckleberry Ridge tuff was deposited. The basalts can still be seen in some places, such as the large overhanging cliff to the northwest above the canyon, but the Huckleberry Ridge tuff was largely stripped by erosion. The broad valley was later cut by a narrower one. This younger valley was then filled by gravels and by interlayered basalt flows and ash beds, the latter dating from about 1.2 million years ago. These rocks have been strongly altered by geothermal activity and are now yellow in color. The younger valley was later recut east of the cliff of yellow gravels and basalts, then filled by glaciation sediments known as Pinedale Till. The valley known as The Narrows, downstream from here, was cut entirely after the Pinedale Till was deposited, less than 12,000 years ago. It is the narrowest part of the Yellowstone Canyon and about 150 m (500 feet) in depth.

About 1.3 km (¾ mile) further down the road is the *Calcite Springs overlook*, from where you can see the activity at the springs and the northernmost part of the canyon. Across from the overlook, the canyon wall shows two exposures of massive basaltic lava flows, each about 1.5 m (5 feet) thick, showing columnar jointing. Between the flows is gravel from glacial outwash and above the uppermost unit is glacial till. The basaltic flows are about 1.5 million years old and came out of vents located to the south on the Yellowstone Plateau volcanic field.

Tower–Roosevelt

Facilities at the village include the Roosevelt Lodge, where horse and stagecoach rides can be arranged. The ranger station gives information on naturalist-led activities that include an all-day hike to the Petrified Forests on Specimen Ridge. Evening programs are offered at Tower Campground Amphitheater.

Mammoth Hot Spring Terraces

Take the road from Tower Junction to Mammoth Hot Springs. The step-like white terraces have been a major attraction of Yellowstone since the early days of stagecoach tourism. Because activity at individual springs varies, the shapes and colors of the terraces change with time. The most striking deposits are bright white, which are fresh travertine formed by the precipitation of calcium and carbonate. As the travertine weathers, it changes to gray. During the building of a terrace, travertine precipitates around the edge of a pool, accumulating at a rate of up to 215 mm (8½ inches) per year.

Because the water cools as it cascades from terrace to terrace, it allows algae to grow, giving it blue-green hues. Bright splashes of colors, in tones of green, yellow, orange, and red, are added by cyanobacteria. The overall effect is truly spectacular. However, there are times when some of the terraces are dry and, while they still look impressive, the beauty of the color effect is gone.

There are two major areas, called Main Terrace and Upper Terrace. Just south of the village is the parking lot for the Main Terrace. A boardwalk and trail (about 2.4 km, 1½ miles) starting from the parking lot lead to the main attractions. Minerva Springs and Terraces is Mammoth's highlight, as it is usually the most colorful and spectacular of the terraces. Another favorite attraction is Liberty Cap, located just to the north of the parking lot. This cone, 14 m (45 feet) high, was formed by the steady flow of hot water from a single source, depositing dense layers of travertine. It was named by the Hayden expedition for its resemblence to the caps worn by colonial patriots during the Revolutionary War.

The Upper Terrace can be reached by a one-way road about 2.4 km (1½ miles) long, located south of the parking lot – look for a turn on the right as you drive towards Norris. A major attraction here is Orange Spring Mound, named for the cyanobacteria which streak the travertine cone-shaped mound.

Mammoth Hot Springs: the Minerva Terrace. (Photograph by David E. Wieprecht; courtesy of US Geological Survey.)

Mammoth is located close to the north entrance of the Park and its facilities include a restaurant and nearby campground. However, if you want lodgings, the closest (inside the Park) is the Roosevelt Lodge back in Tower–Roosevelt, 29 km (18 miles) to the east.

Leg 4: Mammoth Hot Springs to Old Faithful
Follow the road south of Mammoth towards Norris and Old Faithful. After about 6 km (3¾ miles), stop at the Golden Gate parking lot. This marks the north edge of the Yellowstone Plateau. The highway bridge goes over cliffs made of yellow tuff, hence the name Golden Gate. Walk south along the side of the road for a view of Swan Lake on the right-hand side.

Overview of Swan Lake and the Gallatin Range

The Gallatin Range is an uplifted block between normal fault zones. Like the Teton Range, it exposes rocks of Precambrian age through lower Tertiary. To the east of the overlook is the Washburn Range, where glacial ice was at least 670 m (2,200 feet) deep during the Pinedale glaciation. Some of the rock surfaces near the top of the mountains have glacial striations, showing that they were at one time covered by ice. Much of the valley and the low ridges within it are mantled by till and show a subdued topography.

Obsidian Cliff

Look for the parking lot at Beaver Lake picnic area and walk back towards the northern end of the lake for a good view of the cliff. Obsidian Cliff is the west margin of a rhyolitic lava flow that formed about 180,000 years ago, during the post-caldera phase of the last eruptive cycle. Note the large columns at the base of the flow. These features are known as columnar jointing and they form as the flow cools and shrinks. The rubbly zone near the top of the flow formed when the flow's surface cooled and hardened, but the core continued to move, causing the surface to break up. The large ridges near the top of the cliff are pressure ridges formed as the lava buckled and folded. The vent for Obsidian Cliff is located about 1 km (⅝ mile) east of here and is one of a series of rhyolitic and basaltic vents along a narrow zone between Mammoth Hot Springs and Norris Geyser Basin.

Roaring Mountain

Located about 4 km (2½ miles) further down the road, this highly acid-altered fumarole area lets out lots of steam, but no longer makes the roaring, high-pressure discharge noise that it did when it was named in the late nineteenth century. The altered rock is from the Lava Creek tuff. This is worth a brief stop if you have the time. A further 3.9 km (2½ miles) down the road is *Frying Pan Springs*, another location whose name is more impressive than the actual feature. Back on the road, go to Norris Junction and take the west fork towards Madison.

Norris Geyser Basin

This is one of the Park's major attractions. Norris has a large parking lot, a museum, and well-marked, self-guided trails. It is the most dynamic geyser basin in Yellowstone, with changes in the behavior of its geysers and pools happening more often than elsewhere. A major characteristic of Norris Basin is its mixed waters: a major amount of neutral to slightly alkaline chloride-rich water that deposits siliceous sinter, and a lesser amount of acid-sulfate water that produces heavy alteration of the Lava Creek tuff. This basin has the hottest shallow reservoir of all the hydrothermal areas of the Park, as well as the largest geyser in the world: *Steamboat Geyser*. You may not be lucky enough to see it in eruption, since it has periods of dormancy that can last decades. When Steamboat does erupt, the display is spectacular, with water jets reaching up to 115 m (380 feet) into the air.

A short trail leads from the parking lot to the museum, and to two major loop trails. The trail to the north leads to *Porcelain Basin*, a colorful but barren, highly acidic environment. Algae and bacteria cannot flourish here and the colors come from mineral deposits that stain the porcelain-like silica: the pink, red, and orange hues are from iron oxides while the yellow color is from sulfur and iron sulfates. Also to the north are some unreliable geysers, with long periods of dormancy. The south trail goes pass Steamboat Geyser and up to *Back Basin*, where Vixen and Echinus Geysers are often active. *Porkchop Geyser* is now only a bubbling pool, but from 1985 to 1989 it erupted almost continuously. In 1989 it exploded in full view of visitors, sending rock and debris up to 67 m (220 feet) away.

Gibbon Geyser Basin

Located just to the southwest of Norris, Gibbon Basin has small but quite interesting features, but it is unmarked and hard to find. The most unusual are the *Chocolate Pots*, located on the west side of the road about 1 km (⅝ mile) north of the Gibbon Meadows picnic area. The Pots are a small collection of cones about 1 m tall (3 to 4 feet) that, thanks to siliceous sinter deposits, are colored remarkably like chocolate. The Pots are very photogenic, as they have brightly colored streaks running down their sides: green, yellow, and orange, all due to bacteria and algae that thrive in warm water. The overall effect is one of broken chocolate bonbons with their gooey, colorful centers spilling out.

Just down the road are the parking lots for Artist Paint Pots and Monument Geyser. The *Artist Paint Pots* are isolated in the forest at the end of a 0.8 km (½ mile) trail. They are bubbling mudpots that are colored in pastel hues of beige, pink, and gray by iron oxides. Depending on the season, the mud can be soupy and bubbly or so thick that the pots erupt like mini-volcanoes, sending blobs of hot mud up to 4.5 m (15 feet) into the air. *Monument Geyser*, also known as Thermos Bottle, has a cylindrical cone which makes one or the other name appropriate, depending on whom you ask. The cone is 3 m (10 feet) tall with a small opening from where there is a constant hissing sound. This geyser ejects little water and it is known to be now in the process of dying by sealing its vent with internal deposits of siliceous sinter. The hydrothermal features in this area are controlled by a fault and related fractures along the Gibbon River canyon, radial to the Yellowstone caldera.

View of Gibbon Falls

Look for the parking area by Gibbon Falls on the left-hand side of the road. The northwestern caldera wall is just south of here. The falls, plunging over Lava Creek tuff, are a result of headward retreat along the Gibbon River from the inner caldera scarp. Looking to the south, notice the forested surface of the Nez Percé Creek flow, a large rhyolitic flow inside the caldera that erupted about 150,000 years ago.

Madison

Located at the junction between the road to the west entrance and the road south to Old Faithful, Madison offers an information station and bookstore. Take the road going south, passing by Firehole Falls, which plunges over part of the Nez Percé Creek flow. There is a side road called Firehole Canyon Drive that leads to the Falls and parallels the Firehole River. There are good exposures of the rhyolitic Nez Percé Creek flow along the riverbanks.

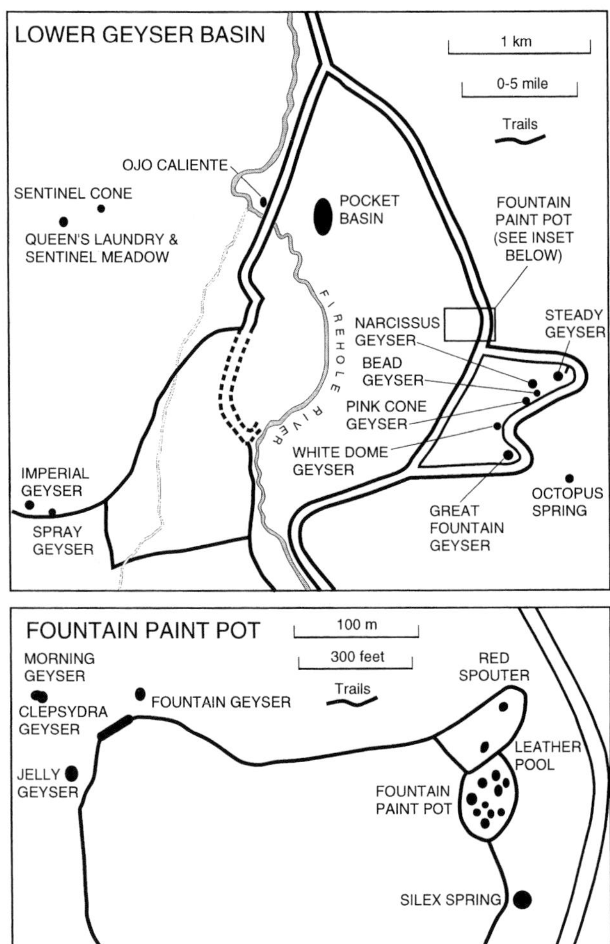

Map of Lower Geyser Basin. (Modified from Schreier, 1992.)

Lower Geyser Basin

One of the major attractions in the Park, this basin is an open valley enclosed by the Madison Plateau on the west and the Central Plateau on the east. Most of the thermal features in this basin are scattered in small groups so it makes sense to drive from group to group. This basin's star feature is without doubt the *Great Fountain Geyser*, located in the southeastern part. Eruptions begin about an hour after the crater fills up with water and overflows onto the terraces surrounding its intricate sinter cone. A few minutes before the eruption, the water surges and boils. Jets usually reach 30 m (100 feet) high. Eruptions usually last about an hour and have several phases, so it is worth watching for a while. The interval between eruptions is long and irregular, lasting from 8 to 12 hours. You can increase your chances of being there at the right time by inquiring at the information station in Madison when the last eruption took place, or else making this your first stop in the Lower Geyser Basin to check if the crater is filling up.

To reach the features on the west side of the basin, take Fountain Flat Drive off the Grand Loop Road. *Ojo Caliente Spring* ("hot eye" in Spanish) is a superheated, alkaline spring surrounded by a heavy shelf of sinter. Walk across the road and take the trail to *Pocket Basin*, formed by a hydrothermal explosion. This trail passes through the basin's ejecta deposits, which are mostly cemented sedimentary rocks. Pocket Basin has the Park's largest collection of mudpots. Stay on the walkways, as this is a dangerous area, as well as a fragile one.

Continue south on the road down to the trailhead for *Imperial Geyser*. This geyser became active in 1927 and its eruptions during the first couple of years reached up to 46 m (150 feet) high. The eruptions were so violent that they probably damaged the geyser's plumbing system and Imperial went into dormancy until 1966. The geyser was again active for a few years, but it entered another dormancy period in 1985. It is still dormant at present, though its beautiful alkaline blue pool continues to boil. You can see activity at the nearby *Spray Geyser* which, although small, has had reliable eruptions for decades. It is one of the few geysers that have a longer eruption time than interval time. It erupts with two main jets of water, one of them spraying at an angle of about 70 degrees, making a truly striking display.

The eastern side of the basin is reached by driving back on Fountain Flat Drive up to the Grand Loop Road, and taking that road towards the south. The first major area worth stopping at is *Fountain Paint Pot*, a collection of bubbling pastel-colored mudpots. A loop trail leads from the mudpots to *Fountain Geyser* which, even though it may be dormant, is worth a look into its beautiful azure pool. Nearby is *Clepsydra Geyser*, a rather reliable gusher as implied by its name, after the ancient Greek water clock. Further along the path is *Jelly Geyser*, small but also reliable, with an azure clear pool that actually looks like blue jello.

Continuing on the Grand Loop Road, take the left fork immediately to the south of Fountain Paint Pot. This road loops around some of the other remarkable features of the Lower Basin. *Steady Geyser* is the largest constantly erupting geyser in the world, having no discernable interval period. *Narcissus Geyser* has an interval time lasting several hours, but signals impending activity: if the water level in the crater starts to overflow the margins, Narcissus will erupt within the next hour. *Bead Geyser* is probably the most regular in

Yellowstone, with an almost constant interval period of about half an hour and an almost constant duration of 2.3 minutes. As the name implies, its crater has an intricate and rather attractive beaded surface formed by deposits of siliceous sinter. Bead Geyser used to be known for its sinter egg-sized beads, but unfortunately these were all collected long ago by tourists who were not environmentally responsible. *Pink Cone Geyser* is named for the beautiful shell-pink color of its sinter cone. *White Dome Geyser* also has a descriptive name. Its sinter cone, which was built over an older hot spring mound, is 6 m (20 feet) high and coated with white deposits. The highlight of this area is the nearby *Great Fountain Geyser*, one of the most spectacular geysers in the Park, which erupts every half day or so. If you can time it right, plan to have a picnic lunch while watching this impressive geyser in action, as each eruption lasts at least 45 minutes.

Midway Geyser Basin

Located just south of the Lower Basin, Midway has a small number of thermal features, but they are quite impressive in size, including two of the largest hot springs in the world: the Grand Prismatic Spring and the Excelsior Geyser. Midway is often called "Hell's Half Acre," a name given to this area by Rudyard Kipling when he visited Yellowstone in 1889. The basin's major feature is the *Grand Prismatic Spring*, the largest hot spring in Yellowstone, the third largest in the world, and one of the most photographed features in Park. Its azure pool measures 75 × 115 m (250 × 380 feet) and sits atop a wide mound which has a series of stair-like terraces. The spring was named by the Hayden expedition in 1871 for its stunning coloration: deep blue in the center that gradually becomes paler towards the edges. Algae color the shallow water near the edge green, and outside the scalloped rim is a band of yellow deposits that change further out into orange and red. Thomas Moran made watercolor sketches of the Grand Prismatic Spring that seemed unbelievable to many who saw it, including geologist A. C. Peale. Seven years later, Peale journeyed to Yellowstone to confirm the colors. Luckily for us, they haven't changed in all these years – Grand Prismatic remains as beautiful as when Moran made his watercolors.

Midway's other great feature is *Excelsior Geyser*, though it no longer holds the honor of being the largest geyser in the world. Its last known major eruptions happened during the 1880s, when the column often reached 90 m (300 feet) high. The violent activity may have caused permanent damage to the plumbing system and the eruptions ceased. In 1985, it woke up for just 2 days, though its eruptions were small compared to those of last century. This activity may have been the geyser's last gasp, but it remains a productive thermal spring, discharging upwards of 18,000 liters (4,000 gallons) per minute of boiling, churning waters. Excelsior seems to have an underground connection to the nearby *Turquoise Pool* because, when Excelsior was active, the level of the water in the pool decreased significantly. Turquoise Pool's blue water over a milky white bottom makes it look like the gemstone after which it was named.

Leg 5: Biscuit, Black Sand, and Upper Geyser basins

Upper Geyser Basin is the Park's most popular location, mostly because it is the home of Old Faithful. Biscuit and Black Sand Basins, located between Midway and Upper Geyser basins, are less known but worth a detour. If you would like to visit an off-the-beaten-track thermal area, make sure to take the trail from the Upper Geyser Basin to Shoshone Geyser Basin, located at the west end of Shoshone Lake. It is also possible to get to Shoshone by taking a canoe trip from Lewis Lake, located near the southern entrance to the Park.

Biscuit Geyser Basin

Named for the biscuit-like sinter deposits that surrounded the crater of Sapphire Pool in the 1880s, this small basin has a small but interesting collection of thermal features. Unfortunately, an eruption from Sapphire Pool in 1959 crumbled the sinter biscuits.

Biscuit Basin is cut by the road and its parking lot, as well as by the Firehole River. There are trails from the parking lot leading east and west. The basin's eastern side has a small group of hot springs, with the only exceptional feature being *Cauliflower Geyser*, named for the cauliflower-like masses of sinter surrounding its crater. *Rusty Geyser*, located right by the parking lot, is a small geyser in a rust-colored basin that erupts every couple of minutes. The color is the result of iron oxides staining the sinter. A trail to the west of the parking lot leads to *Sapphire Pool*, with crystal-clear waters which occasionally boil. *Jewel Geyser* is a short-interval geyser whose eruptions can reach up to 9 m (30 feet) in height. The jet collapses almost immediately but may be followed by up to five other gushes. The geysers in *East Mustard Spring* and *Avoca Spring* are all small but reliable gushers with short interval times. *Shell Geyser*, very irregular in its activity, has an attractive crater, lined with golden sinter, which is said to resemble the shell of a bivalve.

Black Sand Basin

This is an isolated group of thermal features belonging to the Upper Geyser Basin. Black Sand is located on the west side of the road just before it meets the side road leading to Old Faithful. There is a parking lot from where a trail leads to Black Sand's main attractions. This basin was named by early tourists after the small fragments of obsidian that cover parts of the ground. Its small collection of geysers and colorful hot springs include the spectacular *Emerald Pool*. Its deep green color is the result of relatively low temperatures that have allowed yellow bacteria and algae to grow on the lining of the pool. Its orange and brown edges make a particularly photogenic contrast with the emerald water. Algae and cyanobacteria are also responsible for the multi-colored edges of *Rainbow Pool* and for the yellow color of *Sunset Lake*. In the early part of the century, the main attraction of this basin was Handkerchief Pool, located at the southern edge of Rainbow Pool. Tourists used to drop their handkerchiefs into the pool and convection currents would pull them under, making them emerge elsewhere moments later, freshly laundered. Sadly, vandals plugged up the pool and the natural laundromat ceased to work in 1929.

Black Sand also has two geysers, Cliff and Spouter. *Cliff Geyser* has irregular intervals and can be dormant for years but, when it does erupt, the jets commonly reach 12 m (40 feet) in height and the eruption can last up to 3 hours. An indication of a pending eruption is the crater filling up with boiling water – if this happens, wait around, as the eruption is impressive. *Spouter Geyser* is more reliable and its eruptions last for many hours, but they are not particularly spectacular. Overflow from Spouter drains into the nearby *Opalescent Pool* and has, over the years, caused it to flood the surrounding area, killing a number of pine trees. The dead trees that still stand in the blue pool are known as "bobby sock" trees, because the lower part of their trunks has been colored white by precipitated silica.

Upper Geyser Basin

This is the Park's most popular (and crowded) area. Its popularity is not surprising, considering that the basin's approximate 5 km^2 (2 square miles) contain nearly one-quarter of all the geysers in the world. If you only have one day in the Park, this is no doubt the place to come. Boardwalks and trails make for easy access. The area near Old Faithful has numerous facilities including the Old Faithful Inn and Lodge, stores, and a service station.

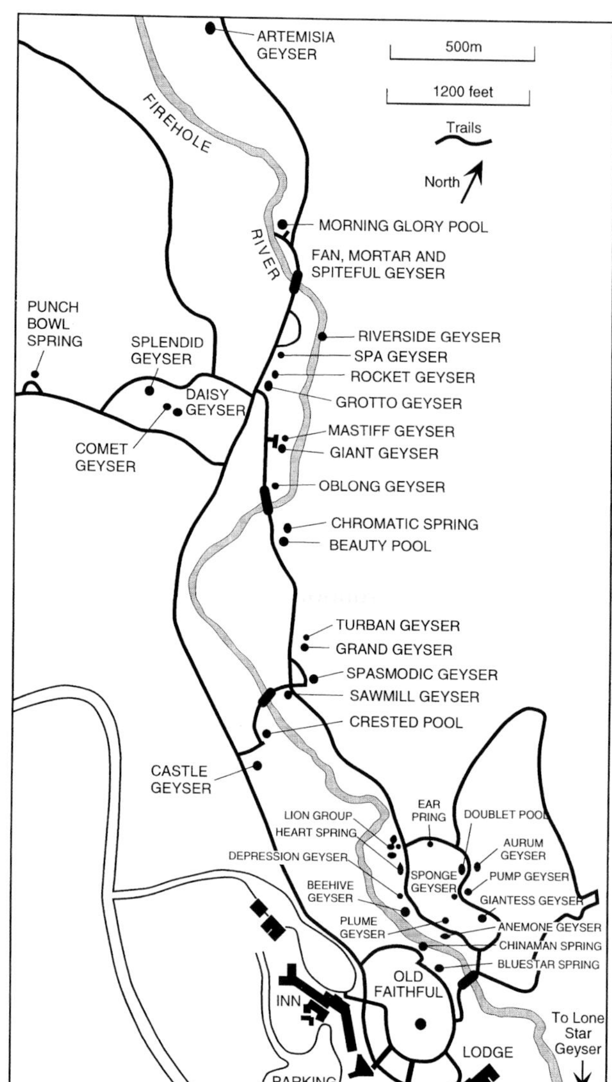

Map of Upper Geyser Basin. (Modified from Schreier, 1992.)

Starting at the north end of the basin, *Artemisia (sagebrush) Geyser* has the largest intricately ornamented crater in Yellowstone, with sinter ridges that look like popcorn. A sudden rise in water level is a sure signal that Artemisia is about to erupt. Further south is *Morning Glory Pool*, which resembles the flower after which it is named because of its deep, funnel-shaped pool with a dark blue center. *Riverside Geyser* is, as its name implies, located on the banks of the Firehole River. The column of water produced during its large eruptions arches over the river, creating a beautifully photogenic effect. *Spa, Rocket, and Grotto Geysers* are located close to one another and are interconnected.

Spa Geyser is irregular, but usually erupts during a long (more than 2.5 hours) eruption of Grotto. Rocket and Grotto have intervals lasting from 1 hour to 2 days and usually erupt simultaneously. Some major eruptions of Rocket have sent water gushing up to 15 m (50 feet) high, but usually the events are far less spectacular. Grotto erupts out of an unusual vent, a grotto-like sinter feature nearly 2.5 m (8 feet) high. It has two modes of eruption: the duration of the short mode is about 3 hours, while that of the long mode is 9 to 13 hours.

A side trail to the west leads to *Daisy Geyser*, the most reliable and predictable of the major geysers in this basin. Daisy's interval is, on average, about an hour and a half, and eruptions lasts under 5 minutes, but give prior warning signs: splashing begins about 20 minutes before the eruption in the largest of the two cones, and 10 minutes before in the smaller cone. Nearby is *Comet Geyser*, which has the largest crater in this group, but its eruptions rarely cause water to splash out of the crater. *Splendid Geyser* had spectacular eruptions when it was named in the 1880s, but nowadays it is rarely active. The last thermal feature in this group is *Punch Bowl Spring*, named after the raised sinter rim of its crater.

Continuing on the path south towards Old Faithful, the first major feature you will see is *Mastiff Geyser*, named because of its "watchdog" position by Giant Geyser. Mastiff has weak, frequent splashing and small, infrequent eruptions. *Giant Geyser* was named for its spectacular eruptions that used to reach up to 75 m (250 feet) in height. Although it has only erupted a few times since 1955, it still splashes and roars out of its impressive broken cone and remains one of the hottest vents in Yellowstone, with temperatures of about 95 °C (203 °F). Just before the path crosses the Firehole River is *Oblong Geyser*, notable for its green-blue pool and sinter formations around the crater rather than for its hard-to-predict eruptions. On the other side of the river are *Chromatic Pool* and *Beauty Pool*, two of the most colorful pools in this basin. Their colors go from deep blue at the center to yellow, orange, and red further out. The two pools are related and, whenever one begins to overflow, the water level in the other drops.

Continuing south, the next group of thermal features consists of four geysers, including the spectacular *Grand Geyser*. This is one of the few major geysers in the Park that has not changed noticeably since its discovery in 1871. Its eruptions are still truly spectacular, with jets reaching up to 60 m (200 feet). Grand is not, however, an easy geyser to predict, as the interval lasts from 6 to 15 hours. The shallow basin does fill with water before an eruption, but this process is slow, lasting about 5 hours. The activity at Grand and at the nearby *Turban Geyser* are linked, and Grand only erupts when activity starts at Turban. As its name implies, Turban has globular masses of sinter that make it resemble the Turkish headpiece. Rising waters in Turban's crater signal that an eruption is about to start at Grand. Both geysers then erupt, and Turban remains active for about an hour after Grand returns to normal. Nearby is *Spasmodic Geyser*, named for its irregular eruptions, and *Sawmill Geyser*, where pulsating jets of water erupt through a pool making a whistling noise.

A short path crossing the Firehole River leads to *Castle Geyser*, a well-known feature because of its sinter cone that reaches 3.7 m (12 feet) and resembles a castle in ruins. Its eruptions have a water phase, lasting about 15 minutes, followed by a 45-minute-long steam phase which makes Castle Geyser look like a locomotive, much to the delight of children. *Crested Pool*, also located along this path, is a superheated pool with water reaching 93 °C (200 °F). Occasionally, surges of hot water from the pool reach 30 m (98 feet) in height.

As the path comes closer to Old Faithful, the concentration of thermal features increases. The single path splits by the *Lion Group*, becoming a loop path that brings visitors close to the major features. The Lion Group is made up of four interconnected geysers: Lion, Lioness, Big Cub, and Little Cub. When viewed from the south, the group does resemble a reclining lion and when it erupts there is a sudden rush of steam that sounds like a roaring beast. Lion Geyser, the most active, is the one with the largest sinter cone. The rest of the Lion family have longer dormancy periods and the whole group sometimes goes to sleep for up to 2 weeks.

Going east along the loop path you will come to *Ear Spring*, popular with visitors because it is shaped like a human ear. The path then curves south, bringing you to *Aurum Geyser*, a small but attractive geyser named for its pastel-colored sinter formations in tones of peach and gold. The sapphire-colored *Doublet Pool* produces a periodic, thumping noise that you can feel, more than hear, if you stand close to the edge. *Pump Geyser*, a small but almost constant geyser, also makes a distinctive thumping sound. Its activity has remained unchanged since its discovery in the late eighteenth century, even though the 1959 earthquake affected many of the other features in the area. Pump's runoff channel harbors a stable microbial community that supports a variety of insect life. *Sponge Geyser*, named for its rounded cone full of small holes, is one of the Park's smallest geysers, discharging columns so small that they appear more like splashing than eruptions. *Giantess Geyser* has long periods of dormancy but when it does erupt the column can reach up to 60 m (200 feet) high.

As the path curves towards Firehole River, it splits, with one fork going towards Old Faithful and the other going north back towards the Lion Group. It is a good idea at this point to head towards Old Faithful and into the Lodge, where the eruption times are posted. Depending on this schedule, stay near Old Faithful or go back across the Firehole River to see the other thermal features near the Lion Group. *Anemone Geyser*, predictable and frequently active, is often used by visiting teachers to demonstrate the principles of geyser eruptions. Anemone has short eruptions every few minutes that shift between its two vents. Just before an eruption, the crater fills up with water while the erupting vent makes a gurgling sound. After the eruption dies down, the water drains back with a remarkable sucking sound. The nearby *Plume Geyser* is a good example of a young geyser. It formed as a result of a steam explosion in 1922, which created an opening on the ground. Plume was irregular at first but since the 1959 earthquake it has erupted about every half-hour. Just before an eruption, water fills the crater and begins splashing and pulsing. The geyser erupts at an angle, towards the west, in one to four bursts.

One of the most spectacular geysers of Yellowstone, as well as one of the largest active geysers in the world, is the *Beehive Geyser*. Its cone, about 1 m (3.5 feet) high, is quite distinct and really resembles a beehive. It is unfortunate that this is an irregular geyser that can be dormant for weeks or even months. However, Beehive does give some warning of an impending eruption: a vent just east of the cone, appropriately called the Beehive Indicator, starts erupting jets up to 3 m (10 feet) high about 10 to 20 minutes before Beehive goes off. An eruption then starts from the Beehive cone, starting with splashes and progressing into a spectacular, rumbling show that sends a straight jet through a narrow opening up to 60 m (200 feet) in the air.

Depression Geyser, named after the appearance of its crater, is another gusher whose activity was triggered by the 1959 earthquake. During intervals, the crater fills up with dark green water until it overflows and the water starts splashing, signaling an eruption. Just south of the Lion Group is *Heart Spring*, a striking feature because of the shape of its crater and its colorful algae and bacteria growth.

Table 8.2. *Predicting Old Faithful's next eruption*

Eruption duration (min)	Interval to next eruption (min)
1.5	45
2.0	52
2.5	59
3.0	65
3.5	70
4.0	75
4.5	80
5.0	86
5.5	89

Take the path to the south, across the Firehole River and towards the Old Faithful area. Near the river are two hot springs. *Blue Star Spring* was named after the star-like sinter formations that surround the pool. In the 1980s, an unfortunate bison calf wandered too close to this pool – its bones can still be seen at the bottom. *Chinaman Spring* was named after an enterprising Chinese laundryman from the 1880s who used the hot water as a clothes boiler. He pitched his tent over the spring and suspended the laundry in the water using a basket. However, the Chinaman was unaware of the mechanics of hot springs and, when he added soap to the water, the spring erupted, taking along with it the tent and laundry. Soap has, in fact, been used by tourists in the past to make geysers erupt, but this practice can permanently damage the delicate plumbing system and has long been stopped.

Old Faithful, the star of Yellowstone, does not have regular intervals, but its eruptions can be predicted quite accurately. The intervals are usually between 45 and 90 minutes and the average duration of an eruption is 4 minutes. If the total eruption time is less than 4 minutes, the next eruption will occur in 40 to 60 minutes. If the eruption lasts longer than 4 minutes, the next interval will be between 75 and 100 minutes. This correlation was found back in 1938 and still works. You can try to predict the next eruption by using Table 8.2. The trick is to time the eruption starting with the first continuous, steady column of water that rises in the eruption, not the splashing that often happens beforehand. The principle behind the predictability is simple: if the geyser has a short eruption,

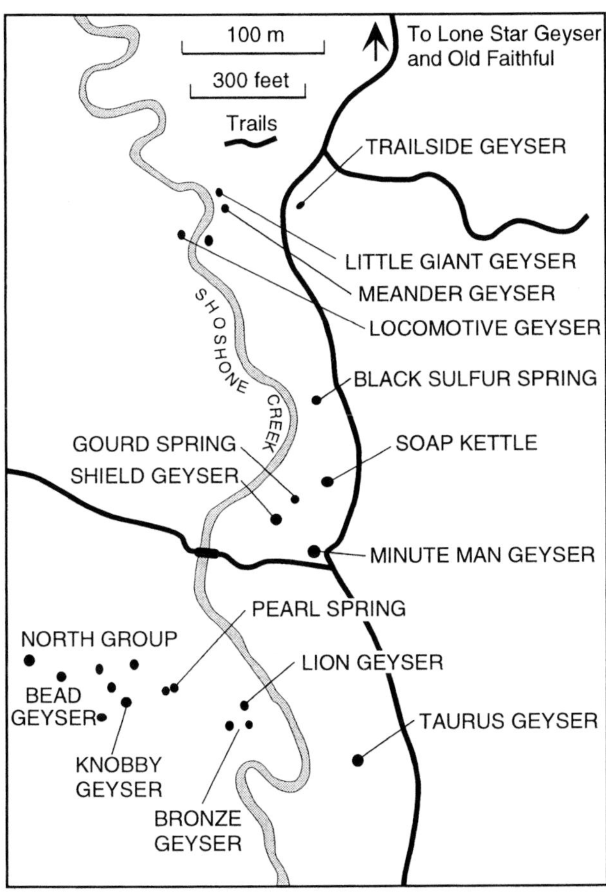

Map of Shoshone Geyser Basin. (Modified from Schreier, 1992.)

it discharges only a partial amount of energy and water and it can recover more rapidly. Old Faithful's show is quite spectacular, with the column of water reaching between 30 and 55 m (110 and 185 feet).

The southernmost geyser in the Upper Basin is *Lone Star Geyser*, named for its isolated location near the Firehole River. To get there, drive southeast on the Grand Loop Road and look for the Kepler Cascades parking lot. From there, a trail 4 km (2½ miles) long leads to Lone Star. The geyser's most impressive feature is its tall cone, reaching nearly 4 m (12 feet) high. Even during its regular 3-hour interval, Lone Star splashes from one of its several vents. This nearly constant splashing has been responsible for building up the tall cone. Lone Star's eruptions have two phases. The first lasts from 3 to 5 minutes and consists of jets that can reach up to 8 m (25 feet) high. After 15 to 25 minutes, the eruption enters the second phase, which lasts about 30 minutes. It begins with splashing, followed by a powerful jet progressing into steam. Considering all the different phases of the activity, it is not hard to catch Lone Star doing something interesting.

Shoshone Lake Trail

Shoshone Geyser Basin, located at the west end of Shoshone Lake, has the advantage of not being accessible by road. It is, therefore, far less crowded than the other basins. Its setting is still unspoiled and its pools and geysers still retain most of their original, intricate sinter formations. To get to Shoshone Basin, take the trail (12 km, 7½ miles) from Lone Star Geyser. The trail eventually parallels Shoshone Creek. Be particularly careful in this area, as there are no boardwalks. There are over 100 thermal features in the basin, with the major attraction being *Minute Man Geyser*, named so because it erupts nearly every minute. A series of short eruptions occurs for several hours before the geyser enters a repose period for a few hours, then the cycle is repeated. Just north of Minute Man is *Soap Kettle Geyser*, which has a distinctive cone lined with gold-colored sinter beads. It is easy to see an eruption from this geyser, as its intervals are at most 21 minutes long. To the northwest of Minute Man is *Gourd Spring*, which has boiling, bubbling water and occasional small eruptions. The trail splits at the Minute Man Geyser, with one fork going south and the other west. The south fork leads to *Taurus Geyser*, which only boils and splashes these days, but is rather attractive looking: deep blue water in the pool's center contrast with orange cyanobacteria lining the edges. The trail west from Minute Man leads to the *North Group*, which consists of a collection of numerous small, boiling hot springs, all of which have colorful basins and runoff channels because of algae and cyanobacteria. Many of the hot springs have intricate sinter borders, some of which are stained red by iron oxides. *Knobby Geyser* is part of this group. Although its eruptions are small and it has long periods of dormancy, Knobby is particularly beautiful because of its sinter formations and square-shaped pool. The white and gray sinter border is rather attractive, forming intricate clusters that resemble roses in bloom.

Other local attractions

Craters of the Moon National Monument

Located in Idaho, this National Monument has a truly lunar-like landscape that has been described as one of the strangest places in North America or, more unkindly, as a desolate and awful waste. In reality, Craters of the Moon is a remarkable volcanic Monument consisting of 215 km^2 (83 square miles) of numerous, overlapping olivine basaltic lava flows that erupted from 15,000 years ago up to about 2,000 years ago. They are surprisingly fresh-looking, making this an ideal location for studying the many different structures found on basaltic flows, both of pahoehoe and aa types. There are more than 60 lava flows that can be mapped individually, some 25 cinder cones, and eight fissures. The main fissure, called the Great Rift, traverses the Monument from the northwest to southeast. It consists of not just one but several "en echelon" fissures that span a swath about 3 km (2 miles) wide.

The Monument is located about 300 km (186 miles) southwest of Yellowstone's west entrance and it is possible to visit most of the attractions there in a day. To get there, take Highway 20 out of Yellowstone and drive south to Idaho Falls. Continue west on Highway 20 until it meets Highway 26. Follow Highway 26 for 29 km (18 miles) past the town of Arco to reach the Monument. The loop road inside the Monument is 11 km (7 miles) long and takes you to the major attractions and to the trailheads for several trails. Make sure to pick up a map at the entrance and follow the suggested itinerary, which consists of seven stops. The first is the Visitor Center, where the geology of the area is explained. Other stops highlight different flows, cones, and features such as tree molds. The highlight of the loop road is the last stop, the Cave Area, where a 0.8 km (½ mile) trail leads to five lava tubes. You will need flashlights for all caves except the Indian Tunnel, one of the largest tubes in the area, some 240 m (800 feet) long and reaching widths of up to 15 m (50 feet). Serious hikers may wish to explore the Craters of the Moon Wilderness, an area south of the Monument that flanks the Great Rift. Permits, available at the Visitor Center, are required for hiking or backpacking in the Wilderness.

Grand Teton National Park

The impressive Teton Range is the youngest of the mountains in the Rocky Mountain system. The Grand Teton reaches an altitude of 4,197 m (13,770 feet), towering over the valley known as Jackson Hole, while 11 other peaks in the range reach over 3,650 m (12,000 feet), high enough to support mountain glaciers. The Grand Teton National Park offers a variety of facilities including four visitor centers and five campgrounds. It is located 90 km (56 miles) from Yellowstone's south entrance and can be reached easily in less than 2 hours by car. There is a lot to do and see in this National Park (a free map will be handed to you at the park's entrance); however, it is worth coming here from Yellowstone even for a half day to see a couple of the major attractions. If you are interested in local history, head for the Colter Bay Visitor Center and Indian Arts Museum to gain a glimpse of Native American life in the nineteenth century. For spectacular views of the entire Teton Range as well as of Jackson Lake and most of Jackson Hole, drive up the Signal Mountain Summit road, starting out from the Signal Mountain Lodge and Campground (8 km, 5 miles). If you have more time, take the Jenny Lake scenic drive for one of the best views of the Grand Teton. The Park also offers extensive hiking trails (over 300 km, 200 miles in total), raft trips down the Snake River, and many climbing and mountaineering routes.

References

Brantley, S.R. (1995) *Volcanoes of the United States*. US Geological Survey.

Bryan, T.S. (1990) *Geysers: What They Are and How They Work*. Roberts Rinehart.

Kene, P.S. (1990) *Through Vulcan's Eye: The Geology and Geomorphology of Lassen Volcanic National Park*. Loomis Museum Association.

Kieffer, S.W. (1981) Fluid Dynamics of the May 18 Blast at Mount St. Helens. US Geological Survey Professional Paper no. 1250. US Department of the Interior.

Parsons, W.H. (1978) *Middle Rockies and Yellowstone*. Kendall–Hunt.

Schreier, C. (1992) *Yellowstone's Geysers, Hot Springs, and Fumaroles*. Homestead Publishing.

Smith, R.B. and L.W. Braille (1994) *Journal of Volcanology and Geothermal Research* **61**, 121–87.

9 Volcanoes in Italy

Southern Italy

Italian volcanoes are among the most notorious and dangerous in the world. Vulcano has given its name to all of the world's volcanoes, the Vesuvius eruption of AD 79 is probably the best-known volcanic event of all time, Strómboli is the longest continuously active volcano in the world, and Mt. Etna is the largest active volcano in Europe. Southern Italy has been the top destination for volcano enthusiasts for centuries and can be considered to be the cradle of volcanology. The historical records of eruptions in Italy go back some 2,000 years, providing a rich source of information on eruption mechanisms, cycles of activity, and on the effects of volcanic eruptions on society. Scientific observations of volcanic activity began on Vesuvius with Pliny the Younger's account of the AD 79 eruption. Vesuvius has the world's oldest volcano observatory, established in 1845, and was the first volcano to be monitored seismically, when the scientist Palmieri measured tremors during the 1872 eruption. Southern Italy's potentially disastrous combination of dangerous volcanoes and prosperous cities has long been a major drive in our need to understand and predict volcanic eruptions.

Most visitors to Italy are attracted to its incomparable historical sites, rich art collections, and splendid scenery. Those who have volcanoes as their focus can find all these elements plus spectacular volcanic scenery and, in the case of Strómboli, an almost guaranteed eruption in progress. The volcanoes highlighted in the following section – Vesuvius, Strómboli, Vulcano, and Etna – are the most active in the country. They have the great advantage of being easily accessible and relatively close to one another geographically, so it is possible to see them all in one trip (allow at least 2 weeks). Considering their diversity, activity, and history, it is hard to imagine a more appealing volcano tour.

Tectonic setting

Volcanism in the Mediterranean stretches from Spain to the Caucasus Mountains and is largely the result of steady convergence between the Eurasian and the African plates. The African plate moves northward at a rate of about 2.3 cm (nearly 1 inch) per year, continuously shrinking the Mediterranean basin. Subduction zones under the Greek islands (the Hellenic arc) and southern Italy (the Calabrian arc) explain the location of the major volcanoes in the region. However, the detailed tectonic story is very complex, with various small plates – called microplates – defying attempts to understand the region in terms of simple models.

The Italian volcanoes are located along the edge of the Tyrrhenian Sea. It is known that the Tyrrhenian basin is an area of high heat flow that has a thin oceanic crust. The Tyrrhenian basin was probably opened by a rifting event some 7 million years ago. Spreading was complete by the end of the Pliocene, about 1.6 million years ago. Since then, the Tyrrhenian Sea has been subjected to tensional faulting, which complicates the tectonic story. On the eastern edge, where Vesuvius and Campi Flegrei are located, volcanism has occurred on the western side of the Apennines, with the exception of the now-dormant Monte Vulture. The southern Italian volcanoes (Roccamonfina, Campi Flegrei, Vesuvius, Ischia, and Vulture) are associated with large faults in the crust, which probably extend to considerable depths and tap magma sources in the lower crust or even the upper mantle. Volcanism in this area started about 2 million years ago and continues today, with Vesuvius and Campi Flegrei being considered the most active and threatening. This whole area is also very active seismically – the Irpinia earthquake of November 23, 1980, killed over 3,000 people and caused considerable damage in Naples.

Sicily forms the southern margin of the Tyrrhenian Sea and the activity of Mt. Etna is linked to tensional

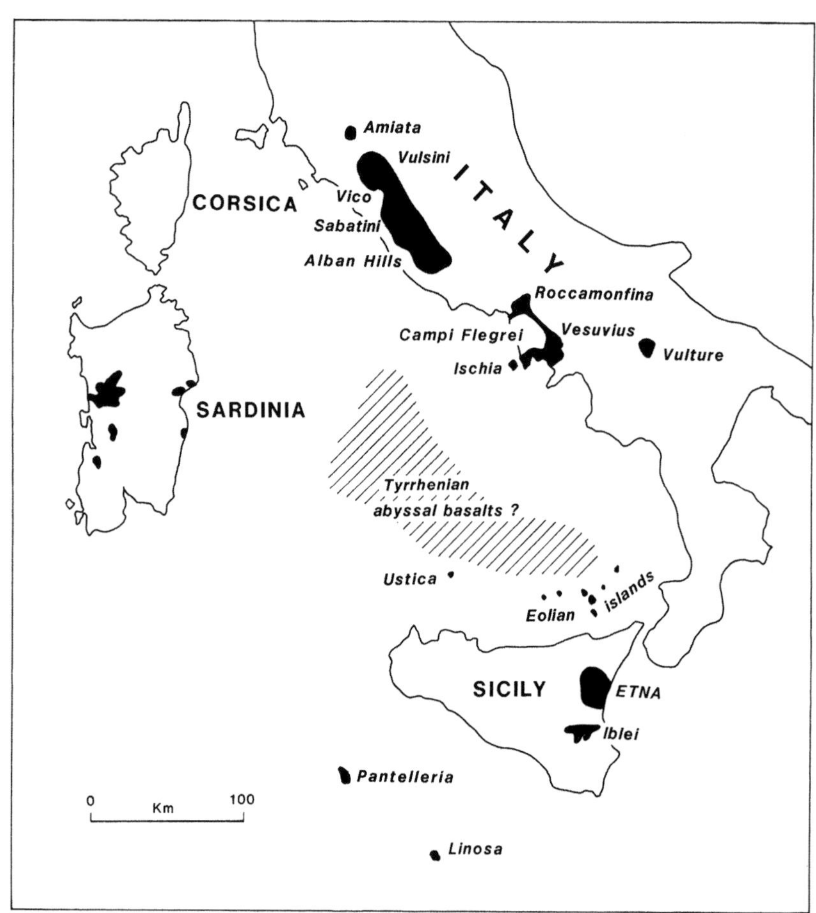

Location of the main volcanic centers of Italy (late Tertiary and Quaternary). (Modified from Chester *et al.*, 1985.)

movements. Etna is located at the intersection of major faults, notably the Messina fault that cuts along Sicily's eastern side and the Alcantara fault running east–west across the island. The activity on Etna has extended from the middle Pleistocene to the present. To the south of Etna lies another volcanic region that is no longer active, the Iblean Mountains.

The most recent volcanic event in the Tyrrhenian Sea was the formation of the Aeolian Islands (Isole Eolie), a chain of seven major volcanic islands that includes Vulcano and Strómboli. The traditional interpretation for the tectonic setting of the Aeolian Islands is that of a volcanic island arc, formed by the subduction of the Ionian sea slab (part of the African plate) beneath the Tyrrhenian Sea. An island arc is the name given to a subducting oceanic region, where one plate is sinking beneath another. However, this interpretation is not entirely consistent with various aspects of the local geology, including the chemistry of the lavas. The magma erupted in recent times – the high-potassium type of basalt known as shoshonite – is expected to be present in island arcs only if they are in a senile stage, that is, if the subduction process has almost come to an end. The problem with this explanation is that the Aeolian Islands are only 1 million years old, which is considered young in tectonic terms. Recent studies suggest that the subduction of the northern margin of the African plate below the Tyrrhenian Sea has indeed ended. Some researchers argue that plate rifting is now happening, but that it affects only the most recently active islands (Vulcano, Lipari, Strómboli).

The relationship between the Italian volcanoes, the Tyrrhenian Sea, and conventional plate tectonics is still being debated. The lack of agreement amongst researchers reflects the limited scientific information we have about this region and also the gaps in our knowledge about the complexities of plate tectonics.

Practical information for the visitor

When to go

The best times to visit the Italian volcanoes are late spring and early fall, when the weather is good and the

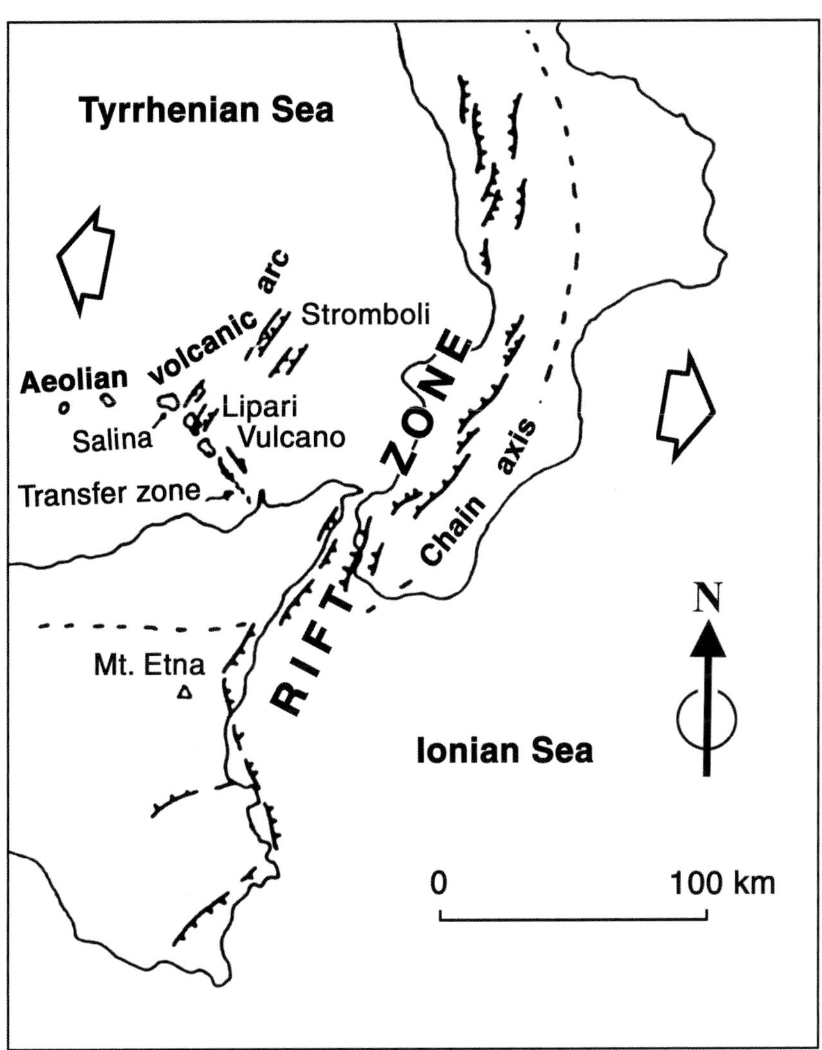

Sketch map showing an interpretation of the present-day tectonic setting of the Aeolian Islands. (Modified from Mazzuoli *et al.*, 1995.)

summits are clear of snow but the summer crowds are no longer there. The Aeolian Islands, in particular, should be avoided during July and August.

Information about volcanic activity

An eruption of Vesuvius or Vulcano would make worldwide news, but those of Mt. Etna and Strómboli occur frequently and tend to go unreported unless they are major events. The best way to check out the activity levels is to consult the S.E.A.N. Bulletin (details in Chapter 5) and look up the homepages devoted to the Italian volcanoes (see Appendix I).

Volcano observatories

The Osservatorio Vesuviano in Naples (Centro Sorveglianza, Via Manzoni 239, 80100 Naples) monitors Vesuvius and Campi Flegrei and carries out research on other Italian volcanoes. In Sicily, the Istituto Internazionale di Vulcanologia (Piazza Roma 2, 95123 Catania) monitors Etna and the Aeolian volcanoes, including Vulcano and Strómboli.

Non-volcanic dangers

Petty crime is common in Italy, particularly in Naples and Sicily, though incidents involving serious violence are relatively rare. It is fair to say that you are more likely to have something stolen in southern Italy than anywhere else mentioned in this book, so take precautions. Don't wear jewelry and keep cameras out of sight. Women should never carry purses – I had two stolen within a 12-hour period in Naples. The advice from a Neapolitan colleague is to carry most of your money in a hidden pocket but to keep some in your wallet (in another pocket) "to make the thieves happy."

Volcano monitoring and emergency services

Monitoring methods are sophisticated and warning of impending eruptions can be expected to the extent that

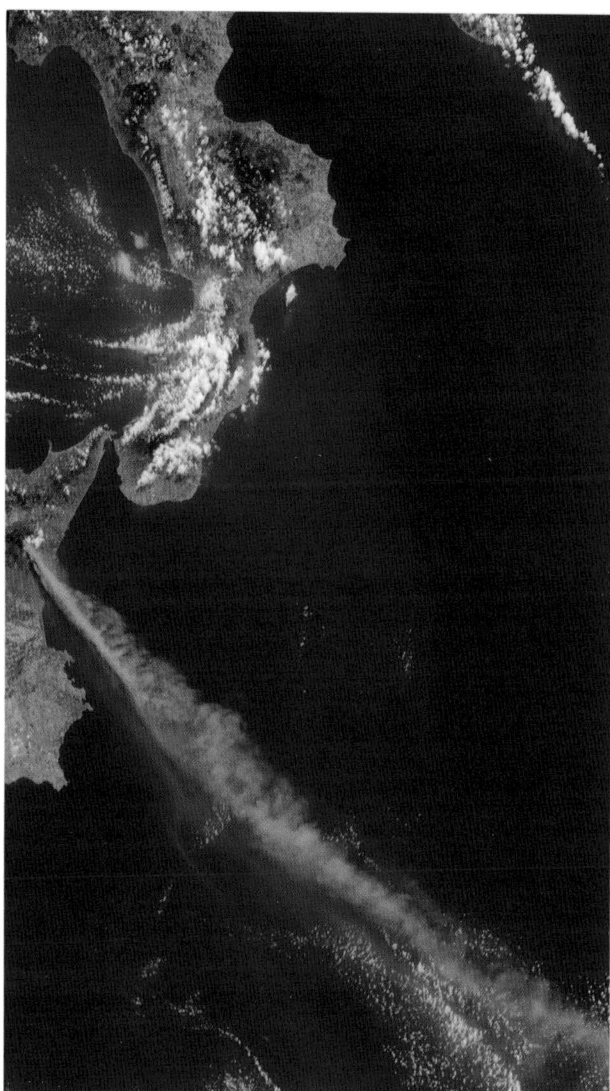

Mount Etna volcano in activity on July 22, 2001. This image was obtained by the Multi-Angle Imaging SpectroRadiometer (MISR) instrument. A large plume of ash and pyroclastics is seen drifting over the Ionian Sea. Ash from the plume caused temporary closure of the airport of Catania. A smaller, whitish plume can be seen near the summit; this one is composed primarily of very fine water droplets and dilute sulfuric acid. (Image courtesy of NASA/GSFC/LaRC/JPL.)

they are possible. However, be aware that little is done to prevent people from entering potentially hazardous areas, unless a full volcanic crisis is going on. It is therefore up to you to keep away from danger. Search and rescue facilities are adequate but not yet up to the standard of those in the USA – do not expect helicopters to be available. Note that the quality of health care in southern Italy's public hospitals can be rudimentary and insurance to cover private care is highly recommended.

Transportation

Public transportation is generally good in Italy. Naples is the nearest airport to Vesuvius and Catania is the nearest to Etna and the Aeolian Islands. The base of Vesuvius is encircled by the Vesuvius railway (Circumvesuviana), with trains linking Naples, Pozzuoli in Campi Flegrei, Pompeii, Herculaneum, Torre Anunziata (site of the Oplontis villa), and Sorrento. Rental cars are useful around Naples and in Sicily, but are not available in Vulcano and Strómboli. Visitors are not allowed to bring their own cars to these islands, but this is not a problem. Vehicles are not necessary or useful in Strómboli, as there are very few roads there. Vulcano and Lipari have limited bus services and sightseeing tours. Taxis, motor scooters, and bicycles are available for hire on both islands. A rental car is almost essential to visit the lower slopes of Etna, but only four-wheel-drive vehicles operated by local tour companies are allowed on the roads to the summit.

Before renting a car in southern Italy or Sicily, remember that driving there is quite an experience – it requires a great deal of nerve and a total disregard for traffic rules.

Getting to the Aeolian islands

Ferries and hydrofoils run all year to the Aeolian Islands, though in winter the services are reduced. The most regular services are from Milazzo in Sicily, but there are also some from Messina and Reggio Calabria. During the summer, daily services connect the islands with Cefalù, Palermo, and Naples. Many of the services go to Lipari and from there connect to Strómboli, Vulcano, and the other islands.

Lodging

Naples is the most convenient base from which to explore Vesuvius, but Sorrento is a good option for those who would rather have a quieter time. Those wanting to stay in Pompeii for longer than the usual day trip can find several hotels and camping sites nearby. Among the Aeolian Islands, Lipari is the one with the most choice in lodging, but there are many hotels and pensioni (guesthouses) on Vulcano and Strómboli. Vulcano's prices tend to be high, particularly in the summer. Camping is forbidden on Vulcano, but tolerated on Strómboli. In Sicily, there is a variety of hotels in Catania, but those wishing to stay on Etna itself can do so – the villages of Nicolosi, Zafferana, and Randazzo offer limited accommodation. An alternative is the Refugio

Sapienza, a no-frills hostel at 2,100 m (6,900 feet) altitude, which is something of a gathering place for the local mountain guides. Unfortunately, it has been damaged by recent eruptions and may not be operational.

Maps

Geological maps of all the volcanoes highlighted here have been published but are not easy to obtain and several are now out of print. They were all published by the Consiglio Nazionale delle Ricerche (see Appendix II) and some are being updated and will be made available again. A topographic map of the Aeolian Islands (at the scale 1:25,000) and a map of Strómboli showing hiking trails can be bought in local bookstores and souvenir shops. The Club Alpino Italiano publishes a 1:50,000 topographic and road map of the Etna National Park (from 1993) and a revised geological map of Etna (1:60,000), last updated in 1990.

Vesuvius

The volcano

Vesuvius is undoubtedly the most infamous volcano in the world. Everyone loves a good catastrophe story, and the demise of the wealthy, sinful cities of Pompeii and Herculaneum has all the ingredients of a tabloid feast. Contrary to popular belief, the eruption did not catch those early Romans partying with no idea about the impending disaster – many had fled during the first stages of activity, before the lethal blasts occurred.

The 1,281 m (4,227 feet) high stratovolcano started its life erupting under the sea, later forming an island, and eventually becoming part of the Italian mainland. Early effusions formed long lava flows that are now superimposed by the summit cone known as Mt. Somma, itself a product of powerful explosive eruptions and voluminous lava flows. Somma's activity ended about 17,000 years ago with a major Plinian eruption that is thought to have caused the collapse of the cone's summit, forming the Somma caldera. The present cone of Mt. Vesuvius was born within the caldera, probably during the same major eruption. The prominent Somma caldera wall has restricted the spread of lava flows from Vesuvius, particularly on the northern side, and influenced the distribution of pyroclastic flows and surges during the AD 79 eruption.

The name Vesuvius is thought to be derived from the Greek "besubios", meaning fire, and the volcano has been dubbed "the unextinguished" – appropriately, as it is still very much active. The last eruption was in March 1944, while the World War II battle of Monte Cassino was being fought some 90 km (56 miles) further north. The volcano has since then enjoyed its longest rest in 350 years, but this only makes us wonder when it will it wake up again. There are some clues from its past behavior: Vesuvius has, so far, followed a series of eruption cycles, each lasting several centuries. A particularly dangerous characteristic of these cycles is that they start with a major Plinian eruption after centuries of inactivity. The start of a violent Plinian eruption, therefore, can come suddenly, as it did in AD 79 – Vesuvius was thought to be extinct at the time. In fact, accounts of the revolt of Spartacus and the Roman gladiators, around 73 BC, describe the volcano as a broad, flat-topped mountain, covered by a dense forest famous for its wild boars. The apparently extinct crater served as the rebels' natural fortress.

Although eruptions of Vesuvius before AD 79 were not documented, studies of older eruption products show that the volcano has passed through six major cycles during the last 17,000 years. Within each cycle, the initial Plinian event was followed by centuries of milder eruptions. The most recent cycle started in 1631 with a violent eruption in which at least 3,000 people perished. The 1944 eruption, a relatively mild event, is thought to represent the closing of this last cycle.

The existence of eruptive cycles may give the impression that the behavior of Vesuvius is predictable. However, the cycles are not identical and, to complicate things further, smaller subcycles can be distinguished within a cycle. Typically, a subcycle starts with minor Hawaiian or Strombolian activity within the summit crater and ends with a strong eruption producing large-volume lava flows and heavy tephra fall. This is followed by a brief repose period, until the milder activity within the crater starts again. The repose periods within the 1631–1944 cycle were at most 7 years. The lack of activity since 1944 has been interpreted as a sign that the volcano has come to the end of one of its major cycles. The question now is whether Vesuvius will remain dormant for many centuries.

Why do volcanic cycles happen on Vesuvius? We know from studying past eruptions that there is a clear relationship between the length of the repose interval and the volume of the materials erupted.

The large-scale Plinian eruptions follow many centuries of repose, the intermediate-size eruptions occur after one or two centuries and the small effusive eruptions after only a few quiet years. This relationship is also reflected in the composition of the magma erupted: the Plinian events erupt magmas that are of the phonolitic or trachytic types. These are akin to andesites or rhyolites in terms of silica content (see Chapter 2), but they have more alkalis. The magma involved in the less violent eruptions is more basic (high-potassium alkali basalts). The interpretation is that Vesuvius has a continuous supply of deep basic magma that is stored within a shallow chamber, probably 3 to 4 km (2 to 2½ miles) down. The magma's composition is altered within the chamber, but fresh batches of magma continue to arrive from below, and the chamber becomes layered. The longer the repose time between eruptions, the more volume there is in the chamber, the more evolved (and potentially explosive) is the magma stagnated at the top, and the greater is the energy needed to break the plug. The Plinian eruption at the start of a major cycle completely empties the chamber and reduces pressure in the system. Later pulses of deep basic magma probably cause the intermediate-size eruptions within the cycles, while the periods of almost-continuous effusive activity are the result of an "open chimney"

Top: Vesuvius seen from Pompeii. (Photograph by the author.) Bottom: Vesuvius and the Bay of Naples imaged by Advanced Spaceborne Thermal Emission and Reflection Radiometer (ASTER) on NASA's *Terra* satellite. Note the densely populated area surrounding the volcano. (Image courtesy of Mike Abrams, JPL.)

system, in which the magma comes up more or less directly from its deep source.

Although the eruption pattern on Vesuvius is understood in general terms, there are still too many unexplained details and variations of behavior to enable us to answer that most important question – when will Vesuvius erupt again? Those who argue that the volcano has come to the end of a major cycle expect a dormant period of many centuries, followed by a catastrophic eruption. Others, however, disagree, saying that the present quiet state is alien to the pattern that Vesuvius has followed during its recent eruptive history. If that is true, the next eruption is expected to be moderate in terms of volume, but it could still be highly explosive and cause major destruction, maybe a couple of centuries from now. What everyone seems to agree on is that the monitoring must be constant, because Vesuvius is expected to give signs before waking up. The big unknown is how long before the eruption the stirring signs will happen – weeks, days, or maybe just hours?

The AD 79 eruption

This Plinian eruption was the most violent that Vesuvius has had in historical times – its volcanic explosivity index (VEI) was a cataclysmic 6. It is not only the world's most famous eruption, but also one of the best studied. The sequence of events during the eruption has been reconstructed from the eyewitness account of Pliny the Younger, as told in a letter to the historian Tacitus, and from studies of the eruption's deposits.

The eruption began on the morning of August 24, or possibly the previous night, with a brief but violent phase of phreatomagmatic explosions. These explosions produced a relatively small volume of fine ash that fell on the eastern slopes of the volcano but apparently did not reach the cities. This initial phase may have caused alarm amongst the local population, but no massive exodus. The two major towns in the region were Pompeii, about 8 km (5 miles) to the southeast, and Herculaneum, some 6 km (3¾ miles) west of the summit. Pompeii's residents numbered about 20,000, while the smaller Herculaneum housed some 5,000. There were numerous smaller communities and isolated villas around the volcano, including the royal villa of Oplontis. The town of Stabiae, 14 km (9 miles) to the southeast of the summit, survives today as Castellamare di Stabia.

Pliny the Younger and his uncle, the well-known

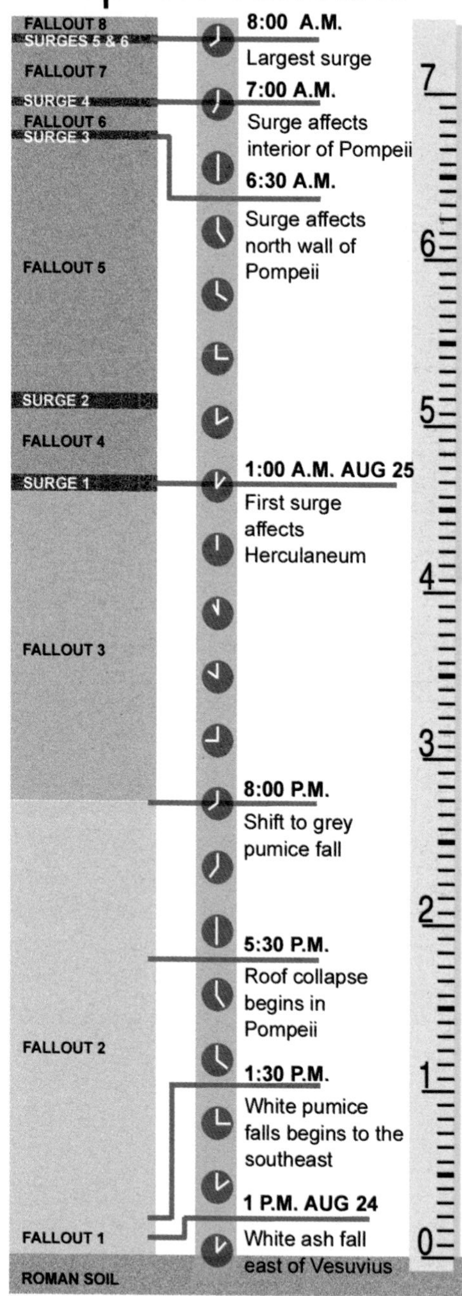

Simulated "scale" showing the sequence and timing of events of the AD 79 eruption of Vesuvius. (Modified from *Kids Discover Pompeii* magazine, published by Mark Levine, USA, 1995.)

statesman and scholar Pliny the Elder, were at Misenum (now Miseno), 32 km (20 miles) across the bay from Vesuvius. Both Plinys witnessed the emergence of a tall eruption cloud above Vesuvius at 1 p.m. on August 24. The column had a narrow lower portion and broadened upwards, resembling a Mediterranean pine tree and, within minutes, it had reached a height of 15 km (10 miles). Pliny the Elder, who was Admiral in command of the Roman fleet at Misenum, decided to depart by boat to investigate the strange phenomenon. As he was leaving his house, he received a message with a plea for help from Rectina, who is thought to have been the wife of the former consul Cascus. She was staying at her "villa under the mountain" – its location is not known, but it was probably within reach of the eruption's first ashfall from the smaller explosions. This presumably alarmed the lady, prompting her to send a messenger asking for Pliny's help.

The emergence of the eruption column at 1 p.m. marks the beginning of the 7-hour long white pumice fall that started to accumulate over Pompeii. Because of unusual wind conditions at the time, Pompeii was the location of the maximum pumice fall. Soon the falling pumice fragments grew larger, and denser fragments begun showering over the city. These dangerous projectiles, some fist-sized, crashed to the ground at speeds of about 50 m (165 feet) per second, no doubt injuring or killing people outdoors. The pumice, accumulating at rates of about 15 cm (6 inches) per hour, probably started causing roofs to collapse during the afternoon of August 24. These events must have been enough to cause a mass exodus from the city, even though traveling was made difficult by the eruption cloud that obscured the sunlight. Pompeii and other areas southeast of Vesuvius must have been in total darkness during this time.

During the evening of August 24, the composition of the erupting magma changed, and gray rather than white pumice was produced. The height of the eruption column increased, reaching 33 km (20 miles). The eruption of gray pumice continued for 5 hours. Around 1 a.m. on August 25, the style of activity changed – Vesuvius produced the first of six *nuées ardentes* that would interrupt the pumice fall during the next 7 hours. Each *nuée* separated into a slower-moving pyroclastic flow and a less dense, more turbulent, faster-moving pyroclastic surge. Because of its internal turbulence and low density, a surge can advance with little regard for topography and move at speeds of about 100 km/hour(62 miles/hour). The pyroclastic surges were to cause most of the deaths during the eruption.

The effects of the first pyroclastic surge were most dramatic in Herculaneum. Because of wind conditions during the preceding 12 hours, the city had suffered only a light dusting of ashfall, accumulating to less than 1 cm (½ inch) thick. However, being only 7 km (4½ miles) away from the summit crater, the residents were no doubt alarmed by the events and many fled the city before the first surge arrived. Those lucky ones probably went to Naples, where it is thought that they established the neighborhood known as the Herculaneum Quarter. The people remaining in Herculaneum must have been on the alert and watching the volcano during the night. They probably saw the first glowing, cascading *nuée ardente* coming down the volcano's flanks. It is estimated that the advancing pyroclastic surge took less than 10 minutes to reach the city. The Herculaneum residents took flight, mostly to the waterfront, seeking refuge inside arched chambers that formed the base of the Sacred Area. The hot ash cloud entered the city at about 1 a.m., running all the way down to the sea and enveloping the people on the beach and in the chambers. The heat of the surge may not have been enough to kill, but the ash would have formed plugs in the victims' windpipes, suffocating them. The pyroclastic flow from the same *nuée ardente* reached Herculaneum shortly after the surge. Because of Herculaneum's relatively elevated position, the flow did not advance through the city, but flowed down a valley along its southern edge and onto the beach in front of the Suburban Thermae (public baths), where it covered the surge deposit before entering the sea.

The surge layer deposited on the beach in front of the chambers and the Thermae contains many human skeletons and, inside the chambers, the layer is crammed with more skeletons. These hundreds of victims were not discovered until 1982, when the waterfront was excavated. Before that, only 10 victims of the eruption had been found in Herculaneum and it was believed that virtually all of the 4,500 residents had fled. Now we know that at least a few hundred stayed behind, waiting nearly two millennia to be uncovered.

Two other *nuées ardentes* were generated during the night, but they did not enter Pompeii. At about 6.30 a.m. on August 25, Vesuvius erupted dark pumice which fell for about an hour. Then, around 7.30 a.m., the fourth pyroclastic surge was generated, overwhelming Pompeii. Most of the residents had already fled, but about 2,000 – about 10% of the population – are estimated to have been killed by the surge. Like the victims in Herculaneum, they were asphyxiated in the hot ash cloud. Shortly after, an even larger surge came down the volcano's flanks. At this time Pliny the Elder was marooned at Stabiae. The previous afternoon he had given up his rescue mission, as the very heavy fall of hot pumice and blocks on his ships, and the thick floating rafts of pumice, made sailing near the southwestern coast of Vesuvius impossible. Pliny the Younger reports (presumably from accounts by his uncle's men) that on the morning of the 25th, accumulation of pumice fall at Stabiae was thick enough to make opening a door onto the courtyard difficult. The buildings shook and swayed and the black eruption plume caused darkness overhead. At about 8 a.m., the sixth and largest surge advanced towards Stabiae, taking the life of Pliny the Elder. Because his companions survived, the cause of his death is debatable. Pliny the Younger's accounts say that his uncle was an overweight man with a weak constitution and that he collapsed in the arms of his two slaves, choking in the dusty cloud. He could indeed have died of suffocation, probably because his often inflamed windpipe became further obstructed by the cloud of fumes and ash, but there are suggestions that he actually died of heart failure.

Back in Misenum, severe earthquakes made Pliny the Younger and his mother decide to leave the crumbling town. During their flight they witnessed "a fearful black cloud" over the volcano and "great tongues of fire" – probably the *nuée ardente* that would shortly cause Pliny the Elder's death. The surge cloud traveled over the Bay of Naples and reached Misenum, though by then it had lost its heat and was much dispersed. Pliny's account does not document the eruption after this, but deposits remaining on Vesuvius indicate that other surges followed, maybe for several days or weeks. The final phase of the eruption consisted of small phreatomagmatic explosions, caused by interaction of groundwater with magma remaining in the conduit. In total, the AD 79 event erupted nearly 4 km^3 (1 cubic mile) of magma, devastated some 300 km^2 (116 square miles) around Vesuvius in just two days, and left behind a tragedy that would never be forgotten. Although this was not the most destructive volcanic eruption in historic times, it is the one that has become the symbol of volcanic catastrophe.

A personal view: Living with the volcano

Like many volcanologists, I consider the events of the AD 79. eruption of Vesuvius to be the most fascinating yet horrifying piece of volcano history. The question of when Vesuvius will erupt again is one of the most urgent in modern volcanology, because the mountain is surrounded by about 1 million people – the largest population ever to live in the immediate vicinity of an active and extremely dangerous volcano.

I had the good fortune to become well acquainted with this most famous of all volcanoes while doing postdoctoral research at the Osservatorio Vesuviano in Naples. Although the volcano itself is incredibly interesting, there is another aspect that caught my attention – how the local population copes with the permanent threat of living in its shadow. They have done this for over 2,000 years, developing their own ways to deal with the danger while exploiting the best the volcano can give them – the beautiful landscape, fertile soils, and, for many years now, profits from the tourism Vesuvius attracts.

Southern Italy is a land of strong traditions, and religion plays a major role in people's lives. Although scientists are accepted and respected, my understanding is that we serve as back-up to the cities' patron saints. Neapolitans in particular seem to be very suspicious of those in earthly positions of authority – often with good reason. Thus the city's protector, San Gennaro, is trusted more than scientists and certainly more than government officials on matters of Vesuvius. Twice a year (May and September), a peculiar ceremony takes place in the Duomo (cathedral), during which two phials of San Gennaro's congealed blood are brought out from the Capella del Tesoro (Chapel of the Treasury). In the crowded Duomo (sometimes in a nearby basilica), the faithful pray for the miracle – liquefaction of the blood – to happen one more time. Sometimes the blood fails to liquefy and this is interpreted as a sure sign of catastrophe. The volatile temper of Neapolitans can erupt if they feel their saint is not cooperating. There are some bizarre accounts of such occasions, when shouts of "porco" (pig) and other epithets filled the vaulted cathedral.

San Gennaro has made an indirect but valuable contribution to the scientific understanding of Vesuvius, because for centuries priests have recorded the occasions in which the saint has been called upon to deliver the people from the volcano's threat. Eruptions that might otherwise have gone unreported are known because of these records, which have been invaluable to pinpoint the cyclic nature of Vesuvius' behavior. The most recent time that the protection of the saint was called for was in 1944, when lava engulfed a great portion of the town of San Sebastiano. The town's priest ordered an image of San Gennaro to be placed in a wine shop in the middle of town but, sadly, even

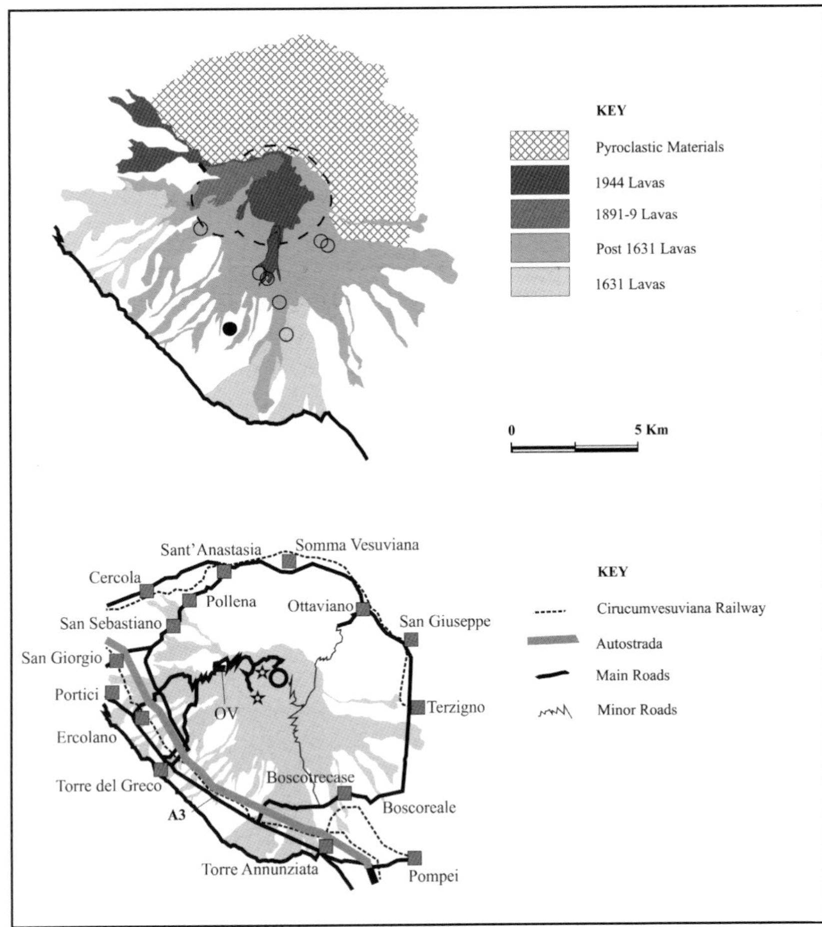

Top: Simplified geological map of Somma–Vesuvius, showing that the northeastern half of the volcano is covered by pyroclastic materials, while the other part has been resurfaced by lava flows as recent as from 1944. Circles show historical flank vents, the filled circle marks the location of the Camaldoli which predates the AD 79 eruption. Bottom: Location map showing locations of major towns, roads (including the A3 highway), and the Circumvesuviana Railway. OV marks the location of the Osservatorio Vesuviano, and ☆☆ the locations of the parking lots for access to the summit crater. Note that the minor road from Boscotrecase to Ottaviano is not open to vehicles and access by foot may also be restricted. (Modified from Kilburn and McGuire, 2001.)

with that precaution the lava destroyed about two-thirds of the town's houses. A quick look at a topographic map of Vesuvius makes it clear that San Gennaro had no easy job – the town is built directly below the Atrio del Cavallo, a valley that channels lavas down from the summit crater. Although San Sebastiano has been destroyed four times since 1822, the people have simply gone back and rebuilt it exactly on the same spot. Tradition still speaks louder than hazard maps.

It is perhaps this deeply rooted sense of tradition and reluctance to abandon one's home that has kept the Vesuvius region populated and prosperous despite the constant danger. In recent years, volcanologists in Italy have been campaigning for restrictions in urbanization around Vesuvius, but their efforts have not yet been successful. It is easy for scientists to criticize the local people for their insistence to live on the slopes of such a dangerous volcano but, at the same time, we have to admire their resilience, faith, and optimism. Many more generations may come and go before Vesuvius reawakens, but we can be sure that many thousands of people – if not millions – will be deeply affected by the event.

Visiting during repose

The summit crater

The summit of Vesuvius is easily reached, as a paved road goes nearly all the way, up to the car park at "quota mille" (1,000 m (3,300 feet) elevation). Buses going up to this point leave from Ercolano and Torre del Greco. Those with private cars can drive on past the turnoff for the car park and get closer to the crater. Driving up to the summit area is a good way of grasping the extent of urbanization on the volcano's flanks and the magnitude of the potential hazard from a large eruption.

There are some interesting locations to stop at on the way up Vesuvius. At the junction of the main road and the road leading up to the old Osservatorio Vesuviano (which is also the road leading to the Hotel Eremo) is an outcrop important for the study of the AD 79 eruption products – it is the nearest place to the summit where those deposits have been found. The old Osservatorio Vesuviano building should not be missed, even though it is currently not open to the public (there are plans to make it into a museum in the near future). The road to the Osservatorio passes just downslope from the Hotel Eremo and leads to a modern building located beside a chapel. Park there, and then walk past the chapel to the old building. The splendid, ornate structure dates back to 1841, when it was commissioned by Ferdinand II of Bourbon and became the world's first volcanological observatory. If you can arrange to visit inside, there are old seismographs and other volcano-related material on display in the richly appointed rooms. The modern Osservatorio, from which monitoring of Vesuvius is carried out, is located in Posilipo, between Naples and Campi Flegrei.

Past the Osservatorio, the road to the summit twists up the 1895–9 lava and cinder cones, after which it cuts the 1944 lava flows. There are good lookout points from which to see the 1944 flow down in the Atrio del Cavallo ("Hall of the Horse"). The Atrio gained its name during the eighteenth century, when it was covered by vegetation and used by visitors to Vesuvius to rest and graze their horses. The Atrio and the Valle dell'Inferno (Hell's Valley), on the eastern side, are troughs, about 300 m (1,000 feet) deep, between the rim of the Somma caldera and the cone of Vesuvius. Both valleys are natural channels for lava flows and are floored by recent lavas, most notably the 1944 flow that invaded the village of San Sebastiano.

The summit car park is surrounded by snack bars and souvenir shops, and by vendors trying to sell minerals that never came from Vesuvius. A 1.5 km (1 mile) steep path leads up to the crater – allow from 30 minutes to 1 hour for the climb. There is a fee to enter the summit area and guides are available, though most people don't need them. You can walk around the impressive crater, which is 450 m (1,485 feet) wide and 330 m (1,100 feet) deep. Individual fumaroles can often be spotted from the rim, giving the crater its still-alive aspect. Lavas from the 1944 eruptions cover the north and eastern part of the crater floor, the rest is covered by older lavas. There is a notable uncomformity (a line separating older from younger rocks) running down the crater wall, resulting from the 1944 crater being carved out from the older, 1906 crater.

Path to the summit crater of Vesuvius (Photograph by the author.)

Most people are content with walking around the rim but, occasionally, someone wonders if it is possible to go down into the crater. It is, but it is not easy to arrange. The best way is to contact the Naples branch of the Club Alpino Italiano, a first-class mountaineering and climbing organization. I'm told it is an exciting trip, but less so than the descent done by Charles Babbage in 1828. The famous inventor of the calculating engine went down the crater when a small cone within it was exploding periodically. Alone on the crater floor, Babbage observed the Strombolian eruptions from a safe distance until he felt confident that there was a consistent rest period of 10 to 15 minutes between explosions. Allowing himself 10 minutes to approach the cone and look down into its bottom, he observed what few had ever seen: a glowing lava lake, from which huge bubbles would rise upwards and threaten to burst. Luckily, his estimate of repose time was correct and the bubbles sank back without exploding. He returned from his adventure unharmed, much to the surprise of the guides who had refused to accompany him.

San Sebastiano and the 1944 lava

San Sebastiano has been destroyed by eruptions in 1822, 1855, 1872, and 1944. Although the town does not offer many attractions, it is a good place in which to see the 1944 lava close up and to pick up a bottle of Lacryma Christi, the famous Vesuvius wine. Some features of the town are quite interesting. For example, several streets near the center are some 6 m (20 feet) lower than the surrounding buildings. This happened because the road was cut through the lava after the houses and apartments had been built literally on top of the lava. In places, one can see the 10 m (33 feet) high lava that invaded the town on March 21, 1944. Many of the residents still remember that fateful day and some can tell stories of their parents and grandparents' houses being destroyed by earlier eruptions.

Pompeii (Pompeii) and Herculaneum (Ercolano)

Whole books have been written about the wonders of these ruined cities (see Bibliography), so the descriptions below are brief and mainly try to place the various sites in the context of the AD 79 eruption.

Most visitors to Pompeii head for the great houses such as Villa dei Misteri, the Casa del Fauno, and Casa dei Vettii. However, the more recently excavated homes have frescos and artifacts in their original places, which makes it easier to visualize what life was like before the eruption. Some well-preserved houses are Casa di Criptoportico, Casa del Sacerdote Amandus, and Casa di Menandro. Visitors should not miss the Amphitheater, the Forum, or a walk down the Via dell'Abbondanza (Street of Plenty), which cuts right through the once-thriving commercial district. The Lupanare (Brothel) is worth a detour and usually has Pompeii's highest density of visitors. Outside the excavation area is the Museo Vesuviano, which contains an interesting collection of artifacts.

The old city of Pompeii was surrounded by a wall with towers and eight gates. The first pyroclastic surge to reach the city swept against the north wall, depositing dark gray ash near the Herculaneum Gate. The city was invaded by the eruption's fourth surge, which was lethal to Pompeii's remaining residents. This surge extended to Bottaro, 1 km (⅝ mile) south of Pompeii, and to Tricino, 3 km (2 miles) west. The victims' remains were found buried by the fifth and, in particular, the thick sixth surge layer. The sixth surge was responsible for most of the destruction of Pompeii's buildings as well as for the death of Pliny the Elder, as the surge's distal edge reached Stabiae. The eruption's sequence of events has been reconstructed in Pompeii largely from the study of deposits near Herculaneum Gate, the Vesuvius Gate, and the Necropolis. Deposits from the surges can be seen in the Necropolis, near the Nocera Gate (Porta di Nocera). As you walk down to the Necropolis, look for the deposits on your left-hand side. Part of the outcrop has unfortunately been covered by a concrete wall, but with luck these important deposits will not be totally entombed in the future. The outcrop shows pumice fallout overlain by the fourth, fifth, and sixth surges.

As for the victims, some of their famous body casts are on display in the Antiquarium near Porta Marina, in the Terme Stabiane, and near the Basilica. The preservation of the victims' remains is very different in Pompeii and Herculaneum, though the cause of death was the same – the effects of pyroclastic surges. However, in Herculaneum the bodies were buried at a level below the post-eruption groundwater table. Therefore, the surge deposit remained wet and, as the victims' soft tissues decayed, the surrounding deposit gradually enclosed the skeletons, and no hollows were left. In Pompeii, the surge deposit containing the bodies was above the groundwater table, hence it remained dry and hard, preserving hollows where the bodies had been. Macabre plaster casts were made out of these hollows, preserving the victims' final demeanor in amazing detail.

Pyroclastic flow and fallout deposits in Pompeii's Necropolis. (Photograph by the author.)

The ruins of Herculaneum, located by the modern town of Ercolano, have not received as much attention as those of Pompeii. They are, however, even more fascinating than those of Pompeii from a volcanological point of view, because the careful archeological digging has recently gone hand-in-hand with volcanological studies. As a result, the catastrophic events have been reconstructed in remarkable detail. As the skeletons were unearthed, the volcanic debris in which they lay were examined and the death of the victims put into the context of the eruption's series of *nuées ardentes*. For example, an overturned boat about 8 m (25 feet) long was discovered on the beach in front of the Thermae. Beside the boat was the skeleton of a male, dubbed "the helmsman." However, careful examination of the pyroclastic layers showed that the male was buried in the first surge deposit, while the boat lay on top of that deposit – therefore, they turned out to be unrelated.

Some extraordinary mosaics survive in the town, the best example being that in the Nymphaeum (fountain and bath) of the Casa di Nettuno e Anfitrite (House of Neptune and Amphitrite). The extensive floor mosaics of the Casa del Atrio a Mosaico (House of the Mosaic

Atrium) were buckled by the weight of the eruption's deposits, and further distorted by earthquakes. The original geometric design is now wavy – a truly dramatic sight. The nearby Casa dei Cervi (House of the Deer) contains a comic statue of a drunken Hercules, who according to legend was the town's founder. As you walk around the town, notice the carbonized wood around window frames and doors. An impressive sight is the carbonized wooden screen in the Casa del Tramezzo di Legno (House of the Wooden Partition).

The Suburban Thermae, the bathhouse of the town's wealthy citizens, is located just above the AD 79 beach and coastline. The arched chambers to the west of the Thermae are where hundreds of skeletons were found – the useless refuge of the people trying to escape from the pyroclastic surge. The Palestra – the magnificent sports arena and public gymnasium – turned out to be an important location in the reconstruction of the eruption's sequence of events. In the center of the Palestra is a large, cross-shaped swimming pool filled with hard volcanic debris that are now hollowed out by tunnels. The layering of the deposits shows how the eruption's events affected the eastern part of the town. For example, the first surge layer in the pool is topped by a massive pyroclastic flow, but this flow filled only the eastern part of the pool, indicating that the flow went around the town but did not enter it. One of the best exposures of the pyroclastic flows that overwhelmed the town is located just outside the entrance to the tunnel leading to the cross-shaped pool. The outcrop shows the fifth pyroclastic flow at the bottom, overlain by the sixth surge and by the sixth flow.

Another important location for volcanological studies is the Theater, built on a hilltop. This splendid structure, which could seat up to 3,000 people, is the only ancient Roman theater so far discovered intact. The statues around the edges were swept away, but all else was preserved, though much of it is still buried. Given that so much of Herculaneum remains unexcavated, finer details of the AD 79 eruption will no doubt come to light.

Villa di Oplontis

This splendid villa was discovered in 1967 in the modern town of Torre Annunziata. It is thought that the villa belonged to Poppaea Sabina, Emperor Nero's second wife – certainly its size and lavish decorations indicate that its owner must have been considerably wealthy. Brilliant murals have been uncovered, and excavations are still going on, though made somewhat difficult by the villa's location in the middle of a modern town. Another villa, known as Villa Crassus Tertius, has been found about 500 m (⅓ mile) east of Oplontis. Both these villas provided important information on the spread of the AD 79 eruption products. The first pyroclastic surge – the same one that killed the residents of Herculaneum – also descended upon Oplontis. No human remains have been found here, so it is assumed that the residents had fled. The pyroclastic layers found in the swimming pool at Oplontis were particularly useful in the reconstruction of the eruption's events; volcanological studies were even able to determine that the pool contained water at the time of the eruption! In fact, the deposit layers found in the pool give a unique example of the passage of

The ruins of Herculaneum are one of the most interesting places to visit in the area. Vesuvius is in the background. (Photograph by the author.)

pyroclastic surges through a body of water. Most visitors to Pompeii and Herculaneum do not get to Oplontis, but the villa is worth seeing not only for its importance in volcanological studies but also for its wonderful architecture and art.

Visiting during activity

Vesuvius is a remarkably versatile volcano in terms of its style of activity. Its eruptions can range from Hawaiian, including lava lakes and fire-fountains within the crater, to Plinian, with deadly pyroclastic flows and surges. If some predictions are right, we can expect a repose interval of at least a couple of centuries, maybe many more. If the volcano surprises us all with an eruption in the near term, it is likely that warning signs will be detected, such as increased seismic events and ground deformation. Historical accounts indicate that these signs were evident before the AD 79 and the 1631 start-of-cycle eruptions. However, it is clear that the evacuation of over 1 million people from the area will be a major problem. The only circumstance in which a visit to an active Vesuvius would be sensible is if the volcano unexpectedly enters a phase of middle-of-cycle mild eruptions.

Other local attractions

Naples (Napoli)

No doubt one of the world's most fascinating cities, Naples is built on a series of old volcanic craters. The summing up of the city given by the 1884 Cook's *Tourist Handbook* is still totally applicable: "Naples is an ill-built, ill-paved, ill-lighted, ill-drained, ill-watched, ill-governed and ill-ventilated city. . . ." The book then concludes that Naples is "perhaps the loveliest spot in Europe." The paradox remains: the city's noise, bustle, and petty theft are offset by its splendid architecture, its fantastic setting, and the warmth of its residents. The sights of Naples are listed in most standard guidebooks, but the *Museo Archeologico Nazionale (Piazza Museo Nazionale)* is worth highlighting, as it is a major Vesuvius-related attraction. It displays many of the buried treasures discovered in Pompeii, Herculaneum, and Stabie, including murals, frescos, and household items. The museum's major crowd-puller is probably its famous collection of Roman erotic art. The graphic paintings and other artifacts were at one time kept in a locked room known as "Objects Cabinet," to which only aged and moral people had access. Aside from those items, one of the best-known displays is of "The Battle of Alexander" mosaic which once paved the floor of Pompeii's Casa del Fauno. More directly relevant to Vesuvius is a painting found in the Lararium of the House of the Centenary in Pompeii: it shows the wine-god Bacchus by a woody, vine-clad mountain identified as Vesuvius. The mountain's single peak on this painting was interpreted by some as evidence that the volcano's twin peaks were formed after the AD 79 eruption. However, wall-paintings from Herculaneum show Vesuvius with twin peaks. Geological evidence agrees with the Herculaneum artists, as it is clear that the wall of the Somma caldera influenced the distribution of deposits from the eruption.

Geologists might like to see the *Museo della Mineralogia* (Via Mezzocannone 8, Naples), a small museum which houses a comprehensive rock collection from Vesuvius. Visitors curious about the miracle of San Gennaro can see the phials of his blood in the *Duomo*, the cathedral dating from the thirteenth century. Photographers should visit the eleventh-century *Castello dell'Ovo*, a perfect place from which to photograph Vesuvius with the Bay of Naples in the foreground.

Ischia

An easy day trip from Naples by hydrofoil, this charming island is famous for its hot springs, "curative" muds, and superb local wines. The volcanological attraction is the 788 m (2,600 feet) high Monte Epomeo, the island's highest point. This volcano last erupted in 1302 and is still considered active. The summit can easily be reached on foot from the towns of Panza or Serrara Fontana – the walk along a footpath takes about an hour and a half. This is a superb place from which to see and photograph the Bay of Naples and Vesuvius, and the view alone makes the trip to Ischia worthwhile.

Campi Flegrei and La Solfatara

Campi Flegrei ("Burning Fields") is a caldera about 12 km (7½ miles) in diameter located along the north shore of the Bay of Naples. About one-third of the caldera is below sea level and forms the Bay of Pozzuoli, named after a fishing town known for its restaurants and its location right on top of the volcano's magma chamber. Campi Flegrei is not just an ordinary volcano – it is considered to be the most potentially dangerous volcano in Italy. This is not reassuring for the 400,000 people who live within the caldera, mostly in Pozzuoli itself. In fact, an eruption from here could be so violent that it could even

The Temple of Serapis in Pozzuoli. The three columns have recorded the rise and fall of the ground level. (Photograph by the author.)

destroy Naples and its residents – about 1 million people.

Campi Flegrei's past history reveals its dangerous potential. The eruption that formed the caldera happened some 34,000 years ago and deposited some 80 km³ (20 cubic miles) of pyroclastics that are known as the Campanian ignimbrite. A smaller explosive eruption occurred about 11,000 years ago, depositing the well-known Neapolitan yellow tuff. Another two major periods of explosive activity happened before the only historical event, which was a week-long eruption in 1538. This eruption was a relatively mild event and formed the 130 m (430 feet) high Monte Nuovo ("New Mount"), a scoria cone located in the northwest corner of Pozzuoli Bay. These past eruptions have had long periods of repose in between, of the order of several thousand years, but this is scant comfort given that the next eruption could be another violent, caldera-forming event. The area is therefore monitored carefully and constantly.

Fumaroles at La Solfatara in Campi Flegrei. (Photograph by the author.)

A volcanic crisis occurred from 1983 to 1984, when hundreds of earthquakes shook the region, severely damaging buildings in Pozzuoli. Was the volcano stirring before waking up? It seemed likely, specially given that the caldera floor rose by nearly 2 m (6 feet) in 18 months. The bulge, centered on Pozzuoli, could be an indication of rising magma. Campi Flegrei's magma chamber is thought to be located right under the town, at a depth of a mere 4 km (2½ miles) or so. The crisis prompted the evacuation of 40,000 people, but no eruption ensued and no further sign of impending activity has been felt since December of 1984. The big question is, for how long will Campi Flegrei be back at rest – and will it give any reliable warning of its awakening? The ground here has gone up and down several times since it was first recorded in 1819 and yet no eruption has happened. This phenomenon (called bradyseism – literally "slow earthquake") is not necessarily an indication of magma movement. However, the periods of faster ground movement that punctuate the general bradyseismic fluctuations are the true causes for concern. Even then, an eruption might not happen for many decades or centuries. What everyone can be sure of is that Pozzuoli is precariously perched on top of one of the world's most dangerous volcanoes.

Pozzuoli and the nearby La Solfatara can be visited easily from Naples. Pozzuoli offers two major attractions. The Great Amphitheater, one of the first of such structures in Italy, is a fine example of the level of sophistication of Roman architecture. The other site has more of a volcanological interest. It is the Serapeum or Temple of Serapis, misnamed because it was in fact a market place. The three surviving columns have recorded the rise and fall of the ground level, acting as a kind of accidental monitoring device for the bradyseismic activity. Between 3.5 and 5.5 m (11 feet and 18 feet) above the base of each column there are pockmarks – unmistakable signs of boring by marine bivalves. This indicates a fluctuation in ground (and therefore sea) level. A close study of the columns has shown that over the last 2,000 years, the columns have moved up and down through about 12 m (40 feet)!

About 2 km (1¼ miles) from Pozzuoli is *La Solfatara*, a 4,000-year-old volcanic crater about 500 m (1,650 feet) across, with low walls and a flat floor dotted with fumaroles and pools of boiling mud. It is given an eerie aspect by the constantly billowing clouds of steam and gases, and by the acrid sulfurous smell so characteristic of active volcanoes. The temperatures of the fumaroles have at times exceeded 200 °C (390 °F). The Romans called the crater Forum Vulcani and used it as a natural steamroom. In the northeastern part of the crater are the remains of a Roman steamroom known as the Grotta del Cane (Grotto of the Dog). La Solfatara's eastern section is particularly interesting, with numerous fumaroles near the pumiceous tuff wall. A trachyte vein, strongly altered by the fumarolic hot gases, is visible on the wall near the Bocca Grande ("big vent"). Still standing on the crater floor is a small abandoned building called the "Osservatorio Friedlaenger" or simply "the old observatory" – an incongruous sight which visitors love to photograph.

The Aeolian Islands

All of the Aeolian Islands are volcanic, and all have been extensively battered, both by fierce winds and by seas that are perhaps the roughest in the whole Mediterranean. The resulting erosion has contributed to the islands' legendary beauty, carving colorful pillars and grottoes along their coastlines. The archipelago got its name from Aeolus, the Greek god of the winds, who was thought, rather logically, to have made these islands his home. According to Homer, Odysseus landed here during his epic voyage and was welcomed by Aeolus himself. The king departed with an invaluable present from the god: a bag of wind, which would speed the long journey back to Ithaca. However, curious sailors opened the bag soon after leaving port, releasing powerful winds which blew their ship right back to the island. Homer's account may be fanciful, but what we know for sure is that Odysseus could not have been the first visitor to these islands, because they were inhabited long before the Trojan Wars started.

Of the seven major islands, the one most popular with tourists is Lipari (see section "Vulcano"), which is also the largest and the most developed and, because of its variety of amenities, makes a good base for exploring the archipelago. The most remote of the islands is Alicudi, which has largely retained the unspoiled character that the Aeolians once had, down to the donkeys that are still the major form of transportation. In contrast, tiny Panarea has become a haunt of the international jet set, with prices to match their lifestyle. Vulcano attracts tourists mostly because of its mud baths, and Strómboli because of its spectacular volcanic activity. Traveling through the Aeolian Islands is relatively easy, as there are many inter-island ferries and hydrofoils. However, services are sometimes disrupted by rough seas, a reminder that Aeolus is still a resident of these islands.

Strómboli

The volcano

Strómboli – known as the "Lighthouse of the Mediterranean" – has been almost continuously active for at least 2,500 years and it is the world's prime destination for those who want to be sure of seeing some action. The frequent explosions, every 20 minutes on

The island of Strómboli seen from the Aeolian Sea. (Photograph by the author.)

average, spray glowing lava and gas hundreds of feet into the air, creating a myriad of trails which are the subject of many a volcano photograph. Strómboli has delighted visitors for centuries and has contributed much to the development of volcanology as a science, including the naming of one of the major eruption types: Strombolian. The earliest records of activity at Strómboli go back to about 300 BC. The volcano was mentioned by Aristotle and Pliny the Elder. The description of Strómboli's eruptions by the Italian scientist Spallanzani in 1788 is considered a landmark in the understanding of the nature of volcanoes. His famous contemporary Sir William Hamilton was another keen observer of Strómboli.

Strómboli has also gained a significant place in the world of fiction. In *Journey to the Center of the Earth*, Julius Verne's characters emerge in the island after their long underground journey from the Snæfellsnes peninsula in Iceland. In *Pinnochio*, Strómboli was the name of the evil puppet-master. The volcano gained the limelight with Roberto Rossellini's 1950 movie *Strómboli*, advertised as "Raging island – raging passions." The climax of this infamous example of the *cinéma-vérité* genre shows an unprepared Ingrid Bergman attempting to make the crater ascent alone. She fails but finds her true self in the process. The movie made Strómboli world famous at the time and people who normally had no interest in volcanoes started coming to visit the island.

Strómboli's major attraction is its activity – the nearly continuous moderate explosions that are almost always safe to see from up close. We can thank the volcano's internal works for the longest-running volcano show on the planet. Strómboli's large, shallow magma chamber is probably fed continuously from a source in the mantle. The activity, however, was not so mild in the past. Strómboli is in fact a stratovolcano, built by violent explosive eruptions as well as by lava flows.

The volcano spent most of its life under water, rising from the floor of the Tyrrhenian sea to build, eruption after eruption, the island we see today. Strómboli is small, only 12.2 km^2 (4¾ square miles) in area and 924 m (3,050 feet) above sea level, and roughly symmetrical about a southwest–northeast trending axis. Most of the volcano's major vents, dikes, and fissures are located along this axis, and so is the sea-rock Strombolicchio, the remnant of an ancient eruptive center. This pattern of dikes and fractures along an axis, rather than radial to the summit, is unusual for a central volcano, and is probably related to local rifting. The pattern has made the shape of the island slightly elongated and the name Strómboli somewhat unsuitable, as it is derived from the volcano's ancient name of Strongyle, meaning "the round island."

The geologic evolution of Strómboli can be divided into six major periods, starting about 200,000 years ago. The only remnant of the first period is the sea-rock Strombolicchio, located 1.5 km (about 1 mile) off the northeastern coast of Strómboli. The first two periods of construction of the main island, called Paleostromboli I and II, started about 85,000 years ago. The activity included not only Strombolian but also violent Plinian-type eruptions that formed pumice deposits and pyroclastic flows. Strómboli's third major period (which has two stages, Paleostromboli III and Vancori) occurred between 35,000 and 13,000 years ago. This period saw a transition in the composition of the magma being erupted, from the calcalkaline to the high-potassium shoshonitic type. During the fourth period, Neostromboli (13,000 to 5,000 years ago), there was a significant decrease in the silica content of the magmas and in the explosivity of the eruptions. A large number of lava flows were produced during this period. Most came from the summit crater, but a few came from vents on the northeastern and western flanks of the volcano. A major event from this period was the formation of the Sciara del Fuoco ("Trail of Fire"), the most prominent feature on Strómboli. This is a horseshoe-shaped, deep and very steep (about 40 degree) scar on the northwestern side of the volcano, which was probably formed by a huge collapse and landslide. The Sciara reaches all the way to the coast, at which point it has a width of 1,500 m (4,950 feet) and is bounded by walls about 300 m (1,000 feet) high. As the name implies, the Trail of Fire is a natural channel along which lava is carried down to the sea. The high walls confine the lava, protecting the villages from damage, and allowing people to live on the flanks of the volcano in relative safety.

The current period of activity, called Recent Strómboli, began 5,000 years ago. Its main products

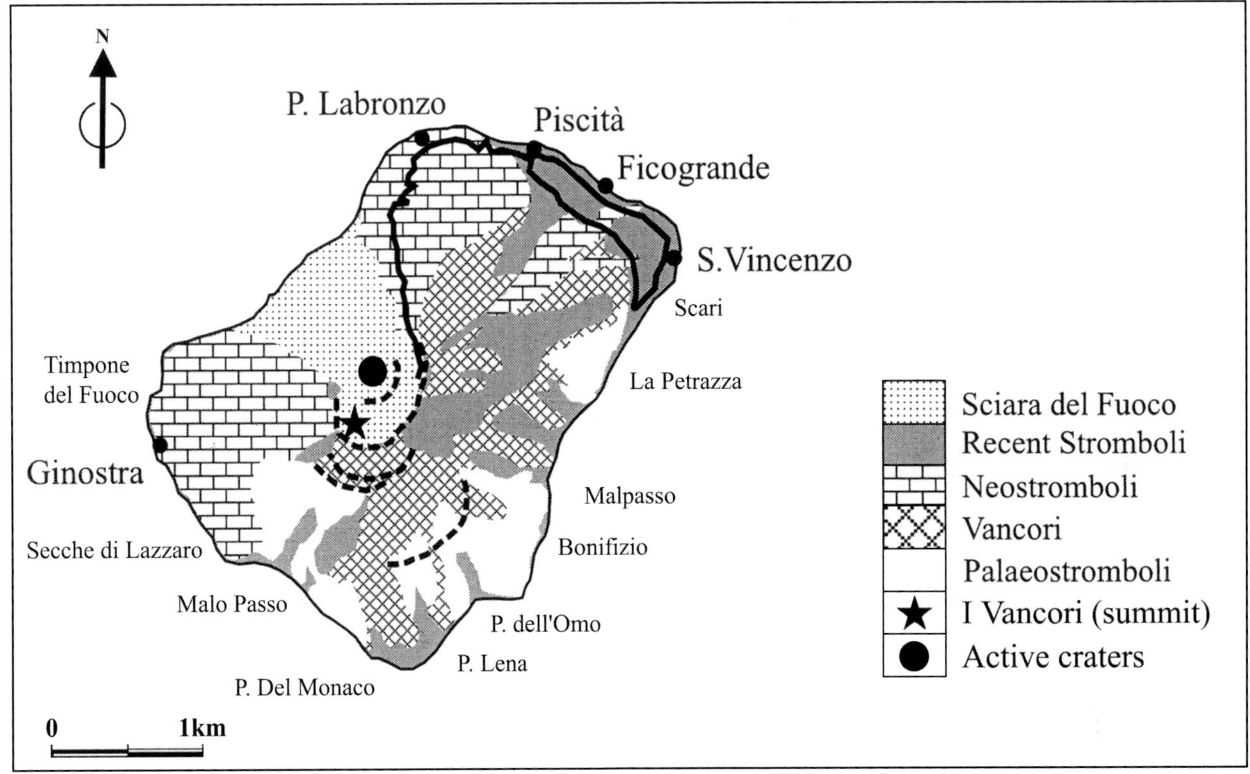

Simplified geological map of Strómboli, showing the four main volcanological units. Dashed lines show rims of collapsed craters. The zone of active craters (black circle) lies just to the northwest of the viewpoint Pizzo sopra la Fossa, located between the two innermost collapsed rims. Strómboli's summit is named I Vancori. The ascent route (black line) skirts the Sciara del Fuoco. (Modified from Kilburn and McGuire, 2001.)

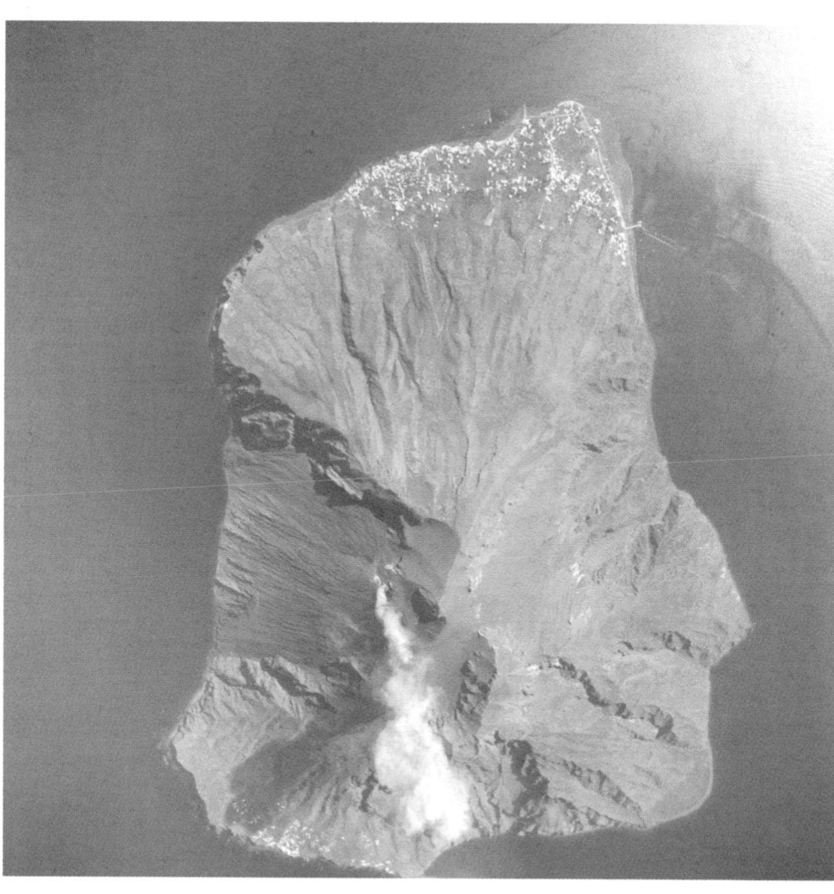

Aerial view of Strómboli taken in July 1986. The Sciara del Fuoco is to the left of the summit and appears darker than the rest of the island. (Image courtesy of Elsa Abbott, JPL.)

are flows made up of shoshonitic lavas and scoria, mostly accumulated at the summit and within the Sciara del Fuoco. During the prehistoric part of this period, Strómboli's activity was probably very much the same as in present times, characterized by relatively mild explosions, occasional lava flows and, about once every 5 years, a period of stronger explosions. The activity has rarely been a threat to the island's small population (about 400), who live mostly in two communities, San Bartolo and San Vicenzo. The two villages are located on the northern side and are often treated as one and simply called Strómboli. Isolated on the southwestern side of the island is the tiny village of Ginostra, which has only about 30 residents. All these communities are safe during the volcano's normal activity, as it does not reach beyond the summit except for the lava flows that run down the Sciara del Fuoco. However, Strómboli's occasional violent events have rained down ash on the villages and, more rarely, large bombs and blocks. Strong explosive activity during 1912 caused ashfalls to reach Sicily and Calabria and blocks to rain down on Strómboli village. Fire-fountains, which are rare on Strómboli, reached 700 m (2,000 feet) in height during this eruption. A few years later, on May 22, 1919, a major explosion again rained ash as far as Sicily and blocks on Strómboli village and on Ginostra. Some of the blocks weighed as much as 10 tonnes. The results of this eruption were tragic: four people were killed, several more were injured, and several houses were destroyed. In addition, the eruption triggered a tsunami that destroyed boats and crops near the shore. Strómboli, however, had yet to demonstrate its full power: on September 11, 1930, it unleashed its most disastrous eruption on record, killing six people and injuring 22 others. Luckily, the volcano has not repeated its 1930 performance, though strong explosions, ashfalls on the villages and even one fatality have occurred in the last few years.

The 1930 eruption

The most violent of Strómboli's known eruptions started almost without warning. Before the morning of September 11, the volcano's activity had been at its normal level for months and there were no impending signs of disaster. At 8.10 a.m., a series of powerful ash emissions caused some light ashfall in the southwestern part of the island, but this phase lasted only about 10 minutes. The volcano seemed to return to normal until 9.52 a.m., when two violent explosions blasted large blocks, made up of old rocks from the summit region, over the northern and southwestern parts of the island. Some of the blocks were over 10 m^3 (350 cubic feet), in volume and they caused considerable damage, destroying several houses in Ginostra and part of the Semaforo ("Lighthouse") Labronzo. These two explosions were accompanied by earthquakes that were felt as far away as Lipari and Calabria. The sea around Strómboli sank by 1 m (3 feet), the effect of which reached Lipari, where boats were swept up onto the shore. It is thought that these first two explosions, which were phreatic in nature, were triggered by the sudden removal of magma from the pipes feeding the main craters. It is not known where the lava was removed to, possibly it was injected into the volcano's walls by dikes or sills or it might have escaped onto the sea floor. The emptied conduits then allowed groundwater to flow in and come into contact with the hot walls, resulting in the two phreatic explosions. Having "cleared its throat" of old material, the volcano started ejecting incandescent fragments and bombs. The new material accumulated rapidly on the steep slopes near the summit, and soon formed two avalanches of still-incandescent materials down the northeastern flank. Unfortunately, this part of the flanks is outside the topographic confinement of the Sciara del Fuoco, which normally channels Strómboli's products down to the sea. The largest avalanche traveled down the Vallonazzo valley and entered the sea just north of the village of San Bartolo. Three people were killed by the avalanche and a fourth person by the boiling water resulting from the hot flow entering the sea. Meanwhile, the main villages suffered from heavy ashfall and one person was killed at Strómboli village by falling blocks. Further damage plus another death were caused by a tsunami with waves 2.5 m (8 feet) high, which reached about 300 m (1,000 feet) inland from the beach of Punta Lena.

The eruption was disastrous to the island, but it was over quickly: the activity subsided after about 10:40 a.m. The final stage of the eruption consisted mostly of lava flows running down the Sciara del Fuoco, and lava emission stopped sometime during the night. In all, Strómboli's uncharacteristic outburst lasted only about 15 hours. However, its effects on the island's life can still be felt. Before 1930, the population bordered on 5,000; it has decreased by an order of magnitude since then. Although economic reasons have greatly contributed to the exodus, the violent eruption is thought to have been its major trigger.

From 1930 to the late 1970s, there were some strong explosions and intermittent lava flows, as well as some quiet periods that sometimes lasted almost a whole year. The historical record of activity is poor around the time of the World War II and during the postwar period. There are scant references in the literature to Strómboli's activity during the 1960s and the records are also incomplete for the 1970s. However, better records have been kept since the onset of a major period of activity in 1985, and Strómboli has been monitored well since that time.

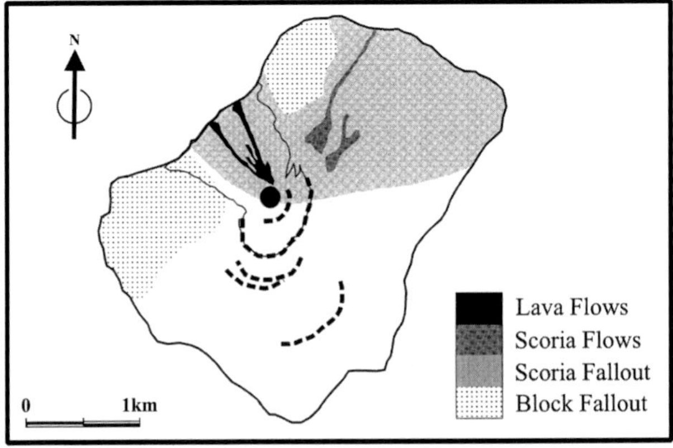

Distribution of material from eruption of September 11, 1930. (Modified from Kilburn and McGuire (2001), geology from Rittmann (1931).)

A personal view: Visiting an unspoiled island

I find Strómboli to be by far the most charming of the Aeolian Islands and sincerely hope that the ever-increasing amount of tourism will not spoil its character. It is delightful to visit a place where the roads are so narrow that the only vehicles on them are motorcycles and the ubiquitous Piaggios, the Italian three-wheeled mini-trucks that seem capable of transporting everything from vegetables to furniture. Other cars stay parked by the port, since the small street there is the only one that can accommodate them. The way things work on Strómboli is simple. You arrive by ferry and walk to your hotel. If you have told them in advance which ferry you will be on, a Piaggio will meet you to transport your luggage. It does not take longer than 10 minutes to walk anywhere in the village so, although a few Piaggio-taxis exist, most people do not need them. The village itself is a pleasure. There are no high-rise buildings, no garish hotels, no traffic lights. The houses have names (such as "the cat's house"), often written on beautiful local tilework. The shops are very casual, some displaying maps and books alongside groceries and wine. The pharmacy, located on the San

Vicenzo church square, sells its own brand of locally made products, which include sea-horse shampoo. Guides to the volcano hang out in their own "shop" but, when business is slow, the establishment doubles as a barber's. The island is peaceful, practically crime-free, and loved by artists. Surprisingly, not many people live here.

Although Strómboli poses less danger to the local people than do Vesuvius, Vulcano, or Campi Flegrei, it is here that the population has dwindled. For a while, there was so much emigration that many of the island's houses had been abandoned, and some visitors described the place as having a certain melancholic feeling to it. This has begun to change in the last few years, because Strómboli has become a fashionable destination. It is not clear yet how this will affect the island's fortunes in the long run, as fashionable is often a short-lived state.

It is true that the island still has its problems, one of them being the total lack of water. Rainwater is still collected on rooftops but now it is also delivered from the mainland, several times a week during the summer. Living with a persistently active volcano is not easy for most people, though the locals are rarely worried by the volcano's explosions, which they call "scoppi" (Italian for "bursts"). These inconveniences were, however, endured by many generations who lived in much worse economic conditions than are found today. The poverty during the 1950s, so well portrayed by Rossellini in his movie, can also be glimpsed from the words of volcanologist Fred Bullard, who visited Strómboli in 1952. His description of the ascent to the summit does not appear to be different from the one visitors make today, except that his local guide had no shoes! Bullard described the island as having "no automobiles, no electricity, no radios, no donkeys, in fact, not even a dog." Yet it had double the population of today.

The violent explosion of 1930 is usually blamed for triggering the exodus from the island, but the reason why this exodus has not completely ceased is undoubtedly economic. In Rossellini's movie, a young housewife dreamt of emigrating to America; in reality, many have done so. Fear of the volcano no longer drives people away and the constant monitoring gives some reassurance that a large eruption will not happen without some warning. It will be interesting to track the island's fortunes in the near future. Maybe Strómboli's newly acquired fashionable state will bring more people to live on the island, cause more hotels and roads to be built, and, 10 years from now, its sleepy charm to vanish.

Visiting during repose

In a few words: this is unlikely to happen! However, Strómboli does have periods of inactivity that can last from months to a few years, so you could be unlucky. The last of these quiet periods started in mid August, 1967, and ended in May 1968. It is definitely worth checking the volcano's current level of activity before setting off to the island. If Strómboli has been unusually quiet, consider postponing the trip. Should you find yourself on the island while the volcano sleeps, it is still worth hiking to the summit to view the craters and perhaps to catch the odd puff from a fumarole. It is also worth taking a boat trip around the island for a good view of the Sciara del Fuoco, and to see the dike swarms exposed by erosion both in the main island and in Strombolicchio. After that, explore the products of past eruptions and come back another time.

One of the most interesting places from which to explore Strómboli's geology, and one that most visitors never get to, is Ginostra. Named after a rampant local plant similar to broom, this tiny, near-deserted village on the island's southwestern coast has the distinction of being the world's smallest port. The narrow harbor is surrounded by lava, which forms a natural protective wall against the frequent storms. The passage is so narrow that boats must enter it one at a time. Ginostra can only be easily reached by boat, as the old mule path connecting it to San Vicenzo has become impassable. It is possible to reach it on foot by hiking around the island from San Vicenzo, but you will be knee-deep in water at times. If seas are rough, Ginostra is completely cut off, and not even the commuting priest can reach it. Apart from its odd and charming setting, Ginostra's attraction to those interested in the island's geology comes from its proximity to Timpone del Fuoco ("Drum of Fire"), a small shield volcano that forms the westernmost part of Strómboli. From Ginostra, you can walk to the 147 m (485 feet) high summit of Timpone and see the crater, only 15 m (50 feet) in diameter, from which the youngest shield lavas were erupted. The lavas are well exposed along the faulted coastline, where they outcrop as a 100 m (330 feet) high vertical wall.

Another parasitic eruptive center, slightly younger than Timpone, is Vigna Vecchia ("Old Vineyard"). It is located at about 600 m (2,000 feet) above sea level on the southwestern part of the island, and its lavas can be seen above Ginostra. Vigna Vecchia represents the last major volcanic event on Strómboli before the formation of the Sciara del Fuoco.

Visiting during activity

Hiking to the summit crater

The highlight of a visit to Strómboli is the climb to the summit. Strombóli's highest point, I Vancori (924 m, 3,050 feet), is the remnant of an older crater, but the 918 m (3,030 feet) high Pizzo sopra la Fossa ("Ledge above the Hole") is the "summit" most people climb to. Once visitors reach that point, they can expect to be rewarded with a fabulous and relatively safe volcano show, with explosions at intervals lasting from 5 minutes to under an hour, each throwing out incandescent magma fragments and hot gas. The activity is normally very moderate in character, with bombs rarely reaching over 150 m (500 feet) above the rim. Strómboli, however, sometimes departs from this normal mode and can become quite dangerous. Strong explosions and even pyroclastic flows have happened in the past.

The path from the village is fairly easy to follow but, according to the tourist office, it is forbidden to climb the volcano without a guide. This restriction is not taken very seriously by most visitors. However, it is very wise to hire a guide if you are unfamiliar with hiking on volcanoes or wish to climb after dusk. The island does not lack experienced guides and they can be contacted through the Club Alpino Italiano office in the village. The usual trip offered is to go up in groups of about a dozen people, starting off at 6 p.m. and returning at 11:30 p.m. This gives adequate time for most tourists to view the activity, but not enough for true volcanophiles and keen photographers, given that the round trip takes about 4 hours.

There are several sensible alternatives to the "tourist climb." The first is to hire your own private guide. This is expensive for one person but fine if you have a small group who wish to make a longer trip. In this case, you can opt for climbing during the late afternoon and returning as late as the guide can be persuaded to stay. Given that dusk is a prime time for good volcano photography, it is a good idea to arrive at the top with enough time to set up equipment and to get a feel for the surroundings and the type of activity. Another sensible alternative is to make the daylight climb on your own and negotiate with the guides in advance to join one of their return parties at night. If you are really keen to stay up all night observing the activity, you might want to wait until the next morning to come back down. In daylight hours it is easy to follow the route on your own. During the summer months, many people stay the night at the summit, so you should have plenty of company. Some bring sleeping bags and even tents, however; sleeping at the summit of an erupting volcano is not a sensible idea, as the activity may change and your camping area may no longer be safe (a French volcanologist recalls waking up to discover freshly erupted tephra close to his tent!). I recommend staying awake and enjoying the spectacular show.

To make the hike on your own, follow the road along the waterfront, past Ficogrande. The path is well marked and easy to follow. Once you go past the village, it heads uphill and after about 20 minutes' walk it deviates to the "observatory" at Punta Labronzo. This is a popular spot from which to see some of the activity, and is also the site of a well-known pizzeria. Above the observatory, the path becomes progressively steeper, zigzagging over rocks, some of which have stripes painted on them to guide climbers. At about 300 m (1,000 feet) above sea level, there is a good viewpoint down the steep scar of the Sciara del Fuoco – notice the recent lava flows and the scoria that run down it. On average, the ascent to Pizzo sopra la Fossa takes 3 to 4 hours. Once you reach the summit, take note of the warning signs and stay well away from the range of the falling material.

It is worth knowing that it is often cool and very windy at the top, with large quantities of fine ash blowing about. One appreciates the ash on the way back, when guides take visitors down a different route. Going down steep slopes of fine ash is one of the truly great experiences on a volcano, making you feel like an astronaut on the Moon, each step carrying you down in an almost effortless way. If you are on your own and don't know the "moonwalk" path, it is best to come down the same way you went up. Few things could be worse than ending up stuck at the bottom, facing a sheer drop into the ocean.

The "tourist path" is not the only way up Strómboli and some climbers prefer to take a more vertical ascent route. However, there have been some fatal accidents, and the route is not recommended for inexperienced climbers. The Club Alpino Italiano guides can offer advice about the condition of the route.

The summit crater area

The somewhat strenuous climb is probably amongst the most worthwhile anyone could make. As British travel writer Eric Newby succinctly put it, Strómboli is the Mediterranean's "greatest free show apart from the Pyramids and even smellier than them." It offers the sights, sounds, and smells of a truly live volcano, as well as some unpleasant effects on the senses, such as the fine ash that seems to get everywhere. Strómboli can be climbed year-round, so long as there is no fog and the visibility is good. It is best to allow a few extra days on Strómboli to ensure suitable weather for going to the summit, particularly during fall and winter.

You may notice that the summit region contains an almost circular depression, about 1 km (⅝ mile) wide. This is thought to be the remnant of an old explosion crater. Strómboli's currently active craters are not located at the summit as one might expect. They lie some 100 to 150 m (330 to 500 feet) below Pozzo sopra la Fossa, at the upper end of the Sciara del Fuoco. The Sciara is bounded by two tall scarps, Filo del Fuoco ("Crest of Fire") on the northeastern side and Fili di Baraona ("Crests of Baraona") on the southern side. From the observation point at Serra Vancura one looks down and across at the craters and, beyond them, at the Sciara del Fuoco. This remarkably located vantage point is one of the reasons why Strómboli's eruptions are truly spectacular to watch.

Since Strómboli is constantly active, some of the geographical details described here may change. At present, there are three craters located at what is called the crater terrace, a feature running northeast–southwest which is gradually growing upwards. Until the middle of this century, the crater terrace was bounded by two spine-shaped rocky promontories. These were called Filo dello Zolfo ("Thread of Sulfur," on the northeastern side) and Torrione ("Tower," on the southwest), though other names were sometimes used. These features have now been almost totally buried by the growing active cones. Torrione can still be barely distinguished on the southwestern side.

The three principal craters have been named simply Craters 1, 2, and 3, counting from the northeast to the southwest. These main craters have been continuously present since at least 1972, and probably much earlier. In 1768, Sir William Hamilton studied all available earlier reports of Strómboli's activity and concluded that the three main vents he had observed had been in existence for at least 150 years. The persistence of the three vents is one of Strómboli's puzzles, since the usual pattern for volcanic vents is to shift and change with time. Strómboli's three craters may represent deep-seated conduits along which magma is being channeled from the chamber. To make matters more complicated, there is often more than one active vent in each of these three main craters, so the smaller vents tend to be designated, for example, 1/1 and 1/2 (meaning Crater 1, first and second vents). Updated maps of the activity at the summit of Strómboli are often published in the Smithsonian Institution's S.E.A.N. Bulletin. As is often the case with continuously active volcanoes, Strómboli's activity alternates between construction and collapse. Constructive activity fills the craters with tephra and occasionally with lava, leading to growth of the cones, which are later destroyed by small collapses and explosions. Larger explosions, which are rarer, can blow away the cones, leaving behind steep-sided, sometimes elongated pits. The largest cone seen during the 10 years of well-recorded observations (1985–95) reached about 30 m (100 feet) in height and was located within Crater 1.

Needless to say, attempting to descend into the crater terrace, or into the craters themselves, is

The NW and NE active craters in a spectacular display of Strombolian activity on August 1, 2002. (Photograph courtesy of Tom Pfeiffer.)

extremely dangerous, though it has been done successfully in the past. In the first part of the twentieth century a seismologist called A. Kerner descended 245 m (805 feet) into one of the craters, earning himself the record for this type of volcanic adventure for a while. However, a more recent attempt resulted in tragedy. In June 1986, a Spanish biologist who attempted to descend into one of the craters was surprised by an eruption and killed by a bomb while attempting to flee for shelter.

Dangers near the craters

Hundreds of visitors climb Strómboli every year to see the activity from the Pizzo and the number of injuries has been very small. However, being on a live volcano always involves some risk. The greatest danger on Strómboli is being hit by bombs or blocks. The most dangerous areas on the summit have now been fenced off and warning signs put up, but these measures are only useful for the usual level of mild activity. A larger than normal eruption on October 16, 1993 threw bombs as far as 600 m (2,000 feet) away from the craters and one person was injured at the summit. It is therefore important to find out what the volcano's current level of activity is, not only prior to arriving at the island (see sources in Appendix I) but also upon arrival. Even if you are planning to make the climb on your own, ask the guides what the volcano has been doing. During most years, Strómboli has prolonged periods of increased activity, with higher frequency of explosions, which may become almost continuous fire-fountaining. While it is normally still safe to climb the volcano under those circumstances, the ejected materials can reach 300 m (1,000 feet) above the vent, rather than the more usual 50 to 100 m (160 to 330 feet), and they can land further away. The Pizzo sopra la Fossa area may be within range of the falling tephra and bombs and may no longer be a safe viewing spot. While an increased period of activity does not mean that a large explosion will suddenly occur, because Strómboli's system is basically open and there is no pressure build-up, it may mean that some areas of the summit are unsafe. It is also important to remember that fountain heights (and hence range of materials) can sometimes increase rapidly, making a quick getaway necessary.

Strómboli's stronger-than-usual explosions, which affect the summit area only, occur sporadically, on average once every 1 to 2 years. This is one of several reasons why sleeping at the summit cannot be considered safe. Although the chance of an individual being there when one of these events happens is small, it is best to be prepared. Follow the safety recommendations for Strombolian-type activity given in Chapter 4 and your own common sense. I strongly recommend adopting the "professional volcanologist look" by wearing a hard hat or helmet while in the summit area. It is true that most of the tourists don't, but it is also true that they don't know any better.

Closure of the summit

During periods of stronger activity, the volcano's summit is sometimes "closed," that is, no guides are available and it is, in theory, forbidden to go up. Although it is, in principle, a good idea to stay off the summit during these times, I recommend asking around. I visited Strómboli when the summit was off-limits, not because of the level of activity but because someone had been injured some 3 months before. The authorities could not figure out what to do, so they "closed" the volcano. Once our group got to the summit, we could see that the activity level was very mild. An Italian colleague, who knew the volcano rather well, admitted that he had never seen Strómboli in such a quiet state.

Prediction of a major explosive eruption

The greatest potential danger of visiting Strómboli's crater is that one of the volcano's occasional violent explosions could occur. If a 1930-size explosion happened today, it is likely that tens of tourists would be killed. Such an event cannot be predicted reliably, because instrumental monitoring of Strómboli's activity has been operational for only a few years. We have no historical information on whether the major explosive events in Strómboli's past had any precursor signs. However, some warning signs related to increase in gas pressure in the magma chamber (or deep in the conduit) may be recognized with the current monitoring instrumentation, which includes tiltmeters and seismometers. While this may not sound very reassuring, it is important to remember that these major events are rare, occurring on average once every 16 years. During these explosions, however, the hazards are not limited to the summit area, particularly if hot avalanches or tsunamis occur. According to a hazard study made by Italian volcanologists, the safest area on the island is just inland from Strómboli village.

Alternatives to the climb

Those who don't feel up to the climb can still see some of the activity in various ways. The easiest is to view the television screen on the village, in a tour shop by the port – a camera mounted at the summit constantly transmits scenes of the activity. (You can even see the images on the World Wide Web, thanks to the Volcanological Institute in Catania.) The next easiest way is to take one of the evening boat trips to view the Sciara del Fuoco from the sea. These trips are usually offered every evening, departing from Ficogrande. A similar trip departs from Lipari. The view is not anywhere as good as from the top, but it is still impressive, with red fiery trails spraying over the Sciara, if the activity is sufficiently strong. Another alternative is to walk part way up the volcano. Many people choose to view the activity from the Osservatorio ("Observatory") at Punta Labrozo, at the northernmost point of the island. During daylight hours the glow of lava is imperceptible, but the picture changes dramatically after dusk. Bring binoculars to see the glowing trails of falling magma. The Osservatorio is also the location of a popular pizzeria, maybe the only one in the world with an almost guaranteed view of a volcanic eruption.

Photographing the eruptions

Strómboli is one of the world's prime volcanoes to photograph, so it worth hauling equipment and plenty of film (and spare batteries) up the mountain. For advice on photographing volcanoes, see Chapter 5. Those with access to the World Wide Web should check out the Strómboli-on-line homepage (see Appendix I). It shows examples of spectacular photographs taken from Strómboli's summit and details on how they were taken. The homepage authors, J. Alean and R. Carniel, recommend setting up the tripod and cameras during daylight hours, and to view at least one eruption through the viewfinder to judge the height and direction of the erupting material. They also offer a particularly useful tip for night-time shots: aim the camera at a crater and expose for half an hour or so. This saves you having to find the cable release after the explosion has begun. If no eruption happens in the half hour, start another exposure, as waiting longer may make the background too bright. Another good tip is to position the camera on the ridge leading towards the summit, at about 800 m (2,600 feet) altitude, from where you can see material sliding down the Sciara following eruptions from Crater 1. A particularly interesting shot can be taken from this position, letting the camera expose for some time after the explosion has finished, so that the sliding material will be caught on film. Keen photographers will want to try a variety of locations and exposures. Don't forget to protect the equipment from the volcano's abrasive ash by using a lens filter (such as a skylight type) and a cloth to wrap the camera body in. Finally, remember to check the lens regularly for condensation which is usually a problem at night, due to falling air temperatures and steaming fumaroles.

Other local attractions

San Vicenzo and San Bartolo

These two parishes, each with a fine church, are the two main components of the village of Strómboli, which also includes the port of Scari. The narrow streets and the simple white houses that characterize the Aeolian style are quite charming. The only red house in the village has a plaque commemorating the fact that Ingrid Bergman and Roberto Rossellini stayed there during the filming of *Strómboli*. Most lodgings and restaurants are located on the village and on the beach at Ficogrande. Shoppers may wish to look for the local white dessert wine, the Malvasia di Strómboli. Although it is considered the best wine in the Aeolian Islands, lack of labor on Strómboli is making it so scarce that its production may soon completely stop.

Ficogrande beach

Literally called "Big Fig," this beach has the distinction of being chosen by Condé Nast *Traveler* magazine as the world's best black sand beach. However, the writers also stated that the volcanic sand feels like a bed of hot coals and is not particularly comfortable. This seems not to bother the 2,000 or so tourists who come to Strómboli every summer, as most of them spend some time here. Probably for this reason, Ficogrande has the island's sole tourist information center, open in the summer months only.

Strombolicchio

This eroded volcanic neck, about 49 m (160 feet) high and bounded by steep cliffs, is the remnant of a pre-Strómboli cone. Although it looks inaccessible, Strombolicchio is topped by a lighthouse, which can be reached by a staircase of over 200 concrete steps. On a clear day, one can see the summit of Mt. Etna from the top. It is easy to arrange for a boat trip to the rock, though most people are content to go around it, catching a good view of the large dikes interspaced with basaltic andesite lavas. It is also possible to arrange a diving trip to Strombolicchio, as the rock is a popular place for underwater fishing.

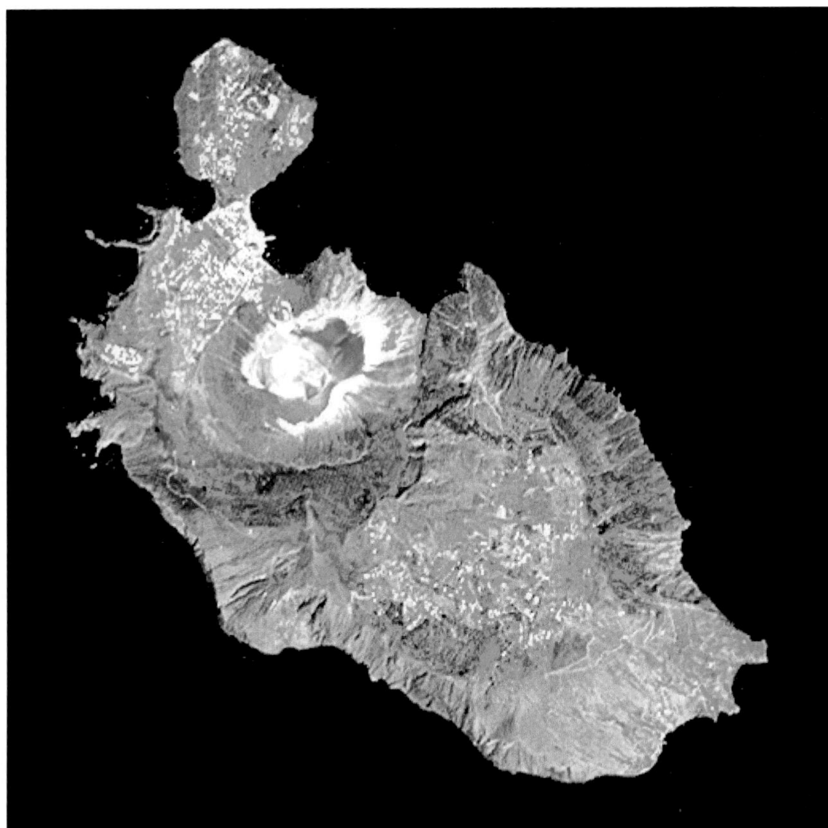

Vulcano island as seen from ASTER (Advanced Spaceborne Thermal Emission and Reflection Radiometer) on NASA's *Terra* satellite. The Fossa caldera is clearly seen as a non-vegetated gray area. (Image courtesy of Mike Abrams, JPL.)

Vulcano

This southernmost of the Aeolian Islands has given its name to all of the world's volcanoes as well as to one of the major types of volcanic eruptions: Vulcanian. During Antiquity, Vulcano was known as Iera (or Hierà), meaning the sacred island. The Greeks and Romans believed that the island was the site of the forge of Hephaestos, the god of fire who was blacksmith to Zeus. Hephaestos was known to the Romans as Vulcan, hence the island's current name. During the Middle Ages, Vulcano acquired a new reputation as the entrance to Hell. People from nearby Lipari credited their patron, San Bartolomeo, with separating their island from Vulcano, though in reality the two islands were never joined.

Vulcano was seldom inhabited, probably because its violent eruptions did not attract visitors, let alone settlers. Most accounts of the volcano's activity were made by observers from neighboring islands and from passing ships. The oldest known descriptions of Vulcano's eruptions are those by Thucydides (in 475 BC) and Aristotle (fourth century BC). In his book *Metereologica*, Aristotle wrote about a particularly violent eruption, during which ashes covered the town of Lipari, 10 km (6 miles) away. This account is regarded as the oldest description of an ashfall in European literature. A few centuries later, the Roman philosopher Pliny the Elder described a "new island" emerging from the sea near Lipari. He was most likely referring to the eruption of 183 BC, which formed Vulcanello in the strait between Vulcano and Lipari.

Vulcano may be lacking in human history, but its geologic evolution has been momentous. The island is made up of four major volcanic centers, which are aligned in a north–south direction, reflecting the tectonics of the region. The composite volcano is 500 m (1,650 feet) high and covers an area of 21.2 km^2 (8¼ square miles). Vulcano is very young – less than 150,000 years old – and it was built during four main periods of activity, which correspond to the four major volcanic structures on the island: South Vulcano, Lentia, Fossa cone, and Vulcanello. During the first period, that of South Vulcano, eruptions of pasty lavas (trachybasaltic to trachyandesitic in composition) and of pyroclastics built a stratovolcano. Around 97,000 years ago the activity was interrupted by the collapse of the Caldera del Piano, a structure about 2.5 km (1½ miles) across. Later eruptions filled the caldera with pyroclastics and lavas, until the activity stopped about 50,000 years ago and Vulcano settled down for a thirty millennia rest.

Vulcanello
Fossa
Lentia
Piano
South Vulcano

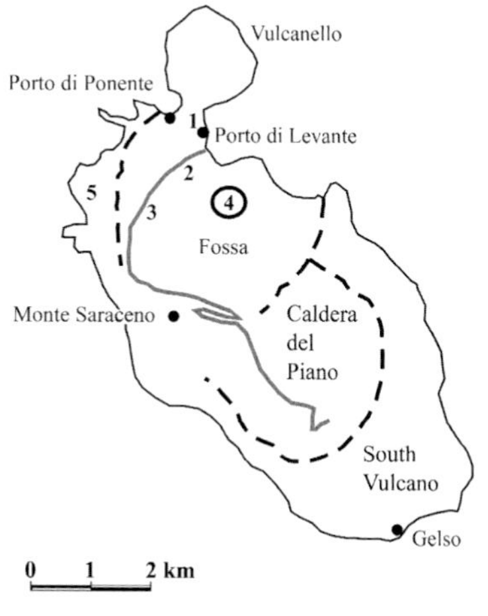

Left: Simplified geologic map of Vulcano island, showing the five major volcanological units and the outlines of calderas. Right: Major locations referred to in the text: (1) Il Faraglione, (2) access to the Pietre Cotte lava flow, (3) start of path for Fossa caldera ascent, (4) summit of the Fossa, (5) main lava domes of the Lentia group. The locations of the main towns and of Monte Saraceno are also shown. (Modified from Kilburn and McGuire (2001), geology from Ventura (1994).)

The second period, Lentia, started about 15,500 years ago with eruptions from three locations: Quadrara in the southern part of the island, Spiaggia Lunga ("Long Beach") in the western part, and from the Lentia lava dome on the northwestern coast. The dome is made up of chunky lava flows, rhyolitic to trachytic in composition. A major event during this period was an explosive eruption somewhere in the strait between Vulcano and Lipari, which deposited the brown tuff that is found over a large area of the Caldera del Piano. Between about 15,000 and 14,000 years ago, Vulcano suffered a second caldera collapse, which this time formed the Fossa ("Pit") caldera. Eruptions within the new caldera produced pyroclastics and lavas, including the Punta Roja ("Roja Point") lava flow which can be seen at the eastern base of the Fossa cone.

The third major period of Vulcano's geologic history started about 6,000 years ago with the formation of the still-active Fossa cone within the new caldera. The Fossa is now Vulcano's only active center and dominates the island's landscape. The geologic evolution of the cone is itself broken down into four major cycles of activity, named after the lava flows they produced:

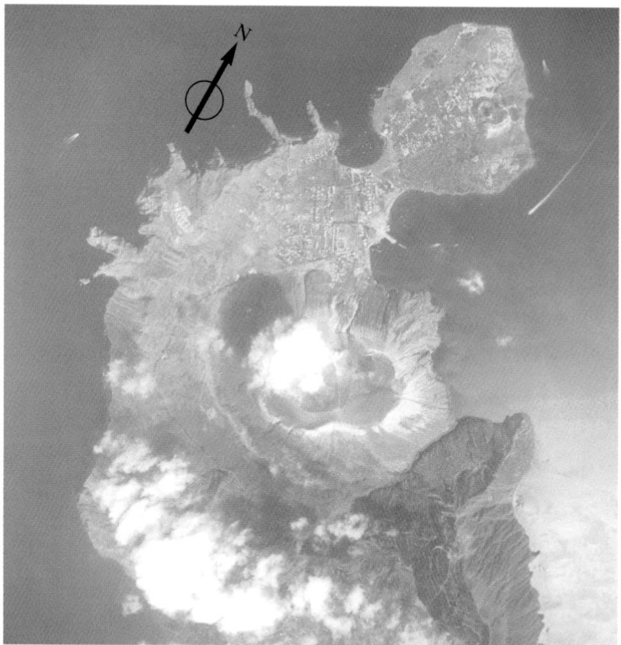

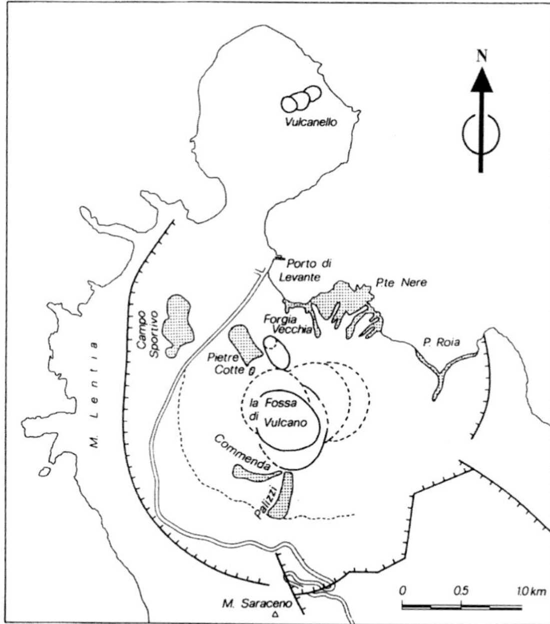

Left: Aerial view of the northern part of Vulcano showing the Fossa and Vulcanello in July 1986. (Image courtesy of Elsa Abbott, JPL.) Right: Map showing lava flows on the northern part of the island, the Vulcanello cones, and other features referred to in the text. The dashed line marks the first part of the Fossa summit path. (Modified from Da Rosa *et al.*, 1992.)

Punta Nere ("Black Point"), Palizzi, Commenda, and Pietra Cotte ("Cooked Stones"). Each of the cycles had a different eruption vent, but a similar succession of eruptions. They all started off with hydromagmatic eruptions in which hot magma and water interacted explosively. As the activity continued, the eruptions became less influenced by external water, so that the final eruptions of each cycle were essentially "dry" (for example, quiet effusions of lava flows). There were long reposes between the cycles, but not within each.

The first Fossa cycle ended about 5,500 years ago with the emplacement of the Punta Nere trachytic lava flow, which forms a delta-like feature on the northern base of the present Fossa cone. The second cycle was similar to the first and ended with the eruption of the Palizzi trachytic lava flow about 1,600 years ago. The flow can be seen as a narrow tongue on the southwest flank of the Fossa cone. There were at least three minor periods of activity between the Punta Nere and Palizzi cycles, during which a flow known today as the Campo Sportivo ("Athletic Field") lava was emplaced, as well as 1 million m^3 (35 million cubic feet) of tephra.

The third major cycle in the Fossa's history, called Commenda, began with a powerful explosion but, unlike the other cycles, it was not hydromagmatic. We know that the Commenda cycle started before the middle of the sixth century, because ash from the 6 AD eruption of Monte Pilato in Lipari covers some of the early Commenda deposits. Later on, the activity did change to hydromagmatic, and produced several surges, both wet and dry (a dry surge consists of a cloud of particles and superheated steam, while the wet type has condensing steam from subsurface water as a third component). The larger of the two craters located on the Fossa's northern flank, known as Forgia Vecchia I ("Old Forge"), was formed during this period by a phreatic explosion. The cycle ended in AD 785 with the eruption of the Commenda rhyolitic flow, which can still be seen on the southwest flank of the Fossa cone.

The last major cycle of the Fossa, called the Pietre Cotte, started in the characteristic way, with hydromagmatic activity followed by wet and dry surges. One of the major events of this cycle was the formation of the phreatic crater Forgia Vecchia II in 1727. In 1739, the stubby obsidian lava flow of Pietre Cotte spilled over the low northern rim of the Fossa cone. This should have been the end of the cycle, if it behaved like the others. However, activity started again in 1771 and continued sporadically until the major 1888–90 eruption. That eruption may indeed have marked the end of the cycle, but we cannot know for sure. Some scientists think that this last eruption may belong to a "modern cycle" which may still be in progress.

Vulcano's eruptions during the last few centuries have not happened only from the Fossa. The formation of Vulcanello in 183 BC marks the last major period in the eruptive history of Vulcano itself. The activity on Vulcanello was at first effusive, with successive lava flows building a lava platform. Explosive eruptions followed, forming the three cones seen today, which are known simply as Vulcanello I, II, and III. More lava flows were erupted, the most notable being the Punta del Roveto trachytic flow that came out of the northern flank of Vulcanello II. Vulcanello's last major eruption was an explosive event from cone III in 1550. By this time, Vulcano and Vulcanello had become connected by sand accumulation in the isthmus between them. The narrow isthmus is today the site of the island's two major ports: Porto di Levante ("Eastern Port") where the ferries and hydrofoils dock, and Porto di Ponente ("Western Port"), used only for pleasure boats. Most of the island's 400 inhabitants live in the village between the ports, known as Vulcano Porto, which is also where most of the hotels are located. Vulcanello has not erupted since 1550, and the fumaroles that were active until the end of last century can no longer be seen.

The Fossa 1888–90 eruption

The reawakening of the Fossa cone started on August 3, 1888. The initial blast ejected only blocks of old material, but later explosions produced incandescent bombs and ash. This eruption is of great importance to our understanding of the characteristics of Vulcano's eruptions, because it is the only one for which good records are available. It was observed by, among others, the Italian scientist G. Mercalli, who pioneered the classification of eruption types and introduced the term "Vulcanian eruption" (see Chapter 3). The main characteristics of Vulcanian eruptions are explosions which eject large quantities of ash, as well as bombs and solid fragments. In general, these eruptions do not produce lava flows.

At the time of this last eruption, Vulcano was not a deserted island. Since the 1870s, it had been the property of James Stevenson, a wealthy but rather dour Scotsman who was known to his own family as Croesus. Stevenson's reason for owning the island was to make use of the sulfur and alum mining operation that had been established earlier that century. Apart from supplying his chemical works in Glasgow with

Giant bombs are one of the potentially lethal products of Vulcanian eruptions. (Photograph by the author.)

Vulcano's minerals, Stevenson also wanted to experiment with harnessing power from the volcano. He set up a chemical plant beside the crater for this purpose, which his family later believed was the cause of the eruption. The locals, who included over 400 remand prisoners exiled from Lipari, also blamed Stevenson for the volcano's awakening, but for a different reason: the Scotsman, a hardline Protestant, had thrown their Catholic priest out of his chapel.

The violent eruption did not cause any deaths, but it devastated the island. Given that so much damage occurred during the first few days of the eruption, it is surprising that there were no casualties. Eyewitness accounts tell of huge explosions that broke windows 10 km (6 miles) away in Lipari, of white ash being deposited over large areas, and of meter-size bombs falling in the Porto area. A characteristic of this eruption was the large number of breadcrust bombs expelled. Breadcrust bombs are rounded or angular bombs that have a smooth, glassy crust broken by cracks. They are formed when lumps of viscous, gas-rich magma are ejected; the outer crust cools quickly as the bomb flies through the air, but the interior remains hot and continues to froth up as the gases are released. Just as in a real bread loaf, the expansion of the interior causes the crust to crack. Some fine examples of breadcrust bomb can still be seen on the rim of the Fossa. The eruption's largest bomb, ejected during a powerful explosion on March 15, 1890, was reported to measure 2.7 m × 1.8 m × 1.8 m (9 feet × 6 feet × 6 feet). It was described as a mass of lava with a 10 cm (4 inch) crust of obsidian. You can still find giant bombs on the flanks of the volcano.

The falling bombs and tephra severely damaged the buildings of Stevenson's sulfur mining operation, located at Porto Levante, and also his house, known as the Piccolo Castello ("Little Castle"). Some of the activity appears to have happened under the sea, because the submarine electric cable that connected Vulcano to Lipari was broken five times. The cable ran about 5 km (3 miles) to the east of Vulcano, where the water was about 900 m (3,000 feet) deep. Contemporary accounts say that each time the cable broke, the sea "boiled" and pumice or scoria was seen near the location of the break. It seems likely that the breaks were caused by submarine eruptions on the submerged flanks of Vulcano. The last cable break occurred nearly 18 months after activity had ceased on land.

The eruption on the island ended on March 22, 1890, after 20 months of persistent explosions. In 1891, Stevenson returned to Vulcano with his cousin John James Stevenson, a well-known architect. John's written account of the trip gives a vivid description of the devastation on the island and tells how the local residents took refuge in caves excavated into rock, which had served as their houses. They were unharmed, but some had to be dug out because bombs thrown from the crater blocked the doors.

The Stevenson cousins and their party landed on Vulcano and climbed the Fossa cone to look into its still-steaming crater, a brave thing to do as they had no way of knowing if the eruption had truly stopped. It seems that James Stevenson was heartbroken at the devastation of his island, and never came back. His cousin John returned in 1907, after James's death, to negotiate the sale of his lands. By then, most of the island was still covered by a layer of black tephra, with only broom and other hardy plants struggling through. John sold the island, at a considerable loss, to a farmer named Giovanni Conti. The farmer's son, foreseeing the island's future as a resort, built the Hotel Conti – the first to cater for visitors seeking the disputable curative powers of the volcano.

The Fossa cone's recent unrest

Since 1890, Vulcano's only activity has been fumarolic, with the exception of four small sulfur flows that occurred between 1913 and 1923, which came out from fumaroles on the external walls of the Fossa. In recent years, the fumarolic activity has been concentrated on the northern rim of the Fossa and on base of the cone, in the beach of Baia di Levanti. The fumaroles give off mostly steam and carbon dioxide, with minor amounts of sulfur and other gases. The temperature of the gases varies; typically it is about

100–200 °C (212–380 °F) but a record temperature of 615 °C (1090 °F) was measured in 1924. Variations in the temperature and composition of the gases could be a sign of impending volcanic activity, so they are monitored frequently.

In recent years, Vulcano has shown some signs of unrest, reminding us that it is still an active and potentially dangerous volcano. The first of these signs was an earthquake of magnitude 5.5 on April 15, 1978, centered about 5 km (3 miles) south of the island. The earthquake caused eight deaths in northern Sicily, though none on Vulcano. In 1985, the gas output of fumaroles increased, their chemical composition showed changes, temperatures rose from 200 to 300 °C (380 to 560 °F), and new fractures and fumaroles opened on the crater rim. In addition to these worrying signs, there were significant changes in ground level indicating that the volcano was inflating. These events caused great concern among the local community and although no eruptions followed, it was enough to strengthen the monitoring program on the volcano.

From late 1986 to early 1987, fumarole temperatures continued to rise gradually, reaching over 400 °C (750 °F). The period between March and June 1988 was marked by numerous small, shallow earthquakes and a more dramatic event: on April 20, a landslide caused a chunk of the Fossa cone to slide into the sea, about 1 km (⅝ mile) south of the populated Baia di Levante area. During the latter part of 1988, fumarole output increased again, with temperatures reaching 470 °C (875 °F). The unrest activity seemed to level off after that and in early 1990 measurements of ground levels indicated that the volcano was contracting. Vulcano seemed to be returning to normal when, in July 1994, seismic activity reached an alarming level and many wondered if the volcano was ready to blow its top again. Luckily, there was no eruption and, since then, Vulcano seems to be going back to a normal, steady state.

The likely explanation for these periods of unrest is that hot gases were injected in the volcano's shallow hydrothermal system, building up the internal pressure and making the ground inflate. Some volcanologists think that the Fossa has undergone a gradual, continuous evolution since 1978, and that the shallow source of heat underneath the fumarolic fields has been revived. However, no magma movement has been detected, and there is no reason to believe that the situation is imminently dangerous. The unrest periods have been worrying, but one must remember that interpretations of Vulcano's behavior are limited by the lack of long-term monitoring, so it is hard to put the recent events into context. Vulcano will no doubt erupt again, but we cannot say when that is likely to happen.

A personal view: A dubious spa

Visitors go to Strómboli to see a live volcano, but they seem to come to Vulcano to feel the effects of one, particularly on their nostrils and skin. It seems surprising that an island with a pervasive rotten-egg smell can attract large numbers of tourists, but Vulcano is proof that it happens. Every summer, about 10,000 people crowd the island, seeking the dubious health benefits of wallowing in warm mud and using noxious fumaroles as natural steam baths. Economically, Vulcano has thrived from the tourist business, with hotels and the inevitable condos cropping up as a result. The present-day Vulcano brings to mind a postcard a colleague sent me from a similarly crowded volcanic island, on which he wrote: "The volcano is magnificent, the resort truly awful. An eruption would fix that."

Having made this point, I must emphasize that Vulcano should be on the must-see list of every volcano lover: the scenery is spectacular, much of the island is still unspoiled, and the steaming Fossa crater is easily accessible. One is constantly reminded of being on an active volcano, thanks to the fumaroles and the beautiful sulfur deposits (not to mention the characteristic smell), but the risk is minimal. Vulcano is colorful and great to photograph, as is nearby Lipari. Geologically, Vulcano offers diverse and fascinating features such as surge deposits and highly viscous lava flows. Despite my earlier comments, I have greatly enjoyed my stays on Vulcano, though from experience

The infamous mudpool near Il Faraglione. Unsuspecting tourists soak up the "healthy" warm mud. (Photograph by the author.)

When hiking up the Fossa, look for the colorful sulfur deposits. Beautiful crystals form by the many fumaroles. (Photograph courtesy of Vince Realmuto.)

I caution against going there during the summer months, when even the Fossa crater can get crowded.

Visiting during repose

Vulcano is, without a doubt, best visited in its current state of mild solfataric activity. The volcano is monitored continuously and it is very unlikely that an eruption could start without warning. Vulcano's star attraction for all visitors, whether or not they are knowledgeable about volcanoes, is the Fossa crater, but the scenery elsewhere is also spectacular. The island is small enough to be negotiated on foot, but scooters and bicycles are available for rent and taxis can be hired. If you are traveling in a group, or can form one, you can arrange a bus tour – inquire at the Gioelli del Mare at Porto di Levante. The port, which is actually Vulcano's major urban center, has a tourist information office, but it is open only during the summer months.

Vulcano della Fossa (also known as Gran Cratere di Vulcano)

The path to the crater starts from Porto di Levante and ends at the 391 m (1,290 feet) high summit. It is an easy 1-hour hike to the rim, but those wanting to see the interesting geology on the way should allow longer. The cone consists mostly of pyroclastics with a small volume of lavas. The pyroclastic deposits were mainly emplaced as dry surges, but there are also multicolored fine-grained tuffs (red, yellow, green, and gray) formed by wet surges, particularly near the crest of the cone. The path passes just below the phreatic Forgia Vecchia craters and, further along, crosses the chunky Pietre Cotte rhyolitic obsidian lava flow. The crater rim has steaming fumaroles and colorful sulfur crystal deposits, as well as numerous breadcrust bombs ejected by the 1888–90 eruption. The scenery is superb to photograph in color – the yellows, ochers, reds, and oranges of the tephra deposits make spectacular pictures. The crater itself is a magnificent sight, about 460 m (1,500 feet) in diameter, 175 m (575 feet) deep, and shaped like a steep-sided funnel. A smaller crater can be distinguished inside the main one, formed by explosions of decreasing violence at the end of the last eruption. There is a path from the rim leading to the crater floor – the more adventurous visitors go down and walk on the flat brown crust. Layers of different-colored tephra are well exposed on the inside walls of the crater. The prominent orange-colored deposit, located about 1.7 m (5½ feet) below the rim, is from the initial phase of the 1888–90 eruption.

You can easily walk all around the crater rim and, on clear days, it is possible to see the other Aeolian Islands from here. The rim is a great vantage point from which to see and photograph Vulcanello and the remnant of the old Caldera del Piano, which is represented by a steep cliff facing the southern part of the Fossa cone. The prominent hill to the south is Monte Sarraceno, a vent formed after the collapse of the Caldera del Piano. From the Fossa rim you can walk south to the crater rim of the Palizzi and Commenda cycles, and further on past the Commenda and Palizzi lava flows. An easy path then leads back to the main road.

Hiking to the rim or to the bottom of the Gran Cratere is normally quite safe, but episodes of intense fumarolic activity could make the crater rim and floor dangerous, because of the hot steam and sulfuric gases. Earthquakes could also make the paths dangerous because of potential rockfalls from the steep cliffs.

Vulcanello

The 124 m (410 feet) high Vulcanello is certainly worth visiting. The problem these days is that so much construction is going on that the paths are getting blocked, sometimes by unfriendly barbed-wire fences. Vulcanello is where the wealthy have built vacation homes and so far the local authorities seem not to be concerned about preserving access to areas of geological interest. Maps showing paths are now out of date, so it is best to ask a local, maybe at your hotel, how you can get to the top of the Vulcanello cones.

The activity at Vulcanello produced three overlapping tephra and scoria cones, with lava flows to the northern, western, and southern sides. The three craters were formed because the pathway used by the magma shifted sideways by about 100 m (330 feet) during successive eruptions. The youngest of the craters dates from 1550 and still shows a perfectly circular rim and steep inner slopes. There is a path going down into the main crater, leading to the colorful "cave" from which alum was once extracted. Vulcanello's eastern side has been deeply eroded by the sea, which has exposed the different layers of lava and pyroclastic

Vulcanello from the ocean. (Photograph by the author.)

materials forming the cone. This natural cross-section can be photographed beautifully from a boat. In terms of composition, Vulcanello's products are mostly leucite-tephritic, but the most recent lava flow at Punta del Roveto is trachytic. This thick flow, which was highly viscous, congealed into the eerie-looking lava formations which are today known as the Valli dei Mostri ("Monsters' Valley").

Laghetto di Fanghi

This is the infamous mudpool in which most visitors to Vulcano wallow at least once. I'm not qualified to comment on the health benefits of bathing in a communal pool of warm mud, but I can say that after spending about 20 minutes here I felt tired, dehydrated, and definitely smelly. It is a good idea to follow the mud soak with a swim in the sea, particularly in the Acqua Calda beach, where hot spring waters bubble from the sand, making the water feel like a natural jacuzzi – a pleasant side effect of an active volcano. The flaky filaments that can give the seabed a milky aspect are deposits of colloidal sulfur. Visitors should know that Vulcano's beaches are (unofficially) clothing-optional – one of my British colleagues, who had chosen to keep his clothes on, had an amusing encounter with his next-door neighbors from back home, who had opted otherwise. Vulcano has been described as a setting worthy of a Fellini movie and around the mudpool it is not hard to imagine why.

Faraglione della Fabbrica and Piccolo Castello

Just north of Porto di Levante is the 65 m (215 feet) high crag known as Faraglione della Fabbrica ("Faraglione of the Factory"), an old pyroclastic cone probably formed during the earlier history of Vulcanello. The colorful material has been altered by fumarolic activity, eroded by the sea, and mined by James Stevenson. The mining operation has long ceased, but it is still possible to see some of the caves from which the materials were extracted. Nearby is a site of historical interest: Stevenson's house, known as Piccolo Castello or Castello Scozzese ("Scottish Castle") because of its peculiar mixture of Scottish baronial and Italian vernacular styles. These days the "castle" is a souvenir shop sometimes referred to as Casa Inglese ("English House"), a name which no doubt would have disgusted the Scotsman. Stevenson's most lasting contribution to the island seems to have been the wine, as he planted the first vines.

Porto di Levante fumarolic area

The fumarolic area of Porto di Levante is particularly interesting. There are mini-volcanoes, small cones with openings from which hissing steam and gases escape. During the dry season (mostly summer) the ground is covered with aluminum sulfate and other fumarolic sublimation minerals, which color the area in photogenic, variegated hues. Unfortunately, these deposits are washed away by rain. Miniature geysers can also be found here, sometimes spouting warm mud. Near the port is a hot spring called Acqua di Bagno ("bath water"), also renowned for its curative powers.

Vulcano Piano

The only road on the island leads to Vulcano Piano, where there are a few summer houses and small restaurants. The road goes past the ruins of the church of Sant' Angelo, which was shattered by an earthquake in the 1950s. It is worth taking the island's bus or a taxi to Vulcano Piano and hiking in the area to see the remains of the Piano stratocone. As the name "piano" ("level plain") implies, the top of the cone is nearly flat. The road stops near the foot of the 466 m (1,540 feet) high Sassara dei Pisani that is one of the two most prominent remnants of the caldera rim. Nearby is the other remnant, Monte Aria, the island's highest point (500 m (1,650 feet) elevation) and the site of a concentration of dikes. There are a few paths from the Vulcano Piano village which are worth exploring. One of them leads down to Faro Vecchio ("Old Lighthouse"), near the southern coast. The coastline is very rugged here and the tiny village of Gelso, on the coast, is best reached by boat.

Boat Trip

This is a good way to see geological exposures which otherwise are practically inaccessible. Excursions are easily arranged, some call in at Gelso for lunch, and most visit the Grotta del Cavallo ("Horse's Grotto"). The highlight of all boat tours is sailing down the Bocche di Vulcano ("Volcano's Mouths"), a 1 km (⅝ mile) wide passage between Vulcano and Lipari that is dotted with numerous volcanic rocks rising sharply from the sea.

Visiting during activity

Vulcano's eruptions tend to be explosive and dangerous and its effects unpleasant, to say the least. If explosive activity starts again, it is unlikely that visitors would even be allowed on the island. According to several researchers, the eruption would probably start with a phreatic blast, as magma will most likely encounter some groundwater when rising to the surface. This steam blast would fragment rock, throwing the pieces out, and would probably be followed by other explosions, as the eruption entered a phreatomagmatic phase (that is, when large amounts of water come into contact with hot magma). This phase could be quite violent and possibly generate wet and dry surges, which could be deadly for the inhabitants. Although it is difficult to predict where the vent area of a new eruption would be, the most likely locations are the Fossa crater or somewhere nearby – maybe to the north or west of the cone, the directions in which the vents have shifted during the cone's history. A new vent on the north side would be very bad news indeed for the inhabitants of Vulcano Porto, as surges are likely to devastate the area. The southern part of the island is considered safe from surges coming from the Fossa cone, but not from falling bombs.

The most sensible place from which to view such an eruption would be Lipari, though ashfall could be heavy there. Several locations on southern Lipari offer good views of Vulcano, the best probably being Quattrocchi on the southwestern side. An alternative would be to hire a boat from Lipari and view the activity from the sea. The Aeolian Islands do not as yet offer helicopter or light plane tours on a regular basis, but it is worth inquiring, as it cannot be long before this type of tourism arrives here.

If Vulcano's activity is effusive, it may be possible to view it from closer range. However, since Vulcano has not been active since tourism arrived, it is difficult to know how local authorities would react. It is likely that any renewal of activity would lead to a complete evacuation of the island, as the potential hazards are serious. Even while in repose, there is a significant hazard on Vulcano due to possible avalanches. Part of the northern rim of the Fossa cone is considered unstable, because of erosion undercutting the rock, which has been weakened by the effects of fumarolic activity. The most unstable area is that above the Forgia Vecchia craters, where numerous cracks can be seen. Blocks could become detached and fall towards Vulcano Porto and, if a major collapse of the rim occurs, the resulting avalanche could bury parts of the village.

Other local attractions

Lipari

The principal island in the Aeolian chain, Lipari is a thriving community of over 8,500 people. Apart from tourism and fishing, Liparesi make a living by exporting pumice as well as by cultivating vines to produce the famed Malvasia delle Lipari, a pleasant sweet wine. Lipari's main town is worth visiting, particularly because of the sixteenth-century Castello, a large citadel with massive stone walls built as a defense against Turkish pirates. Located inside the Castello is a cathedral, a Greek-style theater, a Roman necropolis, and one of the Mediterranean's prime archeological museums. Do not miss the annex to the main museum called "A. Rittman Sezione di vulcanologia," which includes first-class exhibits on the volcanic history of the Aeolian Islands.

The white cliffs of Lipari, where much of the world's bathroom pumice comes from. (Photograph by the author.)

Lipari has proved to be a treasure-trove of archeological findings, which showed that the first settlers on this island came from the Near East at around 3000 BC. There is evidence that the islanders established a successful trade in the volcanic glass obsidian, which was then a highly prized material for making tools. Obsidian from Lipari has been found in France, Spain, and Malta, and evidence of contact with various cultures is also seen in the island's pottery. For reasons unknown, the obsidian-exporting people vanished around 2350 BC. Their successors seem to have flourished thanks to the islands' position on the east–west trading route between the Aegean and the Tyrrhenian seas. The Liparesi fiercely and successfully defended their island against attacks by Phoenician and Etruscan pirates and, in 427 BC, against the Athenians who wanted to use it as a base during their Great Expedition against Syracuse. Lipari became an ally of Carthage and some of the most important naval battles of the First Punic War were fought in Aeolian waters. However, in 251 BC the Romans sent a fleet against Lipari, leaving the island with few survivors. Centuries later, in AD 1544, Lipari and other Aeolian Islands were dealt another nearly mortal blow: the infamous pirate Barbarossa, in a spree of killing and burning, left the islands unpopulated and the settlements in ruins. Emperor Charles V, who considered the islands' strategic position important, ordered that they be recolonized. The invasions ceased, but life was not easy on the Aeolian Islands until tourism started becoming significant in the 1970s.

Lipari is now the most frequently visited of the Aeolian Islands, mostly for its charming setting and historical connections. The island is still classed as volcanically active, though there was only one known eruption in historical times (in AD 729). The only volcanic signs at present are the hot springs at San Calogero in the western part of the island and some low temperature fumaroles. Lipari's northern side has stunning deposits of white pumice stone (called Campo Bianco, meaning "White Field"), which make up the 476 m (1,570 feet) high pumice cone called Monte Pilato, the youngest volcanic center on the island. The cone was built by numerous explosive eruptions of rhyolitic magma, and seawater probably played an important role. On the northeastern side, the cone is breached by an obsidian flow that erupted from the crater in AD 729. The flow is known as Rocche Rosse ("Red Rocks"), because of its coating of iron oxide that resulted from natural aging (weathering) of the surface. If you break a piece with a geologic hammer (using great caution because of the sharp fragments), you can expose the glassy obsidian inside. The flow runs down the side of the white cone and into the sea and is a truly striking sight. Apart from its esthetic appeal, Rocche Rosse is of special interest to geologists because it was here that the rock type liparite (the same as rhyolite) was recognized.

Monte Pilato and Rocche Rosse are easily reached from the main town and are on the itineraries of local sightseeing tours. It is worth climbing Monte Pilato for the best view of the pumice deposits and of Rocche Rosse. The summit crater is impressive, about 1 km (⅝ mile) wide, with steep walls that are 120 to 180 m (400 to 600 feet) high. A good view of the crater can be obtained from the summit of Monte Chirica, the highest point in Lipari (602 m, 1,987 feet), which is a remnant of a larger cone.

The main road around the island passes near the working pumice of quarry of Vallone Fiume Bianco, an impressive sight on the flanks of Monte Pilato. There is an abandoned quarry near the town of Acquacalda that I hope will remain accessible. You can walk right through the blindingly white expanse of pumice and into the caves dug out on the side of a pumice wall about 100 m (330 feet) high. In the not-distant past, men worked in Monte Pilato quarries such as this one in conditions so appalling that the quarries became known as "l'inferno bianco" ("the white hell").

One of the most visited places in Lipari is Quattrocchi ("Four Eyes") on the southwestern coast of the island, which can be easily reached by road. Some say that you need four eyes to take in the magnificent view of Vulcano and Vulcanello to the south, and of Lipari's Monte Guardia and Monte Giardina rhyolitic domes to the southeast.

Mount Etna

Mount Etna is one of the world's most active volcanoes as well as the largest in continental Europe. Its base extends 60 km (36 miles) from north to south and its height – a majestic 3,350 m (11,055 feet) – led navigators of the Classical Age to believe it was the highest point on Earth. The Etna region has been occupied since ancient times and Sicily's tumultuous history of invasions is reflected in the volcano's various names. It was "Gibel Uttamat" ("Mountain of Fire") to the Saracens, who occupied Sicily from the ninth to eleventh century. The volcano is still known to many locals as Mongibello, a Latin–Arabic combination meaning "mountain of mountains." The origin of the name Etna is uncertain – it may have been derived

Mount Etna viewed from the town of Randazzo on the mountain's northern side. (Photograph by the author.)

from the Phoenician word for furnace, "athana," or from the Greek and Latin verbs meaning to burn. Etna's activity has been observed for some 2,500 years and the historical records for the last 1,500 years are considered nearly complete – an invaluable reference for volcanological studies.

Etna's ancient history shows that the mountain grew as a series of broad volcanic structures that were superimposed on one another, starting about 500,000 years ago. The early eruptions, of basaltic (tholeiitic) lavas, took place under water within a vast gulf. Eruptions filling in the gulf, together with the gradual uplifting of this part of Sicily, eventually allowed the lavas to build volcanoes on land. These ancient volcanoes (called Calanna, Trifoglietto I, and Monte Po) were formed between 168,000 and 100,000 years ago and were relatively small. The next phase in Etna's history was the formation of a stratocone, the Trifoglietto, by both explosions and lava flows, between 80,000 and 64,000 years ago. This period was also marked by the formation of the Valle del Bove ("Valley of the Cattle"), a striking keyhole-shaped depression on the eastern flank, some 3.5 km (2¼ miles) across. Exactly how the valley formed is still not known – some say it is the result of several caldera collapses which eventually formed a single massive depression, others propose a giant collapse and landsliding of a sector of the volcano, with the material sliding into the sea. It is known that numerous other caldera collapses occurred during the Trifoglietto period, and little remains of this ancient volcano. The present period in Etna's history, called the Mongibello, started 34,000 years ago. It began with explosive eruptions causing more caldera collapse, similar to the previous period. However, during the last 8,000 years Etna seems to have become less violent, and much more visitor-friendly. The eruptions have been mostly effusive and often very spectacular, producing vast rivers of basaltic hawaiite lavas that pour down the mountain's flanks.

A quick look at Etna shows that it is a stratovolcano with steep upper slopes. The steep top part is the 260 m (860 feet) high summit cone, made up of layers of lavas and tephra. At the top of the cone is Central Crater, which is about 500 m (1,650 feet) across. This is not one but two coalescing craters, called the Chasm (or La Voragine) and the Bocca Nuova ("New

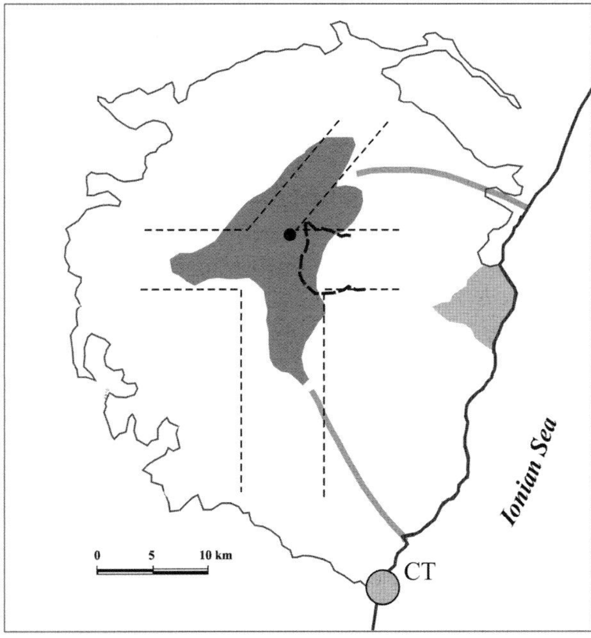

Simplified geological map (top) and structural map (bottom) of Mount Etna. The geological map shows lavas erupted since the sixteenth century (gray) and, in darker gray, the 1669 lava flow towards Catania and the 1991–3 flow towards Zafferana. The outline of the Valle del Bove depression is marked by a dashed line on the eastern flank. The structural map (bottom) shows that Etna has formed at the intersection of regional faults. The regional trends (indicated by the dashed lines) are apparent from the concentration of vents on the volcano: the dark shading shows areas where the concentration of vents is at least one per square kilometer. The pale shading on the eastern side marks the Chiancone formation, consisting of material that came from the Valle del Bove. The area on the eastern flank between the gray lines is slowly slipping into the sea. (Modified from Kilburn and McGuire, 2001.)

Mouth"). The summit cone has two other prominent craters, known simply as the Northeast Crater and the Southeast Crater. The frequent eruptions of Etna are reflected in the complex and constantly changing geography of its summit region – and the fact that mapmakers cannot keep up with the volcano. The most recent eruption (at the time of going to press) lasted from October 2002 until January 2003 and took place from both the northern and southern sides of the volcano, causing considerable changes to the landscape.

Etna's lower flanks are broken up all over by numerous small, scattered cones. These mark the volcano's frequent flank eruptions, in which the magma comes out of vents on the sides of the mountain rather than at the top. Many glowing, gushing torrents of lava have been erupted from these flank vents, often threatening the many towns and villages that otherwise profit from the volcano's fertile soils. It is not surprising that the first known attempt to divert the course of a lava flow happened on Etna. In 1669, an enormous lava flow erupted on the southern flank – the lava would eventually cover over 37 km^2 (14 square miles) of land. It moved inexorably towards the city of Catania, prompting its residents to make an attempt to change its course. Led by Diego Pappalardo, some 50 men attacked the side of a lava stream with picks and axes, using water-soaked hides to protect their bodies from the heat. They succeeded in breaching the chilled margin on the side of the flow, and lava poured out, now flowing away from Catania. Unfortunately, the lava flowed towards another city, Paternò. Some 500 armed Paternesi persuaded the group from Catania to abandon their efforts. The breach on the flow margin quickly healed, and the flow resumed its course, eventually destroying a large part of Catania.

Large, sluggish aa-type lava flows from flank eruptions that occur on average every few years have been the major destructive force in Etna's recent history, but they have rarely killed people. The relatively few deaths that have happened on Etna were caused by unexpected Strombolian or phreatic explosions, mostly from the summit craters. Summit explosions are not usually violent, but they can happen anytime without warning signs. Etna's summit cone is built directly over the volcano's central pipe, so it is the site of almost non-stop activity. At least one of the four major summit craters seems to be open to the feeding pipe at all times. This does not mean that there is constant activity – it is common to have many quiet months in a row – but there is always the potential for activity.

Even during quiet times the summit area is dangerous, because of the frequent ground collapses that continue to enlarge the craters. It is sobering to know that the Bocca Nuova started its life in 1978 as an 8 m (26 feet) diameter hole, but 15 years later it had become a crater about 300 m (1,000 feet) across. Summit collapses can also make the craters deeper – this tends to occur when flank eruptions are taking place, because they drain the magma column in the central pipe. During some flank eruptions, the depths of the Chasm and the Bocca Nuova have reached about 1 km (⅝ mile).

Prior to a fatal summit explosion in September 1979, tourists were routinely able to stand on the edge of the Chasm and the Bocca Nuova peering into the steamy clouds to see their near-vertical, apparently bottomless walls. In July of that same year, a small lava lake, about 80 m (265 feet) in diameter, was active at the bottom of the Chasm, which was then about 150 m (500 feet) deep. Tourists and volcanologists marveled at the rare view of Strombolian activity – the lake bottom rose every few minutes as one huge bubble. When it burst, glowing fragments were sprayed high into the air, but not so high as to threaten the safety of spectators at the top. Loud bangs filled the air, thrilling hundreds of visitors. I was on my first volcano field trip at that time and was delighted that Etna had put on such an unforgettable show. The Southeast Crater also erupted that same summer, its first explosion catching some members of our team by surprise as they collected gas samples inside the crater. Luckily, they weren't hurt, and we were able to witness spectacular activity later that evening, when nearly continuous Strombolian explosions ejected glowing fragments more than 100 m (330 feet) into the air. Late in that 1979 summer, Etna seemed to go totally quiet. None of us expected that the volcano would soon take the lives of nine of the people who had come to simply view its craters, like thousands had done that same summer without harm.

The 1981 and 1983 eruptions

It is interesting to contrast these two eruptions rather than highlight a single one because they are good examples of the different types of eruption that are common on Etna. Both eruptions produced long lava flows that threatened towns, but the 1981 event was over in a few days, while the 1983 lava flowed for months. Although we don't exactly know why, Etna's lava flows seem to either come out in a quick-and-furious manner over a short period of time (hours to days), or dribble out slowly over months or years, with few eruptions in between. Part of the work I did on Etna was to recognize the existence of these two major types of lava eruptions, and the fact that they produce flows of different shapes. The flows that come out fast are narrow relative to their lengths, while the others are wide. The reason behind this is the rate at which lava comes out of the volcano, and how it changes during the eruption. The rate tends to be high at first, producing a long narrow flow. Some eruptions stop there quite abruptly, after hours or days – this is the first major type. Other eruptions decay slowly, with the lava coming out at a much reduced rate. New flows form alongside the first one, fattening the flow-field, and a wider area is covered.

The March 17, 1981, eruption is an example of the first type. Lava broke out at 2,550 m (8,410 feet) above sea level, about two-thirds of the way to the summit, and rapidly flowed down the northern side of the volcano. At first it was not clear what type of eruption this would be. Local scientists phoned my then-thesis advisor in London, John Guest, who headed the UK's Volcanic Eruption Surveillance Team. The British team decided to come out to study the eruption and we left just over a day later. At the time there were no direct flights from London to Catania and delays in internal Italian flights were common. Unknown to us, the eruption turned out to be of the fast and furious kind, and the lava stopped while our team waited for a flight at Rome airport.

Feeling extremely frustrated, we got to work on the volcano the next day, as we could at least collect fresh samples and map the extent of the flow. We learnt that this quick eruption had been the costliest in 10 years, as the 7.5 km (4⅔ miles) long flow covered about 4 km² (1½ square miles), including prime farmland, went over main roads and railway tracks, and killed hundreds of sheep and cattle. The eruption had been

Mount Etna's 1981 eruption. (Photograph by the author.)

spectacular – within 6 hours, lava came out of fissures extending 4.5 km (2¾ miles) down the volcano's northwestern flank, etching a fiery pattern complete with fire fountains reaching 200 m (660 feet) in height. Luckily, few houses were destroyed, as the flow passed in the 3.5 km (2¼ miles) wide gap between the town of Randazzo and the village of Montelaguardia.

We were soon to learn that the eruption had not yet completely stopped. John Guest and I walked up to one of the fissure vents, where a 15 m (50 feet) high new spatter and cinder cone had grown. We started climbing onto the cone when we heard a noise – a rustle, like broken pieces of glass rubbing against one another, the characteristic sound of aa lava on the move. Following the sound, we soon came across a small lava flow – maybe 10 m (30 feet) long – that was barely moving. "This thing is not dead yet," John said. As if on cue, the cone exploded, throwing glowing bombs high into the air. We looked up, taking a few steps back, and then he shouted: "Run!" For a fraction of a second, I hesitated – I knew that taking one's eyes off the falling bombs and running away was the wrong thing to do. But I reasoned with myself that my advisor had been around active volcanoes far longer than I, and this was not the time to argue with him. When I quizzed him later about his strategy, he explained that the bombs had not come very far above the cone, and he judged that we could get out of range before they landed. Under these circumstances, it made sense to run. I must stress, however, that it takes a lot of experience to be able to judge correctly the range of bombs. If in doubt, I would follow the standard practice and stay put, sidestepping if one came my way.

Once the eruption was truly over, we calculated that the average rate at which lava was discharged was 75 m^3 (2,650 cubic feet) per second, that is, it could have filled 100 Olympic-size swimming pools in 1 hour. In contrast, the next time Etna unleashed a lava flow, in 1983, the eruption lasted 131 days and the lava came out at only about 9 m^3 (315 cubic feet) per second. It is interesting that both eruptions produced flows of nearly equal length and area, though more volume came out in 1983, reflected in the thickness of the final lava field.

The 1983 eruption also started during the month of March, but this time the lava came out on the southern flank, initially at an altitude of 2,300 m (7,600 feet). Soon the lava destroyed tourist facilities and homes, and severely damaged the Rifugio Sapienza – our team's usual lodgings and the base station for the aerial cable lift to Etna's popular southern side skiing area. It became apparent that the eruption would be long-lived and that the economical damage could be vast, particularly if the lava reached the town of Nicolosi and other towns downslope. Civil authorities decided that something had to be done to prevent the disaster and the only solution was to attempt a lava diversion – the first since 1669.

Lava was flowing in a complex system of channels and tubes but most of it headed to the southeast towards the Sapienza tourist complex. Construction of a barrier on the eastern side of the flow began on May 1 and, at the same time, work started on an experiment to breach the flow higher up, using explosives. The idea was to make more of the lava head towards the west, to lessen the threat to the tourist complex. The operation involved about 200 men, including the two blast engineers who devised the plan. Bulldozers were used to take away part of the flow margin, to make drilling the holes for explosives easier. The initial plan was to drill three rows of 20 holes each on the now-thinned flow margin and pack each hole with dynamite. However, the temperature of the lava wall turned out to be much higher than 200 °C (380 °F), which is considered the maximum that is safe for Gel-A dynamite. The engineers' solution was to cool the holes with water and pack them with dry ice. This did not work, as the temperature was not lowered enough to be safe, but the procedure did chill the flow margin, making the lava channel narrower. This, in turn, caused overflows – which buried many of the lowermost drill holes. After much frustration, the engineers eventually solved the problem by using pneumatic hoses to inject all the charges into the flow, sending them through steel pipes that stood out from the lava margin. The charges were all injected at the same time and the workers had a scant 30 seconds to get clear.

The explosion blasted the lava wall on May 14, but the resulting breach – and its effect on the lava channel – were smaller than anticipated. In less than 2 days, the breach had healed. However, the experiment worked in an unexpected way, as debris thrown out by the explosion was blown into the flow, creating a major blockage in the channel some 500 m (1,650 feet) downstream. The blockage achieved what the flow breach had not, as it succeeded in diverting lava to the southwest. Within a few days, about 65% of the lava was being rerouted in the direction of a newly built barrier near the Monte Vetore cone. Another barrier was built near the Rifugio Sapienza – it held up until the end of the eruption on August 6, saving the remaining tourist facilities. Although the lava diversion did not work quite as expected, it was considered very successful. At a cost of about 3 million dollars, the effort is estimated to have prevented at least 5 million dollars' worth of damages, and possibly much more.

Lava diversions can be attempted on the type of flow the 1983 eruption produced, and thus hazard management is viable for those long-lasting, relatively sluggish eruptions. Although the volume of lava produced by the 1983 eruption was about three times larger than that of the 1981 event, it is clear that the 1981 type of eruption can be far more damaging, as there is usually no time to do anything but evacuate the areas at risk. Lava diversion was attempted again on Etna in 1992, having some more contestable benefits this time, and again in 2002 to save some facilities near the Sapienza tourist complex. However, even if the flow cannot be stopped, it is clear that we learn a lot from each attempt and this alone can make the cost worthwhile in the long run.

A personal view: My first volcano

Etna is special to me because it was the first volcano I visited as a volcanologist-in-training. Its frequent eruptions and unexpected explosions taught me valuable lessons about safety on volcanoes and, more importantly, gave me a perspective on both the wonders and the tragedies of volcanic eruptions. My first field season, in the summer of 1979, was truly my best and worst. As part of a team of about a dozen volcanologists from Britain, I helped with surveying and sampling, and quickly got to know the volcano. My own project was to study the shapes of lava flows to find out how they varied depending on the different ways the lavas had flowed: fast or slow, over flat ground or steep slopes, and so on. Eventually I would be able to use these studies to estimate how lavas on the surface of Mars had flowed many millions of years ago. My project took me all over Etna, as I had to make measurements of many of its older flows. The work was interrupted in the best possible way by an eruption from the Southeast Crater in July. We carried out a study of this Strombolian eruption, measuring the size and range of the bombs and the frequency of explosions. I learnt how to approach the erupting crater, ready to dodge the incandescent fragments, and to be ready for unexpected events. It was invaluable training, but the real lesson that Etna would teach me happened later that summer.

By early September, the volcano was quiet and our more mundane work proceeded smoothly. Our team, by then of only five people, was due to leave on the 14th of the month, and we had left our "leveling traverse" to do last. This consisted of surveying along a line crossing the top part of the volcano – the precise measurements would tell us about any inflation of the ground. The traverse was slow and tedious work, requiring nearly a week to complete. When we finished for the day on September 11, we drove up to the summit to view the steaming craters. There had been internal collapses in the Bocca Nuova, and now and then the crater would spew out billowing clouds of dust. Our brief visit was uneventful – we had been there, luckily, almost exactly 24 hours too early.

In the morning of September 12, we picked up the traverse again, this time starting out on the road just below the Central Crater, some 400 m (1,320 feet) away from the Bocca Nuova. Volcanologist John Murray, our team's most expert surveyor, set up the instrument and tried to obtain the first reading. After a few minutes, he gave up. Although we couldn't feel it, the ground was vibrating up and down like a boat on choppy seas – the motion was apparent when we looked through the surveying telescope at the measuring rod located a few meters away. Ground tremor is a common, but by no means certain, precursor to volcanic activity. Our team was always hoping for an eruption to happen, and John Murray half-jokingly predicted that one would start very soon. We did not have any means to make a scientific prediction at the time, or to know what the almost imperceptible ground tremor signified. We carried on with our work, starting our measurements further away from the crater, where the ground appeared to be stationary.

At 5:47 p.m. that day our group was about 2 km (1¼ miles) away from the Central Crater when suddenly we saw a black ash cloud rising rapidly over the Bocca Nuova. We did not hear an explosion, but knew immediately that one had taken place. We quickly packed up and drove towards the summit, soon encountering one of the tourist Land Rovers coming down. The driver, one of the mountain guides, stopped to tell us that a violent explosion had happened and many people were hurt or dead. His Land Rover, full of panicked tourists, had a gaping hole in the roof where it had been hit by a bomb. About 150 tourists had been near the crater at the time of the explosion and many needed help getting down the mountain. Our team leader decided to drive up alone, so he could ferry the remaining tourists down. I remember standing by the roadside, trying to figure out what had happened. The billowing black cloud continued to rise, by now reaching hundreds of meters above the crater. Frightened

The tragic explosion of September 1979: plume above the crater. (Photograph courtesy of Christopher Kilburn.)

tourists were running or walking down to the safety of the lower slopes. Some stopped to tell us about a sudden explosion that had showered the whole summit area with large bombs.

When we finally returned to our lodgings in the Rifugio Sapienza, we found the place full of police, army personnel, and reporters. Several of the tourists who had been at the summit were also there. Some had lost loved ones – parents, a spouse, friends – and their pain made this freak explosion a personal tragedy, not just another scientific event to study. In one evening, Etna taught me that the work of a volcanologist is not all science and adventure. Our science had failed these people, because we still know so little about how volcanoes work.

The explosion was a one-off, unexpected event, and not even a true eruption – there was no new magma involved. The bombs had, in fact, been old blocks of rock, and most were less than 25 cm (10 inches) across, but they were ejected at speeds up to 50 m (165 feet) per second and were deadly to those in their path – up to some 400 m (¼ mile) away from the crater. The big question was, and still is, what caused the explosion. We think that the Bocca Nuova had become blocked because of repeated collapses in the previous couple of weeks. The explosion was probably triggered by a build-up of gas or steam, either from the magma underneath the blocked vent or from water mixed in with the collapsed material. When the build-up reached a critical state, the Bocca Nuova simply coughed out the obstructing material. It was a minor volcanic event, but a tragic one. The ground tremor we detected may have been associated with the gas build-up, but it is unlikely that we will ever know for sure.

The next morning was the last for our field season. Army personnel blocked the road to the summit, but they let us through to finish our work. We had to restart our leveling traverse close to the summit crater and were quite aware of the danger. We considered giving up, but that would mean not accomplishing one of the major purposes of the trip. We devised a safety plan: one person would drive our Land Rover along, staying close to the group. In the event of another explosion, we knew the interior of the vehicle would not provide adequate shelter – we had seen the gaping hole in the roof of a Land Rover the previous day. However, we thought that the chassis of the vehicle would protect us from most bombs, so our plan was to dive under it in the event of an explosion. It was not the best safety plan, but might have been adequate. Luckily, we didn't need to put it to the test.

As we drove up the mountain, we met a caravan of army trucks coming down. It was only then that I knew how many had been killed, as I counted the coffins on the backs of the trucks. With a mere hiccup and not a drop of fresh magma, Etna had claimed nine victims, more than any of its spectacular eruptions have done in recent times.

Visiting during repose

In 1987, the Parco dell'Etna was established, making much of the volcano, including the whole summit region, part of a national park. The facilities for visiting Etna have been steadily improving since then. The Club Alpino Italiano has done much to facilitate hiking on the volcano. It maintains several rifugios (mountain shelters) and publishes a geological and hiking map that is invaluable for visitors (see Appendix I). Etna is a lively ski resort in the winter months, though volcano enthusiasts will fare better

The summit of Mt. Etna. (Photograph by the author.)

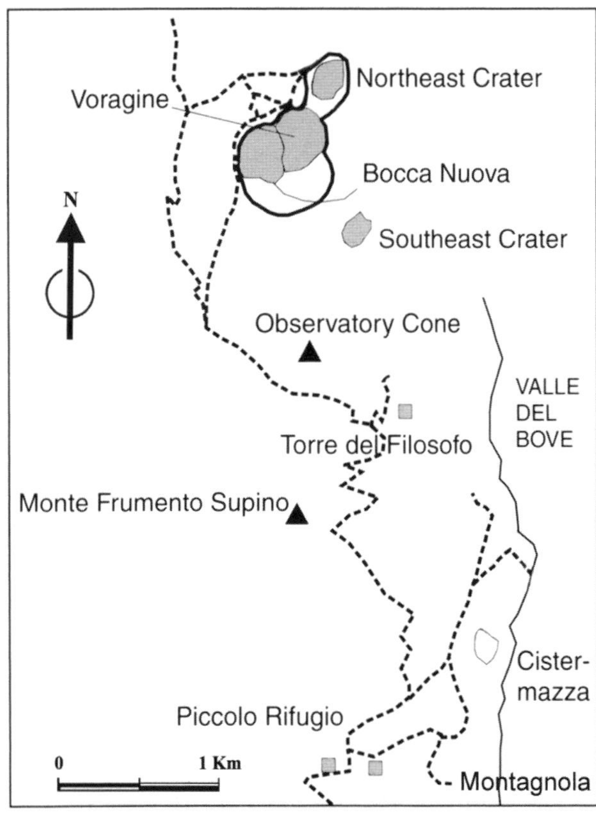

Sketch map of the summit area of Mt. Etna showing the summit craters, the main pyroclastic cones (triangles) and the wall of the Valle del Bove. Buildings or shelters are shown as squares. Tourists are usually taken to Piccolo Rifugio (top of the cable-car station) and, depending on conditions, to the Torre del Filosofo or summit craters. The path is shown as a dashed line. (Modified from Kilburn and McGuire, 2001.)

when snow does not obscure the lava formations and summit access is easier. Etna is rarely crowded, so anytime from May through September is a good choice. Etna erupts often, and therefore some of the information given below may be soon outdated. At the time of going to press, access to the summit was severely restricted following the 2002–3 eruption, and buildings and roads had been destroyed. I recommend checking websites about Etna, and making some phone calls, for the most recent information (see Appendix I).

Access to the Summit

There are two major routes, one from the south and another from the northern side, both closed in winter. Because of the fatalities in and after 1979, groups of tourists are no longer routinely taken up to the Central Crater. The present rules about summit access are unclear, changeable, and at times about as effective as red traffic lights in Naples. The best way to arrange to go up is to make inquiries at local tourist offices or at the Rifugio Sapienza tourist complex, the starting point for the southern route. The Rifugio Sapienza is located at 1,910 m (6,300 feet) altitude, and the tourist complex includes a cable car, shops, and restaurants. Keep in mind that the cable car and the Sapienza and other buildings have been threatened and damaged by eruptions several times, most recently in 2002, so there is no guarantee that they will be operational or even standing when you visit.

To get to the Sapienza tourist complex, follow the Strada dell'Etna from Catania or take the daily bus from the city's Stazione Centrale. The Sapienza is operated by the Sicilian Alpine Club and, when operational, is a favorite hang-out for volcanologists and Etna guides. It was severely damaged by the 2002–3 eruption, so its future is uncertain. Before then, its rooms were basic and shared, but it was a convenient place from which to explore the volcano's south side and summit.

The ascent from Etna's northern side is considered a secondary route but it is worth trying if guides operating the southern side tours are not making the trip – the policies of the two groups often differ and of course much depends on the condition of the roads. The best place from which to pick up the north side summit tour is Linguaglossa, a ski resort town of alpine character. The town's name, meaning "glossy tongue," probably refers to the lava flow of 1566. Tours are arranged at the tourist office, which houses a mini-museum of Etna's minerals, flora, and fauna. The tourist office can advise you about hiring your own four-wheel-drive vehicle and guide to go up the mountain. The northern road to the summit has particularly good views of the Northeast Rift, the sector of the volcano with the maximum density of eruptive vents.

Ascent from the southern side

The guided summit tour (assuming it is running) is an easy trip, as four-wheel-drive vehicles can take groups nearly all the way up (if conditions allow), following a private road that starts by the Sapienza. Alternatively, if the cable car is running, groups can go up to the 2,500 m (8,250 feet) level by cable car and pick up vehicles from there.

The tour's first stop is usually at the Piccolo Rifugio ("small shelter"). Located at the 2,500 m (8,250 feet) level, near the top of the cable-car station, this ruined building was once a mountain shelter. During the 1983 eruption, fractures damaged the building, and in 1985 Etna finished the job. More fractures opened up, and lava poured out from a fissure just tens of meters uphill. Construction on Etna's flanks is often short-lived; in fact, the previous cable car was destroyed by the 1971 lava. Walking on the lava, just uphill from the Rifugio, it is possible to see small lava channels, tunnels, and other interesting structures.

Before the 2002–3 eruption, the second stop used to be by the Torre del Filosofo ("Philosopher's Tower"), but the eruption covered up the building and damaged the road so when tours resume they will probably stop elsewhere. The situation before the 2002–3 eruption was that visitors had two choices from the cable-car station. The first was to go up with the other visitors in one of the four-wheel-drive buses and the second was to walk up. The buses often stopped by a hut near the then-standing Torre del Filosofo building at 2,919 m (9,632 feet). The original tower was gone long before the last eruption and replaced with an ugly but serviceable building. Local history says that the Tower was constructed to celebrate a visit by the Roman emperor Hadrian. The "philosopher" designation comes from Empedocles (494–432 BC), a Greek scholar who spent several years living near the summit, observing Etna's activity. It is said that he died either by falling into the summit crater or jumping into it, though historians refute this popular story.

A stop near the destroyed building provides a good view of the Southeast Crater and of the summit cone. Depending on the level of activity at the summit, it may be possible to see some tephra from explosions, and hear the loud rumblings. The guides are quite careful to make sure that all the visitors they take up to this point come back down again. This means that if you want to risk going all the way to the summit, it is best to walk up from the first stop, by the top cable-car station. The road was easy to follow, but it has been cut by the recent flows and it is uncertain when it will reopen. The hike, even on a good road, is a long one – allow at least 4 hours up. The old road stopped below the Central Crater and from there a steep path led up to the steaming summit craters. Given the changeable situation at the summit, it is extremely important to hire a guide or at least make enquiries locally at the time of your visit. Etna can be a dangerous and unforgiving mountain.

The Southeast Crater

Located on the side of the summit cone at an altitude of 3,050 m (10,065 feet), this impressive crater, now about 400 m (1,320 feet) in diameter, started its life during the 1971 eruption as a pit less than 100 m (330 feet) across. It became active again in 1978 and since then has erupted more often than any of Etna's other craters. Its Strombolian explosions can be dangerous: in 1987, two people were killed and ten others injured. Since 1989, the activity of this crater has been particularly impressive. This has been linked to a huge batch of magma being intruded under Etna and draining mostly into the conduit feeding the Southeast Crater.

Getting to the Southeast Crater used to be fairly easy, either by hiking up the road from the cable-car station,

The Valle del Bove, a giant depression on Etna's eastern flank, channels many of the volcano's lava flows. (Photograph by the author.)

or simply taking the tourist bus. As for the summit, make local enquiries and find out if tours are taking visitors that far. If you are able to go all the way up to the summit, you can hike down from the Central Crater to the Southeast Crater. However, it is not advisable to do this without a local guide and a gas mask, as it often involves walking through the summit crater's plume and it is dangerous terrain.

The summit

The Chasm and the Bocca Nuova are two impressive steaming craters, each over 250 m (825 feet) in diameter, separated from one another by a sharp, narrow crest. Both have near-vertical walls and, although their actual depth varies, they look like bottomless abysses most of the time. Mild Strombolian activity is often happening at the bottom of at least one of the craters, and the loud bangs coming from deep down are very impressive indeed. Billowing clouds of steam make it unpleasant to stand downwind; it is advisable to have a gas mask handy at all times in case the wind direction shifts.

In 1974, mountain guide Antonio Nicolosi descended about 130 m (430 feet) into the Bocca Nuova using a rope ladder and did not reach the bottom. Wearing a protective suit, he withstood temperatures of over 100 °C (212 °F) before being forced to abandon his daring attempt due to lack of oxygen. A look over the edge of the steaming crater makes you realize why others have not repeated Nicolosi's adventure.

The highest point on Etna is the Northeast Crater, separated from the Central Crater by a narrow saddle. It opened up as a vent in 1911 and since then has often been the site of spectacular Strombolian eruptions, though its level of activity has decreased since the Southeast Crater opened in 1978. The Northeast Crater is a short walk from the Central Crater. Depending on conditions at the time, you can climb the crater's side to gain a good view of the summit region. However, the rim is often unstable due to collapse. If you are on the summit without a guide, it is unwise to take the risk.

It is prudent to remember that unexpected explosions do happen on Etna's summit, hence it is a good idea to wear a hard hat and to plan escape routes. A year after the fatal 1979 explosion, the Bocca Nuova surprised me with another sudden explosion, just as I was standing on the crater's edge. I watched the black bombs come up a few meters above the rim and, thankfully, go back down again, not one of them falling outside the crater. Like the 1979 explosion, this was not the start of an eruption – just a minor but potentially deadly volcanic cough that can happen anytime.

Valle del Bove

This spectacular valley, 7 km (4½ miles) long and 3.5 km (2¼ miles) wide, can easily be viewed from both the northern and southern roads to the summit. Guided tours to the summit usually stop at a good lookout point, such as the "Belvedere" on the southern side at 2,760 m (9,100 feet). The 1991–3 lava completely covered the floor of valley's southern part. Many of Etna's lava flows go down this valley, covering one another, so lavas may not stay exposed for long. A few cinder cones stick out here and there from the sea of congealed black lava. Anyone looking at the valley floor can imagine what a spectacle it must have offered when the lavas were active and glowing red at night, such as in 1989.

It is possible to hike down into the valley, but not easy and it has become much more difficult since the 1991–3 eruption. The valley is remote, has no water, and there is always a danger of fog coming in. Venturing into the valley when flows are coming down into it is very hazardous indeed. If Etna is quiet and weather conditions are good, it is worth going down. Since eruptions can cause major changes in the hiking route (and Etna erupts often) I recommend that you hire, or at least talk to, a local guide on arrival. Given the conditions of the valley, hiring a guide is highly recommended. At the very least, go with another person. There is much to see inside the valley: fresh lava flows, pyroclastic cones, and numerous dikes and faults, which offer a unique view into the volcano's complex feeding system. The Valle del Bove is a prime location for studying Etna's ancient activity, as exposed volcanic products date back to the Pleistocene. Near the southern wall are deposits from Etna's Trifoglietto period, and near the southeast end of the valley is Monte Calanna, from the period of the same name.

Southern side attractions

There are many interesting sights and hikes along the road from Catania to the Rifugio Sapienza. Just north of the town of Nicolosi are the Monte Rossi ("Red Mounts"), a double-peaked pyroclastic cone rising to 200 m (660 feet), which was the vent of the huge 1669 lava flow. The lava itself can be seen along the Catania–Nicolosi stretch of the road. Once you get to the Sapienza area, you can hike onto the 2002 flow which practically destroyed the building. Many visitors walk east from the Sapienza and climb onto the red-brown Monte Silvestri cones of the 1892 eruption. Looking to the north, you see a dark peak which some mistake for Etna's summit. It is, in fact, La Montagnola, the site of Etna's 1763 lava flow and vent area. This is Etna's thickest historical lava, reaching 100 m (330 feet) in places. Energetic visitors climb the steep ridge to catch one of the best views of Etna's southern side. To see another unusual flow, drive from the Sapienza towards the town of Zafferana: the road winds through the 1792 lava, one of Etna's rare pahoehoe lava flows.

The 1989 lava flow. In this photograph the sky was overexposed so that details of the dark lava would be visible. (Photograph courtesy of Christopher Kilburn.)

The 1991–3 lava flow front

The front of the 1991–3 lava flow can be seen in Piano dell'Acqua ("water plain"), about 500 m (1/3 mile) northwest of the town of Zafferana. To get there, follow signs from the town that say "Colata lavica 1992." Despite the construction of four barriers to try to stop the flow, lava invaded houses on the outskirts of town in 1992. After the eruption stopped, a small park was built by the flow front, complete with a statue of the Madonna. Visitors come to walk on the congealed lava and to see houses poking out from the flow. It is evident from the graffiti on some of the houses that their owners were not impressed by the government's efforts to stop the lava.

Visiting during activity

Etna erupts often, typically two to four times per decade and sometimes more. In addition to its spectacular flank eruptions, the summit region is often active, so it is sometimes hard to define what an "eruption" here really means. Since the volcano is easily accessible and often active, it is potentially a very good place in which to see and photograph an eruption. Most of the activity consists of Strombolian explosions and large, chunky aa lava flows which, in theory, make Etna's eruptions ideal to visit. However, the Strombolian activity can quickly turn violent, phreatic explosions can occur without warning, and eruptions can start rather suddenly. If volcanoes had personalities, Etna would be a trickster. Unlike many other volcanoes, it does not kill by having devastating, violent eruptions – it is the surprise element that causes tragedy. A recent example was a strong explosion near the Sapienza complex during night of December 17, 2002, which injured 32 people. The explosion was not directly caused by the eruption, but rather by the vaporization of oil and water inside the building while it was overrun by lava.

Apart from the well-publicized recent fatalities, such as in 1979 and 1987, historical records show two other tragic events happening: in 1843, 36 spectators from the town of Bronte were killed when the front of a lava flow exploded suddenly as it encountered wet, marshy ground. During the eruption of 1928, three men allegedly decided to spend an extra night in their threatened homes near the town of Mascali. By morning, the houses and, presumably their inhabitants, had been consumed by lava. These events epitomize Etna's greatest danger: the false sense of security one can easily be lulled into. Etna's activity straddles a fine line between mild and violent and is often hard to predict which way it will go. Luckily, much of the activity can be watched in relative safety, and tourists join locals to watch sluggish aa lava flows consume the lower slopes. Flank eruptions are the easiest to watch, and you are likely to have plenty of company unless the flow is in a hard to access place.

Activity on the summit, the Valle del Bove, or the adjacent Valle del Leone is difficult to get to and potentially very dangerous. A local guide can be invaluable in these circumstances. Even if they cannot take you close to the activity, they will know about vantage points from which to view the eruption safely. During the 1991–3 eruption, many went to the lookout point on the road to Monte Zoccolaro, as it offered a great view of the southern floor of the Valle del Bove. Unfortunately, two overly eager spectators fell to their deaths from there.

Those wishing to see the activity from the air can contact the Catania airport. Sightseeing flights are becoming more popular in the area, but are not yet offered routinely.

Etna is a big mountain, so it is particularly important to note where the eruption is before making plans to see it. The time of year also needs to be considered, as many places on Etna are hard to access during the winter months, and snow remains at high altitudes for much of the year. Going to see an eruption on Etna may be as easy as walking to an active flow in the company of many, or as hard as mounting a costly expedition for a brave few.

Other local attractions

There are many interesting towns as well as geologic attractions in the Etnean area. The best way to get around the mountain is by car, though there are buses to nearly every town and the Circumetnean railway serves many of the towns in the lower slopes of the mountain.

Aci Castello

Named after the river Aci and a castle dating back from 1297, this town offers one of the area's prime geologic attractions – a spectacular exposure of pillow lavas which form the "rock" that the castle is built on. This is probably the most scenic pillow lava exposure to be found anywhere: a neck some 50 m (165 feet) high overlooking the sea, topped by a Norman castle. The lavas exposed are some of Etna's early tholeiitic lavas, which were erupted under water in the beginning of the volcano's history about half a million years ago. The total extent of these lavas is unclear, as they are covered virtually everywhere else by younger products. A staircase leads from the road down to the bottom of the volcanic neck, which is surrounded by a platform also made up of the ancient lava.

Aci Trezza

Located along the coast, just north of Aci Castello, this is a charming town made famous by the ancient lavas that form the rocks offshore called the Ciclopi. They were named after the one-eyed monsters from Homer's *Odyssey*, who are said to have hurled rocks into the sea as Ulysses and his crew made their escape. It is possible to take a boat trip to examine the Ciclopi from up close. The lavas here are also ancient tholeiitic basalts, about the same age as those of Aci Castello.

Taormina

Perched on a hillside just north of Etna is this major resort town, one of the best places from which to photograph Etna. Taormina is famous for its beautiful Greek amphitheater, which is still used for a variety of artistic events. Come here in the early morning or late afternoon for that perfect photograph of the volcano. Taormina has a rich history and wonderful architecture, but the town's beauty is marred by crowds and the nearly infinite number of souvenir shops.

Bronte

History buffs may wish to visit this charming town, once the dukedom that Ferdinand III of Naples bestowed upon Lord Nelson. Until recently, Nelson's heirs still owned the Castello Maniace, where the Admiral planned to retire to with his beloved Lady Hamilton. The castle now belongs to the municipality and is undergoing renovation, but can be visited by applying in advance – check with the tourist office in

Catania. While in the area, it is worth seeing the Museo dell'Antica Civilità Locale (Museum of Old Local History), which reconstructs a typical rural scene of ancient days in the setting of an Arab–Norman building from around AD 1000.

References

Chester, D.K., A.M. Duncan, J.E. Guest, and C.R.J. Kilburn (1985) *Mount Etna: The Anatomy of a Volcano*. Chapman and Hall.

Da Rosa, R., G. Frazzetta, and L. La Volpe (1992) An approach for investigating the depositional mechanism of fine-grained surge deposits: the example of the dry surge deposits at "La Fossa di Vulcano." *Journal of Volcanology and Geothermal Research* **51**, 305–21.

Kilburn, C.J. and W.J. McGuire (2001) *Italian Volcanoes*. Terra Publishing.

Mazzuoli, R., L. Tortorici, and G. Ventura (1995) Oblique rifting in Salina, Lipari, and Vulcano Islands (Aeolian Islands, southern Italy). *Terra Nova* **7**, 444–52.

Rittman, A. (1931) Der Ausbruch des Strómboli am 11 Sept. 1930. *Zeitschrift für Vulkanologie* **14**, 47–77.

Ventura, G. (1994) Tectonics, structural evolution and caldera formation on Vulcano Island (Aeolian archipelago, southern Tyrrhenian Sea). *Journal of Volcanology and Geothermal Research* **60**, 207–24.

10 Volcanoes in Greece

Greece

Greece is not a country known for active volcanoes, but it does have one of the world's most famous – the beautiful Santorini, in the Aegean Sea. Around 1600 BC, Santorini exploded with the energy equivalent to some 3,000 atoms bombs, in what was one of the most cataclysmic eruptions of all time. The Minoan eruption, as it is often called, had a tremendous impact on Greece and the Aegean, but just how much is a subject of considerable debate and tomes of research. Did the effects of the eruption cause the downfall of the Minoan civilization? Some refute this, but none doubt that the eruption changed Santorini forever. The original volcano, a stratocone, fragmented into the broken circle of small islands that we see today. Since then, several eruptions have occurred in these islands but all have been considerably less violent than the Minoan. At present the only activity is fumarolic, concentrated on the island of Nea Kameni – the "Young Burnt Island" in the center of the Santorini archipelago. Visitors are often surprised to find out that this island could erupt again anytime.

Greece has a few other volcanoes still considered active, on the islands of Methana, Milos, Nisyros, Yali, and Kos. Of these, only Methana and Nisyros have erupted during historic times. All activity has been mild in the last few hundred years, consisting of small explosive events. People don't come to Greece to see volcanoes in action, though the modest eruptions in recent times have been enjoyed by those lucky enough to be there at the right time. Christos Doumas, Greece's most celebrated archeologist, tells of a colleague welcoming him on his first visit to Santorini with the words, "If you are lucky, sir, there may be a little eruption for you to admire. It really is a marvelous sight."

Since the local economy of Santorini and much of Greece is sustained by tourism, visitors will find traveling extremely easy. Greeks are friendly and hospitable people and crime is not usually a problem. There is much to see in Greece, and visitors of all tastes can find something to entice them, though those interested in ancient history are likely to have the most enjoyable experiences. Others may find that exploring Santorini and the products of its Minoan eruption is an ideal way to become acquainted with Greece's rich past.

Tectonic setting

Santorini is part of the Aegean arc of volcanic centers which, like the Tyrrhenian Sea and its volcanoes (see Chapter 9), formed as a result of the complex collision between Africa and Europe. The African plate has been diving beneath Crete and the southern Aegean for about 26 million years, but volcanism of the current island arc begun only about 3 to 4 million years ago. The volcanic islands along this arc, which cuts the Aegean from the northwest to the southeast, are Aegina, Methana, Poros, Milos, Santorini, Kos, and Nisyros.

Santorini's growth has been strongly controlled by the tectonic faults in the region. Santorini's main island, called Thera, is crossed by two major fault systems, both trending northeast–southwest, called the Kameni and Columbus lines. The Kameni Line crosses the islands of Nea and Palea Kameni ("Old Burnt Island") at the center of Santorini's caldera and manifests itself on the surface as faults and as lines of vents and fumaroles. A brief look at the geologic map of Santorini shows that the oldest ("basement") rocks are not seen on the northern part of Thera. This is because faulting along the Kameni Line has effectively split the island into two and the northern part has is now below sea level.

The second major fault system, the Columbus Line, crosses the northern part of Thera but with more subtle effects. The only obvious signs of its presence are two cinder cones about 70,000 years old, called Megalo and Kokkino Vouno. The Columbus Line was named after its most famous feature, the submarine volcano Columbus Bank, located to the northeast of Thera. It erupted in 1650, but has remained quiet since.

The Kameni and Columbus lines are major controls on the evolution of Santorini and give us some idea of where future eruptions are likely to start. The odds favor somewhere along the Kameni Line, with the young island of Nea Kameni being the most likely site for new vents.

Practical information for the visitor

When to go

Like all Greek islands, Santorini has crowds of tourists in the high season (July and August). Prices are more reasonable from September through June, but between October and May many tourist establishments are closed and ferry services are considerably reduced. Temperatures are pleasant year-round and there is little rainfall, but strong winds blowing the abundant fine ash can be a problem, particularly during the cooler months. May is probably the best month to be in Santorini, as the contrast between the blooming wild-flowers and the dark volcanic landscape is truly striking.

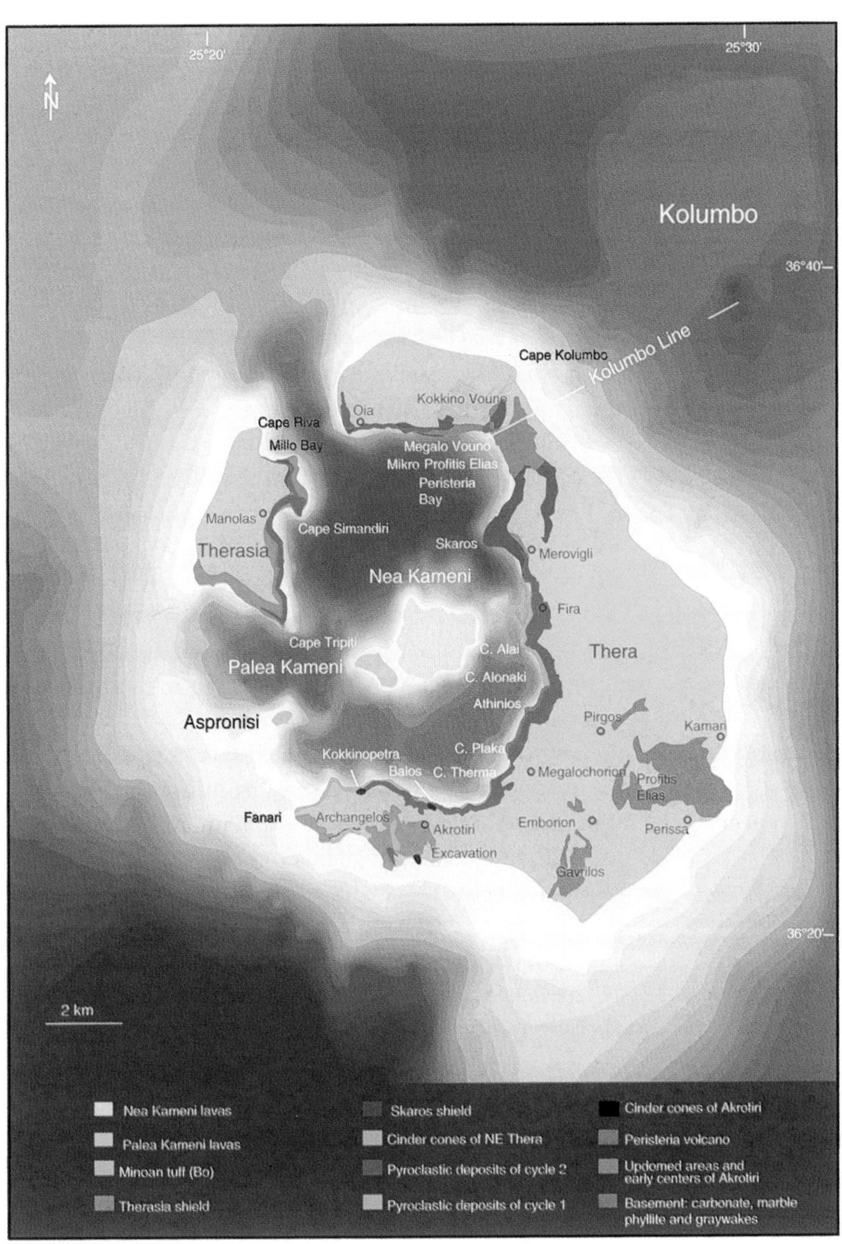

Simplified geologic map of Santorini showing also the bathymetry and the location of Columbus Bank (Cape Lolumbo). (After Druitt *et al.* (1998), reprinted from *Fire in the Sea* by Walter L. Friedrich, Cambridge University Press.)

Information about volcanic activity

The solfataric activity at Nea Kameni is usually steady, but any changes in this or in seismic activity would be reported in the S.E.A.N. Bulletin (see Bibliography). The local volcano observatory is the Institute for the Study and Monitoring of the Santorini Volcano (see Appendix I)

Volcano monitoring and emergency services

Nea Kameni, Santorini, and nearby islands are monitored by a modern network of dozens of different types of instruments that are sure to record any precursor signs of an impending eruption. In its current state, Santorini offers no special dangers, though you should be careful near active fumaroles. Emergency medical service is free in Greece, but visitors carrying medical insurance get better care. Language is not usually a problem as most physicians speak English.

Transportation

Santorini has a major airport and frequent flights to Athens and other European cities. Ferry travel is cheaper and very convenient, linking Santorini (Ormos Athinios port) to Athens (Piraeus port) and to other Greek islands. It is no problem to find a boat from Santorini to Nea Kameni. Most visitors get around Santorini by public bus, taxi, or rented mopeds. Rental cars are also available, and you can bring cars to the island on the ferry, but driving (and parking) in downtown Fira is often a hassle.

Lodging

Santorini has a wide variety of hotels and guest houses, most located in Fira. Camping is only allowed at designated sites, though it tends to be tolerated elsewhere. There are official camping sites near the towns of Perissa and Kamari on the southeastern coast of the island.

Maps

A simple 1:40,000 "road tourist map" is widely available for sale. The map barely shows the topography, but gives heights of major hilltops and shows the locations of the most popular attractions. A color geologic map of the islands Nea Kameni and Palea Kameni is included in the popular *Santorini: Guide to the Volcano,* published by the Institute for the Study and Monitoring of the Santorini Volcano. The book is widely available on the island. A new geological map of Santorini was published in 1999 in a Memoir of the Geological Society (see Bibliography).

Santorini

Santorini is one of the world's most fascinating volcanoes, not only because of its superb geologic scenery but also for its rich and intriguing history. The island has been occupied for some 5,000 years. Before the Minoan eruption, Santorini was a stratocone that may have looked like Strómboli, as both islands were known by the name Strongyle, meaning "the round one." Santorini's alternate ancient name was Kallisti, "the fairest one", reflecting its still unquestionable beauty. The Minoan eruption and its associated caldera collapse changed the stratocone into the group of five islands that we see today, with Thera, Therasia, and Aspronti arranged in a broken ring around the central islands of Nea and Palea Kameni. They are known collectively as Santorini, a name given by thirteenth-century Crusaders after the chapel of Agia Irini (Saint Irene), whose location in the islands is no longer known.

The Santorini caldera is in fact a complex of four flat-bottom structures, which were the result of at least four caldera collapses. The steep cliffs marking the walls of the various calderas are truly striking, reaching 300 m (1,000 feet) in height. The largest of the calderas is that on the northern part of the island, extending down to Fira, and delineated by Therasia on the western side. The northern basin appears to have been the focus of the deepest caldera collapse during the Minoan eruption. Its current depth below sea level is 390 m (1,285 feet).

Santorini's momentous geological history left its record on the steep cliffs that come down to the sea and give the island its characteristic rugged look. It is easy to see that the cliffs are made up of layer upon layer of volcanic deposits. The succession of eruptions is revealed by these layers, each telling a story many thousands of years old. Santorini is an ideal volcano for those people interested in learning about volcanic stratigraphy, that is, how to use the different layers to reconstruct the volcano's long history. There are many places along the cliffs where the succession of layers from different eruptions is truly striking, showing variations in color and texture. Santorini's paucity of trees or tall vegetation is a bonus in this case, as there is nothing to obscure the view.

Geologists have been able to trace Santorini's beginnings back to a pre-volcanic island in the Triassic period. Crystalline limestones and schists from this ancient island can still be seen today, particularly on Profitis Elias, the highest point of Santorini. The first known volcanic eruptions on Santorini were

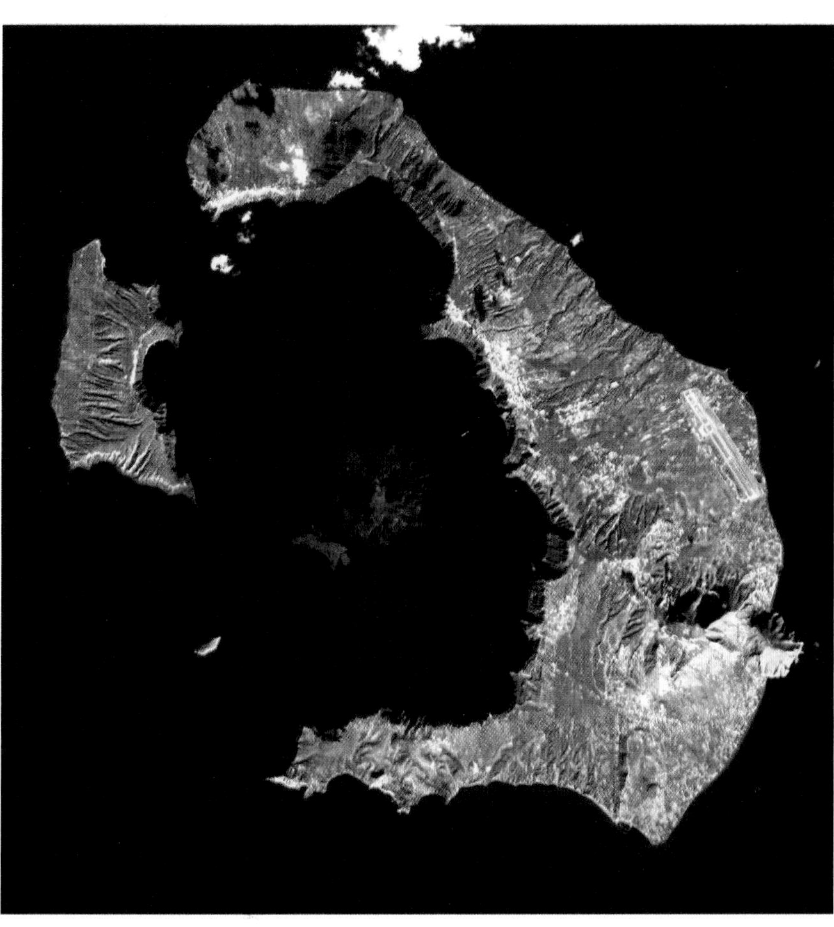

Satellite image of Santorini obtained by Advanced Spaceborne Thermal Emission and Reflection Radiometer (ASTER) on NASA's *Terra* satellite on November 21, 2000. The larger island is Thera, the smaller is Therasia. The Kameni islands (which appear dark in the image center) were formed after the main caldera and have erupted as recently as 1950. (Image courtesy of Mike Abrams, NASA/JPL.)

submarine and happened during the late Pliocene. After that, the evolution of the volcano is best described as six separate stages. The first, called Early Centers of the Akrotiri Peninsula, followed the submarine eruptions and ended about 580,000 years ago. Santorini erupted pasty, high-silica lavas of rhyodacitic composition from vents in the southern part of Thera. We can still see lava hills near Akrotiri that were formed during this early stage. The rocks from this period can be distinguished easily from others because they have a high amount of the mineral hornblende.

The second stage, called the Peristeria Volcano, lasted from 530,000 to 430,000 years ago. Vents on the northern part of Thera erupted lavas and pyroclastics, mostly andesitic in composition, which eventually built up a large stratocone. Its remains can be seen today in Megalo Vouno and Micros Profitis Elias. Around the same time, activity started again in the south, forming cinder and spatter cones in the Akrotiri Peninsula, some of which can still be seen.

Around 360,000 years ago, Santorini entered what is called Cycle 1, the first of two major eruptive cycles focussed along the Kameni Line fault. Lavas and pyroclastics from Cycle 1 can be seen along the caldera wall in southern Thera, where visitors interested in the volcano's stratigraphy can have a fine time unraveling the various deposits with the aid of a geological map and a stratigraphic column. The succession of eruptions is only summarized here, but the dating and mapping of their deposits has kept many geologists busy.

The first eruption in Cycle 1 – called Cape Alai – produced andesitic lavas that form a layer 60 m (200 feet) thick at the base of the cliffs about 1 km (⅝ mile) south of Fira. The next eruption, called Cape Therma 1, was Santorini's first major explosive event and left thick deposits of dacitic lavas in southern Thera. Cape Therma 2, another major explosive eruption, left a pure-white pumice fall layer, about 2.5 m (8 feet) thick, that is easy to spot in the cliffs of southern Thera. The eruption also extruded pasty rhyodacitic lavas that can be seen on Cape Alonaki. The Cape Therma 3 event produced a gray scoria flow deposit that is widespread on the Akrotiri peninsula. Cycle 1 culminated with a pair of spectacular explosive eruptions that are known by the rather dull names of Lower

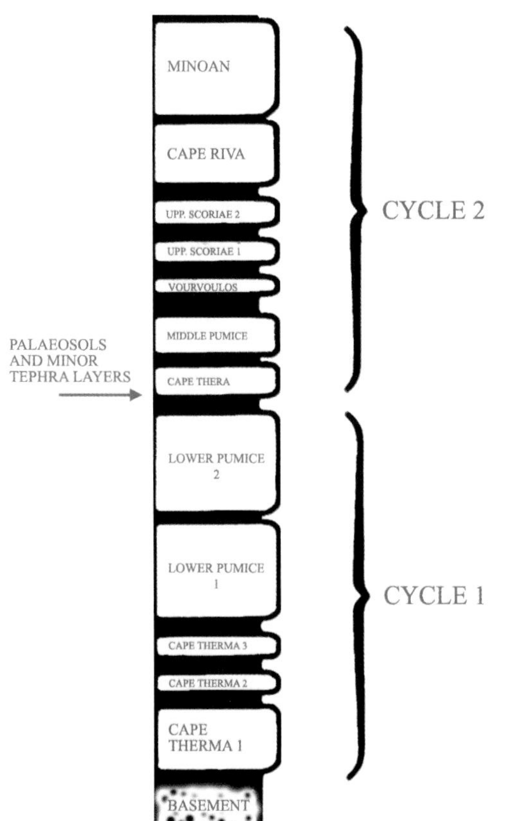

Stratigraphy of the pyroclastic eruption deposits in Santorini, showing Cycles 1 and 2. The black bands represent minor eruption tuffs and palaeosols. (Modified from Druitt *et al.*, 1996.)

Pumice 1 and 2. Both eruptions started with a Plinian phase from vents near the present Nea and Palea Kameni islands and left deposits all over southern Thera. Cycle 1 ended rather cataclysmically around 180 thousand years ago, with a massive caldera collapse.

Cycle 2 of Santorini's evolution begun about 170,000 years ago with the eruption of the Cape Simandiri andesitic lavas. These formed a lava shield in northern Thera, parts of which can still be seen at the base of the Therasia island cliffs. A series of explosive eruptions followed, with four being considered major events, as each discharged at least a few cubic kilometers of pyroclastics. The "big four" are called Cape Thera, Middle Pumice, Vourvoulos, and Upper Scoria 1, but those who cannot remember their names simply refer to the whole lot as "Middle Tuff." Their deposits consist of ash, pumice, scoria flow, pyroclastic flows and surges, and other products. You need a good eye to distinguish the various layers and, if you are among geologists, the discussions about the origin of each can get quite lively. During the same period as these Middle Tuff eruptions, the cinder cones of Megalo Vouno and Kokkino Vouno were built over the old Peristeria Volcano, in the northeastern part of Thera. The Middle Tuff phase ended with another major caldera collapse. Cycle 2 then entered a quieter phase, during which the Skaros lava shield grew inside the caldera. It is estimated that Skaros reached 350 m (1,150 feet) above sea level, with a diameter of about 9 km (5½ miles). Remains of the 70,000-year-old shield can be seen along the cliffs north of Fira and also on Therasia island.

Three more major eruptions occurred before Santorini was settled by unsuspecting humans. First came the Upper Scoria 2 eruption about 55,000 years ago, in which a vent near the present Nea Kameni island sent scoria flows down the flanks of the Skaros shield. Next came the construction of another shield volcano, called Therasia, part of which still survives as an island. The cliffs of the present day Therasia show the layered succession of domes and flows, which in places reaches over 200 m (660 feet) in thickness. Most of the Therasia shield was destroyed by the next eruption, the Cape Riva, which also caused the collapse of the Skaros shield. This explosive eruption took place about 21,000 years ago, starting with a Plinian phase which sent pumice falling all over – these deposits can be still seen in northern Thera. The Plinian eruption column collapsed, and pyroclastic flows poured out of multiple vents, mantling most of Santorini. The volcano then entered a quiet period lasting some 17 millennia, during which settlers started to arrive.

The Minoan eruption, described below, was the last major explosive eruption to date. Santorini is still very active but, luckily for the present residents, since 197 BC it has been in a fairly quiet phase, during which dacitic magma has leaked out slowly from the Kameni line and built a shield volcano inside the caldera. The islands of Palea Kameni and Nea Kameni are the surface tips of the 3.5 km (2¼ miles) diameter shield, which stands 500 m (1,650 feet) above the submerged caldera floor. Only nine of the shield-building eruptions have been reported or deduced from historical records. The oldest known took place in 197 BC and was recorded by the Greek geographer Strabo. He mentioned the existence of a new island, Hiera ("holy island"), which is thought to be the present-day Bankos Reef on the northeastern side of Nea Kameni. The next known eruption was from AD 46 to 47 and built the island of Palea Kameni. The volcano woke up again with a major blast in AD 726, which sent pumice all over Asia Minor. This eruption was marked by a major collapse, which formed the rather steep northeastern cliffs of Palea Kameni. Activity from 1570 to 1573 formed the island of Mikri Kameni, which later became united with the present Nea Kameni, the "new" island born during the eruption of 1707 to 1711. The intervals between the shield's eruptions appear to have become shorter during the last two centuries, but this may just be our impression, as the records of activity are more complete. Eruptions from 1866 to 1870 and from 1925 to 1928 extruded lava which added land to Nea Kameni, but also had explosive phases with spectacular eruptions columns reaching 2 to 3 km (1¼ to 1¾ miles) in height. The activity from 1939 to 1941 started with a submarine explosion, but it soon shifted to the center of Nea Kameni, building four lava domes. The last eruption, in 1950, begun with phreatic explosions, followed by the extrusion of the Liatsikas lava flow – the newest land in Greece.

The activity on Nea Kameni has been relatively mild, but a violent eruption happened in 1650 on the Columbus Reef, about 6.5 km (4 miles) away from Thera. After 2 years of quiet submarine lava extrusions, a series of strong explosions above sea level produced large quantities of ash and fumes. The nasty mixture drifted to Santorini, causing excruciating pain to the residents and blinding many for 3 days. The explosions were soon followed by a large tsunami that wrecked Santorini and other areas within a 150 km (95 miles) radius, killing at least 50 people and 1,000 animals. The island cone formed at Columbus Reef has since been wiped out by the sea, but bathymetric surveys show a caldera 3.5 km (2¼ miles) in diameter and 500 m (1,650 feet) deep, which may one day come to life again.

Tourists today come looking for Santorini's sun and sand, but many take a side trip to Nea Kameni, which is known locally as "the volcano." Visitors are often surprised to find out that its last eruption was as recent as 1950. When Santorini's tourist boom started in the 1970s, it was not "the volcano" that drew the crowds, but Santorini's superb natural beauty, with craggy cliffs and black sand beaches, and the picturesque towns of Fira and Oia with their white-domed houses. Santorini is primarily a holiday island for both Greeks and foreigners. However, visitors interested in volcanoes, ancient history, or both, are the ones who can fully appreciate the island. The steep cliffs tell the story of a series of prehistoric and historic eruptions and of the cataclysmic succession of events during the Minoan eruption. The excavations at Akrotiri uncovered a wealthy Minoan city overwhelmed by the eruption but remarkably well preserved. Like Pompeii and Herculaneum in Italy, the excavated city reveals a way of life totally destroyed by a volcano, though here the residents had ample warning and managed to escape – no bodies have yet been found.

Santorini's residents today are well aware of the volcano's violent past but do not expect repeat performances of the same kind. They know that a new eruption from Nea Kameni is the most likely threat and, if it is as mild as the ones in recent history, the event will thrill rather than frighten most visitors. A more serious danger is that of a strong earthquake, such as that of magnitude 7 which devastated the island in 1956. Another of these earthquakes could cause many casualties and would be disastrous for the island's thriving tourist industry. Residents live with the possibility, praying that it will not happen, least of all during the crowded summer months. Most visitors are blissfully unaware of the danger.

The Minoan eruption

This cataclysmic event is one of the world's best-studied eruptions and also one of the most debated, as many aspects of it are sources of controversy. For a start, there is no agreement on the actual date for the eruption, so it is often just referred to as "Minoan." The conventional archeological date for the event is sometime between 1500 and 1550 BC, but recent radiocarbon dating of materials excavated from Akrotiri suggest that the eruption took place about a century earlier. This agrees with evidence taken from much further afield. It is expected that an eruption of this magnitude would have left its signature in the atmosphere (increasing the amount of sulfur) and in the climate (lowering average temperatures). An ice core from Greenland shows an increase in the amount of sulfuric acid around 1645 BC. Moreover, studies of tree rings from high-altitude bristlecone pines from California and oaks from Ireland show frost damage to these trees, implying that a large eruption took place in 1627 BC, with an estimated error of only 2 years. While it is possible that another eruption could have caused these environmental effects, the mounting evidence points to a date for the eruption between 1600 and 1650 BC.

It is not known how long the eruption lasted, but the sequence of the various events has been reconstructed from the deposits that mantle most of Santorini today, as well as from archeological excavations at Akrotiri. It is known that earthquakes rocked the island well before the catastrophe, causing the residents to leave their homes. This is why neither bodies nor precious artifacts have been found in the ruins. More violent earthquakes followed, destroying many buildings. However, before the eruption took place, there must have been a long period of quiescence, because some residents returned and started repairing the damage to the buildings. It was at this time that the main eruption started, but its opening salvo – probably relatively small phreatic or phreatomagmatic explosions – still gave residents time to flee. Fine pumice fell over the whole island, creating a layer about 2 cm (¾ inch) thick. The eruption must have halted for some time, sufficient for this pumice layer to suffer oxidation. Luckily, it seems that the people did not return, as the next explosions showered the island with pumice, creating deposits over a meter thick.

It was the next event, however, that was truly catastrophic. A Plinian eruption ejected huge quantities of ash into the atmosphere and hurled huge boulders onto Akrotiri, some of which can still be seen today. The eruption column reached 36 km (22 miles) in height, and pumice fragments showered the whole island, leaving a light gray mantle that is over 30 m (100 feet) thick in places. Directly on top of the pumice fallout we can see fine-grained surge deposits – a type of diluted but rapidly moving, deadly pyroclastic flow. We know that the Plinian fallout continued, as there are pumice fragments mixed in with the surges. Studies of the deposits show that seawater entered the vent, probably after several hours of Plinian activity, when collapse of the caldera began and fractures spread from the vent area towards the northeast and southwest. Water and hot magma mixed, causing violent explosions and sending powerful mudflows downhill. Blocks of lava up to 10 m (33 feet) in diameter were carried down in the torrential mixture of water, ash, and magma fragments.

Following the mudflows, the eruption continued its destructive course, now sending out hot pyroclastic flows, estimated to have been between 200 and 400 °C (380 to 740 °F). These deposits – called ignimbrites – form an apron up to 40 m (130 feet) thick, all around the outer coasts of Thera, Therasia, and Aspronisi. Within and over the ignimbrites, there are enigmatic deposits called flood breccias. These are thought to have formed when the pyroclastic flows ripped up portions of the mudflow deposits, smearing out the debris downstream.

After the eruption stopped, nature did not quiet down for long. The ignimbrite deposits in the area of the Akrotiri excavations have been eroded and covered by alluvial deposits, showing that powerful flash floods probably happened soon after the end of the eruption.

It is estimated that the Minoan eruption discharged some 30 km^3 (7 cubic miles) of magma and the island lost some 83 km^2 (32 square miles) of its original area. Ash from the eruption drifted east and today can be found on the islands of Kos and Rhodes, and as far as western Turkey. The vast pumice deposits on Santorini eventually became the island's major source of income, until mass tourism started and the quarries were shut down because of environmental concerns. The locals like to point out that pumice from Santorini was at one time exported all over the world and that cement made from the pumice lines the Suez Canal.

It is not known exactly how much time elapsed after the eruption until people started returning to Santorini. Fragments of Mycenaean vases suggest resettlement by the end of the thirteenth century BC. The Lacedaemonians, who settled here in the ninth century BC, named the largest island in the group Thera, in honor of their leader Theras. The island has been inhabited ever since, prized first for its location on the trading routes and fertile volcanic soil, and more recently for its exquisite scenery and rich archeological sites.

A personal view: Where myths begin

Almost everyone who starts delving into Santorini's history ends up gripped by the lore and mystery surrounding the Minoan eruption. The far-reaching and sometimes far-fetched effects of the great catastrophe have launched thousands of publications, from the purely academic to the downright sensational. Was Santorini the "Lost Atlantis" described by Plato? Did the eruption cause the downfall of the Minoan civilization? How about Egypt's plague of darkness and the parting of the Red Sea by Moses? Such questions may seem extraordinary and their answers may never be known, but there is no doubt that the Minoan eruption had vast effects at the time, and it could have easily have inspired fantastic legends.

The large quantity of ash injected into the atmosphere must have had marked effects on the Aegean Sea and beyond, and probably caused a temporary global climactic change. It is quite possible that the biblical account of Egypt's plague of darkness could have been inspired by the drifting ash. A link between the parting of the Red Sea and Santorini's eruption is harder to argue. The suggestion comes from Hans Goedicke of Johns Hopkins University, and is based on an Egyptian temple inscription carved around the beginning of the fifteenth century BC. Pieces of the text seem to parallel the biblical account of the Exodus and suggest that a sudden flood engulfed the army pursuing the Israelites. For the account to fit, the Israelites would have been traveling along the north coast of Egypt, which is actually a more direct route to Palestine than the Red Sea. The fugitives would have assembled onto high ground to prepare their defense against the army when the flood – presumably a tsunami – struck, washing away their pursuers. A big question is: did the Minoan eruption generate a tsunami? If so, could it have traveled about 850 km (530 miles) to the southeast, causing a devastating flood onto the desert?

Tsunamis are frequently triggered by eruptions of island volcanoes. In the case of Santorini, it is possible that a large enough wave could have crossed the Mediterranean and, within a matter of hours, reached the northern African coast. The problem with this theory is that one would expect the tsunami to leave some geological evidence of its existence and so far none has been identified for sure. Another great source of uncertainty for any links between biblical events and the Minoan eruption is timing: the dates for the Exodus are even more poorly constrained than those for the eruption. In the absence of any real evidence, these links must remain in the realm of interesting conjecture.

A more fiercely debated theory is whether the eruption caused the eventual downfall of the Minoan civilization. This idea was initially championed by the famed Greek archeologist Spyridon Marinatos, who discovered the buried city of Akrotiri in 1967. While there is evidence that the eruption, or earthquakes associated with the magma movement, did great damage to the cities and palaces in Crete – the major Minoan center – most scholars now agree that the eruption coincided with an already existing decline in the quality of civilization in Crete. It is interesting that archeological and geologic studies of Crete show that the entire history of the Minoan civilization was punctuated by earthquakes. Each phase of development of the Minoans appears to have been halted, at least locally, by destructive earthquakes. So, while geologic forces no doubt played their part in the Minoan's decline, the role of Santorini's eruption may have been only that of a *coup de grâce* to an already dying empire.

As for the most fantastic suggestion – that Santorini is Plato's Lost Atlantis – it seems unlikely, even if we assume Plato's account was based on true facts. Santorini's major Minoan center, Akrotiri, was largely destroyed and abandoned before the eruption took place. This does not agree with the legend of Atlantis, which says that a great center of civilization was suddenly and violently destroyed by the sea. Besides, Santorini is an unlikely candidate for the legendary continent: the island is too small, and its naval power and standard of civilization were far below those of the omnipotent Atlantis. Minoan Crete, in fact, fits the descriptions of Atlantis far better than Santorini. The connection between Santorini's eruption and the "Lost Continent" comes from the accepted date for the destruction of Atlantis, around 1500 BC. If we assume that the Atlantis catastrophe was an historic fact, the Minoan eruption could be its most probable cause, particularly if a tsunami was generated.

A sensible interpretation of the Lost Atlantis story was given by Spyridon Marinatos. Plato's account of Atlantis was the first to be written, but the story came to him orally and is first attributed to Egyptian priests around 590 BC. Marinatos argued that the priests were confused when they attributed the destruction of a civilization (the Minoan) to the sinking of the island where the people flourished (Crete). In reality, it was Santorini that "sunk," though not completely. Given that about 900 years had passed since the eruption, it is not surprising that the priests had gotten some facts wrong. Plato himself may have embellished them more, as he used the story as a paradigm of an ideally organized society that flourished until it became arrogant and decadent, invoking the wrath of the gods. Whether or not there is any truth behind the legend of Atlantis, we can be sure that Santorini's cataclysmic eruption could indeed have destroyed an island empire "in a single day and night."

Visiting during repose

Santorini's most likely state is far from disappointing. Nea Kameni shows visitors that the volcano is still much alive. There are fumaroles, warm rocks, and that unmistakable, sulfurous volcano smell. The major volcanic attractions of Thera are the products of old eruptions, the Minoan in particular, and the archeological excavations at Akrotiri, which let us glimpse what life was like before the cataclysm.

Nea Kameni

Called simply "the volcano" by the locals, this small island is a major tourist attraction. Boats to Nea Kameni depart from Thera's port of Athinios and the

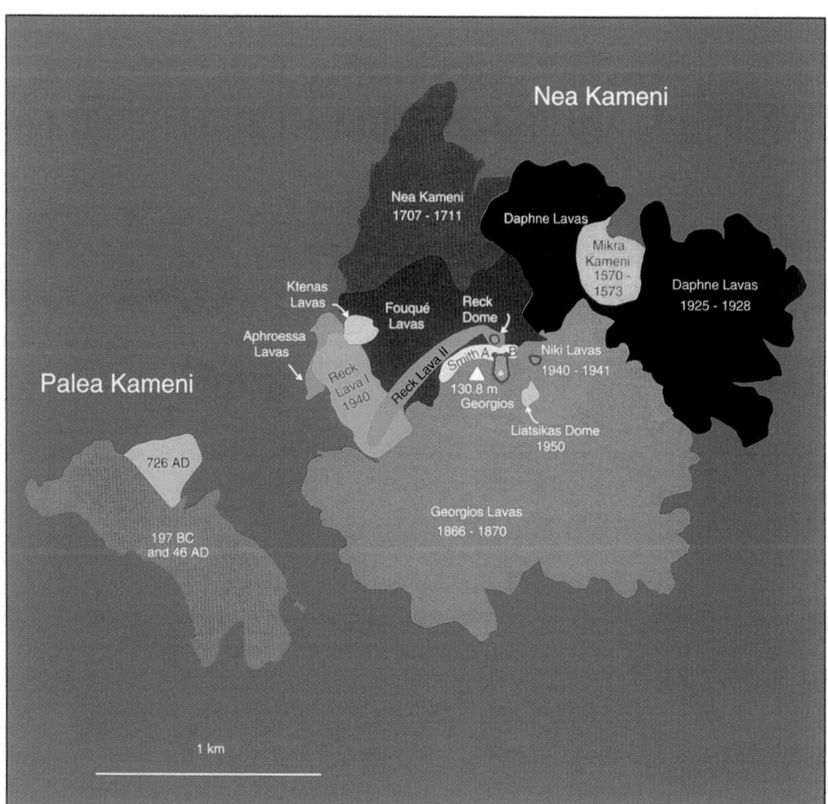

Geological sketch map of Nea Kameni and Palea Kameni (After Georgalas (1962), reprinted from *Fire in the Sea* by Walter L. Friedrich, Cambridge University Press.)

trip affords some spectacular views of the caldera cliffs. Boats land on Nea Kameni's small harbor of Kato Fira (called Yialo by the locals), which is flanked on both sides by lobes of blocky flow from the 1925–6 eruption. The harbor is itself set on the oldest lava exposed on the island, from the 1570–3 Mikri Kameni eruption. A path to Nea Kameni's summit starts by Kato Fira and crosses the black, blocky Dafni lavas (from the 1925–6 eruption) for a few meters before climbing the ash-covered slopes of the Mikri Kameni dome. Look to the sides of the path and you will see breadcrust bombs from that eruption. Nea Kameni has produced an abundance of these bombs, and it is worth detouring to take a close look at them. The path continues to the top of the Mikri Kameni dome and on the left-hand side is the crater from that eruption. Looking south from the crater, you will see one of the several seismic stations used to monitor the island's rumblings. The path continues down to where the Dafni lavas flow over Mikri Kameni, then forks and goes around the rim of the main Dafni crater, coming together again by the edge of the 1938–41 lava. Before climbing to the summit of the island, take a short detour to the right of the path, to view closely the lava domes of Fouqué, Reck, and Smith – named after the French, German, and American scientists who studied the island's eruptions. The slopes of these domes are littered with breadcrust bombs, including some as large as 2 m (6 feet) in diameter. You can climb to the top of the domes, from where there are excellent views of the 1939–40 lava flows. Another dome, called Niki, is located to the left of the path.

The path continues to the two spectacular craters of 1940, which formed over the Georgios lavas from the eruption of 1866–70, named after the reigning King of Greece, George I. The craters are dotted with active fumaroles, complete with the characteristic rotten-egg smell, and brilliant white and yellow sulfur deposits. Gases seep out at temperatures between 93 and 97 °C (199 to 206 °F), and the ground is warm in many places. Several of the fumaroles are located on the eastern rim and you can put your hand above them to feel the heat. The highest point on the island, at 127 m (419 feet), is marked by an obelisk located southwest of the two craters.

Most people end their visit here, but it is worth continuing along the path that leads down the slopes of the Georgios dome to the Taxiarhes bay harbor. If time is short, walk just the first 200 m (660 feet) to see on your left the 1950 Liatsikas lava butting against the Georgios dome. This small flow, named after an eminent Greek geologist, is the youngest piece of

Nea Kameni, seen from Santorini. (Photograph by the author.)

Greek real estate. You can take a detour from the path and walk down the steep slope to the blocky, black flow. The fresh lava has large crystals of feldspars, pyroxenes, and olivines. From here the path skirts around the Niki ("Victory") lavas, from the 1940–1 Niki dome, and continues over ash and lava from the Georgios dome. The last part of the path is steep and rough as it descends to Ormos ton Taxiarhon ("Archangels Cove").

Most organized excursions to Nea Kameni do not allow enough time to explore the island, so it is better to hire a boat to take you there and to pick you up later – allow at least 4 hours on shore. When making arrangements for the boat, be sure to ask for a stopover at Palea Kameni and for a tour around both islands, to see their otherwise inaccessible coastal cliffs.

Palea Kameni

This small island, about 3 km (1¾ miles) in length, is famous for the bay of Agios Nikolaos. Many excursions to Nea Kameni make a stop here so that visitors can swim in the bay, which is heated up by hot springs to an ideal 36 °C (97 °F). The site is very picturesque, as the iron oxides brought up by the hot springs stain the shoreline and seafloor with orange hues. The entrance to the bay is demarcated by a sharp contrast between the warm, green bay water and the cold blue Aegean Sea.

It is worth landing on the island and taking the steep path to the 98 m (323 feet) high summit, from where the views of Nea Kameni are spectacular. You will also be able to see the submerged Islets of May in the narrow strait between Palea and Nea Kameni. They were formed during the 1866–70 eruption and are now about 1 m (3 feet) below sea level.

Most of Palea Kameni's lavas date from the AD 46–7 eruption and are known as Thia lavas. Like those of Nea Kameni, they are dacitic in composition. There are some deep fissures in the Thia lavas, some about 30 m (100 feet) deep, which are often home to wild doves. The "new" part of the island, formed during a violent eruption in AD 726, is located north of Agios

Nikolaos. You can take a path to the eruption crater – an elongated structure about 250 m (825 feet) in diameter, made truly photogenic by its green lake. The eruption sent pumice and ash drifting across the Aegean Sea and as far as Asia Minor.

Palea Kameni is surrounded by spectacular cliffs best viewed from a boat. Look out for the lava dome on the southeastern shore of the island. The dome was sliced in half by subsidence of the coastline and its interior is now exposed, an onion-like structure of lava layers. Further around on the southern shore are two lobes of lava flows entering the sea. Most of the coastline shows signs of the fragmentation and subsidence that have, in fact, gradually made the island smaller.

Be sure to take advantage of the return boat trip to Santorini to take a good look at the different deposits on the caldera cliffs – the immensity of the volcano's past eruptions is immediately brought home.

The "donkey steps" link the city of Fira to the port. The walk is a great way to see the many volcanic layers forming the cliffs, but watch out for those donkeys. (Photograph by the author.)

Fira and "the steps"

Thera's capital, Fira, is a dazzling collection of white cubes and domes, between which narrow streets wind and twist. The place has a bazaar atmosphere and the tiny shops sell everything from gold jewelry to pumice – and even small pieces of concrete masquerading as "volcanic rock." The town is precariously perched above the caldera cliff and offers a truly breathtaking panorama of Santorini's other islands, with Nea Kameni down in the center. Fira has a port below the cliffs, these days used mostly by cruise ships. There are two ways to get down to the port, and both provide great views of the layered deposits from past eruptions – the modern cable car and the older "donkey steps" that zigzag down from the city to the port. The walk down the steps – actually a steep path cut by shallow steps every one donkey length or so – offers great opportunities to get close to the many volcanic layers forming the cliffs. However, as the name implies, the right of way belongs to the morose donkeys who carry tourists up and down. Stay close to the edge – I didn't and was literally carried down a few steps by a wave of donkeys, two of which stuck their heads under my armpits to help me along. Another word of caution: the steps are very slippery and, yes, your shoes will remind you of the donkeys for quite some time.

Those who are not put off by this description can make their way from Fira down to the port, keeping track of the number of bends in the path which are the best way to pinpoint the locations of interesting rock exposures. The deposits range in age from the Middle Pumice eruption (part of Cycle 2) which are found near the port, to Therasia shield lavas which are close to the top. All those eruptions happened before the Minoan cataclysm.

The first interesting stop is on the second bend, on the left-hand side (as you face the sea). You will notice a dark lava flow with a base of reddish oxidized rubble. The lava is a dacite flow from the Therasia shield. It has some interesting structures such as bands and folds, and scattered nodules of brown andesite, called inclusions, which represent older magma fragments that got mixed in and carried along by the flow. It is rare to see such good examples of inclusions on lavas. After the fourth bend on the path, look to your right and you'll see pink-brown pyroclastic deposits from the Vourvoulos eruption that happened during Santorini's Cycle 2. At the base of this deposit there is a 30 cm (1 foot) layer of pumice, topped by layers of pyroclastic surges that look like wavy dunes in cross-section.

As you continue down, you will pass through the 80 m (265 feet) thick lithic breccia deposit from the

Middle Pumice period. Geologists who come to Santorini talk a lot about lithic breccias. Simply explained, they are the result of pyroclastic flows that carried older blocks of material ("lithics") along with them. The best place to look at this deposit is just after the seventh bend on the path, to the right. You can see lithic blocks up to 2 m (6 feet) across. The larger blocks are formed of orange tuff, some smaller ones are brown obsidian. Analysis of the obsidian shows that it was part of the deposit underneath, which was laid down as airfall. The fragments were still hot when they were ripped up by the pyroclastic flows and carried along by them. On the left-hand side of the path you can see the airfall deposit formed by the Plinian phase of the eruption. Unlike most airfall deposits, which consist of loose pieces of pumice plus ash, this is what geologists call "welded airfall." As the name implies, the fragments were so hot that they became stuck together. It is estimated that the temperature of the pumice fall, as it was emplaced, was over 500 °C (920 °F). If one follows this deposit south, away from the vent, one can see that it grades into a normal, non-welded airfall – logical as the pieces had time to cool before reaching the ground.

When you reach the port, walk to its northern end and look at the exposed rock there – the deposit from scoria flows emplaced during the Cape Thera eruption. A scoria flow is a type of pyroclastic flow (Chapter 3) – its main characteristic is the large quantity of vesicles or "bubbles" in the magma fragments. These flows ponded inside the caldera, so they are particularly thick here.

At this point, I recommend a drink by the port and the cable car back up to Fira – the ride only lasts a few minutes, so have your camera ready, as the views are quite spectacular.

The quarries

Santorini's abandoned quarries are the best places in which to see cross-sections of the immense deposits from the Minoan eruption. There are two particularly good quarries. The first is on the southern end of Fira and the second just off the Akrotiri to Fira road.

The south Fira quarry is located on the caldera rim. The entrance road, opposite the Daedalus Hotel, leads you down the white pumice Minoan tuff until the quarry floor, where the airfall deposits from the older Middle Pumice eruption are exposed – the same deposits that you can see in the Fira steps walk. There is a proposal to make this quarry into an open-air geology museum, with diagrams at various locations explaining the history of eruptions in Santorini. I hope this comes to fruition but even in its present empty, dusty state, the quarry is worth visiting. Here one can easily grasp the enormous size of the deposits laid down by various eruptions.

The Minoan deposits are beautifully exposed in both quarries, but they are particularly impressive in the Akrotiri road quarry. To get there, drive or take the public bus to a dirt track located on the western side of the road, 2 km (1¼ miles) north of Akrotiri. Walk down the dirt track to the quarry floor. The white walls of the quarry, 20 m (65 feet) or so in height, show spectacular cross sections through the first three phases of the Minoan eruption. To distinguish between the three phases, first look for the brown layer of soil underneath the white pumice layer. Geologists call this

The immense deposits from the Minoan eruption, seen in the disused pumice quarries of Santorini. The Akrotiri road quarry is particularly spectacular. (Photograph by the author.)

palaeosols and it is the actual layer of soil from the Minoan days. One of the most thrilling things one can do in Santorini is to look for tiny pieces of broken pottery in this ancient soil. I found a small handle from a vase in this quarry; it was amazing to hold it and realize it had been buried there for some 3,500 years. For the sake of conservation, try to put back anything you can't resist to dig up from the palaeosols and remember that it is strictly illegal to take any antique artifacts out of Greece.

Just above the palaeosols is a very thin layer of white ash, and then a layer of white pumice, about 2 m (6 feet) thick – both are part of the phase 1 deposit. The ash and pumice are airfall – they were thrown up in the air during the Plinian eruption and eventually fell down to the ground, piling up as a white blanket. The pumice fragments in this layer are coarse, typically one to a few centimeters (½ to 2 inches) in diameter. The next layer up is called phase 2 deposit – it has fine particles and often a wavy structure, resembling a series of sand dunes in cross section. This layer was formed by the series of pyroclastic surges that followed the pumice airfall. In some parts of the quarry, you can see how the surges – which form many fine layers – are deformed by "bomb sags," caused by the impact of bombs in the soft material. In some sites around the quarry, one can still see the bombs.

Above the surge layer is the ignimbrite deposit, laid down by a denser form of pyroclastic flow than the surges. The buff-colored ignimbrite layer – called phase 3 deposit – is seen as a veneer over the white deposits.

While walking around the quarry you may notice arch-shaped cavities close to ground level that look like doors. They are reminders of the old methods of pumice quarrying that were used here. The workers never used dynamite, they excavated tunnels in the deposits instead. This rather dangerous practice required the workers to rely on experience to know when a tunnel was about to collapse – and to make a quick getaway.

Akrotiri

More than any place else on the island, this archeological marvel truly conveys the effect of the Minoan eruption on history. Excavations begun in 1967 and the town, buried by pumice and ash, is slowly being revealed. There is much to still be discovered in Akrotiri; as so far less than a dozen buildings have been fully explored. The town's architecture is sophisticated, with some three-story buildings, cobbled streets and squares, and an underground drainage system.

Akrotiri is located on the sheltered southern coast. The fertile plains to the east were used for agriculture and just west of town was a small harbor, which has not yet been excavated. The items recovered so far include pottery, milling equipment, large pithoi (containers) for storing oil and wine, and even some food remains such as dried fish and flour. Many loom weights were found, suggesting that weaving was a major occupation. However, Akrotiri's treasures are the numerous wall frescos that are among the most beautiful ever found in the world. They are now exhibited in the Archeological Museum in Athens (see below).

Much of the damage to the buildings was done by earthquakes prior to the eruption. Excavations showed that part of the rubble had been cleared and sorted into piles before the onset of the eruption. Once the eruption entered its more violent phase, pyroclastic surges entered the town, causing walls and roofs to collapse. Some huge blocks were thrown out by the eruption and traveled 15 km (9 miles) to land on the town. One particularly large block, about 1.5 m (5 feet) in diameter, can still be seen near the "room of the women." By the end of the eruption, the town had been completely buried and would remain so for nearly 3,500 years.

The excavations have board signs to guide the visitor and give brief explanations about the various buildings, but anyone with more than a passing interest in the town should purchase one of the several guidebooks that are sold at the visitors' entrance. A particularly well-illustrated guide is that by archeologist Christos Doumas. Akrotiri is one of the world's prime archeological sites and it is worth spending at least half a day here, getting to know what life was like before Santorini exploded.

Visiting during activity

The likelihood of another eruption on Nea Kameni is reasonably high and monitoring is carried out continuously. The next eruption is likely to be similar to the last few, starting out with explosions caused by the interaction between magma and seawater, followed by extrusion of pasty dacitic lavas. The major hazard is from the explosive activity, but it is doubtful that people on Santorini (Thera) would be in danger. In fact, the view of the activity would be outstanding from there.

Judging from the past behavior of Nea Kameni, an impending eruption will probably give plenty of warning signs, mostly in the form of earth tremors. The

An ancient working place in Akrotiri shows a jar, a bench, and, on the floor, a stone mill. (Reprinted from *Fire in the Sea* by Walter L. Friedrich, Cambridge University Press.)

island would then be off-limits to visitors, maybe for several months before an eruption started. It is hard to tell whether visitors would be allowed on Nea Kameni while activity was taking place, even if the eruption had entered a quiet phase of lava flows. However, it is likely that boats would be allowed to circle the island at a safe distance and that the locals would want to take advantage of the volcano's spectacle and organize boat tours. Since Santorini has an airport, flights over the active island could also be arranged, though at present there are no operating sightseeing flights.

It is possible that a violent eruption could happen again, even if it is not of Minoan proportions. If a major eruption starts, all the islands around would be evacuated and no one should try to enter the endangered area. Santorini should not be underestimated, as it still has the potential to unleash a cataclysm.

Other local attractions

Ancient Thera

Located on the limestone rock of Mesa Vouno are the ruins of a Lacedaemonian city known as "ancient Thera." Although it lacks the appeal of Akrotiri, this city played an important role in the Aegean from about 9 BC to Christian times. Excavations have revealed ruins of a theater, an agora (market place), temples, and various civic buildings and private residences dating from Hellenic times.

National Archeological Museum, Athens

Those who want to see the extraordinary wall paintings of Thera, or learn more about Aegean history, should plan to visit this museum, which is undoubtedly one of the world's best. The museum is still the treasury of ancient Greek art, but visitors should be aware that there are plans to house various collections in regional museums, close to their original sites. The Thera frescos are due to be moved to a site in Santorini in the future, so it is a good idea to check with the Greek tourist board on the status of the move. The beautiful frescos currently in the museum include a spring landscape with clumps of lilies and flying swallows, a naked fisherman carrying strings of fish, and a naval expedition that shows the island volcano as it might have looked before the cataclysm. The fine Minoan art is well represented by the frescos and seeing them allows us a glimpse of the life and level of sophistication of this culture. It is hard to believe that each fresco was found as thousands of little pieces on the ground in Akrotiri and painfully reconstructed.

There is a small archeological museum in Fira, but at present it contains only a limited collection of artifacts. Visitors who are really keen on Minoan art and culture should also plan a visit to Crete to see the splendid palaces of Knossos, Phaistos, and Malia and the Iraklion Archeological Museum, where Crete's treasures are kept.

Nisyros

This island, 16 km (10 miles) east of Kos, is a typical cone-shaped stratovolcano. Phreatic explosions happened here as recently as 1888 and the volcano still shows strong fumarolic activity. Boatloads of visitors come from Kos every day to hike up to the steaming crater, a bowl about 250 m (825 feet) across and some 30 m (100 feet) deep. Sometimes the gray mud at the bottom bubbles and sputters, much to the delight of onlookers. The strong smell of sulfur permeates everywhere, and the sides and floor of the crater are streaked with brilliant yellow. Nisyros is still a very quiet island by Aegean standards, thanks to its lack of sandy beaches. It is a delightful place in which to stay and sample traditional Greek hospitality and cuisine.

Reference

Druitt, T.H., L. Edwards, M. Lanphere, R.J.S. Sparks and M. Davis (1998) Volcanic development of Santorini revealed by field, radiometric, chemical, and isotropic studies. In: R. Casales, M. Fytikas, G. Sigvaldason, and G. Vougioukalakis (eds.). *The European Laboratory Volcanoes*, European Commission.

Druitt, T., M. Lanphere, and G. Vougioukalakis (1996). *Field Workshop, Santorini; Guide and Excursion Booklet.* IAVCEI Commission on Explosive Volcanism Workshop.

Friedrich, W.L. (2000) *Fire in the Sea*. Cambridge University Press.

Georgalas, G.C. (1962) *Catalogue of the Active Volcanoes of the World including Solfatara Fields*. International Association of Volcanology.

11 Volcanoes in Iceland

Iceland

God may have been thinking of geologists when He created Iceland. Here the outcrops are unrivaled, unobscured by trees, and set amongst unspoiled, uncrowded, and breathtakingly beautiful wilderness. Anyone seriously interested in volcanoes should have Iceland in their must-see list. Located atop the mid-Atlantic ridge, which cuts across the island from the southwest to the north, Iceland has all of the tectonic and volcanic ingredients of a typical mid-oceanic ridge, including spectacular gaping fissures that result from the spreading of the Earth's crust. The ridge, however, is only part of the story: Iceland is also placed above an oceanic hot spot, which results in a much greater yield of molten magma than is typical for mid-oceanic ridges. In fact, Iceland has produced one-third of all the lava erupted on Earth during the last 1,000 years. The interplay between hot spot and mid-oceanic ridge volcanism is responsible for Iceland's very high rate of magma production, which over millions of years has constructed a richly varied landscape. Here one finds not only basaltic shield volcanoes, typical of mid-oceanic ridges, but also features formed by silicic magmas (mainly rhyolites and andesites), which are usually associated with subduction zones. These include pyroclastic flows and widespread ash deposition from Plinian eruptions. The silicic magma beneath Iceland is produced by partial melting of the lower basaltic crust, caused by the high temperatures above Iceland's hot spot.

Iceland is often called the land of fire and ice, an interaction that is also reflected in the country's geology. About 10% of Iceland is covered by glaciers, but during the Ice Age the glacial coverage was complete. The country's volcanic activity has been strongly affected by overlying ice and its meltwater. Spectacular features were formed by eruptions under ice, such as table mountains and mounds of pillow lavas. Part of Iceland is still covered by ice-caps, and the largest – Vatnajökull – is the world's third largest. A subglacial eruption under Vatnajökull that started in late September 1996, melted through 500 m (1,650 feet) of ice within 30 hours. Heat from the eruption melted

Declining activity in the Grímsvötn eruption on December 27, 1998. Grímsvötn lies largely beneath the vast Vatnajökull icecap. The geothermal area in the caldera causes frequent jökulhlamps (glacier outburst floods) when melting raises the water level high enough to lift its ice dam. The 1998 eruption was a phreatomagmatic basaltic eruption within the Grímsvötn caldera. Surtseyan explosions such as this one occurred throughout the eruption. (Photograph courtesy of Magnus Tumi Guðmundsson.)

more than 3 km^2 (1 square miles) of ice and the meltwater accumulated in a subglacial lake, which later drained in a catastrophic jökulhlaup (glacier burst). Jökulhlaups are relatively common in Iceland. Some of the most voluminous have been from Katla, one of Iceland's most active volcanoes, located beneath the Myrdalsjökull ice cap in the southern part of the country. Katla has discharged jökulhlaups where the water has reached flow rates of over 100,000 m^3 (3.5 million cubic feet) per second. Because of the large quantities of sediments carried by the water, this activity has extended the country's southern coastline by several kilometres in the last four centuries.

One of the most fascinating aspects of Iceland's volcanoes is how they have shaped the lives of the population since the country was colonized by the Vikings 1,100 years ago. Icelanders have suffered major losses as a result of volcanic activity. The most serious disaster was the death of 24% of the population, mostly from starvation, caused by the effects of the "Laki Fires" eruption of 1783 destroying crops and killing livestock. However, the people not only have been willing to accept the dangers of living on such a volcanically active land, but they also have learnt to take advantage of it: a major example is geothermal power, which heats 85% of Icelandic homes. Much of the power used in the country comes from geothermal and hydroelectric sources. The lack of pollution from fossil fuels is one of great gifts from the volcanoes and glaciers.

Iceland is a remarkably pleasant country to travel in: it is safe, uncrowded, clean and, provided the weather is good, offers remarkably crystalline air and magical light for photography. The best months to visit Iceland are July and August, with the last 2 weeks in July being considered the best. The best way to explore the country is by renting a four-wheel drive vehicle, even though car hire is very expensive in Iceland. However, most of the locations described in this book can be reached using a standard car or even public buses. When using public transportation it is best to be prepared to do some camping and backpacking as well. An efficient way to see the country's geologic highlights and other interesting places is to take a bus or hiking tour. There are many of these tours available in Iceland and some are even guided by local geologists. Flights on light aircraft are a good way to get an overview of some of the country's most spectacular areas: they are available from Reykjavík and, although not cheap, are well worth their cost.

English is widely spoken in Iceland, and the people are friendly and helpful. There is very little crime and women can travel alone safely. As for personal safety on volcanoes, erupting or not, it is your own responsibility. Iceland offers everyone a remarkable degree of personal freedom but this means that visitors are expected to use their common sense and be responsible for their own safety. One of the ways that this concept of personal responsibility is manifested is in the lack of fences or barriers in potentially dangerous places, such as the edges of cliffs, craters, boiling hot springs, or waterfalls.

Visitors to Iceland are expected to help preserve nature and rock collecting is strictly forbidden except for research purposes (a special permit is required). The Icelandic flora and fauna also need to be protected as they subsist at the limits of tolerable conditions and are, therefore, very fragile. Iceland has beautiful wildflowers and first-class birdwatching, as well as some of the most interesting volcanic rocks to be found anywhere. Visitors should be careful to help preserve this unique environment for future generations to enjoy.

Tectonic setting

Iceland can be considered an anomaly in the North Atlantic region, because of its high rate of volcanism and the fact that, compared to other mid-Atlantic islands, it is substantial in both size and topography. Iceland owes its existence to an underlying hot spot, which enables high rates of volcanism to be maintained and, therefore, the island's substantial landmass to be built. Without the hot spot, Iceland would probably be just a series of small volcanic islands built by typical oceanic-ridge volcanism.

Iceland's tectonic structure is more thus complex than is typical for mid-oceanic ridge volcanoes. The active volcanic zones of the country consist of a series of elongated fissure swarms, each ranging from tens to over 100 km (60 miles) in length. Each swarm has a large volcanic complex in the center and these central complexes are the major pathways for magma to reach the surface. The fissure swarms trend northeast–southwest in southern and central Iceland, but north–south in the northern part of the country, where Krafla is located. The fissures in each swarm are arranged in *en échelon* (parallel) manner.

The tectonics in Iceland are dominated by the spreading of the mid-Atlantic ridge, caused by the pulling apart of the North American and Eurasian tectonic plates. Hence Iceland is being slowly split into two and gradually widened as new magma comes to the surface. The country's tectonic structure is far from simple when looked at in detail and there is no simple,

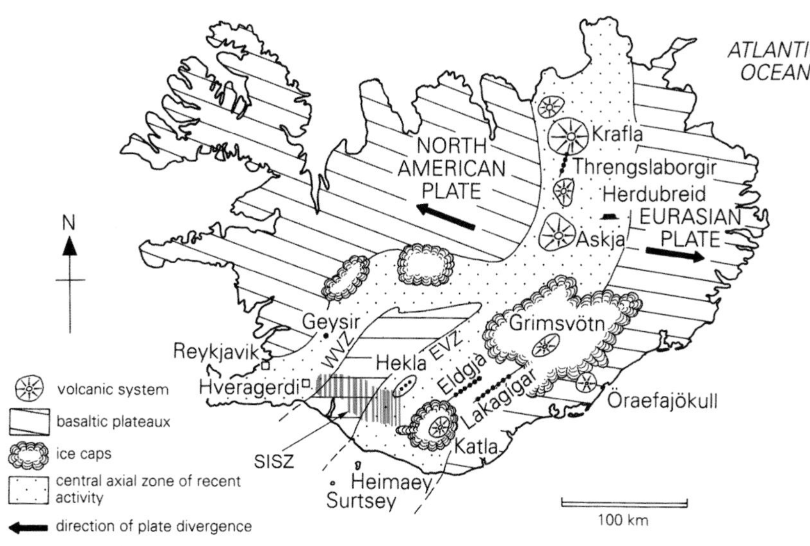

The major volcanic features of Iceland. The central axial zone of recent activity splits in the south into the Western Volcanic Zone (WVZ) and the Eastern Volcanic Zone (EVZ). These two zones are linked by the South Iceland Seismic Zone (SISZ), which is schematically represented in this figure. (Modified from Scarth, 1994.)

linear split. The mid-Atlantic ridge plate boundary is defined by the Western Volcanic Zone in Iceland's southwest and by the Eastern Volcanic Zone in the north and east of the country. The Eastern Volcanic Zone is the most active, containing Hekla, Katla, and Grímsvötn volcanoes, as well as the Vestmannaeyjar ("Westmen islands") further south.

The two volcanic zones can be considered as distinct portions of the ridge, where spreading is taking place at an average rate of 1.6 cm (⅔ inch) per year – about the same rate as the growth of our fingernails. These two portions of the ridge are linked together by a transform fault in the region called the South Iceland Seismic Zone, which is about 15 km (9 miles) wide and crosses the lowlands of southern Iceland from east to west. The transform fault represents the lateral slipping of one block of oceanic crust past another, offsetting the ridge. The fault ends abruptly at both ends, where it meets the two portions of the oceanic ridge. Shallow earthquakes are characteristic of transform faults. Hence, Iceland's major earthquake activity occurs in the South Iceland Seismic Zone and in the Tjörnes Fracture Zone to the north of the country, which is also a transform fault.

Earthquakes in Iceland can be destructive but major events only occur on average about once per century. Earthquakes tend to occur in swarms, both in terms of location and time. Known earthquakes in the South Iceland Seismic Zone took place in 1732–4, 1784, and 1896. An earthquake sequence in Vatnafjöll in 1987, where shocks reached up to magnitude 5.8 in the Ritcher scale, could represent the start of a new seismic episode. Even if this is not the case, there is still a 90% chance that an earthquake of magnitude 6 or larger will occur in the South Iceland Seismic Zone sometime in the next 20 years.

Eruptions in Iceland are as varied as the landscape they create. Hekla's eruptions can be highly explosive; some have been of Plinian type. Strombolian activity has also been common on Hekla and several other volcanoes. Eldfell volcano (Heimaey) was built by Strombolian explosions. However, the most characteristic eruption type in Iceland is a fissure eruption, appropriately referred to as Icelandic (see Chapter 3). Krafla has had two remarkable historical eruptions of this type, the latest ending in 1983. The fissures are formed by the tectonic plates moving away from each other. The motion stretches the crust for some time – 100 to 150 years – until a section rifts. Magma invades the fissure, though it does not reach the surface and cause an eruption every time. When an eruption happens, it is spectacular and relatively safe to watch, as attested by the many visitors to Krafla during its last awakening. When an eruption ends, magma clogs the fissure, making another dike. Lava-filled dikes are characteristic along the spreading and rifting axis of the country. However, lava dikes are not all the same. Recent studies have shown that many Icelandic dikes are not caused by vertical movement of magma, but rather by lateral movement: the dikes are fed by magma chambers underneath other volcanoes. The magma travels sideways along the dike, and can be erupted a long way from its source. This is why the magma that comes out in some fissure eruptions is different (more evolved) than what one would expect if it came straight up from the mantle: the magma had the opportunity to change while in residence at a chamber somewhere else. A classic case of this lateral migration is the famous Laki eruption in 1783: it is thought that the magma came from the Grímsvötn volcano, about 70 km (45 miles) away (see section "Hekla"). Understanding this sideways magma movement has been very important in Icelandic volcanology. Much progress in these studies was made during the last Krafla eruption, where seismic monitoring showed lateral flow of magma in the fissure swarm, forming dikes.

Practical information for the visitor

Transportation

Reykjavík is Iceland's capital and major airport. The nearest airport to Krafla is in the town of Mývatn, and daily flights from Reykjavík are available during the summer. Flights to Heimaey operate at least once a day from Reykjavík City Airport during the summer. It is also possible to fly to Heimaey from several locations in southern Iceland by using a charter service, Leiguflug Vals Andersen, based on Hella. The airport in Heimaey is less than 2 km (1¼ miles) outside town. A cheaper alternative to get to Heimaey is to take the car ferry from Þorlákshöfn, on the south coast. This trip, which takes about 3 hours, has the benefit of sailing past dramatic cliffs on arrival to Heimaey. The disadvantage is that the ride can be extremely rough if the weather is bad.

Car rental can cost a lot in Iceland. A cheaper alternative is to take local buses. There is an extensive bus service to most parts of Iceland, including the highlands. Schedules vary, but in the summer it is usually possible to reach many interesting areas by bus. For more information contact BSÍ Travel (Vatnsmýrarvegur 10, IS-101, Reykjavík). In Heimaey, the local taxi service is a good alternative to renting a car, though many visitors find that the island is small enough to be negotiated on foot.

Those wishing to go further afield and rent a car to explore Iceland's rugged interior should prepare their trip carefully, as this is not an activity to be undertaken without planning. All mountain and interior roads have loose gravel surfaces, which means four-wheel-drive vehicles are almost a necessity. The roads tend to be windy, narrow, and to go across shallow unbridged rivers. Traveling as groups of two or more vehicles is advisable. A helpful booklet called *Mountain Roads* can be obtained from the Iceland Tourist Board. Foreigners are often surprised to know that most roads are impassable and closed for most of the year, opening only in early July or later if the weather has been bad. By mid September, many of the roads are closed again.

Tours

Geological tours of Iceland are offered regularly by Icelandair's European Vacations and occasionally by geological societies or other specialist groups. Some geological tours are also available in Iceland. The Icelandic Touring Club (Mörkin 6, IS-108, Reykjavík) organizes some 200 hiking tours every year and they are open to non-members. The club owns over 30 mountain huts in uninhabited parts of the country

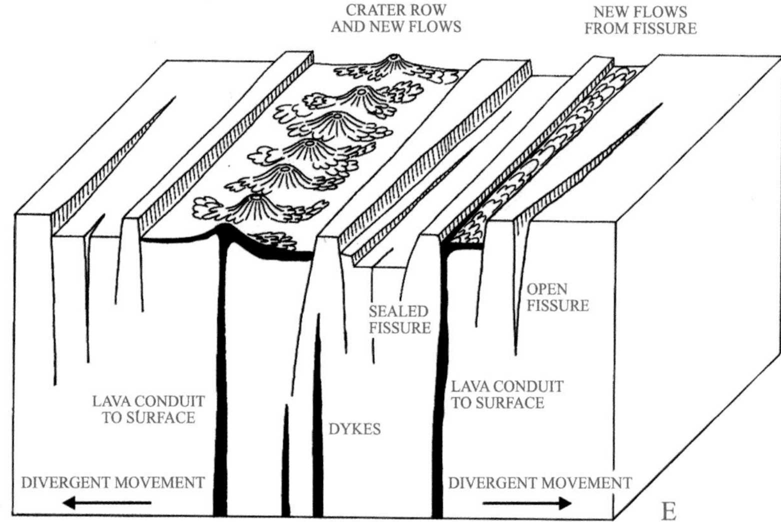

Sketch diagram of typical Icelandic rift and fissure eruptions. (Modified from Scarth, 1994.)

that are used as lodgings. There are a number of Icelandic tour operators who offer "adventure tours," some of which include hiking on volcanoes and, in the past, have included watching eruptions. Information on these locally arranged tours can be obtained from Tourist Information Centers, located in the capital (Bankastræti 2, IS-101, Reykjavík) and in major cities.

Local bus tours of Heimaey are generally available in the summer and can be booked through local hotels. Information on tours is also available at the Tourist Information Center in town. Eldfell is one of the major attractions in these local tours, but most do not allow people enough time to climb and appreciate the volcano. More interesting tours are those that take you around the island on a boat. These tours usually take a couple of hours, during which you will see the spectacular klettur (escarpments) where birds nest, and sea caves.

A visit to Krafla is included in the Grand Mývatn tour, which departs daily from the lakeside town of Reykjahlíð. Day tours to Krafla are also available from Reykjavík, flying into Akureyri. Although these trips are easy to arrange, most visitors seriously interested in the volcano will want to spend more time at Krafla than these tours allow. Local tours to Hekla depart from the town Hella – ask at the Tourist Information Center in Hella or at the Hekla Visitor Center. Some climbing tours can be arranged on an individual basis.

Sightseeing flights

Numerous flights are available from Reykjavík airport and they can be easily booked by local travel agents or directly at the Reykjavík city airport. Some helicopter tours are also available. Sightseeing flights of the Mývatn and Krafla areas depart from the Reykjahlíð airfield.

Lodging

Iceland has a variety of hotels and guest houses, but interesting alternatives are hostels and farmhouses. Lodging in the Mývatn area is plentiful in the town of Reykjahlíð on the northern shore of the lake and in several lakeside villages, including Mývatn. The Edda Hotel in Stóru Tjanir is further away, but is exceptional value for money. Edda Hotels are operated by the Iceland Tourist Bureau; most of them are school dormitories that are used as hotels during the summer months only.

The most popular place to stay near Hekla is the youth hostel (Gistihúsið) at Leirubakki, just 10 km (6 miles) away from the volcano on Route 26. It offers a variety of facilities, including private rooms, sleeping-bag accommodation in a dormitory, and outdoor camping. They also rent horses or, rather, Icelandic ponies, which are unique in having a fifth gait. (Icelanders are so concerned about the purity of the breed that ponies that go abroad are never allowed back in the country.) Those who don't feel energetic enough to try the ponies can soak in one of the two outdoor hot tubs that offer a fantastic view of Hekla.

A lot of tourists come to Heimaey in the summer, so a variety of lodging choices are available then. These include hotels, guest houses, and a youth hostel (Faxi Youth Hostel) which tends to fill up quickly. There is also a very popular campground in Herjólfsdalur, to the west of town. I stayed at the Hótel Brædraborg: the accommodation was nothing to write about, but the restaurant did serve a tasty puffin steak.

Camping is an alternative used by many visitors. Iceland has over 100 campgrounds that are normally open from early June to end of August or middle of September, depending on location. Camping outside the official grounds in not permitted. Reservations can be made through the Iceland Tourist Board (Lækjargata 3, Gimli, IS-101, Reykjavík). The Lake Mývatn camping site is located in Reykjahlíð. Bear in mind that Mývatn means "Midgewater" – a well-deserved name but, fortunately, not all the time. There are photos of the lake that show black clouds of these pests hovering over the water, but most visitors are not likely to encounter them in such great numbers. Most of the midges do not bite, but it is a good idea to protect yourself with insect repellent and to take along a head net, just in case they become too much or too many to bear.

Safety and emergency procedures

Medical care and rescue services are very good in Iceland. Only citizens of the UK and Scandinavian countries are covered by the national health insurance, so all others need to have appropriate medical insurance. Almost everyone in Iceland can speak English, so communication is not a problem.

Maps

The Icelandic Geodetic Survey (Laugavegur 178, IS-125, Reykjavík, phone 354-5334000) sells a wide variety of maps, including geologic maps. Copies of aerial photographs can also be ordered here. Bookstores in Reykjavík and Akureyri generally stock many of the survey's maps. A geologic map for the Mývatn area (Norðausturland) is available in

1:250,000 scale and a topographic map showing lavas of 1975–80 is available in 1:100,000 (Húsavík/Mývatn). Topographic maps are also available in scales 1:250,000 and 1:50,000.

The 1:250,000 General Section map of southern Iceland (sheet 6 in the series) is useful for an overview of the Hekla region. At the scale of 1:100,000 (with contour interval of 20 m) there are two maps available for Hekla. The Þórsmörk/Landmannalaugar Atlas sheet map is the most up-to-date, showing several lava flows including those of 1991. It covers Hekla and the area to the south and east of the volcano. The Hekla map (sheet 57 in the series) is part of an older series last revised in 1987. There is also a map of Hekla in the topographic series (1:50,000 with contour interval 20 m), but it was last revised in 1988 and does not show the 1991 lava. A geologic map of the region is available in 1: 250,000 (sheet 6, south Iceland), but this was also made prior to the last two eruptions. Since the Survey is updating its maps, more recent versions may soon become available.

Maps of Heimaey are also available at several scales. The 1:50,000 map is the most useful for visitors: one side of the sheet has a topographic map of Heimaey and the other shows the Vestmannaeyjar, a map of the town, and even a geological map.

Krafla

The volcano

Krafla is not so much a volcanic mountain but a volcanic region associated with a major fissure swarm 100 km (60 miles) long and up to 10 km (6 miles) wide, trending north–south. The Krafla region has been active for about 200,000 years, since the last stages of the Ice Age. There is a small, 800 m (2,640 feet) high mountain within this region which is specifically called Krafla, but it has not erupted in post-glacial times. This small Krafla is made up of palagonite, which is basaltic glass that cements together with time and turns brownish.

The region known as the Krafla central volcano is a low, broad shield some 25 km (16 miles) across, dominated by a 10 km (6 miles) wide caldera, inside which is a much-visited high-temperature geothermal field. About 5 km (3 miles) to the south of the caldera is another geothermal field very popular with visitors, Námafjall. These colorful geothermal areas, with sulfur deposits giving the ground tones of yellow, orange, and brown, are contrasted by black lava fields made up of tholeiitic basalts, dotted with fumaroles and spatter

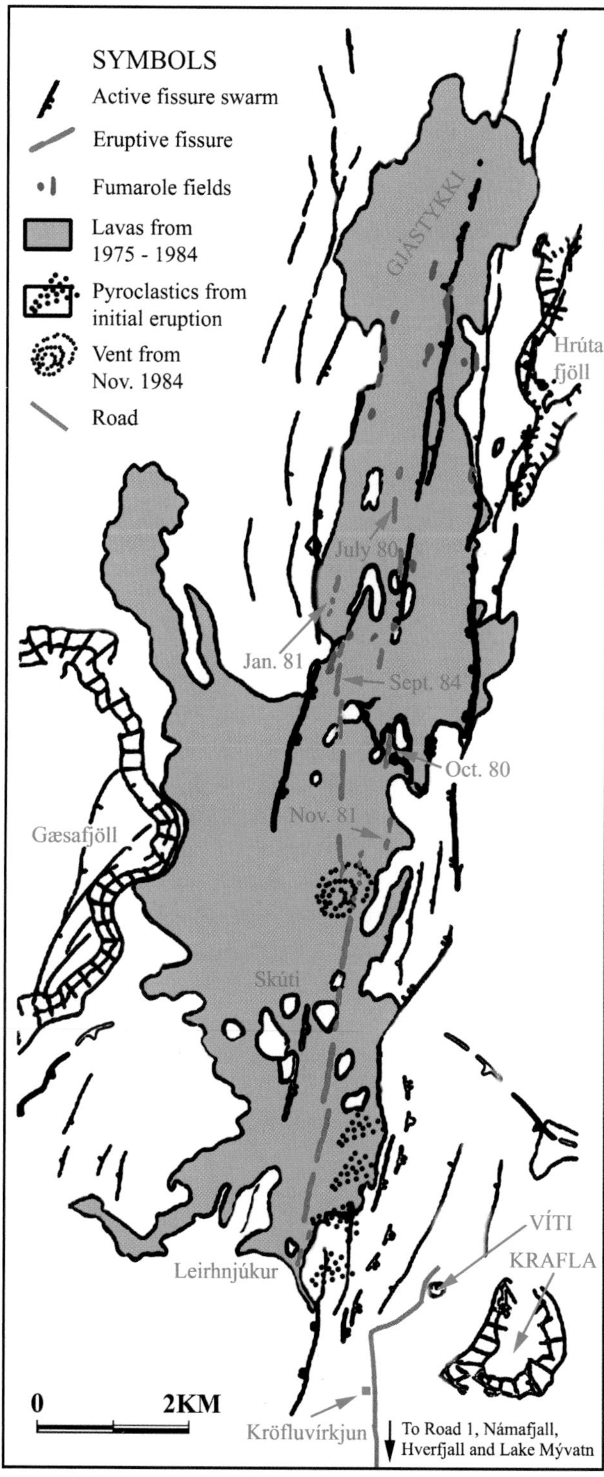

Sketch map of the Krafla region showing lavas and fissures from the Krafla fires eruption of 1975–84. The locations of Víti crater (near parking lot) and of the Leirhnjúkur geothermal field are marked. Lake Mývatn, Námafjall, and Hverfjall are located further south. (Modified from map on the website of the Nordic Volcanological Institute, Iceland.)

cones, which are amazing to hike on. Krafla offers some of Iceland's most spectacular volcanic sceneries and has the added advantage of being easily accessible. Predictably, it has more visitors than any other Icelandic volcano, but one could hardly call it a crowded place.

The magmas erupted in the Krafla region have been of basaltic and more silicic compositions (rhyolites and andesites). A wide variety of volcanic features can be found here, including both subglacial (erupted under ice) and surface eruption products, such as pillow lavas, lava flows, and craters formed by explosive activity. The caldera that dominates the landscape was formed by a large explosive eruption of silicic magma near the end of the last interglacial period, which also produced pyroclastic flows up to 15 m (50 feet) thick. During the last glacial stage, the caldera was partially filled in by eruptions.

Krafla's first post-glacial activity is known as the Lúdent episode, which began in the eastern part of the volcanic system and ended about 8,000 years ago. This episode produced the lava shield Gjástykkisbunga ("shield of the fissure area") as well as lavas flows and tuff rings on the main fissure. After the Lúdent episode, the region was invaded by an unusually large basaltic lava flow, known as the Older Laxárlava. The source of the lava was the Ketildyngja ("kettle shield") volcano, 25 km (16 miles) south east of Krafla. The flow dammed up the Laxá ("salmon") river and created the lake basin which today contains Lake Mývatn. In spite of its unappealing name, Lake Mývatn is one of Iceland's most picturesque regions and is very popular with Icelanders for summer vacations, as well as with birdwatchers from all over the world. Thanks largely to the lake's abundant supply of food in the form of the annoying insect life, Mývatn is home to thousands of birds, including more species of breeding ducks than anywhere else in Europe.

The next major period in Krafla's history occurred about 3,000 years ago and is known as the Hverfjall ("crater mountain") episode. The activity, which had by then shifted to the central axis of the Krafla system, began with a large phreatomagmatic eruption caused when magma from the southern part of the fissure entered the lake. The explosive eruption produced a series of craters, the largest of which is known as Hverfjall. This is a basaltic explosion crater created by the interaction of water and magma, and has been classed by geologists as a tuff ring. Hverfjall is one of the Krafla system's most interesting places to visit.

Krafla's activity continued shortly after the Hverfjall episode, when a major fissure rifted the surface and produced the lava flow known as the Younger Laxárlava. The lava covered an area of 170 km^2 (66 square miles) and, in places where it flowed over wet ground or into shallow lake waters, it produced steam explosions that fragmented the lava forming small cones. These cones are known as pseudocraters, because they are not "true" craters, that is, they were not formed by magma exploding over a vent area. Iceland has many examples of pseudocraters, because of the prepondence to wet ground, and some good examples can be seen in the Mývatn area. Other interesting features formed by the Younger Laxárlava are the lava pillars and other strange formations at Dimmuborgir, another of the region's popular attractions.

The most recent eruption prior to that of 1975 started in 1724 and lasted until 1729, with minor activity continuing as late as 1746. Known as the Mývatnseldar ("Mývatn Fires"), it began with an explosive phreatic eruption that formed the Víti ("Hell") crater, one of the points of interest on a visit to Krafla. In 1727, large quantities of basaltic lava outpoured from a fissure 13 km (8 miles) long, covering an area of 33 km^2 (13 square miles), with a total volume of 0.25 km^3 (9 million cubic feet). The last lava flow, erupted from Leirhnjúkur ("clay mountain") crater in June 1729, destroyed or damaged several farms, including the parsonage of Reykjahlíð, and entered the lake on the northern side. In Reykjahlíð, the lava surrounded but did not destroy the wooden church, an event seen as miraculous by many. The church is no longer there, but another has been built on its place and a wooden carving inside it depicts the lava flow surrounding the original church. After 1729, Krafla was to remain quiet for over 200 years.

The 1975–83 eruption

Known as the "Krafla Fires," this volcano's most recent series of eruptions begun in December 1975 and lasted until September 1984. The spectacular activity, which included rows of fire-fountains and great outpourings of fluid basaltic lava, presented relatively little danger to sensibly placed observers and was one of the world's most photographed eruptions.

Krafla started to give warning signs in mid-1975, in the form of earthquakes that reached magnitude 4 in the Ritcher scale. The first but short-lived eruption started where the Mývatn Fires had ended: the crater Leirhnjúkur. A small lava flow came out from the crater on December 20, 1975. More significant was

Krafla in vigorous activity, 1984, during a classic fissure eruption displaying fire-fountains and lava flows emerging from a fissure. (Photograph courtesy of Eysteinn Tryggvason.)

an increase in seismic activity: it quickly propagated from the confines of the caldera to the northern coast of Iceland and towards Lake Mývatn, a total distance of 60 km (37 miles). Immediately after the eruption, the fissure swarm in the region subsided, sinking by up to 2 m (6 feet) and widening by up to 1.5 m (5 feet). The caldera floor also subsided. A few months later, the caldera floor gradually inflated again.

This deflation–inflation pattern turned out to be typical during the next few years and the regularity of the pattern allowed the forecasting of deflation episodes, when lava was likely to erupt at the surface. For weeks or months prior to magma breaking out at the surface, the ground would slowly rise as the underlying magma chamber was being invaded and expanded by new magma. The rate of inflation at the center of the caldera floor was about 7 to 10 mm (⅓ to ⅖ inch) per day. When the inflation reached a critical level, the seismic activity increased, until rapid deflation happened. The critical level was reached when the pressure in the chamber exceeded the strength of the surrounding rock. The deflation that followed, signaled by an abrupt drop in the level of the ground, meant that magma had migrated sideways from the chamber into the fissure zone north and south of the main caldera and that a small portion of it was moving upwards to the surface. From 1975 to 1979 most of the magma moved sideways within the crust, with minor surface eruptions. However, from 1980 to 1984, fissure eruptions at the surface became the dominant activity, with large-volume lava flows. The total lava volume extruded by the Krafla Fires was about 0.25 km^3 (9 million cubic feet), covering an area of 35 km^2 (13½ square miles) – very close to the estimates for the earlier Mývatn Fires.

Spectacular as the eruption was, with fire-fountains reaching 50 m (165 feet) above ground, the most fascinating aspect of it happened below the ground: it was what the typical pattern of inflation and deflation was able to tell us about volcanism at the mid-Atlantic ridge. Krafla is a surface expression of the rifting of the two tectonic plates either side of the ridge and observations of this last series of eruptions has been of considerable value in understanding how plate rifting happens. We know that eruptions do not happen frequently at Krafla, as the Mývatn Fires are the only other historic activity. However, when eruptions do happen, they come in clusters. The explanation for the centuries-long intervals between active periods, as well as for the periodic nature of the eruptions once they start happening, lies in the way that the tectonic plates under Iceland are rifting and moving apart. During the volcano's quiet intervals, plate movements are building up stress that is finally released by short bursts of rifting and volcanic activity. Krafla's lesson has been that the plates are not moving apart smoothly and gradually, but rather in fits and starts. The average rate of plate spreading under Iceland is about 1.6 cm (⅔ inch) per year, but across the Krafla fissure swarm it can be as much as 5 cm (2 inches) per year. Moreover, because of the often abrupt way in which the movement occurs, it is sometimes possible to see overnight changes at Krafla – astonishing manifestations of the rifting of the Earth's crust.

A personal view: Harnessing the power

My introduction to Krafla came in the last few days of a 2-week geological field trip to Iceland, by which time all of us in the group had already seen so much breathtaking geology that the word "amazing" was beginning to sound commonplace. I had, therefore, wondered if this topographically unimpressive volcano would be disappointing. This was far from the case: for a start, on Krafla we felt that we were walking on a truly active volcano. Even though the "Fires" no longer rage, there is plenty of steam making parts of the black lava fields feel like Turkish baths, which can be rather pleasant under Iceland's often inclement weather. The sulfurous smell, so characteristic of active volcanoes, is another reminder that Krafla is very much alive, even if its spectacular eruptions do not often happen. We were also thrilled to see Krafla's colorful geothermal fields, which are among the best in Iceland, and the Hverfjall crater, a superb example of a tuff ring.

Geologic wonders aside, Krafla offered me a fascinating glimpse into the interaction of Icelanders with their volcanoes. The fact that Iceland sits atop a hot spot has provided great benefits to the people in the form of cheap geothermal energy, of which Icelanders use more per capita than anyone else in the world. Visitors appreciate the fact that even the simplest lodgings are pleasantly warm and often have an outdoor spa for relaxing in after a day of hiking.

Krafla has two high-temperature geothermal fields and it is not surprising that Icelanders decided to exploit this resource by building a geothermal power station there. Unfortunately, their timing was wrong. Building work, which started in 1974, was soon interrupted by the 1975–84 eruption. Some Icelanders like to say that the power station became the world's most expensive tiltmeter: it turned out that the 50 m (165 feet) long roof provided a great bench for tiltmeters, which were set up to detect the inflation and deflation of the ground during the eruption.

It soon became clear that this unfortunate power station has been the subject of some ridicule, maybe because it is known that political pressures forced construction to begin before the results of the test-drilling program were completed. The eruption was not the power station's only problem: the hot water tapped by the wells was about 350 °C (660 °F), nearly double that expected. Besides that, pulses of carbon dioxide and sulfur dioxide gas corroded metals from well casings, making several wells unusable. But despite all the problems, the power station survived and started operating in 1977. It now produces 19 megawatts using steam from 11 boreholes and one of the two installed 30-megawatt geothermal turbines. The station, called Kröflustöð, is not open to visitors, but it can be seen from the road (number 863) going to Krafla.

Another local operation that suffered from its proximity to Krafla was the diatomite factory, an unusual local industry. Lake Mývatn's floor is covered by a 3 to 6 m (10 to 20 feet) thick layer of diatomite-rich mud. Diatoms are the remains of a single-celled type of algae that builds shells out of silica, which is brought into the lake by flowing warm water from the geothermal fields. Diatomite is used for filtering, as a filler in fertilizers, and in cosmetics. It is extracted by pumping mud from the lake bottom and drying it in a plant near the lake's shore. Given the proximity of Krafla, it is logical that geothermal heat should be used for drying the mud, hence the plant has steam wells. In 1977, Krafla played havoc with the operation by shooting lava through one of these steam wells. Icelanders like to joke about the well being the world's smallest volcano. This volcanological curiosity no longer exists, but the diatomite plant (called Kísilgúrverksmiðja) has been in operation again since 1980. The holding tanks full of turquoise-blue water are a conspicuous sight from the bypass road to the north of Highway 1. Further along the road, in the area called Bjarnarflag, one finds another example of how locals use volcanic heat: they have built an ingenious underground bread oven where hverabrauð – a cake-like loaf – is slowly baked for over 20 hours. One can see small glass doors into the ground and the hverabrauð (literally "hot spring bread") can be bought in Reykjahlíð. Apparently, farmers used to grow potatoes in these fields, but gave up when the geothermal activity increased and they found themselves digging up half baked potatoes. The heat, however, works very well for the hverabrauð, a delicacy with special appeal for those of us who have cooked (and burnt) dinners on active lava.

Visiting during repose

The most interesting areas of Krafla are not far from roads. Road number 863 takes you to a parking place near the Víti crater. There is a path up to the rim of Víti and another path up through the 1975–80 lava to the Leirhnjúkur solfataras and crater. A walk on the new flow, past steaming fissures and spatter cones, is not to be missed. Away from the main Krafla area, the most interesting locales are the tuff ring Hverfjall, the hot springs at Hveraröng, the fields of pseudocraters near Lake Mývatn, and the ancient lava formations at Dimmuborgir.

Víti crater

This impressive 320 m (1,060 feet) wide crater containing a photogenic blue lake is located within the Krafla caldera, but breaching its western rim. Víti was the explosive vent that marked the beginning of the Mývatn Fires eruption of 1724. The series of eruptions that formed the crater lasted for 5 years, starting with a phreatic explosion on May 17, 1724. The erupted material consists mainly of greenish, fine fragments that came from highly altered (changed) rocks, mixed with silicic pumice and basaltic scoria. The ejection of great amounts of altered rocks tells us that the explosion involved the hydrothermal system of the volcano, as exposure to hydrothermal activity causes such changes in rocks. Geologists visiting this site may want to look for the large xenoliths of granophyre (an

A layer of ice clings to the sides of Víti ("Hell") crater in July 1995, proving that in Iceland, Hell does freeze over. (Photograph by the author.)

intergrowth of the minerals feldspar and quartz) that were also ejected by the explosion and are common on the crater rim. Xenoliths are fragments of pre-existing rock which are carried to the surface in the magma and can often be seen as round nodules of crystals set into lava. (Xenoliths are described in more detail in the "Hualalai" section.)

Víti is no longer active, so it does not merit its name, which is Icelandic for "Hell." However, during the Krafla Fires (which began in 1975) Víti gained a smaller companion, formed when a borehole exploded in the vicinity, creating a mini-crater. Icelanders were quick to name this essentially man-made crater Sjalfskaparvíti, meaning "self-inflicted hell."

Leirhnjúkur

One of the major attractions of the Krafla area is Leirhnjúkur hill and its solfataras. Leirhnjúkur was another of the Mývatn Fires vents and quite a spectacular one. In January 11, 1725, this hill split open, spouting lava. Large lava flows were erupted from here during the 1724–9 Fires, including the one that threatened Reykjahlíð. Leirhnjúkur woke up again in 1746, with a powerful but short-lived eruption. Several of the lava flows from the Krafla Fires eruption came out from fissures north of Leirhnjúkur. Leirhnjúkur itself was active as recently as 1984. It is interesting to see the contacts between the younger lavas, which are very dark, and the older ones, which tend to be dark brown rather than black in color. You can get a good view of the lava fields from the rim of Leirhnjúkur. From the bottom of the hill it is easy to make your way along the fissures. The 1984 lava is still steaming and you can walk on places that give the feeling of a natural steam bath.

The Leirhnjúkur solfataras can also be easily reached from the parking lot. The pastel colors of the deposits surrounding the mudpots and steaming fumaroles are very attractive, but the area has its dangers: stay on the paths, as there may be hidden fissures and holes covered by thin deposits.

Hiking on Krafla in 1995. Plenty of steam still comes out of fissures reminding visitors that this is an active volcano. (Photograph by the author.)

Gjástykki ("fissure area")

Located at the northern end of the Krafla fissure swarm area, this impressive red hill, surrounded by lava flows, was the vent area for some of the Mývatn Fires lavas and was also a very active vent area for the Krafla Fires. It can be visited as a day hike from Leirhnjúkur.

Hverfjall

This beautifully symmetric, 163 m (540 feet) high basaltic tuff ring is a prominent landmark in the region. It can be reached in a day by walking from Reykjahlíð but it is also possible to drive to the crater following a dirt track. The crater itself is just over 1 km (⅝ mile) wide and its rim can be easily reached by a path that cuts diagonally across the flanks on the northern side. Hverfjall is the largest of the craters formed by the phreatomagmatic explosive activity that took place along a 25 km (16 miles) fissure some 3,000 years ago. Be sure to walk around the rim and examine the tuff beds, most of which dip away radially from the centre of the crater, but some dip 15 to 35 degrees towards the centre. Some of the beds were formed by wet surge flows and it is possible to see some flow structures on them. In some places one can see accretionary lapilli, which are small "mudballs," formed when particles thrown up by the explosion gathered ash from the air as they fell. You may also notice some larger blocks interbedded in the tephra layers: these were older blocks thrown out by the more violent steam explosions. While walking around the rim, you can see faults parallel to the rim, cutting across the crater, which are evidence that the activity took place on a fissure swarm. Inside the crater is a smaller ash cone. A curious feature of Hverfjall is the graffiti on the floor of the crater, written using light-coloured rocks. This is a remnant of the days when the crater floor used to serve as a hikers' message board. However, because of the erosion caused by hikers descending from the rim, the crater floor is now off-limits to visitors.

Lúdent

This crater was chosen as a training ground for astronauts, including Neil Armstrong, during the days of the Apollo missions to the Moon. Lúdent is the largest of the Lúdentsborgir crater row east of Mývatn, which were formed by explosive eruptions between 6,000 and 9,000 years ago. Lúdent itself is nearly 1 km (⅝ mile) in diameter and 70 m (230 feet) deep, topped by an almost perfect oval crater. It can be reached by a track that goes around the southern base of Hverfjall and continues some 5 km (3 miles) to the southwest through the Lúdentsborgir row. The oldest crater in the area is Lúdentskal, formed about 9,000 years ago. On this crater's northwestern flank is a lava flow of the volcanic rock known as Icelandite, which is similar to andesite but has less aluminum and more iron. Those wishing to walk for a further 2 km (1¼ miles) to the south beyond Lúdent will come to the fissure swarm known as Þrengslaborgir, also worth a close look.

Dimmuborgir

The "dark fortress" is one of the most popular attractions in the Mývatn area. Located to the southwest of Hverfjall, Dimmuborgir was formed some 2,000 years ago by the Younger Laxárlava, which is mostly of the aa type, when it ponded in the area. The 20 m (65 feet) deep lava pond eventually breached a ridge to the west and was drained, leaving behind pillars formed by lava which had begun to solidify. The pillars show ledges or benches which represent "tide marks," indicating the different stages of lava drainage. Notice also the vertical marks formed by the crust scratching cliff walls and pillars when it subsided. There are several marked paths for visitors to walk along in Dimmuborgir, which are described in leaflets one can pick up by the parking lot. One of the most interesting routes is Kirkjuhringur (the Church route), which takes you to the lava cave Kirkjan (Church). The cave was given this name because of its resemblance to a vaulted gothic cathedral. All the paths meander through interesting lava formations, including lava channels and tubes, and spectacular pillars showing lava cooling patterns. Because of the large number of visitors and the potential for environmental damage,

Hiking on Námafjall geothermal field. (Photograph by the author.)

wandering off the paths is not allowed. Be aware that the paths are rough, the terrain uneven, and that numerous small but deep cracks are to be found all over the area.

Námafjall

This striking pastel-colored palagonite ridge, with numerous steaming fumaroles and hot springs, used to be the site of an important sulfur mining industry. Its name actually means "mountain with a mine" and the sulfur found there was exported and ended up as gunpowder, fuelling many European wars. Námafjall is located 6 km (3¾ miles) east of Reykjahlíð, south of the ring road. A trail leads from the road to the top of the ridge and the view from the top makes the easy 30 minute climb well worth the effort. The ridge, formed by a fissure eruption, sits atop the mid-Atlantic rift, as does its neighbor ridge to the north of the highway, Dalfjall. The spreading of the mid-Atlantic rift is gradually splitting Dalfjall into two, as shown by a growing notch on the ridge. Námafjall itself sports a series of cracks, the largest of which was formed in 1975. Very high near-surface temperatures have been recorded at Námafjall: during exploration drilling for the Krafla power station, the temperature measured at 1.8 km (1⅛ miles) depth was 260 °C (500 °F). This is another area that visitors should be particularly careful walking on, as the crust near some of the boiling mudpots can be only a few millimeters thick.

Hverarönd ("hot spring line")

This geothermal field just east of Námafjall is Iceland's best known area of mudpools and solfatara activity. It has numerous pools of hot, blue-gray bubbling mud, as well as hot fumaroles and even small fuming cones which look like mini-volcanoes, formed by spatter from the bubbling mudpools. Hverarönd can be easily reached by road. As in all geothermal areas, visitors should stay on paths and avoid walking on light yellow or white colored terrain, which may be unstable.

Pseudocraters

Many of these "fake" craters are found in Iceland. They differ from "real" craters because they don't represent actual vents from where magma came out. Pseudocraters are formed by explosions when lava comes into contact with waterlogged ground, causing

The author tests the temperature of a "mini-volcano" in the Námafjall geothermal field. (Photograph courtesy of Armando Ricci.)

trapped subsurface water to turn explosively to steam. The steam explosions break up the lava and form small scoria cones with craters within, which look remarkably like spatter cones formed over vents. One striking difference between the two crater types is that "real" craters tend to occur either isolated or, when they are erupted over fissures, aligned in rows. Pseudocraters, however, tend to occur in clusters. The most striking pseudocraters around Mývatn are located on the lake's western shore, east of Vindbelgjarfjall and they can be more than 300 m (1,000 feet) across. Smaller examples, some only a couple of meters (6 feet) across, can be found along the southern shore, and as islets within the lake. The most easily reached swarm of pseudocraters is the Skútustaðagígar, located near the village of Skútustaðir on the southern shore. This village is famous because of its association with saga-times villain Vigaskúta (Killer Skúta), who was partial to murdering his neighbors. From the village, it is easy to visit the pseudocrater field near the pond called Stakhólstjörn, where there are well-marked trails wandering over the craters.

Visiting during activity

How likely is Krafla to erupt again in the near future? After the 1984 event, the caldera begun to inflate again, but inflation ceased in early 1985, with some minor and intermittent inflation occurring in October 1986. Since 1989 the caldera has shown a steady and slow deflation, unlike the abrupt patterns that foretold of eruptions during the Krafla Fires. Although Krafla has not shown signs of impending activity, it could certainly erupt again anytime. However, a potential eruption should not be a cause for concern. The volcano is monitored and warning signs of an eruption, such as earthquakes and ground movement, are likely to be detected days or weeks before the event. Also, Krafla's eruptions tend to be relatively mild, consisting mostly of long fissure vents spouting fire fountains and lava flows. However, violent phreatic explosions have occurred in the past. There is also the danger of simply being in the wrong place at the wrong time. When the last eruption started, a Danish tourist was standing quite close to the area where the ground split, rifted apart, and lava started coming out. Luckily, the Dane got away quickly, but he was in significant danger, as the low-viscosity lava was moving quite fast – up to 10 m (33 feet) per second, which is faster than people can run.

As there have been only two major documented eruption events at Krafla, it is hard to say how an eruption would begin. If we take the two events as typical, then we can say that the next eruption is likely to start as localized eruptions, perhaps preceded by earthquakes and ground movement. The continuing activity is likely to be dominated by lava flows and fire-fountains, of limited danger and of immense interest to visitors. Icelanders like to watch eruptions and they flocked to Krafla during the Krafla Fires. Another eruption will no doubt attract many visitors and it is reasonable to expect that viewing areas will be designated.

An eruption from Krafla is likely to be spectacular and well worth a special trip. However, local facilities are limited and are likely to be in short supply as people rush in to see the fireworks. It would, therefore, be important to book accommodation and transportation. The time of the year is also important, as it can be difficult to travel in Iceland during the winter and the hours of daylight are few. If past Krafla eruptions are typical, the activity is likely to last for a long time, so you may consider waiting for good weather, though pauses in the activity could happen.

If you make your own way to Krafla during activity, study the safety rules for Hawaiian-style eruptions but remember that steam explosions have often taken place on Krafla. Do not get close to lava flows going over waterlogged or icy ground, or entering a lake or pond. Follow the directions and recommendations given by the locals and keep in mind that Icelanders expect you to take responsibility for your own safety: they are unlikely to expend significant resources to make sure that visitors don't get into trouble.

Those who want to see and photograph the eruption from the air can arrange a sightseeing flight from the Reykjahlíð airfield or from Reykjavík. Some of the most spectacular photos from the last eruption were taken from the air. Finally, remember that an eruption is likely to cause significant disruptions in the area and that many local attractions, such as the old lava fields, may be off-limits. For example, during the last eruption the thermal activity in Námafjall increased so much that the area had to be closed to the public. Krafla is, in fact, a volcano that should be visited both in repose and in activity.

Other local attractions

Reykjahlíð

This is the Mývatn area's principal town, with about 600 residents. Many tours are available from here, not only to Krafla and Mývatn, but also to the Askja caldera, the ice caves of Kverkfjöll and the stunning Dettifoss ("falling falls"), Europe's most powerful waterfall. An interesting sight in Reykjahlíð is the Leirnjúkur lava of 1729, which surrounds the church and shows good examples of ropy lava and tumuli. The present church was built in 1972, but it is located on the same site as the old wooden church. Some say that the old church was miraculously spared in 1729, when the lava parted into two and flowed around it (the fact that the church was located on a low rise no doubt helped). Something else worth seeing while in town is *The Volcano Show*, which is also available in Reykjavík. The movies last a couple of hours and have some of the most spectacular footage of volcanic eruptions available anywhere. The show usually includes the award-winning *Birth of an Island*, about the eruption of Surtsey. The owners of *The Volcano Show* have been filming eruptions in Iceland for over 40 years, so this is a way of seeing what the more recent eruptions of Krafla and Hekla were like.

Askja

The Askja caldera is located in the remote region of central Iceland, but getting there is worth the effort. There is a road going nearly all the way to the caldera's edge but it is an Icelandic interior road that requires a four-wheel-drive vehicle, and even then is sometimes impassable. An easy way to see Askja, though briefly, is to take a tour from Reykjahlíð, which is available daily in the height of summer. Askja is the central part of the 20 km (12 miles) diameter Dyngfufjöll collapse caldera. Askja's caldera – which is large at 50 km^2 (19 square miles) – formed very recently, during a violently explosive eruption that started in 1875 and lasted for 30 years. The main caldera was born early in the eruption, but towards the end of the activity a smaller crater was formed, about 11 km^2 (4 square miles) and some 300 m (1,000 feet) below the rim of the original. This younger crater became filled with water and now contains the 217 m (715 feet) deep lake Öskjuvatn, the deepest in Iceland. When the ice melts in the summer, the lake waters are extraordinarily blue. There is a smaller lake at the bottom of Víti, a tephra crater located near the northeastern corner of Öskjuvatn. The lake at Víti is deliciously warm and attracts many swimmers. It can be reached by a somewhat slippery path descending from the rim. Askja was last active in 1961, when it formed a row of explosive craters named Vikraborgir, which can be seen near the road entrance to the caldera.

Akureyri

Its name means "meadow sand pit," but this is actually Iceland's third largest town (with a population of 14,000!). The town is located near the end of a major fjord, the Eyjafjördur ("island fjord"), and sits on a gravel and sand terrace built up by alluvial deposition. The 1144 m (3,775 feet) high mountain Súlur, a silicic volcanic center, rises above the town and can be climbed in a day. Akureyri is a good place in which to shop for books and maps as well as for the traditional Icelandic goods. Sunny skies are frequent in this area and the weather is considered the best in the country. Visitors are often surprised to find African and Mediterranean plants growing outdoors in the town's famous Botanical Gardens (Lystigarður Akureyrar), because the gardens are located just a few kilometres south of the Arctic Circle. A major nearby attraction is one of Iceland's most beautiful waterfalls: Goðafoss, the waterfall of the gods. The falls were formed by the glacial river Skjálfandafljót cutting through the 8,000-year-old Bárðardalur lava flow. The flow originated from the crater Trölladyngja in the south of the country, near the Vatnajökull glacier.

Heimaey

The volcano

Heimaey ("Home Island") is a 16 km^2 (6 square miles) volcanic island about 9 km (5½ miles) off Iceland's southern coast, which captured the world's attention with its unexpected eruption in 1973, its first in historic times. The island has two major volcanic cones: the 227 m (750 feet) high Helgafell ("Holy Mountain"), which dominated the landscape before 1973, and Eldfell ("Fire Mountain"), built by the 1973 activity to its present height of 221 m (729 feet). Although neither can be con-

The port of Heimaey, with Eldfell and Helgafell in the background. (Photograph courtesy of Armando Ricci.)

sidered particularly impressive as a volcano, the island is well worth visiting, because the easily accessible products of the 1973 activity represent more than just interesting volcanic landforms: they are also the grounds of a riveting battle between Eldfell and a small contingent of volunteers who would not surrender Heimaey to the destructive power of the new volcano. Their efforts have gone down in history as the first truly ambitious attempt to minimize and control the damage caused by an eruption. What is most remarkable is that their effort was successful. After the eruption was over, the people returned, cleaned up the ash, rebuilt where necessary, and went back to fishing. The eruption gave them some benefits: the island became 20% larger, the harbor was improved, the cooling lava turned out to be a source of inexpensive heating, and it is even reported that the warm crater became a favorite meeting place for lovers. The eruption made Heimaey world famous, not because it was particularly noteworthy as far as volcanic activity goes, but because of the courage and ingenuity of the people who fought against it.

The 1973 eruption was unexpected. There were no recognizable precursor signs and Heimaey had not had an eruption for several thousand years. Carbon dating on Helgafell indicates that the last eruption happened around 3450 BC. However, there had been two, possibly three, historical eruptions elsewhere in the Vestmannaeyjar ("Westmen islands") archipelago, of which Heimaey is the only inhabited island. The best known of these previous eruptions is Surtsey, the new island which was formed between 1963 and 1967, about 40 km (25 miles) to the southwest of Heimaey. All the Vestmannaeyjar are part of Iceland's Eastern Volcanic Zone (see section "Iceland" for a description of Iceland's tectonic regime), thus they are located on the same major fissure system as Hekla. The Eastern Volcanic Zone is propagating to the south of Iceland and advancing into relatively thick (15 to 25 km, 9 to 16 miles), old lithosphere. The Vestmannaeyjar are part of a mostly submarine volcanic complex about 40 km (25 miles) long and 30 km (19 miles) across, with Heimaey's position being central. The volcanic complex has 18 islands and about 80 vents which have been active in the last 10,000 years.

Heimaey itself was built by several volcanoes that at one time formed three separate islands. One of the islands is the present northwestern part of Heimaey, Norðurklettur, where the volcanic centers of Dalfjall and Stóraklif are located. The second island is now Heimaey's eastern extreme that acts as a shelter for the harbor and consists of Heimaklettur, Miðklettur, and Ystiklettur. Those two islands were formed about

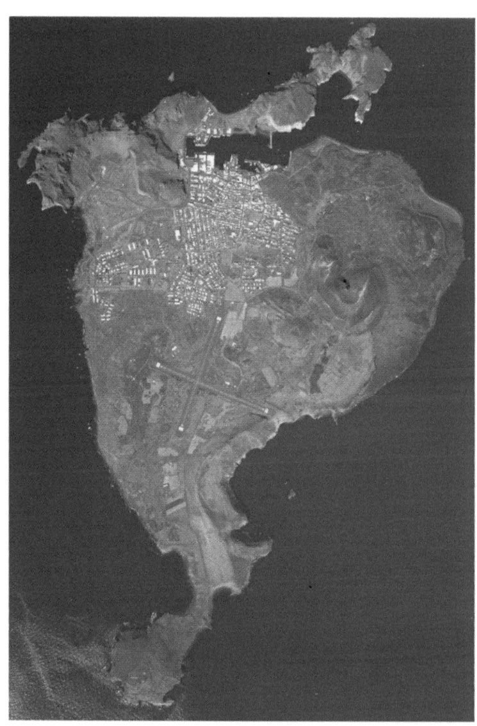

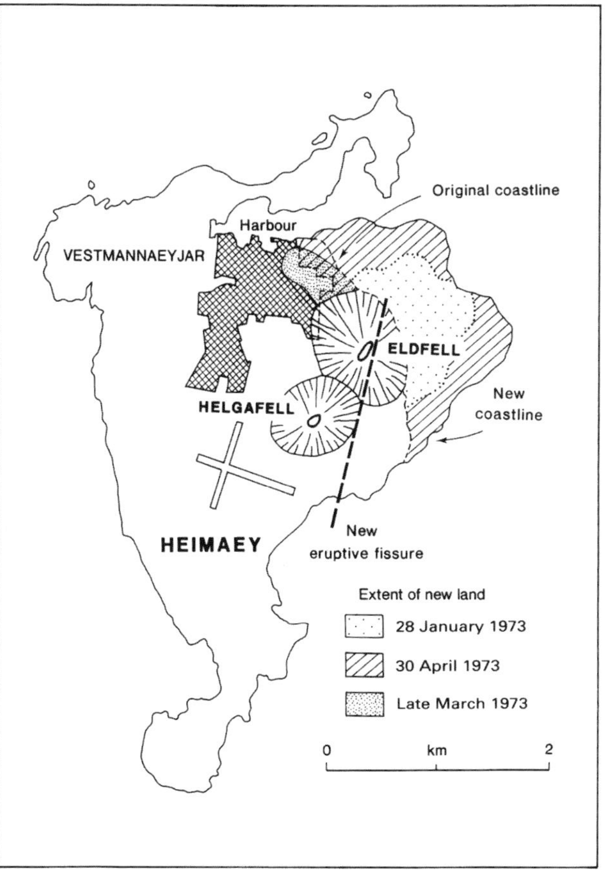

Top: Aerial view of the Heimaey island. (Photograph courtesy of the Icelandic Geodetic Survey.) Bottom: Map of the island, showing the locations of the new volcano (Eldfell) and the older volcano (Helgafell). The town is shown in cross-hatch pattern. (Modified from Chester, 1993.)

In 1973 a new volcanic cone, Eldfell, erupted right by the town of Heimaey. There had been no eruptions on the island in some 5,000 years and residents were not expecting to be awakened one night to find an eruption in their backyard. Locals and volunteers from the mainland were able to save the harbor from lava flow damage, but many houses were destroyed. (Photograph courtesy of Ralph White.)

10,000 years ago. The volcanic centers that form the present-day southern extreme, Stórhöfði and Sæfjall, erupted between 6,000 and 5,000 years ago. All these separate centers consist of hyaloclastite or palagonite tuff cones, which means they formed by the rapid breakup of chilled magma as it came into contact with seawater. These cones were joined together by the flows of alkali basalt lava from the eruption of Helgafell, which occurred about 5,000 years ago. This was the last activity on Heimaey until 1973.

The 1973 eruption

When the residents of Heimaey went to sleep on the night of January 22, they had no idea that they would soon wake up in a most unexpected way. Although there had been some faint seismic tremors starting about 30 hours before the eruption, these were barely noticed by the islanders, who were not expecting their volcanic island to wake up after a 5,000 year slumber. At 1:55 a.m. on January 23, when almost everyone was asleep, an 1,800 m (6,000 feet) long fissure opened on the eastern slopes of Helgafell, traversing the island from one shore to another. Fire-fountains over 200 m (660 feet) above the ground burst forth along the fissure, forming a continuous and spectacular curtain of fire. The fissure was barely 1 km (5/8 mile) from the center of Vestmannaeyjar, the island's only town.

The reaction of Heimaey's people was the most

amazing feature of the eruption. They did not panic, perhaps because volcanoes are so much a part of life in Iceland. Some of them even lingered to take photographs, though most packed their essential belongings quickly and went down to the harbor. Thanks to inclement weather, Heimaey's fishing fleet was not at sea, and evacuation was swift. Aircraft and boats also came from the mainland to transport Heimaey's residents. The Icelandic State Civil Defense Organization had a contingency plan for such an emergency, which they quickly put to work. In 3 hours, 4,000 out of the 5,300 islanders had been evacuated, and most of the islanders were safely away within 6 hours of the start of the eruption. There was not a single casualty that night. Only one person, a man, died during the 6 months of activity – his death was caused by breathing poisonous gases. The islanders' good fortune on the night of the eruption was in part due to the weather. Not only were the fishing boats in the harbor but, more importantly, the wind was blowing opposite to its usual direction and ash was carried out to sea rather than towards the town.

The new volcano's initial activity was fire-fountaining along the fissure that soon became concentrated in a small area. Within 2 days a cinder and spatter cone over 110 m (360 feet) high had been formed. Later it was christened Eldfell, the fire mountain. The eruption was Strombolian in character and moderate in terms of its explosivity (the VEI was 3). Had this eruption not taken place in Heimaey, it might have been easily forgotten. But the proximity to the town and the threat to the harbor made the birth of Eldfell a potential economic catastrophe. Within a few days, strong easterly winds carried tephra over the town, and the houses closest to the new volcano were totally buried. The major threat, however, came from a massive lava flow that advanced towards the harbor in early February, and began to narrow its entrance. The lava's course meant that it would eventually fill and destroy the harbor.

Vestmannaeyjar is no ordinary harbor. Because of its location and natural protection, it is Iceland's most important fishing port. Though the island houses only 2% of Iceland's population, it produces about 12% of the country's fish catch. Given that Iceland's exports are 75% fish, one quickly realizes the value of the threatened harbor. It was clear that something had to be done to try to stop the lava flow. Several Icelandic scientists recommended spraying seawater on the lava to cool and harden it, so as to stop the advance of the flow. Volunteers started fighting the flow's advance by aiming water hoses at its front. This had some encouraging local effect, as the heat required to evaporate the water caused the lava to cool to near solidus temperatures, thus forming a wall of chilled lava which acted as a dam against the hot flow behind it, causing the lava to flow in a different direction. However, it soon became clear that to stop the flow this effort would have to be attempted on a much larger scale. There were two major problems to tackle: one was to halt the advance of the main lava towards the north, which threatened the harbor, and the other to stop another flow that was advancing to the northwest, towards the town and its fish-processing factories.

The next stage of the fight against the flows involved a twofold approach: to stop the advance of the main flow, a system of large water pumps in the harbor was used, together with a network of pipes to carry the water to the flow. The northwestern flow towards the town was to be halted by constructing a lava barrier and bulldozers were to be used to move the chilled lava. This became the most ambitious program that had ever been attempted to control the damage of a volcanic eruption. It was a truly heroic effort. The pumping of water onto the hot lava meant that there was dense, warm fog everywhere on that battleground. Author John McPhee, who wrote a vivid account of the events after interviewing many of the islanders, reported that bulldozers had to operate without seeing their blades, with volunteers walking alongside to help guide them. The pumping crews used the noise of the volcano to orient themselves and carried bottles of water so that they could cool down their boots. Five hundred volunteers at a time participated in the effort, of which about 75 were in the pumping crews. Some were residents who had never left, but people came from all over Iceland, helping as they could. They lay down more than 31 km (19 miles) of pipe, which by early April were delivering seawater at rates of up to 1,000 liters (265 gallons) per second. From the time the operation started, in early February, until if ended on July 10, over 1 million m^3 (35 million cubic feet) of seawater were pumped onto the lava flows, converting an estimated 6 million m^3 (210 million cubic feet) of molten lava into solid rock. Temperature measurements made using boreholes in various parts of the flow indicated that the parts of the flow which were sprayed with seawater cooled between 50 and 100 times more rapidly than those left to cool naturally.

Lava kept coming and ponded behind the barrier that halted the flow towards the town. On March 18, lava breached the barrier and the flow now known as the "town lava" engulfed about 200 houses. However, volunteers were able to save many other houses by clearing the roofs of accumulated ash before they collapsed. The islanders eventually won their fight against the flow. The harbor was saved: some lava did enter, but its effect was to make the harbor's entrance more protected from the rough seas. The eruption stopped in early July. One of its more lasting effects was the addition of about 2.5 km^2 (1 square mile) to the size of the island.

The 6-month Eldfell activity was predominantly Strombolian in character, with heavy ashfall, which piled up high enough to cover houses. The massive but slow-moving aa lava flows reached over 100 m (330 feet) in thickness in some places, advancing several meters a day towards the town and the harbor. Large blocks of solidified lava broke off from the main cone and were carried along with the flow, some as big as 200 m^2 (1,500 square feet) and standing 20 m (65 feet) above the lava surface. The largest of those was nicknamed Flakkarinn ("the wanderer"). It was a piece of Eldfell that slid off its northern side and remained afloat on the lava, traveling over 0.5 km (⅓ mile) in 2 weeks. It is reported that some people climbed up Flakkarinn for "rides." However, the wandering block soon became a major threat, as it headed straight for the harbor. A plan was developed to stop Flakkarinn: an area of lava in its path was chosen for cooling, and every available pump was aimed at it. About 115 million liters (30 million gallons) of water were poured into the area. Eventually, Flakkarinn crashed into the cooled lava with mighty force and broke up into smaller pieces that soon stopped moving. Its remains can still be seen protruding from the lava flow.

The lava flowed from Eldfell as thick basaltic flows. The lava's composition changed during the eruption, from mugearite in the initial phase (with about 3.5% magnesium oxide) to hawaiite at the end (5% magnesium oxide). This compositional change implies that the magma chamber was zoned, that is, made up of horizontal layers of the different magmas, with the upper part richer in alkalies and silica. Petrological studies indicate that the separation of the magma happened at a depth of about 18 to 20 km (11 to 12½ miles). Seismic data showed that earthquakes during the eruption were located about 15 to 25 km (9 to 16 miles) down, and also that the source of the Heimaey magma was much deeper than is the case for most Icelandic eruptions.

The 1973 eruption was not wholly confined to Heimaey, as it included some submarine activity that lasted from February to late May. Most of the activity was near the island, but on May 26, a boat captain reported a short-lived submarine eruption about 6.5 km (4 miles) northeast of Heimaey, only about 1.5 km (1 mile) from the coast of Iceland.

After the end of the eruption, the residents who had fled started coming back to tackle the hard job of rebuilding the town and the island's economy. Houses buried under ash were dug out, the streets cleaned, and, by sheer persistence and willpower, the islanders' lives eventually returned to normal. It was slow work: the ash on the flanks of Eldfell had to be stabilized, as wind continued to sweep great quantities of it into the town. During the next 2 years, volunteers worked to move ash and to sow grass seed by hand. During the summers, students came from the mainland to help out. The tenacity of the people ensured the town's rebirth and, by 1975, 80% of the residents had returned. As one walks through the streets of Vestmannaeyjar today, it is hard to imagine that this sleepy town was recently the scene of the most fearless and famous battle between a community of people and a volcano.

A personal view: The struggle for survival

Like most visitors to Heimaey, I knew about the 1973 eruption and the lava diversion long before I got to see the island, and therefore had already concluded that the people who make their homes in Heimaey are a resilient lot. As I delved further into the island's history, I was surprised to find out just how resilient: life has been tough in the Vestmannaeyjar from the very beginning of the settlement. This rugged island has been so marked by misfortune that I wondered why people kept living there.

The "Westmen islands" were, in fact, named after the five runaway Irish ("westmen") slaves who became their first residents, though their stay was brief. According to the *Landnámabók* – the revered Icelandic *Book of Settlement* – the slaves belonged to Hjörleifur Hrodmarsson, the brother of Iceland's first settler Ingólfur Arnarson. The slaves murdered their master in south Iceland and fled to the uninhabited, inhospitable islands, in the hope of escaping pursuit. But the Vikings were not about to let the deed go unpunished and, 2 weeks later, they tracked down and killed the fugitives. Heimaey was eventually settled by Icelanders who wanted to escape the feuding in the mainland. However, life was not peaceful for long. In 1413, a bishop gave away the islands to the king of Norway as a debt settlement and the unprotected Heimaey then became an easy target for pirates. It was taken over by the British and later by the Danish. In

1627, the island was invaded by 3,000 bloodthirsty Moroccan pirates, who murdered, pillaged, and torched, then carried off over 200 people. The king of Denmark paid a ransom for the islanders, but only 13 of those taken away returned home: many had died or had been sold into slavery, while others chose not to come back. The islanders never forgot the barbaric invasion and tales about the pirates are still told to misbehaving children.

Natural disasters also plagued the Vestmannaeyjar. One of the effects of the catastrophic 1783 Laki eruption (described in the "Hekla" section) was to poison the seawater, killing all the fish around the islands. Even when the islanders had enough to eat, their diet lacked fruits and vegetables, so scurvy was rampant. Shortages of fresh water compounded the environmental problems and fierce storms caused the deaths of many fishermen. Given all these misfortunes, it is perhaps easier to understand why the islanders chose to fight the lava flow rather than to abandon their home. When viewed in the context of history, the eruption of Eldfell seems just like one more trouble to be overcome by a tenacious people who had seen worse.

I found the history of the island – including the accounts of the lava diversion – to be its most engrossing aspect. Eldfell is certainly interesting from a geological point of view but, as volcanoes go, it is not particularly impressive. What is stunning is its location: Eldfell sprang up practically in people's backyards. When you see it looming behind the town it is hard to believe that so many of the houses still standing existed before the volcano was born. And you cannot help imagining what it would have been like to be here on that cold January night, waking up to see a curtain of fire right behind your house.

Visiting during repose

Eldfell

The base of the cone can be reached by a dirt road and the view across the lower lava fields is spectacular: one can see that much of Heimaey is covered up by the 1973 lava – about one-sixth of the island. The 221 m (729 feet) high Eldfell can be quickly climbed following one of several paths. The view is even more impressive from the top, particularly to the north, where the crater is open and dark lava flows extend towards the sea. Rising from the lava flow is the block known as Flakkarinn ("the wanderer") which was carried by the lava. The ground at Eldfell is still warm and the summit has steaming fumaroles and brilliantly colored sulfur deposits. The distinctive smell is a reminder that this volcano was recently active. There is a path down the west side of the crater which leads to the foot of the verdant Helgafell. The easiest route to the summit of the older cone is up the southwest slope and the view from the top is worth the short climb.

Kirkjubæjarhraun ("church farm lava")

This is the 1973 lava field that forms the eastern part of the island. A white pillar placed on the flow marks the location of the Kirkjubœr farm, now buried underneath 100 m (330 feet) of lava. The pillar is a memorial stone to the Revd Jón Þorsteinsson, a former resident who was killed by the invading pirates in 1627. The stone was rescued before the lava buried it, and re-erected on the same spot after the eruption ended. Several roads have been cut into Kirkjubæjarhraun and geothermal units have been built to provide cheap, natural heating for the town. The lava is thick and blocky and not particularly easy to walk on, but dedicated volcano enthusiasts will not

Eldfell crater, with town in background. (Photograph courtesy of Armando Ricci.)

want to miss the chance to wander through the still-fresh lava field. Heimaey lava flows are mostly hawaiite, and have a high alkali content relative to the tholeiitic basalt lavas in mainland Iceland. This means that the viscosity of the lavas in Heimaey is higher, and the thicknesses of the flows therefore generally greater. This is apparent when you look at the edges of the flows, which tend to be tens of meters thick.

Vestmannaeyjar town

The town is spread around the harbor and the setting is spectacular: the rugged cliffs of Heimaklettur formed by hyaloclastites and pillow lavas on one side, the volcanic cones of Eldfell and Helgafell on the other, and, on the harbor, a variety of colorful boats docked along the natural windbreak formed by the 1973 lava. When you walk away from the harbor and into the town, along the street called Kirkjuvegar ("church path"), you can see the base of the lava wall which acted as a barrier to halt the flow in 1973. This really brings home the fact that the eruption occurred virtually in people's backyards. The ruins of one house are still standing, perhaps as a monument to the many which were totally destroyed. A striking reminder of the eruption is a colorful mural on the side of a building near the harbor: it was painted by local schoolchildren and depicts their view of the morning of Eldfell's awakening.

View of a house partly destroyed by the Eldfell eruption. (Photograph courtesy of Ralph White.)

Heimaklettur ("home rock")

Located just north of the harbor, this hyaloclastite hill surrounded by steep cliffs can be climbed without special equipment for a great view of the island, but it should not be attempted by the faint-hearted. Ladders, ropes, and steps bolted into the rock allow you to get up the steep scarp and onto the grassy steep upper slopes where sheep roam and puffins flock. Many people wonder how the sheep managed to get up there – the only explanation I was given is that they are Icelandic sheep. From Heimaklettur it is possible to walk across to Miðklettur ("middle rock") by picking your way carefully along the ridges, but it is not possible to reach the top of Ystiklettur ("outer rock") without climbing equipment. A hike up Stóraklif (where the radio tower is) leads to another great view. There is a track from the harbor and ropes and ladders allow you to scramble up the steep scarp. You will also have to climb a large scree slope. It is possible to cross that slope and hike west to reach Dalfjall, but again this hike is best left to those who can identify with the Icelandic sheep.

Herjólfsdalur

This is a grassy amphitheater open to the south, formed by the crater of the old volcano Norðurklettur. Excavations on this site uncovered the remains of an old farmhouse, which many think belonged to Herjólfur, the island's first intentional settler. However, carbon-14 dating of charcoal found in the site has yielded dates in the AD 700 to 800 range – that is, prior to the settlement of Iceland in AD 874. These results are intriguing but generally viewed with suspicion. Most visitors to Herjólfsdalur come here for its campground and golf course, and for the 3-day "People's Feast" in early August, which commemorates the granting of Iceland's constitution by Denmark in

1874. Those who don't like crowds should avoid Heimaey during the festival, when the island's population usually doubles.

The seaward side of Herjólfsdalur has cliffs formed of columnar basalts, and there are some good hikes along sheep tracks down the rugged cliffs to lava caves, which were formed by erosion. Hundraðsmannahellir ("one-hundred-man cave") and Teistuhellir ("razorbill cave") are two of the best known.

Stórhöfði

This is the southernmost volcano on the island. It is linked to the rest of Heimaey by a low narrow isthmus. The road to Stórhöfði leads to the 122 m (400 feet) high summit, where a lighthouse and an important weather station are located. Since this volcano is heavily vegetated, it is of limited interest from a geological point of view. However, the cliffs around it are a good place from which to watch puffins and their hunters. Like many animal-loving foreigners, I was uneasy about eating the island's delicacy until I realized that there are literally millions of puffins on the Vestmannaeyjar. An unwritten rule amongst the hunters requires that only non-breeding birds can be taken, otherwise they must be released. Hunters use a butterfly-type net on a long pole to catch the puffins in mid-flight, a skilled and sometimes dangerous pursuit, as the cliffs are steep and treacherous.

Sæfjall

This tuff ring open to the ocean on the island's southeastern coast is, geologically, one of the most interesting places in the island. Here you can see classic examples of cross-bedding: layers of fine material that cross one another, rather than lie neatly on top of each other. When explosions produce fallout – that is, material is deposited as it falls from the air – the beds representing different explosions produce distinct layers. However, when the explosion pulses send out tongues of material (surges), these can go in different directions, producing cross-bedding in the deposits. The surge deposits are poorly sorted, that is, the particles of different sizes are not layered as one would expect from an ashfall. The surges that produced these deposits were probably occurring every few seconds.

Sæfjall is also a classic site for bomb sags. Looking up from the beach, you can see "sags" in the layers of the tuff ring, where bombs landed. Some of the bombs have fallen all the way down to the beach. For those who are more interested in animals than rocks, this is a good spot from which to watch orcas (killer whales).

Bomb sags in tuff layers near the beach at Sæfjall. (Photograph courtesy of Armando Ricci.)

Also on the eastern coast, between Sæfjall and Stórhöfði, is Ræningjatangi, where the pirates swarmed ashore in 1627. Many of the islanders tried to escape by hiding in the lava caves of the area, which are interesting to explore.

Visiting during activity

Eldfell is considered to be a monogenetic volcano, that is, the type that erupts only once. This does not mean that there couldn't be another eruption on Heimaey, when a new volcano would be born probably in a similar way to Eldfell, but the likelihood of it happening in the next few decades is remote. Eruptions in the Vestmannaeyjar archipelago have been rare. However, the fact that two occurred in a period of 10 years (Surtsey and Eldfell) is worth noting. Several studies indicate that Iceland's activity is shifting to the islands and this could lead to more frequent eruptions in the area. The next eruption could happen in Heimaey again, or elsewhere along the chain, perhaps forming a new island much in the same way as Surtsey emerged.

The prospects of a visitor being able to watch an eruption in the Vestmannaeyjar would depend very much on where the activity was. During the 5 months of the Eldfell eruption, there was ample opportunity for visitors to see the activity, and even for volunteering to fight the flow's advance. The Strombolian eruption could be watched in relative safety and, despite the risks taken by the crews pumping water on the lava, there was only one fatality: a man entering a cellar was overcome by toxic gases that had accumulated there.

Many visitors and residents of the islands were also able to catch great views of Surtsey's series of eruptions during the island's 5-year birth. If a similar event happens in the Vestmannaeyjar, it should be possible for visitors to view the activity from boats or from the air. It is, however, unlikely that anyone who is not on official duty would be able to alight on the new island. Icelanders would probably repeat the very successful experiment they are carrying out on Surtsey and not allow tourists to come ashore, even after the eruption is over.

The major danger of the likely volcanic activity on the Vestmannaeyjar would be explosions caused by the interaction of magma and water, as these can be quite violent. If you watch a Surtseyan-type event from a boat or plane, you will need to rely on the captain or pilot following the local authorities' recommendations for safe distance. Other dangers from an Eldfell-type eruption come mostly from the bombs thrown out by the explosions – staying out of range is the most important consideration in this case. The lava flow from Eldfell was slow-moving and did not represent a real danger; however, the flow produced some local explosions when it entered the sea. In retrospect, the volunteers working against the lava flow were lucky, because there were some close calls. In fact, those who worked on the live flow were at one time known as "the suicide squad."

Other local attractions

Museums in Heimaey

The Folk Museum is worth a visit, especially for its display of a relief map of Heimaey before the 1973 eruption, which helps to visualize what the island was like without the Eldfell cone. Other displays include Icelandic stamps, currency, and art. The Natural History Museum has a geology room with a remarkable collection of "landscapes in stone," in which stones such as agate and jasper are arranged to give the effect of landscape paintings. The museum's other highlight is an aquarium displaying exotic Icelandic fish. The Icelandic catfish, with piranha-like teeth, is particularly frightening. John McPhee wrote about a famous French volcanologist who was asleep in the building one night during the 1973 eruption. During a heavy tephra fall, he woke up to the terrifying sight of an Icelandic catfish swimming up to the glass and opening its mouth. According to some Icelanders, this horrifying encounter was the real reason behind that scientist's extremely pessimistic (and erroneous) predictions for Heimaey's doom.

Volcanic Film Show

This is not the same as the *Volcano Show* in Reykjavík (see section "Krafla"), and the two films normally shown here are not particularly spectacular. However, they provide a way of learning more about the 1973 eruption and island life in general. The show consists of two movies, though the program may change: one of the movies is about the salvage operations after the end of the Eldfell eruption. The other is about the amazing story of Gudlaugur Fridthorsson, a Heimaey fisherman. In 1984, he survived the sinking of his boat, and then stretched all limits of expected endurance by swimming for about 6 hours in the icy sea and then walking barefoot for another couple of hours over sharp, rugged lava fields. Medical studies concluded that Gudlaugur's life was saved by his seal-like layer of subcutaneous fat. Watching this movie may persuade you to order a second helping of puffin.

Surtsey

Located 18 km (11 miles) southwest of Heimaey, Surtsey is off-limits to visitors, because its natural colonization by plant and animal life is being studied by a variety of scientists. A distant view of Surtsey can be glimpsed from the Þorlákshöfn to Vestmannaeyjar ferry, but even through binoculars the view is disappointing. Visitors intent on seeing the island should take an air sightseeing tour, either from Heimaey or from Reykjavík City Airport. It is cheapest to go from Heimaey – ask at the airport. The Icelandic Geodetic Survey publishes a 1:5,000 topographic and geological map of Surtsey which is useful to take along on the sightseeing flight. Surtsey is 1.57 km^2 (⅔ square mile) and its highest point is 154 m (508 feet) above sea level. The island has a "sandy" point to the north, made up of fine tephra, and a narrow boulder beach

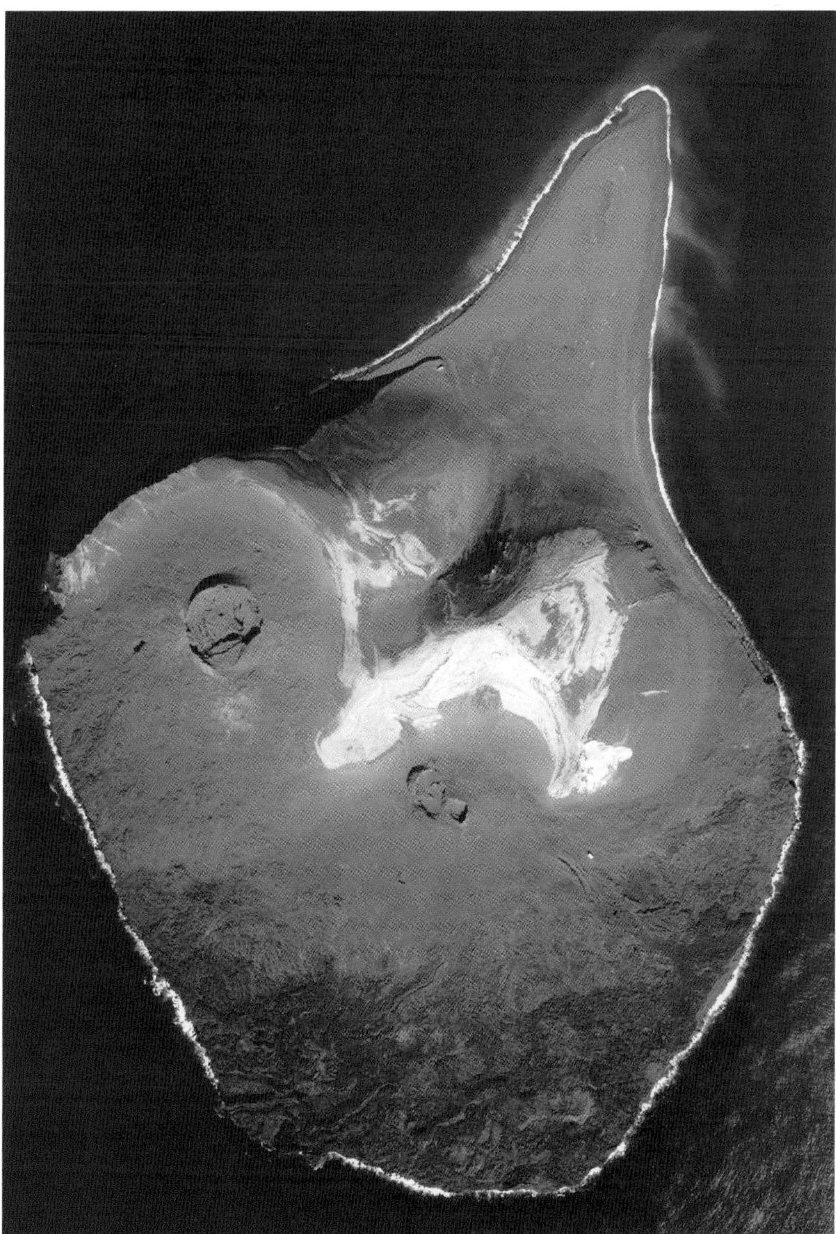

Aerial view of Surtsey island. The west and east craters can be seen near the center of the image. Tephra from 1963 and 1964 accumulated to the north of the craters and lavas from 1964/5 and 1966/7 covered the southern part of the island. (Photograph courtesy of the Icelandic Geodetic Survey.)

fringing the rest of the island. There are steep lava cliffs on the eastern side and compacted tuff on the west. Surtsey's southern half is dominated by lava flows, of both aa and pahoehoe types. From the air one gets a magnificent view of the two semicircular tephra cones opening to the south, located about halfway down the island. The craters within the tephra cones are called Surtur I and Surtur II. The tuff material is gradually becoming palagonite, that is, the basaltic glass is cementing together and turning into the brownish vitreous material known as palagonite. Icelanders call palagonite tuff móberg, meaning literally "brown rock".

The island was named after Surt, the Norse giant who sets the world alight at Ragnarok, the Nordic equivalent of doomsday. The birth and evolution of Surtsey have been extremely instructive to scientists. For volcanologists, the eruption demonstrated the role of magma–water interaction in the formation of tephra during basaltic volcanism and has served as a model for the formation of the other Vestmann islands. Biologists have been studying the gradual colonization of the new island which started life devoid of living organisms, thanks to the magma temperatures of 1000 °C (1890 °F). Black-backed seagulls were the first birds to nest on the new island, back in 1970. Since then, 60 bird species have been observed on Surtsey, and six regularly nest there. During the early days of research in the island, some of the birds were sacrificed so that the grit contained in their stomachs could be analyzed, to locate where the birds had come from – a useful but rather grim geological research project.

Hekla

The volcano

Hekla is Iceland's most famous volcano and also the most active, erupting on average once every 10 to 30 years. Hekla's eruptions can be very dangerous and some have been catastrophic. However, the most recent eruptions, such as those of 1991 and 2000, have been only moderately explosive. The 1991 eruption was relatively easy to watch and practically emptied Reykjavík, as hoards of volcano chasers came to see the fireworks.

Hekla is 1,491 m (4,920 feet) high and is the central feature of a fissure system 40 km (25 miles) long and 7 km (4½ miles) wide. The volcano's location atop a fissure caused it to grow to the shape of an overturned canoe or, as early-twentieth-century author Murat Halstead put it, "a beautiful mass of whiteness, formed like a hen's egg lying on its side." Hekla's snowy summit is cut by a 5.5 km (3½ mile) long fissure, Heklugjá, which is aligned northeast–southwest and reaches down both shoulders of the mountain to a level of about 850 m (2,800 feet). In typical Hekla eruptions, such as that in 1947, this fissure erupts, pouring out lava. Hekla looks like a shield volcano (or a hen's egg) when viewed at right angles to this fissure line, but it shows a dramatic change when viewed along the fissure. The concave profile seen this way is similar to that of stratovolcanoes such as Mt. Fuji. Hekla's eruptions have, indeed, produced large quantities of both lava and ash, so it is classified as a stratovolcano rather than a shield volcano.

Although Hekla's copious lava flows have eaten up many square kilometers of land, the heavy ashfalls are the volcano's greatest potential source of damage. Some of Hekla's eruptions have been of the Plinian type, notably that of AD 1104 which buried everything within a 50 km (30 mile) radius underneath 1 m (3 feet) thick layer of ash and magma fragments. In 1947, Hekla spewed out a cloud of ash and pumice that rose more than 27 km (17 miles) above the volcano and, within hours, turned day into night. Two days after the

The imposing, snow-capped Hekla in 1995. (Photograph by the author.)

onset of the explosive activity in 1947, the heavy ashfall reached as far as Finland, 2,800 km (1,750 miles) away, meaning an average speed of 50 km/hour (30 miles/hour). Fortunately, this Plinian awakening of the volcano was soon replaced by relatively quiet outpourings of lava when Hekla cracked open along the main fissure direction and poured lava on both sides. The 13-month eruption turned out to be one of the most voluminous in Hekla's history: it is estimated that over 1 km^3 (35 million cubic feet) of magma came out as ash, tephra, and lava. The greatest damage was done by the ash and tephra fallout, which destroyed two farms and damaged many more.

Even when the eruptions are not particularly violent, Hekla's ashfalls can cause a lot of damage, as they tend to have a nasty effect on livestock. The volcano's magma is unusually rich in fluorine, which can contaminate the groundwater system and, in addition, adhere to tephra particles, which then contaminate grassland. Grazing animals ingest the fluorine in the water and grass with lethal results. Fluorine poisoning has been a persistent problem of Hekla's eruptions: in 1970 alone, 7,500 sheep were killed this way. Some livestock also perished by wandering into hollows near the base of the volcano where carbon dioxide gas had accumulated. It is interesting that, in some of those cases, men tending the sheep suffered no ill-effects simply by virtue of being taller, as their heads were above the rim of the carbon-dioxide-filled hollows.

Hekla has erupted over 20 times during the 1,100 years of Icelandic historic record and the fissure system associated with the volcano is known to have erupted another five times. It is possible that only violent eruptions were documented in earlier times. More eruptions are known to have taken place this century than in earlier times, but may not represent a true increase in the frequency of eruptions. Perhaps the less violent events, such as those of 1970, 1980–1, 1991, and 2000, were simply ignored in past centuries.

Hekla's eruptive history can be traced back into Iceland's last glacial episode. The earliest known major Plinian eruption of silicic magma occurred about 7,000 years ago and a tephra layer from this event can still be seen over a large area of Iceland. Since this time, Hekla's pattern of activity has been mostly one of large Plinian eruptions in which silicic magmas were involved, alternating with effusions of lava flows mostly of andesite and basaltic andesite composition. Many of these eruptions have discharged large volumes of magma: it is estimated that the long-term magma production rate of Hekla is 10 million m^3 (350 million cubic feet) per year. For comparison, Iceland's total magma production rate is estimated to be between 50 million and 100 million m^3 (1,950 and 3,500 million cubic feet) per year.

A particularly interesting feature of Hekla's activity is the variety of magma compositions that can be erupted during the same episode. The first eruption products are often highly silicic in composition, sometimes being rhyolitic magmas, but these are followed by a rapid transition to basaltic andesites. The fact that both types of magma are present in the same eruption raises questions about the volcano's magma source and reservoirs, such as: are the silicic and basaltic magmas connected in origin? Since Hekla is located over the spreading-dominated, basalt-producing mid-Atlantic ridge, where is the silicic magma coming from? The general understanding that has been gained from both chemical and geologic studies is that there is an extensive reservoir of basaltic magma at the base of the crust underneath the Hekla region. This layer of basaltic magma feeds fissure eruptions outside the main Hekla volcano. During periods of repose, the basaltic magma acts as the heat source for melting the overlying crust, which then produces the silicic magmas (dacites and rhyolites). The basaltic layer is also thought to be the "parent" magma that produces the andesite and basaltic andesite magmas. The mechanisms are magma mixing and fractional

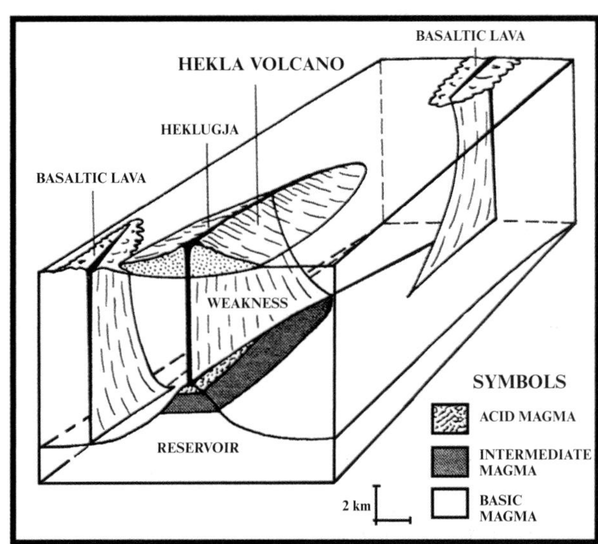

Proposed model of the interior of Hekla. The magma in the reservoir is layered, with acid and intermediate magmas on top of basic magma. When Hekla erupts, the acid and intermediate magmas are erupted via the zone of crustal weakness. Eruptions from the marginal regions of the reservoir tap the basic magma and produce basaltic lavas. (Modified from Guðmundsson *et al.*, 1992.)

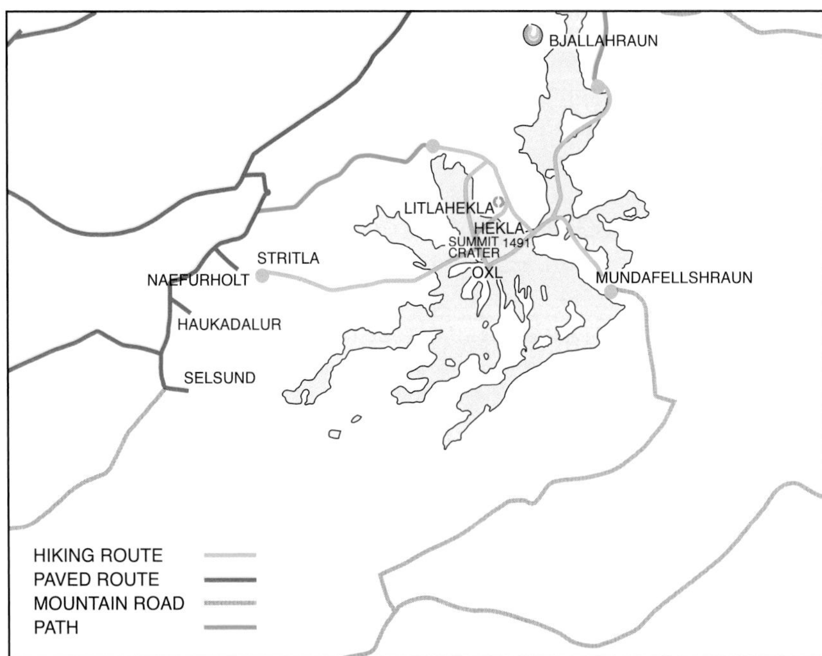

Sketch map of the Hekla summit area. Recent lava flows (prior to 2000) are shown in pink. The main lavas from the 2000 eruption flowed from the summit to the southwest (toward Vatnafjöll and Trippafjöll) and to the northeast (toward Skjólkvíar) but did not affect the climbing routes. (Modified from Þórðarson and Snorrason, 1994.)

crystallization. Magma mixing happens when the molten magma in the reservoir causes the rock around it to melt and be mixed in. Fractional crystallization is the partial crystallization of the molten magma as it cools in a reservoir, which can cause the magma's composition to change as some of its constituents form crystals and sink to the bottom. The magma reservoir beneath Hekla, which is about 7 to 8 km (4½ to 5 miles) deep, is layered: the lower density silicic magmas accumulate in the upper part and this is why they are erupted first.

Hekla's relation to Iceland's tectonic setting is also important for understanding this volcano. Hekla is situated at a critical juncture where the Eastern Volcanic Zone (a segment of the propagating mid-Atlantic rift) joins up with the transform fault system of the South Iceland Seismic Zone. Hekla's location at this juncture is probably one of the factors that causes the volcano to erupt so frequently. According to one model, Hekla takes up a lot of the tensile strain associated with the propagating rift and this, in turn, implies that the reservoir under the volcano has a high frequency of dike injection.

Earthquakes are also common in the Hekla region. The South Iceland Seismic Zone was given this name because of its high frequency of earthquakes. Major events happen about once per century; the last one occurred in 1896 and left hundreds of farmhouses in ruins. Seismologists estimate that there is a 90% chance of a magnitude 6 or larger event happening in the region in the next 20 years. In contrast, the minimum known eruption interval for Hekla has been just under 10 years (between 1991 and 2000) and the average interval (after the 1104 eruption) has been 55 years. From these numbers it looks like visitors to Hekla during the next decade are more likely to feel an earthquake than to see an eruption. However, since both these phenomena are notoriously hard to predict, the opposite could easily happen.

The 1991 and 2000 eruptions

The 1991 eruption was the third eruption in 20 years and represented a great increase in the frequency of Hekla's activity compared with the preceding few hundred years. The eruption started on January 17, when a Plinian-type eruption column rose to nearly 12 km (7½ miles) above the ground in only 10 minutes, producing a plume that reached the north coast of the country in about 3 hours. Hekla gave almost no warning about this eruption. Precursor signs from seismic and ground deformation monitoring were detected at 4.32 and 4.36 p.m. and the first explosion happened around 5 p.m. At 5:05 p.m. the ash and tephra column was seen from two nearby farms. A few hours later there was a second major explosion that produced a similarly large eruption column. Tephra fall peaked at about 6:20 p.m., but continued for the next 3 days, affecting an area of 20,000 km^2 (7,200 square miles). The composition of the tephra is basaltic andesite, the same as the lava flows that started to come out from fissures shortly after, or perhaps simultaneously with, the explosive activity. These fissures were mostly in the southern part of the volcano, below the summit. The height of the fire-fountains from these fissures was probably about 300 m (1,000 feet), but the spectacle had no witnesses.

During the first 2 days of the eruption, when most of the activity occurred, the effusion rate of lava flows reached at least 800 m^3 (28,000 cubic feet) per second and some estimates are as high as 2,000 m^3 (70,000 cubic feet) per second. After this, the activity became minor and concentrated on a southwest-trending fissure, where a scoria cone was built. The eruption stopped on March 11, by which time an area of 23 km^2 (9 square miles) had been covered with basaltic andesite lava.

Hekla erupted again on February 26, 2000. A fissure nearly 7 km (4½ miles) long opened along the southwest flank of the Hekla ridge, from which lava erupted as a discontinuous curtain of fire. An ash plume reached 11 km (7 miles) high in minutes. The eruption was at its most vigorous during the first hour, then started to decline. Lava flows came out the next day and covered a large part of the southeast flank, while in the evening flows to the north were observed to be advancing at a rate of several meters per hour. Three craters in the southermost part of the fissure were particularly active, sending forth lava flows and Strombolian-type explosions. The eruption was short-lived and by February 29 the activity from the summit stopped, but continued from the southermost part of the fissure with Strombolian activity and flowing lava. Four main vents and three smaller vents produced Strombolian explosions every few minutes on March 1 and the lava flowed towards the east. Large steam clouds were seen the next day but the interval between explosions had increased to about half an hour. Glowing lava flows were seen coming down the southwest flank. Bad weather made some observations difficult, but on March 5 the southwest lava could be seen still flowing. At sunset, people in the town of Selsundsfjall, 15 km (9 miles) away, could see a red, pulsating glow from the uppermost craters in the southwest-trending fissure. Scientists on a reconnaissance flight on March 6 observed the fissure still steaming, but all activity appeared to have stopped. Minor tremors were detected until the morning of March 8, when the eruption was declared over. There were no casualties and the problem of fluorosis in grazing animals did not arise because of the time of the year, as the animals were being kept indoors.

A personal view: Hekla as a time-marker

From a volcanologist's standpoint, I found the most fascinating aspect of Hekla to be its importance in the success of a method of dating eruptions which Icelanders have developed to a fine degree: tephrachronology. This involves analyzing the different layers of tephra that are deposited by successive eruptions and matching the layers with the historical records of eruptions. Tephra layers can then be used to constrain the dates of, for example, lava flows and archeological remains found between the different layers. Iceland's lack of trees helps the preservation of

Aerial view of the Hekla eruption of 1991. This was Hekla's third eruption in 20 years – a great increase in the frequency of the volcano's activity compared with the last few hundred years. The eruption happened during the winter but numerous volcano watchers braved the cold weather to view the volcanic fireworks. (Photograph courtesy of Oddur Sigurðsson.)

these tephra layers, as there are no deep root systems to upset them. Because Hekla has been the source of Iceland's largest explosive eruptions, it has been an important contributor to the tephra record. In fact, some of Hekla's eruptions have deposited tephra all over Iceland, and these layers are important chronological markers. Prehistorical layers are also important: by dating them using carbon found in the layers, the long-term eruptive history of Hekla can be reconstructed and traced over large parts of the country. Five major eruptive cycles have been identified, each having a similar tephra pattern that starts with a Plinian eruption. The first cycle identified happened about 6,600 years ago, and the tephra layers are known as Hekla 5. The latest cycle, Hekla 1, started in AD 1104.

The backbone of tephrachronology in Iceland is the analysis of the content of iron (FeO) and titanium (TiO_2) in the layers: when these two quantities are plotted against each other in a graph, the tephra samples from different volcanoes fall into distinct groups in the graph, making their identification easier. Hekla, having widespread tephra layers, is Iceland's most useful volcano for chronological dating. Historical records, including the famous Sagas, have helped to reconstruct the sequence of eruptions since the country's settlement. A particularly interesting aspect of the tephra layers is that they have recorded the major ecological and historical change to occur in Iceland: the settlement of the land by the Vikings. The layer called Settlement tephra, a product of a major fissure eruption in southern Iceland, is very widespread and thought to have been deposited around AD 900. Soils above this layer show evidence of agricultural and other human activities: for example, the proportion of birch pollen is seen to decrease (due to the trees being cut down), and the proportion of grasses and cereals increases.

Much as the intellectual pursuit of tephrachronology fascinated me, I have to admit that climbing Hekla was the highlight of my visit. I was lucky with the weather on that day, as the summit was clear and not at all doing justice to the mountain's name, which in Icelandic means something between "the shrouded" and "the hooded." Hekla is a beautiful, interesting mountain and quite an easy one to climb if one stays on the main route. When Hekla is quiet, there is nothing particularly frightening about this volcano. Therefore I was rather surprised to find out that, as far as we know, nobody had successfully climbed the mountain before 1750. The reason Hekla remained unexplored for so long is that, in medieval Europe, the volcano was widely believed to be the entrance to Hell and most people did their best to stay away from it. But when pioneer naturalists Eggert Olafsson and Bjarni Pálsson made their way to the top in June 1750, all they found there was snow and ice. Their feat helped to demystify Hekla and its possible effects on those who attempted to climb it. It is likely that an unknown Icelander had climbed Hekla in the late sixteenth century but, according to writings by the Bishop of Skálholt, the man returned "so terrified that he got to his home almost out of his mind and did not live long afterwards." True or not, this account may have discouraged potential climbers for the next two centuries.

Hekla is thought to have acquired its bad reputation back in the fourteenth century, when Caspar Peucer wrote of the volcano: "Out of the bottomless abyss of Heklafell, or rather out of Hell itself, rise melancholy cries and loud wailings, so that these may be heard for many miles around. Coal-black ravens and vultures flutter about. There is to be found the Gate of Hell and whenever great battles are fought or there is a bloody carnage somewhere in the globe, then there may be heard in the mountain fearful howlings, weeping and gnashing of teeth." Even Hekla's snow-white summit seemed to fulfill an evil purpose. It was said that the Devil would lay out the souls of the tormented on the ice, to cool them off whenever they showed signs of getting used to the flames of Hell! "Go to Hekla" was a popular curse in sixteenth century Europe, and the word "heck", meaning Hell, may have been derived from the volcano's name.

The contrast between the Hekla of myth and the placid-looking, dormant volcano is quite fascinating. I found the real Hekla to be beautiful and serene, totally silent, and so cold on the summit that coming across the gate to Hell began to sound attractive. The view from the top was magnificent, but the wind was so fierce that I could not hold my camera still. The descent, however, was pleasant and picture-perfect, with sunshine making the contrast between exposed dark lava and snow quite breathtaking. The weather seemed very changeable and the sunshine did not last for long. As we drove away from the mountain, clouds moved in and hugged the upper slopes, turning Hekla into "the hooded" again.

Visiting during repose

The magnificent Hekla fully dominates the landscape around it. A good way to get a feel for the size and shape of the mountain is to drive the local roads and see the volcano from several directions. For example, Hekla looks magnificent on a clear day viewed from the Þjórsárdulur river valley. It is a good idea to plan to spend several days in the Hekla region, not only to explore its other attractions but also to increase your chances of catching a fine day for climbing the volcano.

Hekla Visitor Center

Located in Brúarlundi, near Leirubakki, this small museum and visitor center is entirely dedicated to Hekla. You can see a movie about Hekla and its eruptions, eat in the coffee shop, buy slides and other souvenirs, and learn quite a bit about the volcano from the exhibits. When I visited, there was even a selection of the many scientific publications about Hekla on display. This is a small, low-tech, but rather charming and informative visitor center.

Climbing the mountain

Although Hekla is a relatively easy mountain to climb, it presents some hazards, particularly as the weather can be rough. Some common-sense recommendations should be followed. Do not climb alone, unless you are an experienced mountaineer. Stay on the climbing routes, particularly near the summit: Hekla has a glacier at the top, complete with menacing-looking crevasses. As can be expected, it is quite cold near the summit and weatherproof clothing (jacket and pants) is essential, as are hat, gloves, and the usual cold-weather gear. If you follow the main routes, no ropes or other special equipment are needed. The northeast climbing route is the easiest and the one that locals recommend. It starts at Skjaldbreið hill, which is on the Landmannaleið dirt road, about 8 km (5 miles) east of the intersection with the main road. If you have a four-wheel-drive vehicle, you can drive down the path from Landmannaleið to the edge of the 1970 lava, depending on the condition of the track. You can look at the 1970 craters and think about how great that eruption must have been for photographing, given that visitors were able to drive this far to watch the fireworks. If you choose to walk along the track rather than drive, you will be rewarded with some interesting findings. Along the edges of the track there are lava "snowballs", 2 to 3 m (6 to 10 feet) across. They were formed when still-molten lava cascaded down the steep slopes. It is also worth starting to look for xenoliths along this path – some xenoliths on Hekla are as large as 1 m (3 feet) in diameter. The climb to the summit along the northwest route takes from 4 to 6 hours for those who are of average fitness levels, and about half that time to come back down.

Those who may not feel energetic enough to climb to the summit should consider coming this far and visiting the photogenic crater known as Rauðaskál (Red Bowl). It is also well worthwhile to walk partway up the route, where it goes past the 1970 lava, to see flow features, spatter cones, and ramparts.

It is actually possible to get to Hekla by public bus, or at least to get close enough to the mountain to be able to climb it during a summer day. There is a bus from Reykjavík to Skaftafell (the Fjallbak route) which travels along the main road northwest of Hekla. You

Visitors descend Hekla at the end of a long day. The hike to the summit is not difficult, but be prepared for snow and strong winds. (Photograph by the author.)

can ask the driver to drop you off at the junction with the Landmannaleið, just to the east of the table-top mountain Búrfell. From there, walk along the Landmannaleið, until the beginning of the climbing route. Be warned that Landmannaleið, like many Icelandic interior roads, is cut by a stream that requires wading across. Taking off one's boots and pants and walking into the chilly water is not particularly pleasant, but it is often done in Iceland.

Visiting during activity

How likely is Hekla to erupt again? The known intervals between eruptions have ranged from 10 to 121 years, but recent intervals have been in the lower range. A sobering thought is that this might happen during your visit, in which case you can expect little warning. If Hekla's past precursor behavior is typical, there will be no more than a few seismic tremors in the hour preceding the first blast – and some eruptions, such as that of 1947, appear not to have given any discernible warning at all. Although the probability of being on the mountain when an eruption starts is very small, it is still worth thinking about. It is advisable to climb the mountain in clear weather when the visibility is good, and to take along a good map in case another return route is needed. An earthquake may signal the start of an eruption and should be taken as a warning to get down and away from the mountain as fast as possible.

Some of Hekla's milder eruptions, such as those of 1991 and 2000, are quite safe to watch, as attested by a large number of Icelanders, and the activity can be very photogenic. As with every eruption, follow the recommendations of local authorities and use your own knowledge and common sense about safety. It is worth remembering that Hekla can be extremely dangerous, not only because it is capable of producing large, Plinian-type explosions but also because of the large quantities of snow and ice packed on its slopes. In 1947, warm mudflows (hlaups in Icelandic) were formed during the first 20 minutes of the eruption, as superheated steam, incandescent bombs, and ash blasted laterally from the base of the erupting column and melted away snow and ice. It is estimated that 3 million m^3 (100 million cubic feet) of water came down the mountainside in about 6 hours. Most of the hlaups went towards the north and eventually into the Ytri-Rangá river without causing major losses. Hlaups of this kind seem to be a fairly common feature of large Hekla eruptions, as floods of the Ytri-Rangá river are also known to have occurred during the eruptions of 1766 and 1845. It is probable that floods also happened during earlier eruptions, of which we have only scanty records.

Despite the potential dangers, those who have been fortunate enough to watch the mighty Hekla in action seem to agree that it is an awe-inspiring, once in a lifetime event that should not be missed. Explorer Uno Van Troil, who climbed Hekla in 1772 as a member of an expedition led by Joseph Banks, expressed the feelings of most visitors to the dormant mountain when he wrote: "There was not one in our company who did not wish to have his clothes a little singed for the sake of seeing Hekla in a blaze."

Other local attractions

Stöng

Hekla's devastating eruption of 1104 destroyed some 20 farms, including one called Stöng, located to the north of Hekla, in the Þjórsárdalur ("Bull River Valley"). The farm belonged to a warrior called Gaukur Trandilsson, whose name is well known in Reykjavík thanks to the popular pub Gaukur á Stöng. Gaukur was important enough to have his own saga but, sadly, it has been lost. He is, however, mentioned in the famed *Njáls Saga*. According to the story, Gaukur survived the eruption but was later killed by his foster-brother, who disapproved of Gaukur's affair with one of his kinsmen. The information given in the saga helped archeologists to locate the buried farm and, in 1939, Stöng was excavated. The farmhouse walls were still standing and today Stöng is referred to as the "Northern Pompeii." Although the comparison is rather strained, the ruins of Stöng are definitely worth a detour, particularly if you can also visit its reconstruction a few kilometers away. In 1974, Icelanders decided to faithfully reconstruct Stöng, as part of the celebrations marking 1100 years of settlement. Known as the Reconstructed Medieval Farm in Þjórsárdalur, the farm was built using only the materials that would have been available at the time, and meticulously recreates the layout, furnishings, and so on. One quickly realizes that country life in Iceland, even in a relatively wealthy farm such as Stöng, was very harsh indeed. The turf roof and walls, broken up by only a few small windows, ensured that the house was dark, humid, and grim. It is sobering to learn that many Icelanders still lived in these conditions in the early part of the twentieth century, as described vividly in the novels by Iceland's Nobel Prize winner Halldór Laxness. I recommend visiting the Reconstructed

The Laki crater row, the site of Iceland's most deadly eruption. (Photograph by the author.)

Medieval Farm as much as the original Stöng, because of its value as a lesson in Icelandic culture and history. Getting to either farm is not easy without a car, but they are both on the route of some day excursions from Reykjavík.

Lakagígar

This fissure and crater row was the source of a catastrophic eruption in 1783, which poured out the Earth's largest historic lava flow. The large explosive and effusive eruption, whose VEI is estimated to have been 4, did not kill anybody directly, but ended up being a major environmental disaster for Iceland. The eruption's huge emission of sulfur dioxide and the formation of some 100 million tonnes of sulfuric acid as atmospheric aerosol led to the widespread destruction of crops, the poisoning and starvation of livestock, and, ultimately, the nightmare of the Moþuharþindi: the period of famine that caused the deaths of 9,000 people. The effects of the eruption were also felt beyond Iceland. In southern England, a month-long "smoky fog" turned the midday sun red. Crops were destroyed in Scotland and the sulfurous fumes were felt in Holland, over 1,000 miles away. The bluish haze observed all over Iceland traveled to Europe and then to Asia, eventually reaching the Altai Mountains in China. The winters following the eruption were particularly severe in Europe. The connection between the Laki eruption and climatic changes was first noted by Benjamin Franklin, who was US Ambassador to France in 1783. In a paper in 1789, he discussed the blue haze that drifted across Europe and the fact that the winter of 1783–4 was particularly severe, attributing both these phenomena to the Laki eruption. Since Franklin's pioneer work, research into the climatic and environmental pollution effects of large eruptions has become one of the most important issues in volcanology and the impact of the Laki eruption remains a major topic of study.

The lava flows produced by the Laki eruption were also truly unusual. The eruption started on June 8, with low-viscosity tholeiitic basalts gushing out of a fissure into the Skaftá river valley, hence the eruption is also known as the Skaftáreldar ("shaft river fires"). During the first 48 days, four to five successive lava surges were produced, having an average discharge rate of about 2,200 m^3 (77,000 cubic feet) per second. The maximum advance rate of the lava was about 17 km (11 miles) per day. The 8-month-long eruption consisted of 12 major episodes. On July 28, the start of the sixth episode, the activity shifted, with lava pouring into the Hverfisfljót river gorge, east of Skaftá. In total, the Laki eruption produced nearly 15 km^3 (3½ cubic miles) of tholeiitic lava from a fissure system 27 km (17 miles) long, with over 100 craters. The area covered by the lava was 565 km^2 (218 square miles) and, in addition, the eruption produced about 0.4 km^3 (14 million cubic feet) of tephra.

The fact that the Laki lava was both voluminous and came out at a very fast rate makes it a possible analog for the high-volume, long lava flows on the Moon, Mars, and Venus, which several studies imply were also erupted at particularly high rates. This eruption is also of prime interest to the study of terrestrial flows, because it is the only historical eruption that can be considered to be in the category of a flood basalt eruption.

Apart from its importance in volcanology, Lakagígar should be visited because it is one of the most spectacular volcanic fields anywhere. The fissure is located to the southwest of Grímsvötn volcano and the similarities in magma composition of these two centers are so strong that they imply a common source. Grímsvötn volcano is itself located underneath the Vatnajökull ice sheet, in Iceland's Eastern Volcanic Zone. There is evidence that the Lakagígar magma did in fact come from the shallow reservoir beneath Grímsvötn, traveling laterally along a dike.

Lakagígar can be reached by a road suitable only for four-wheel-drive vehicles starting west of the village of Kirkjubæjarklaustur ("church farm convent"). Like most Icelandic interior roads, it is only passable in the summer, has shallow rivers to ford, and it is not for timid drivers. However, it is possible to take a tour bus from Kirkjubæjarklaustur or Skáftafell during the tourist season (mid July to end of August), which allows a few hours to explore the Laki crater and surroundings. The parking lot is just at the foot of the main Laki crater, which is easy to climb. The view from the top is breathtaking: looking along the fissure you can see a series of tephra cones on the crater row, many covered with soft moss; while further in the distance are magnificent glaciers such as Síðujökull. On the day I climbed Laki, the weather was so clear that I could see the 2,119 m (6,993 feet) high ice-capped caldera Öræfajökull, the highest peak in Iceland, some 75 km (50 miles) away.

Those who have the time should try to walk along the path that meanders through the crater row, exploring the numerous tuff and tephra cones, some of which are covered in grayish-green moss while others are still bare and gray. The vegetation is quite remarkable. The moss is so soft and spongy that one is tempted to take a nap on it. Patches of gray tephra are punctuated by clusters of white and pink wild flowers. One the whole, Laki is one of the most remarkable volcanic areas of Iceland, as well as one of the most visually attractive. Getting there is definitely worth the effort.

Geysir

Some of the most spectacular manifestations of Iceland's volcanism are its geysers and the most famous is Geysir, which has named all the world's geysers (see section "Yellowstone" in Chapter 8 for discussion of how geysers erupt). Iceland's Geysir is located about 40 km (25 miles) northeast from Hekla's summit. It is possible to visit the Thingvellir National Park (Iceland's old parliament), the Geysir geothermal area, and the spectacular Gullfoss ("golden falls") waterfall in the same day. Geysir's first documented eruption was in 1294 and it eventually became one of Iceland's major tourist attractions but, sadly, it no longer performs well for visitors. In the early part of the twentieth century, the eruptions became infrequent and ceased by 1916. This is thought to have happened because the surface area of Geysir's bowl became much larger resulting in water

A visitor pitching coins into a rift zone lake at Thingvellir on a rare clear day. (Photograph courtesy of Jim Garvin, NASA.)

cooling very quickly as it came to the surface. Since then, new eruptions have been induced by various methods. Tourists in the early part of the twentieth century poured gravel into the bowl to lower the water level and force an eruption. Another favorite method was to pour soap into the water, which decreases the surface tension and facilitates superheating of the water, leading to an eruption. Occasionally, Icelanders use soap to induce an eruption on special days, such as Independence Day, but even then there are no guarantees that Geysir will perform. If it does, the eruption can be spectacular, reaching 60 m (200 feet) in height.

Visitors who come to Geysir will not be disappointed, as the nearby geyser Strokkur ("the churn") performs reliably every 5 minutes or so. Strokkur's jet reaches about 20 m (66 feet) and is quite impressive, though it is over quickly. Also impressive is the whole geothermal area that these geysers are part of. Well-marked trails take you by numerous steaming vents and mudpots. Stay on the trails, as the ground around these formations can break away easily under your weight. During the summer months, an average of one tourist per day is scalded in some way, mostly from putting their hands in water pools to see how hot they are.

Strokkur geyser, near the famous Geysir. (Photograph by the author.)

References

Chester, D. (1993) *Volcanoes and Society*. Edward Arnold.

Gudmundsson, A. *et al.* (1992) The 1991 eruption of Hekla, Iceland. *Bulletin of Volcanology* 54, 238–46.

Scarth, A. (1994) *Volcanoes*. Texas A&M University Press.

Þórðarson, B. and J.K. Snorrason (1994) *Hekla*. Töðugjöld.

12 Volcanoes in Costa Rica

Costa Rica

An oasis of peace in troubled Central America, Costa Rica is a country of magnificent active volcanoes set amidst glorious rainforests. It is the ideal destination for those who want to see a variety of volcanoes within a short time, including the very active Arenal. Aside from active volcanoes, Costa Rica offers tropical rain and cloud forests, rivers for whitewater rafting, pristine white, black, and pink sand beaches, and an abundance of wildlife and exotic plants. Eco-tourism could be a term invented for this country, or maybe it was here that the concept was proven. Rather than profit from the destruction of rainforests, Costa Ricans ("Ticos") decided to use them as a magnet for the emerging tourism market. About 25% of the country's land is now protected as National Parks and forest reserves. There is an increasing number of privately owned reserves, some bought with charitable donations from all over the world. The reserves showcase the country's remarkable flora and fauna, which include over 1,000 types of native orchids and 850 species of birds.

Costa Rica has over 200 volcanic centers, most of which are less than 3 million years old. Five volcanoes are known to have been active in historical times: Rincón de La Vieja, Irazú, Arenal, Poás, and Turrialba. A few others, such as Miravalles and Barva, still show fumarolic activity and are thought to be potentially active. Unfortunately, much remains unknown about the activity of all the volcanoes, because written records in Costa Rica only go back about 250 years.

Archeologically, Costa Rica is considered to be the frontier territory between the Mesoamerican and the Intermediate cultural areas. Findings from excavations show that hunter–gatherers were living here about 11,000 years ago. Unfortunately, oral history is scant and only a few legends make references to volcanoes. Even after Columbus discovered Costa Rica in 1502, recorded history continued to be meager for quite some time. The first Western settlement, Cartago, was not established until 1562 and the new land remained sparsely populated. Costa Rica played a minor role in the Spanish conquest of the New World because, although its name means "rich coast," no riches such as large amounts of gold and silver were found. The indigenous population, potential slaves for the Spanish, was small, so most explorers directed their attention elsewhere.

It is interesting that there is no record of any volcanic activity between 1502 and 1723. It is possible that no major eruptions happened during that time, but more likely that the events were simply not observed and written about by Westerners. Even during the last 250 years, some volcanic activity probably was never recorded.

Costa Rica begun to change in the eighteenth century, when coffee was introduced and thrived on the fertile volcanic slopes. The country's wealth grew and, towards the end of last century, vast banana plantations were established along the Caribbean coast. Coffee and bananas remained the most important wealth until the 1980s, when tourism boomed and replaced both as the country's biggest industry. Early explorers could not have predicted that this new land's future riches lay in its rainforests, rugged volcanic landscapes, and incredible diversity of flora and fauna.

Tourism undoubtedly has been helped by Costa Rica's peaceful, democratic government. The first free elections in Central America were held here in 1889 and the military was abolished following a short civil war in 1948. Past president Oscar Arias was awarded the Nobel Peace Prize in 1987 for his efforts to bring peace to his country's belligerent neighbors. Ticos are extremely proud of their half-century of democracy, their lack of military, and their high standards of education and health care. All this has attracted not only

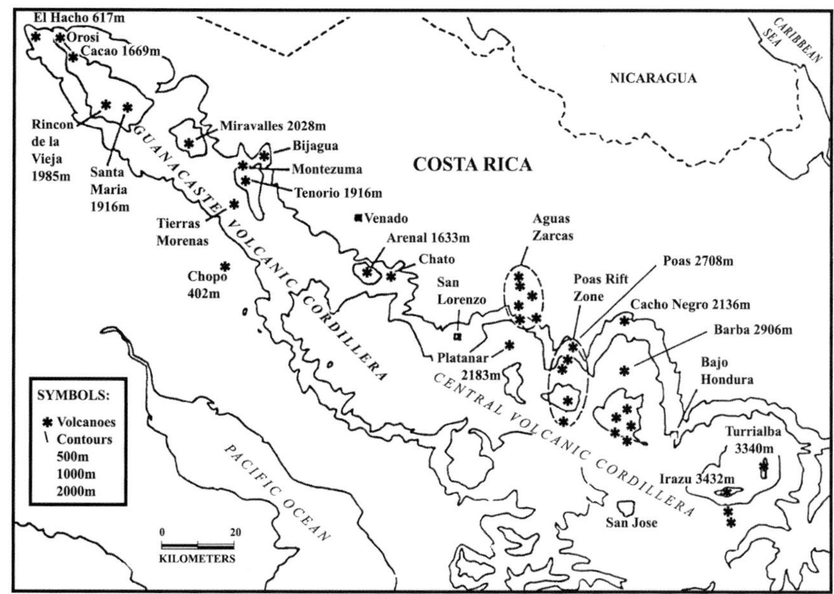

The volcanoes of Costa Rica are located along the Guanacaste range and the Central range. Most of the country's volcanic centers are less than 3 million years old. Rincón de la Vieja, Irazú, Arenal, Poás, and Turrialba have erupted in historic times. Heights of some of the major volcanoes are given in meters. (Modified from Viramonte *et al.*, 1997.)

tourists, but also immigrants, many of them retirees from the USA.

The volcanoes of Costa Rica have been, on the whole, kind to the population, as eruptions have killed only a small number of people. Most of the population do, in fact, live a safe distance away from the most active volcanoes, making their home in the Meseta Central (Central Valley), one of the fertile valleys that lie amongst the mountain ranges (cordilleras). Four major cordilleras – Guanacaste, Tilaran, Central, and Talamanca – cut the country from northwest to southeast, separating the Caribbean and Pacific coasts and giving the country its rugged appearance. Cordillera Talamanca, unlike the others, is not volcanic in origin. It is a granite batholith, that is, it consists of intruded igneous rocks that formed under great pressure and have since been uplifted. Arenal and Rincón de la Vieja are located in the Cordillera Guanacaste, while Poás, Irazú, and Turrialba are part of the Cordillera Central. Irazú is considered the most hazardous of the historically active volcanoes, as it can potentially threaten the capital city, San José.

Tectonic setting

The *Jurassic Park* story may have been set here, and the forests do look primeval, but Costa Rica was still under the sea when real dinosaurs roamed the Earth. Costa Rica emerged as part of a bridge of volcanoes linking South and North America, a direct result of the subduction of the Cocos plate beneath the Central American plate. The subduction occurs in the Middle America Trench off the west coast of Central America. The trench reaches depths of over 6 km (3¾ miles), but beneath Costa Rica it is about 4 km (2½ miles) deep. During the subduction process, the Cocos plate bends down to an angle of 25 degrees and is then pushed beneath Central America. The rate of subduction under Costa Rica is very fast – about 9 cm (nearly 4 inches) per year. The dragging of the Cocos plate under Central America causes convective overturn ("churning up") of the wedge of mantle that is sandwiched between the two plates. This brings hot mantle to shallower depths, and ultimately magma is released towards the surface.

The subduction beneath Costa Rica is complex, because the angle of subduction is steeper in the northwest (about 65 degrees) and shallower underneath the central part of the country (about 35 degrees). The change in angle occurs at what is known as the Quesada Sharp Contortion, below Platanar volcano east of Arenal. The change is related to the varying age of the crust of the Cocos plate. North of the Quesada Sharp Contortion, the Cocos plate that is being subducted is old and is at a greater depth (about 100 km, 60 miles). The Cocos plate being subducted to the southeast of the Contortion is younger, more buoyant and at a shallower depth (about 80 km, 50 miles). This explains the differences in the Guanacaste and Central volcanic ranges. The northern Guanacaste range is relatively narrow (7 to 14 km (4½ to 9 miles) across) and has smaller volcanoes, while the Central range is wider (about 30 km, 19 miles) and its volcanoes are larger. The Quesada Sharp Contortion is also a geochemical boundary. Magmas underneath the Central volcanic range are similar in composition to typical ocean island basalts, while magmas in the Guanacaste range are more like a mid-oceanic ridge depleted mantle source. The variations in chemistry are related to the relative importance of subducted sediments in the generation of magmas and in the melting of the subducted slab and overlying continental crust.

Costa Rica has rare outcrops of ultramafic rocks (ophiolites) in the peninsulas of Osa, Nicoya, and Herradura in the western part of the country. Called the Nicoya Complex, the rocks represent a succession of submarine volcanic and deep-marine sedimentary rocks that are mostly Jurassic to early Cretaceous in age. The early volcanic rocks are similar to those from oceanic hot spots and it has been suggested that they originated in the Galápagos hot spot, and formed the foundation of the Caribbean plate.

Practical information for the visitor

The National Parks and Reserves

All of the National Parks and most of the reserves are open to the public, usually for a small fee. Arenal, Poás, Irazú, and Rincón de la Vieja are located within National Parks. The National Park regulations are similar to those from all over the world, such as requiring visitors to stay on marked trails and forbidding the collection of rocks or plants. The quality of the trails is generally good and improvements continue to be made.

Transportation

Public transport is reasonable in Costa Rica, but buses tend to go to towns rather than to volcanoes. Renting a car is the best option. Many of the car rental companies are American and they tend to offer a cheaper rate to people who book in advance. Four-wheel-drive vehicles are available at a higher price and give you greater flexibility, but are only necessary when visiting a few of the places described here.

Those who want to travel by public transport can get to Irazú and Poás by bus from San José, though at present only on weekends. Enquire at the bus terminal in San José (known as the Coca-Cola Bus Terminal), which is actually a collection of small bus stations near one another. The nearest town to Arenal is La Fortuna, to Rincón de la Vieja is Liberia, and to Turrialba volcano is Turrialba. It is possible to take buses from San José to these towns and then taxis to your lodgings. Some lodgings will provide transportation from a nearby town at little or no cost – check first, because taxis are not cheap. If you are staying at Turrialba Lodge, or at Rincón de la Vieja Mountain Lodge during the rainy season, arrange for the owners to pick you up unless you have a four-wheel-drive vehicle.

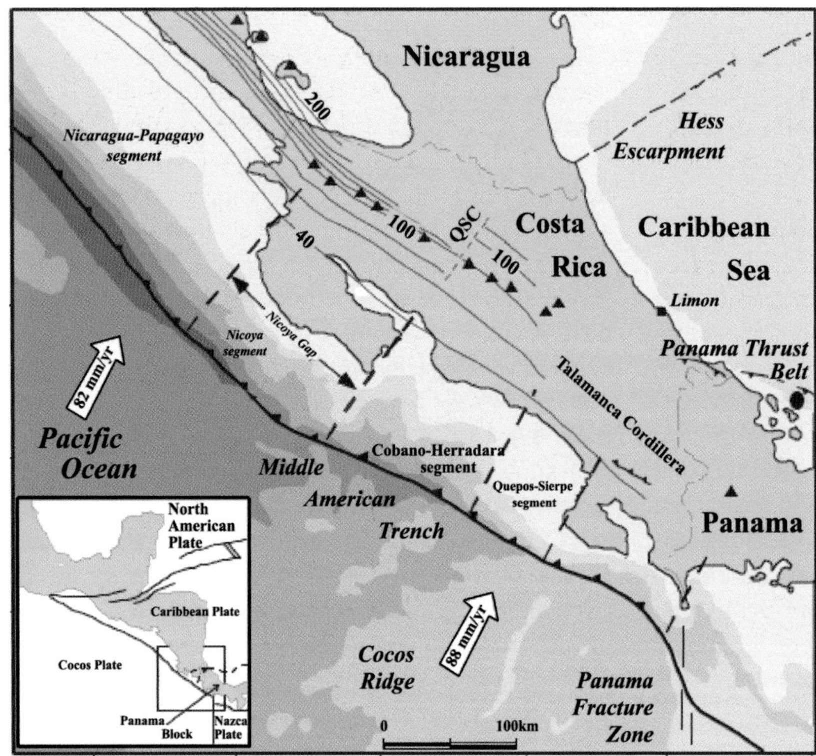

The tectonic setting of Costa Rica. The Cocos plate is subducting underneath the Central American plate. The subduction is steeper in the northwest and shallower underneath the central part of Costa Rica. The change in angle occurs at what is known as the Quesada Sharp Contortion, labeled here. (Modified from Viramonte *et al.*, 1997.)

Tours
Numerous tours to Arenal, Poás, and Irazú are offered by local companies; however, they are typically half-day or 1-day trips that are very limited in scope. Some local tour operators cater to "adventure trips" for small groups, offering tailor-made itineraries and a van, driver, and guide if required. The Sun Tours Company is a good choice for Arenal, as they own Arenal Observatory Lodge. Most of the time, these companies don't provide four-wheel-drive vehicles. Magic Trails offers horseback tours to the south slopes of Irazú and one of the tours takes riders all the way to the summit.

Tours from the air
A few light planes tours (including one entitled "Dance on the Volcanoes") and helicopter tours are now available in Costa Rica. The best way to find out what companies are offering air tours is to check advertisements in the *Tico Times* (the English-language local newspaper) and in *Costa Rica's Best Guide*, usually available at hotels. Most of the companies are based at the Tobias Bolanos airport west of San José.

Lodging
There is a trend in Costa Rica towards small "nature lodges" and haciendas (farms), which seem more ecologically friendly than big hotels or resorts. These lodges are not necessarily cheaper than hotels and may seem quite expensive for the amenities they offer – it is their magnificent locations you are paying for. Fussy travelers should beware that eco-friendly generally means that guests commune with nature – meaning no air-conditioning – but this is not a problem in the cool volcanic areas. Lodges that are particularly convenient for the volcanoes described here are the Poás Lodge, Arenal Observatory Lodge, Rincón de la Vieja Mountain Lodge, and the Volcán Turrialba Lodge (see Appendix I for addresses). Irazú is within easy driving distance from San José or Cartago, which is the country's old capital but unfortunately does not offer much choice in lodgings or food. Visitors to Poás have a variety of hotel choices in the nearby town of Alajuela, though the stone-built Poás Lodge is truly special. The Arenal area also offers a number of choices, but the Observatory Lodge, originally built for volcanologists, cannot be beat for the view and atmosphere. The Rincón de la Vieja Mountain Lodge is very rustic but cheap, and runs an even cheaper campground. Tours to the volcano and sulfur springs are available from the Lodge. The choices around Turrialba are very limited, as the area is not yet developed for tourism. The Volcán Turrialba Lodge is the closest to the volcano and has the convenience of offering horseback tours to the crater. The Lodge's location is magnificent and truly rural and remote, but prices are high considering the facilities offered. Since tourism is on the increase in Costa Rica, new establishments become available often – check a yearly guidebook to Costa Rica to find out what is new. Camping facilities are available inside the Rincón de la Vieja National Park, but at present not inside the other parks.

Safety and emergency services
Medical care in Costa Rica's public hospitals and emergency services in urban areas are of very high standard. However, volcanoes are far away from hospitals, the roads in the country are remarkably slow, and rescue services are not sophisticated (don't expect helicopters). Personal safety is generally not a problem in this peaceful society, but take sensible precautions. Driving can be dangerous, as most of the roads are narrow and windy – do not expect to get anywhere fast. Poisonous snakes are common in Costa Rica, so watch out for them while hiking. Strong earthquakes can occur; the last to date was in 1991 and measured 7.4 on the Richter scale.

Maps
It is not easy to find maps once you leave the major cities and it is best to buy them in San José. Topographic and other maps are available from Lehman and Universal bookstores (Central Ave, near the Grand Hotel) and from Instituto Geográfico Nacional (IGN), located in the Supreme Court Building. The IGN publishes a 1:200,000 map in nine sheets, and 1:50,000 topographic maps. A 1:500,000 geologic map by J. Tournon and G. Alvarado (in Spanish with English abstract) is published by Editorial Tecnologica de Costa Rica. For free maps of San José, stop at the Costa Rican Tourism Institute booth at the airport.

Arenal

The volcano
Arenal is at present one of the world's most active and spectacular volcanoes. Few sights are comparable to the night-time view of red-hot lava tumbling down the volcano's steep flanks and it seems unbelievable that we can watch it in almost complete safety from a number of vantage points. During the daylight hours, the handsome stratovolcano shape is fully revealed

Arenal is at present Costa Rica's most active and dangerous volcano. Arenal was a dormant volcano until 1968, when it started erupting continuously. Strombolian eruptions still occur at intervals from minutes to hours. (Photograph by the author.)

and Arenal appears steep, barren, and gray, contrasting with the surrounding lush, green rainforests. Arenal's size is modest – it stands only 1,100 m (3,630 feet) above the surrounding terrain (1,633 m (5,389 feet) above sea level) and has an area of 33 km^2 (12¾ square miles). The volcano has attracted and delighted visitors by being almost continuously active since it woke up from a five-century slumber in 1968. It has become one of the most sought-after sights in Costa Rica and its popularity is definitely on the rise. The Costa Rica Tourism Institute depicts Arenal as the country's logo, promising "no artificial ingredients." The Laguna ("lake") Arenal at the foot of the volcano has become one of the country's top tourist destinations, and new hotels and lodges with views of the fiery cone continue to be built. Like Kilauea and Strómboli, Arenal is an ideal destination for those who want to be sure of seeing an eruption. However, Arenal, a basaltic andesite stratovolcano, is potentially much more dangerous than either of the other two. To climb Arenal, you need to either be truly fearless or have a death wish. Luckily for those of us who don't qualify, it is possible to view and hear the activity from the lower flanks, well beyond the reach of bombs and of tumbling blocks of hot lava.

Arenal has been continuously active since 1968, producing frequent Strombolian explosions, lava flows, gas jets, fumarolic activity, small pyroclastic flows, and a lava pool that has been active since 1974. This eruption may be typical for Arenal. In its 4,000 years of activity, the volcano has produced many lava and pyroclastic flows, made up of basalts, andesites, and dacites. What we don't know for sure is whether the long duration is typical, as there are no past records of Arenal's eruptions.

Incredible as it may seem, Arenal was not even listed as one of the world's active volcanoes until it erupted in 1968. The only clues to its earlier activity came from legends of the Melekus (or Guatusos) indigenous people, who believed that the mountain was the home of their god of fire. It is now known that the legend was based on reality, as prehistoric Indians who lived near the volcano suffered numerous times from the effects of eruptions. However, the European colonizers who came to Costa Rica thought that Arenal was just another, ordinary mountain. It became known as Pan de Azúcar ("Sugarloaf") because its steep shape, and later as Cerro Arenal ("Sandpit Mountain"), because of the vast quantities of ash on its flanks.

Evidence about the early history of Arenal is slowly being gathered. Radiometric carbon-14 dating of charcoal found below an old lava flow indicates that the last major eruption before 1968 happened in AD 1525 – before the first Western settlement had even been established. Other radiocarbon dates indicate that major eruptions occurred in AD 1450, AD 800, AD 1, 800 BC, and 1800 BC. A large eruption, dated at about 1000 BC, produced about 0.4 km^3 ($^1/_{10}$ cubic mile) of material, about six times the volume that the present eruption has produced so far. It formed a Plinian deposit of dacitic pumice and ashfall, attesting to the volcano's potential for large, violent eruptions.

Archeological sites can also help to reconstruct Arenal's history. A thick layer of ash and tephra fall was found at the Sojo archeological site, excavated during the construction of the Arenal dam. The volcanic layer covered artifacts that were dated between 500 BC and AD 500, so the eruption must have occurred sometime during that period. Among the findings at Sojo were human bones, ceramics, jade, and stone tools. Unfortunately, the site can no longer be visited, as it was flooded after the construction of the dam.

The studies of past eruptions suggest that Arenal may have a cyclic behavior, with a major explosive event of andesitic magma opening a cycle, followed by lava flows of basaltic andesite that come out slowly for decades. When the lava flows stop, the volcano enters a repose period that may last for centuries. Since the current eruption started in 1968, it may not last much longer – and it could be a once-in-several-lifetimes occurrence.

The 1968 to present eruption

Arenal did not officially become a "volcán" until it blew up in 1968, but by the late 1950s its volcanic nature had started to become known. The first clue came from some aerial photos taken by Costa Rica's geographical institute as part of a survey. Active fumaroles were spotted on photos of Arenal's summit crater but, even then, the mountain was not thought to pose a threat. About 10 years before the present eruption, a local author wrote that the volcano had been "extinct for thousands of years."

Before 1968, visitors came to Arenal to climb its steep flanks up to the "Crater of the Finches." Thousands of these birds nested in the forest-clad summit crater and the morning symphony was described as truly enchanting. A journalist wrote that it was "as if the very soul of the volcano had turned to music." The crater is now known by the dull name "D crater" and, sadly, the finches are long gone.

The first signs that activity might happen again were noticed in 1965, when the Instituto Costarriquense de Electricidad (ICE) made the first topographic survey of the volcano. ICE was preparing to build the Arenal hydroelectric complex, which was completed in 1983. The hydroelectric plants use water from Laguna Arenal, the 88 km^2 (34 square miles) dam near the foot of the volcano. During the initial survey of the area, assistant geologist Hugo Taylor reported several unusual phenomena. Perhaps the strangest was the rapid drying up of the Cedeño lagoon, which caused the death of a large number of fish. In 1967, the temperature of the Tabacón River increased so much that the cattle could no longer drink its water. Señor Taylor concluded quite

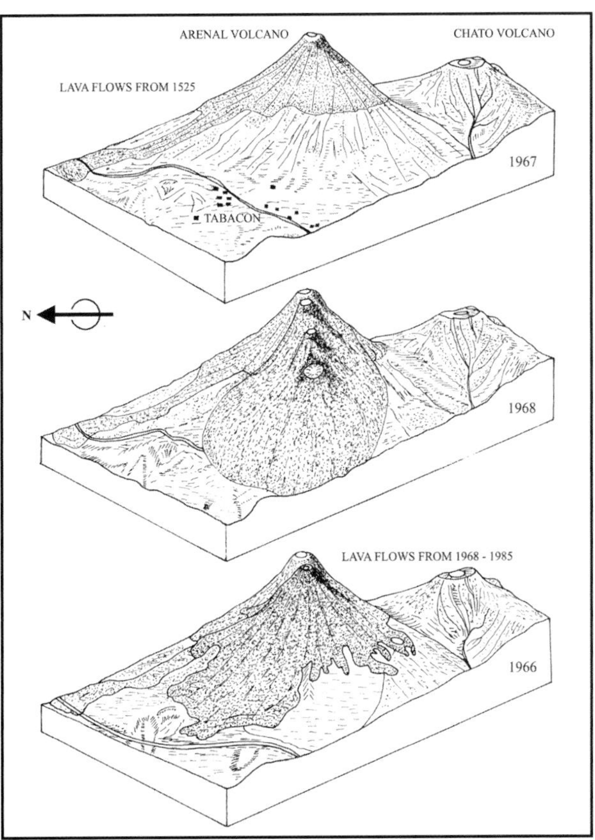

Schematic diagram of the recent evolution of the Arenal volcano. The top figure shows the volcano before the current eruption started in 1968. The arrow in the middle figure shows the areas devastated by pyroclastic flows and bombs in 1968. Since then, lava flows have continued to cascade down the steep flanks. (After Alvarado Induni, 1989.)

correctly that Arenal was entering a new eruptive phase.

Predicting volcanic eruptions is always difficult and, in the case of Arenal, this was specially true because so little was known about its past activity. The first warning sign came as a series of seismic events during May of 1968, but this was not enough to scare people away from the area. At 7:30 a.m. on July 29, after 10 hours of intense seismic activity, Arenal woke up with a violent explosion, the first of several. Bombs up to 10 m (33 feet) in diameter were thrown out, forming craters up to 60 m (200 feet) wide as they landed, as far away as 5 km (3 miles) from the summit. Three new craters were formed on the volcano's western flank, which became known as "A," "B," and "C" craters, aligned along a roughly east–west zone. According to eyewitnesses, the three craters were all formed during the first explosion.

Several villages and farms stood in the way of Arenal's blast. Pyroclastic flows (block and ash flows) reached two of them, Pueblo Nuevo and Tabacón, leaving no survivors. Eyewitnesses tell of thunder and lightening, earthquakes, and darkness. At Tabacón, the ground was completely peppered by craters formed by falling bombs. Studies of the burnt and scarred vegetation indicate that temperatures reached up to 400 °C (750 °F). The official death toll was 78 people, which included some missing and presumed dead. One of them, a young boy orphaned in the tragedy, was found many years later living in a local farm. Another miraculous survivor was a parrot, who was still on its perch when it was found by rescue workers.

Dozens of the injured were rescued, even though the conditions made rescue work nearly impossible. Over 1,200 evacuees came to Tilaran, quickly exhausting the drinking water supply. The government sent water trucks, declared a state of emergency, and soon Ticos from all walks of life volunteered to help. Sadly, the tragedy was not over. On July 31, the last of the major explosions took the lives of eight residents of Ciudad Quesada, 2 km (1¼ miles) away from the volcano.

Some deposits from this tragic first phase of the eruption can still be seen, telling the story from the volcanological point of view. The lowest deposits were produced by the initial phreatomagmatic blast, as hot magma came into contact with groundwater. Pieces of uncharred wood are found in this layer, indicating that the temperature was not high enough to burn wood. The blast deposit is overlain by several pyroclastic units in which the fragments did not weld to one another, meaning that they were not very hot when they were deposited. The next unit is a surge deposit and, on top of these are deposits from muddy fallout, block and ash flow deposits, and ashfall. The total volume from the initial eruption is about 25 million m^3 (875 million cubic feet).

During the following few months, Arenal continued to emit ash and vapor, but was otherwise quiet. People began to return to their lands and to rebuild their homes. The damage had been considerable, estimated at over 1 million dollars at the time. Between September 14 and 19, renewed explosions ejected small bombs and scoria, but these were low-energy, minor events. On September 19, the lower crater ("A") was filled by viscous lava, which eventually spilled out, forming the first of many thick, blocky lava flows. The flow advanced slowly down the valley of Quesada Tabacón, burying forever much of the devastated zone. Lava continued to spill out from the "A" crater (at 920 m (3,036 feet) elevation) until 1973. In March 1974, activity migrated to the "C" crater at 1,460 m (4,820 feet) above sea level, from where lava flows descended towards the southwest, eventually filling up the "A" crater and covering the western and southwestern flanks of the volcano. The "A" and "B" craters no longer exist, but it is still possible to find where they once were by looking for breaks in slope on the volcano's western side.

A new eruptive phase began in June 1975, with strong explosions that produced great quantities of ash and pyroclastic flows down the Tabacón valley. Fortunately, this time there were no casualties. Crater "C" continued to be the center of activity, building a small cone around it that now detracts from Arenal's otherwise symmetrical shape.

As I write, the eruption continues, but we don't know for how much longer. If it carries on at the same level, the typical scenario greeting the visitor will be mild explosions and sluggish, blocky lava flows, most of which stop at altitudes between 1,000 and 1,200 m (3,300 and 3,600 feet). On average, Arenal produces four to six lava flows per year, each lasting for several weeks. Although the magma output is small in volume, it has been steady for 30 years, making Arenal an ideal destination for people who want to see a volcano in action.

Now and then, Arenal surprises us with more dramatic events. In 1993, a wall containing the lava lake in the "C" crater breached, creating three pyroclastic flows that came down the steep slopes to 550 m (1,815 feet) altitude level. One of the flows came towards Tabacón, but stopped about 300 m (1,000 feet) away

Arenal's spectacular eruptions can often be witnessed from the safety of the Arenal Observatory Lodge, much of the delight of visitors from all over the world. The Lodge is one of the world's prime locations from which to photograph an erupting volcano, providing the weather cooperates. This photograph was taken by the late Fred Aspinal, owner of the Lodge. (Photograph courtesy of William Aspinal.)

from the famous Tabacón hot springs. Another came up to one of ICE's tiltmeter stations, but did not destroy the metal hut. Volcanologists found a trash can by the hut with burnt garbage inside and, by carefully examining the type of garbage, concluded that the flow was accompanied by a hot blast or surge with a temperature of about 400 °C (750 °F).

In September and October of 1995, large explosions caused plumes to rise up to 1 km (⅝ mile) above the crater, and bombs fell down to 1,000 m (3,300 feet) elevation. In 1998, the activity neared its 30th anniversary with no signs of decline. In the early afternoon of May 5, part of the crater wall collapsed again and the volcano began discharging pyroclastic flows that traveled several kilometers towards the northwest. Several hundred people were evacuated, though authorities reported some trouble getting tourists to move away. Luckily there were no fatalities, but Arenal proved its fatal capabilities on August 23, 2000. A series of thunderous explosions started at about 10 a.m. local time, soon followed by pyroclastic flows rushing down the mountain. One of the pyroclastic flows rushed toward the Los Lagos resort, inflicting severe burns on a local guide and two tourists who were hiking on one of the resort's trails. The guide was able to get the injured tourists to safety, but died of third-degree burns later in the day. This deadly event, considered the most serious since 1968, served as a sobering reminder that Arenal's threat is far from over.

A personal view: Volcanoes that go bang in the night

I consider my first visit to Arenal a good example of how not to visit a volcano. It was, frankly, quite dismal and could have been dangerous. My only excuse is that I had not intended to go to Arenal at all during my first visit to Costa Rica. I was in San José for an international conference, and a field trip to Poás was offered as a bonus. I thought one volcano would satisfy me and, in any case, I would not have a free day to get to Arenal. All this was right, but then I met José, a self-appointed tour guide whose only credential was probably the ownership of a van. He told me that I could indeed visit the notorious volcano – if I had a free night.

"Volcano very beautiful at night," he said emphatically, and then moved in for the kill: "You can see red lava." Unable to resist those magic words, I went on to persuade seven of my colleagues to join me in a misadventure they would not soon forget. Our group included my office-mate, my boss, and an internationally renowned scientist who was well past his 70th birthday. Our plan was to leave San José at dusk, travel to Arenal, and return "around midnight," according to our expert guide.

It rained as we left San José, and the rain followed us all the way. At around 10 p.m., we arrived at what is now the entrance to the National Park, but we could have been anywhere. The rain and fog made it impossible to see the volcano, let alone any red lava. Having gone that far, we did not want to give up. I had heard of the Arenal Observatory Lodge, so I suggested that we drive up there and wait a while – maybe the weather would improve. Everyone agreed, and José drove us on towards the Lodge.

The dirt road to the Lodge crosses a shallow river and, back in 1991, there was no bridge. José's van did not have four-wheel drive but he thought it could cross the shallow water. He inched the van towards the edge and, in the gleam of the headlights, we saw a bank halfway across, around which the river forked. The bank was only about 15 to 20 m (50 to 65 feet) from the river's edge and we knew the water was shallow. Still, the current seemed strong and in the dark, the river looked ominous. Down into it we went, with water reaching halfway up the wheels. I was relieved when we began to climb out onto the bank but, just then, the engine died. José bemoaned our fate in Spanish and, as if orchestrated, the volcano exploded just then with a very loud, reverberating boom.

"Was that thunder?" asked one of my anxious-looking traveling companions, "or the volcano?"

Since I was the only volcanologist in our unfortunate little group, everyone turned to me. I said it sounded like thunder, reasoning with myself that it was the best answer under the circumstances. My quick assessment of the situation was that I didn't like it. I knew Arenal was very active, and potentially lethal, but I hadn't done my homework. I had no idea where we were, or how dangerous the volcano was at the time. I knew Arenal discharged pyroclastic flows down its river valleys and we were stuck in one – not only inside the valley but in the river itself, in nearly two feet of rushing water. And it was a dark, stormy night.

My intellectual chain of thought was mercifully interrupted by José, saying that we had to get out of the van and push. Off came the boots and socks and we did as we were told, except for one member of the party, who seemed frozen to his seat. I chose to believe he was frozen with fright. Our septuagenarian scientist, a real trouper, did not hesitate to get into the water. I'd take him to a volcano anytime.

Once the van was onto the sandy bank, the next problem was how to get it out of there. My boss decided to cross the second half on foot and walk to the Lodge to get help. He soon turned back – the water was too deep and the current too strong. José, meanwhile, was taking the engine apart. To make matters worse, the next boom we heard could in no way be blamed in the weather. We strained our eyes in the direction of the sound, but the rain and fog had gotten even worse. We were at the mercy of the volcano, but it would somehow have helped to be able to see how far away the fuses were being lit.

After what seemed a long time, we heard the most wonderful sound – the purr of the engine that José had somehow fixed. We cheered, then remembered we were still surrounded by water. The only way back was across the river. Somehow this time the engine held on, and we crossed safely. We were back on the road and feeling rather cheerful. José, however, felt bad. He had promised us red lava.

We had one more option. We went back to our first stopping place, now the entrance to the National Park, and José advised us to take a path. He told us that, after about 100 m (300 feet), we would find a sign saying we should go no further. "Stop there," he said. "Volcano very dangerous." Taking our flashlights, we set out on the path. We may have taken the wrong path, or just failed to see the described sign. Soon we were walking along a shallow gully, then a deep one, with walls some 5 m (15 feet) high. It would have channeled a pyroclastic flow very nicely. Just as I was about to tell the others that we should turn back, the clouds broke and blurry red lava flows appeared as if by magic, delineating the steep cone. I stopped to look. The others charged towards the volcano like bulls at a cape.

It is difficult to judge distances at night and, although we seemed to be a few kilometers from the lava flows, I knew that we could get into trouble. The steep slopes of Arenal allow lava blocks to detach themselves from the end of flows and roll downhill at great speed. We had some fantastic views of some such blocks which, fortunately, did not roll directly towards us. We could hear explosions but saw no bombs. Questions started flashing in my mind, but I didn't have the answers. What was the topography between the volcano and us? Could blocks roll towards us? What state was the volcano in? Could a pyroclastic flow come right down the gully we were walking on? Most pressing of all – where exactly on the volcano were we?

One thing was clear – we could not get out of that gully in a hurry. The walls were very steep, blocky, and hard to climb. I told the others we shouldn't go any further, but several of them did not seem inclined to listen to me. I could not be very convincing, since I had no hard facts to present. I promised myself to do my homework before any future trip to a volcano.

When you can't reason with people, you have to trick them. I suggested stopping for a group photo against the glowing volcano. Somehow none of those highly qualified scientists seemed to realize that lousy weather and no tripod meant a photo was a really dumb idea. But I stopped their momentum and, after the photo was taken, everyone seemed content to stand and watch for a while. We then started to head back.

Our adventure had not ended as we lost our way and got into waist-high grass. Years later, I walked on that terrain and am amazed that none of us that night fell into a ravine or deep hole hidden in the grass. Our eldest colleague tripped and fell, but luckily was unhurt. We decided to retrace our steps and try to find our original way. We eventually got back to the road, helped by a very worried José who guided us by flashing the van's headlights. Exhausted but unhurt, we arrived back at San José at 7 a.m., just in time for breakfast before joining the "official" field trip to Poás. I felt very proud of my group of non-volcanologists, as all of them gamely went along to see the next volcano – probably because I was not guiding that trip.

Visiting during repose (or bad weather)

Nobody knows when the present eruption is likely to end and, even if Arenal is active, the weather can be quite bad, covering the volcano with clouds. It is possible to spend several frustrating days waiting for the weather to improve, though hearing the loud booms from the volcano reminds you that it is very much active. It helps to know that nearly everything that you can do during a visit to Arenal does not depend on the volcano being cloud-free. Some cloud and rain can even make the jungle trails more dramatic, giving you a real rainforest experience. If possible, keep your travel plans flexible, so you can wait for clear nights with the fantastic red fireworks. Arenal's weather is very localized, so forecasts for the country or state cannot be relied upon.

Visiting Arenal is easy. The local infrastructure has greatly improved in the last few years, particularly since the National Park was established in 1995. Tourism has grown, but not to the point of detracting from the natural wonders. The trails and places described below can be enjoyed whether the volcano is active or not.

Arenal Volcano National Park

Because the park is so new, its services are limited, but are likely to improve in the near future. The main entrance is reached by driving west of La Fortuna for 15 km (9 miles), then making a left turn into a 2 km (1¼ miles) dirt road leading to the Park. The rangers at the entrance hut can sell you a trail map but, in general, are not very forward with information. A short trail (1.3 km, ¾ mile) starts by the entrance and leads to the volcano lookout point, beyond which fences and signs warn visitors not to go into the danger zone. The lookout point is one of the best places from which to watch and photograph the activity.

There are three other trails inside the Park, and visitors are not allowed to deviate from them. The Sendero las Heliconias ("Trail of the heliconias," a type of tropical plant) and the Sendero Colada ("lava flow trail") run parallel with one another and overlap. The most notable part of the Colada trail is where it climbs onto the 1992–3 blocky lava flow. You can also get to the 1992–3 lava by following the Sendero los Tucanes ("The toucans' trail"), which starts near another Park entrance, on the south side. A fourth trail, Sendero los Miradores (1.2 km, ¾ mile), starts from the road opposite the main entrance to the Park and goes up to the Arenal hydroelectric dam.

Sendero las Heliconias

This is an easy, 1 km (⅝ mile) long trail that starts by the main entrance to the National Park. The purpose of the trail is to show visitors the plants, mostly ferns and small bushes, that have sprouted since the

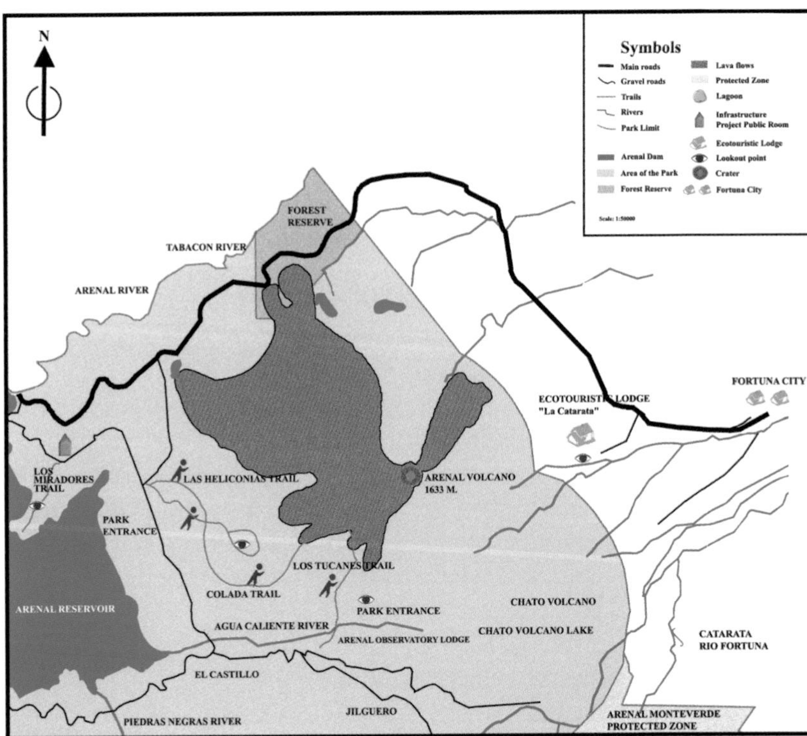

The Arenal National Park has numerous senderos (trails). It is easy to spend several days exploring the slopes of the volcano, which are rich in wildlife. Ascent of the volcano is extremely dangerous and should only be attempted with an experienced guide. (Modified from Arenal National Park's brochure.)

Trails through the Arenal National Park not only provide spectacular views of the volcano but also the opportunity to see tropical flora, birds, and animals. (Photograph by the author.)

devastation of the 1968 eruption. Although of limited interest to volcanophiles, it is worth the short walk, as you are likely to see colorful hummingbirds and butterflies along the way.

Sendero Colada

This is another easy trail that starts from the main entrance to the Park. It goes over fairly flat ground covered with ash for about 2.8 km (1¾ miles), up to the edge of the 1992–3 blocky lava flow. The trail climbs onto the large blocks of the flow and continues near the flow's edge, up to where it joins the Tucanes trail. There are particularly good views of the Chato volcano and of the Arenal lake from the top of the flow. Look away from the volcano and notice the difference in vegetation between the old jungle further away from the flow, where tall trees still stand, and the shorter grasses nearer to the flow, where the forest was devastated by the 1968 activity.

Sendero los Tucanes

This beautiful trail through the jungle starts by the southern entrance to the Park, near the Agua Caliente ("hot water") river. You can get to the southern entrance by following a short trail from right outside the gate for Arenal Observatory Lodge, which is located 1.3 km (¾ mile) from the lodge itself. The Park entrance is located just across the Agua Caliente river.

The toucans' trail is mostly flat and easy, but you may have to go over or around fallen trees. The flora and fauna are remarkable. Apart from the toucans, you may see howler monkeys, armadillos, wild pigs, sloths, and coatis. Go early in the morning and walk quietly to have the best chance of seeing some of these animals. The trail ends by the 1993 blocky flow, where it joins the Colada trail.

Lake Arenal resort area and 1996 flow

This collection of resorts around the edge of Arenal Lake offer sailing, windsurfing, fishing, and a variety of other sports. The Jungla y Senderos ("jungle and trails") Los Lagos resort, located on the northern side of Arenal and 5.5 km (3½ miles) east of the town of La Fortuna, is noteworthy for its trails. The resort and its trails suffered some damage from the pyroclastic flow of August 2000, which killed a local guide on one of the trails. Because of this tragic event, some of the trails are currently closed off, but will probably re-open (check the website given in Appendix I to find out current status). Of particular interest here is the Rio de Lava ("lava flow") trail up to the 1996 blocky lava flow. The trail is about 800 m (½ mile) long from the nearest parking lot, and starts near one of the resort's two lakes (pick up a trail map at the entrance to the resort). There is a small admission fee to the resort and, for a little extra, you can rent horses to take you most of the way to the flow. It is, however, a very easy trail. Once it gets to the flow margin, the trail continues alongside the flow. You can easily climb onto the flow, which is about 12 m (40 feet) thick, and catch an awesome view of the volcano's northern slopes. This flow cut right through the jungle, but did not cause much damage beyond the margins of the lava. Mosses and ferns have already started to invade the cooled flow and, in as little as 10 years, it may be impossible to see where the lava once was. If you can, take time to explore some of the other trails of Los Lagos, where howling monkeys and toucans abound. There is a campground at Los Lagos, as well as lodgings in casitas (bed and breakfasts).

Volcán Chato and lagoon

Chato ("flat") is a small volcano located about 3 km (1¾ miles) to the southeast of Arenal's crater. Like Arenal, Chato's lavas are basaltic andesites and andesites, but the two volcanoes belong to separate systems. Chato is no longer active, and radiocarbon dates indicate that its last eruption occurred at about 1550 BC. The summit, at 1,140 m (3,760 feet) above sea level, is capped by an explosion crater about 500 m (1,650 feet) in diameter. The crater is now filled with a rainwater lake and native frogs can often be seen by the edge of the emerald-green water. To climb to the summit, take the Chato trail (round trip 6 km (3¾ miles), about 5 hours) starting from the Arenal Observatory Lodge. It is a fairly arduous trek, but rewarding because of the jungle setting, the stunning crater lake, and the views of Arenal's flanks. A small cone, called Chatito, can be seen on the southern flank of Chato.

Tabacón hot springs

The Balneario Tabacón resort is built around the natural hot springs of the River Tabacón. It is located about 12 km (7½ miles) west of La Fortuna and is popular with foreigners (who come by the busload), as well as with locals. The admission fee is not cheap, but

A popular Arenal attraction is the Balneario Tabacón, where visitors can take advantage of natural hot springs of the River Tabacón. Locals recommend coming here at night for the best views of the erupting volcano. (Photograph by the author.)

the lovely gardens and the experience make it worth the money. I advise skipping the artificial swimming pools and heading straight for the less crowded, natural pools in the river, where the temperatures range from about 27 °C to 39 °C (about 50 to 100 °F). There are lovely views of the volcano from here and the locals recommend coming at night for the most spectacular scenery. The main attraction may be the thrill of knowing that you are bathing right under the volcano, in one of the places at the highest risk from pyroclastic flows.

Visiting during activity

The suggestions below are those that are best done while Arenal continues to erupt, though climbing the volcano will be far safer once the activity ends.

Arenal Observatory Lodge

Located inside the National Park, the lodge started out in 1987 as a private observatory where volcanologists from all over the world stayed while studying the ongoing eruption. It is set on a ridge about 2 km (1¼ miles) away from Arenal and the views of the volcano, plus accompanying sound effects, make staying here an almost unique experience. Non-residents can enjoy the grounds and views for a small fee. Book well in advance if you want to stay here, because although the Lodge has expanded over the years to several buildings, it still only has a total of 24 rooms. Some rooms have volcano views and there are several "observations decks" where visitors can sit and watch the nightly fireworks. One of the Lodge's buildings, the Smithsonian block, houses a small museum and a working seismograph. To get there from the main building you have to go over a dramatic suspension bridge.

The Lodge offers daily guided hikes, including Volcán Chato and the Sendero los Tucanes. Their "red lava walk" at night is said to offer great views of the volcano's glowing flanks. Several trails start by the Lodge, the most interesting being the Cerro Chato and the Old Lava trails. The New Lava Trail is really a trail to the south side entrance to the National Park, where you join the Toucans trail up to the 1993 flow.

Climbing the volcano

I haven't done this but would like to. Given the present state of activity, many of my volcanologist colleagues wouldn't do it, and some would say that you'd have to be crazy to want to. It is possible to climb the active Arenal – provided that the only guide I know of who does this trip is still alive and doing this kind of work. Gabino Hidalgo lives in La Fortuna and has been taking visitors to Arenal for many years, so far without losing any of them. Hidalgo takes advantage of the fact that the eastern ("D") crater has been inactive for years and he hopes it will remain that way. From the rim of this crater, there is a great view of the active "C" crater.

His route goes through the jungle-covered eastern slope of the volcano up to a "safe rock" a few hundred meters below the summit. Safe is a relative term, but this rock serves as a crude shelter. The climbers stay there until they hear an explosion, which Hidalgo knows happens about every 30 minutes. From his years observing the volcano, he believes that the interval between explosions is at least 15 minutes. As soon as the bombs fall from the last explosion, he and his group make a mad dash up to the rim of "D" crater. They marvel at the view of red lava in the "C" crater, take a picture or two (automatic winding helps in this case), then make another mad dash down to the "safe rock." From there, the party walks back down the eastern slope. Hidalgo always stops on the way and uses a mirror to signal his mother in La Fortuna, to say that all is well. I assume he also has a secret "send rescue" signal worked out.

Hidalgo claims his method is 90% safe, whatever that means. My opinion is that his way to climb the volcano is probably the safest possible, but still involves considerable risk. Volcanoes do not behave like clockwork and explosions can occur at unexpectedly. A dormant crater in an active volcano could wake up again anytime. However, if you are determined to climb, go with Hidalgo (ask locals for his whereabouts). His record is good, he knows the mountain very well, and we hope his luck will last.

Other local attractions

Volcán Rincón de la Vieja

This active, complex stratovolcano is the main attraction of the Rincón de la Vieja National Park. The volcano's odd name means "old woman's corner" – it is said that an old, hermitic woman lived at the foot of the volcano long ago. Rincón is not heavily visited because of its remoteness, but there are interesting sights and good trails for those who are willing to detour from Costa Rica's more beaten path. The volcano is a complex of at least nine vents forming an elongated ridge, covering an area of 400 km^2 (150 square miles) and reaching an altitude of 1,895 m

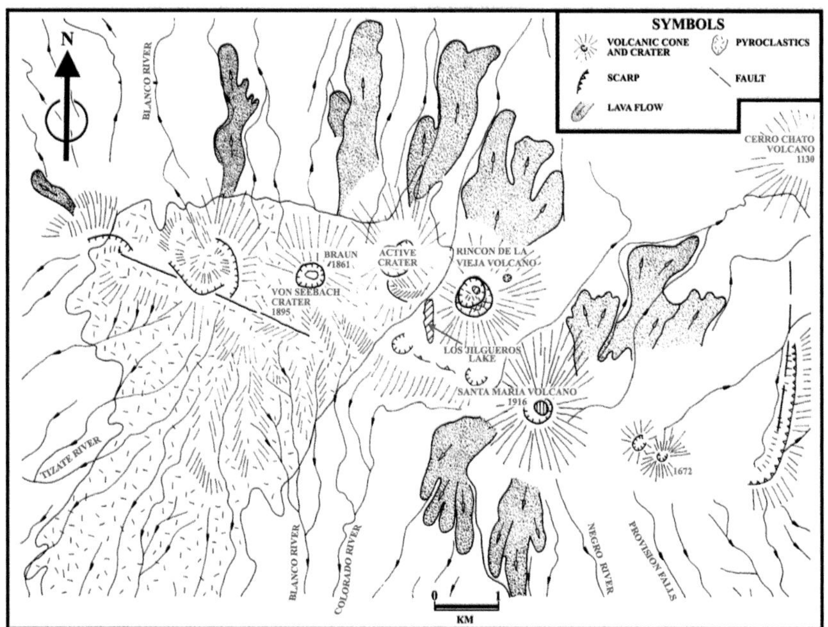

Sketch map of the Rincón de la Vieja volcano. Rincón de la Vieja is an active volcano that is little visited because of its remote location, about 50 km (30 miles) south of Lake Nicaragua. There are nine craters in the Rincón complex, but only the main crater is active. The volcano is located within a National Park that has several trails worth exploring, including one leading to the summit crater. (After Alvarado Induni, 1989.)

(6,250 feet). The overlapping cones that form the complex are of different ages, and indicate that activity has migrated west to east. The volcano has often been active in historical times and during the nineteenth century was known to sailors as a "natural lighthouse." Eruptions during the last few decades have mainly consisted of mild to moderate phreatic explosions from the summit crater, which contains a lake. The explosions have often triggered mudflows and noticeable ashfall.

Because of its remoteness and the few population centers located nearby, Rincón has not been a priority for monitoring or intensive study. As far as we can tell, it does not pose a great threat at present, as recent eruptions have not brought up to the surface any juvenile (new magma) material. Local scientists visiting the crater during November of 1995 described what is thought to be a typical eruption. Explosions produced jets of very dark, wet ash and hot water. The cypresoidal jets fell outside the crater, forming mudflows along the drainage of rivers. Blocks and wet ash fell as much as 1 km (⅝ mile) away from the crater. From this description, it is clear that Rincón can be dangerous if you get too close to the active summit crater. Before attempting to climb to the summit, stop by the ranger station at Casona Santa Maria and ask the rangers about the volcano's current activity. The Casona, located 25 km (15 miles) northeast of the town of Liberia, is an interesting nineteenth-century ranch house that apparently once belonged to US President Lyndon Johnson. Note that the road from Liberia is poor and requires four-wheel drive during the rainy season.

If you are staying at the Rincón de la Vieja Mountain Lodge, near the Las Pailas mudpots, it is the best to take a different road. To find it, take the Interamericana from Liberia and drive about 5 km (3 miles) north up to a turnoff onto a gravel road. The sign may say Mountain Lodge, Las Pailas, or Curanbade, the name of a small village along the way. Beyond the village the road deteriorates and requires a four-wheel-drive vehicle during the rainy season. The road goes up to Hacienda Guachipelin, which is now also open as a lodge. To proceed to the Mountain Lodge, you need to pay a small fee for the use of the Hacienda's private road, which is accessible during daylight hours only.

There are several trails within the park. From the Santa Maria ranger station, a 3 km (2 miles) long trail leads to sulfurous hot springs where visitors can bathe and supposedly benefit from the volcano's therapeutic properties. The trail continues for 3 km (2 miles) to the Las Pailas boiling mudpots and for a further 2 km (1¼ miles) to the Las Pailas ranger station, which is not always open. From the Las Pailas mudpots, another trail leads north for about 2 km (1¼ miles), then forks. The left-hand fork continues for another 2 km (1¼ miles) to the small fumaroles of Las Hornillas ("the kitchen stoves"). The right-hand fork takes you to the summit, about 8 km (5 miles) from Las Pailas. Below the summit is the Laguna de Jilgueros, known to have many tapirs and other wildlife.

Poás

The volcano

Volcán Poás is one of the most active volcanoes in Costa Rica, and one of the most easily visited. Located only 30 km (19 miles) northwest of San José, the 2,708 m (8,885 feet) high volcano is accessible by a picturesque road that crosses through a rainforest and then goes all the way to the summit. Poás is very popular with tourists because of the spectacular beauty of its two craters. The active crater is usually filled with a hot-water, turquoise-colored lake, while the dormant – or perhaps extinct – Botos crater contains a dark blue, cold-water lake.

Poás is a complex stratovolcano with an area of about 300 km^2 (116 square miles). It has an irregular shape with a broad and relatively level summit cut by faults. Together with Platanar, Barva, Irazú, and Turrialba volcanoes, Poás forms the Cordillera Central of Costa Rica. Poás volcano itself consists of three main structures, roughly aligned northwest to southeast. At the northwestern end is the von Frantzius crater, about 1 km (½ mile) in diameter, which is considered extinct. The active crater, about 1.5 km (1 mile) in diameter, is located in the middle. The smaller Botos crater, last active in prehistoric times, is located at the southeastern end.

The volcano is built over silicious ignimbrites and lava flows dating back to the Plio-Pleistocene. The oldest lavas from the summit were emplaced about 7,500 years ago, around the same time as Botos was last active. Poás erupts mostly basaltic andesites and andesites, though there are some basalts. Some of the older lavas are particularly rich in aluminum and titanium. Chemical analysis of the hot crater lake has shown a very high content of dissolved solids (60 to 350 g/kg) and sulfuric acid content of as much as 4 g/kg. Despite its attractive color and jacuzzi-like temperatures (50 to 70 °C 112 to 158 °F), the lake would be a nasty place in which to take a dip.

Poás has had a long history of violent eruptions. Reports on the activity go back about 200 years, and most describe "geyser-like" phreatic eruptions and fumarolic activity. The earliest account is from 1828, by a Señor Miguel Alfaro, who described ash and blocks that burned in the air with blue flames. The blue color has been seen in recent years, and is due to the burning of sulfur. The next report of activity at Poás is from 1834 and describes ash being carried by winds to Esparza, a town on the Pacific side of the country. In 1860, the German scientist Alexander von Frantzius went down to the bottom of the active crater and measured the lake temperature, reporting 39.1 °C (about 102 °F). Another scientist, Dr. Henry Pittier, went down to the lake in 1888 and again measured its temperature – this time it was higher, at 55.5 °C (about 132 °F). Severe earthquakes during that same year caused a landslide that created the Fraijanes pond, located close to the present road to Poás.

A spectacular phreatic eruption occurred in 1889, during which water and mud columns from the crater reached heights of about 70 m (230 feet). Temperature measurements during this time showed the lake water to be at about 70 °C (about 160 °F). By the early twentieth century, it became clear that phreatic eruptions are the principal modus operandi of Poás: these eruptions happened in 1903, 1904, and 1905. On July 18, 1905, violent explosions ejected materials that landed as far as 500 m (1,600 feet) away from the crater's rim. Reports from the time mention that activity continued sporadically until 1907.

One of the major eruptions on Poás happened on January 25, 1910, when an immense cloud of ash rose 8,000 m (26,000 feet) above the volcano, causing ash to fall all over the Central Valley. Bombs were ejected with so much force that impact craters up to 1 m (over 3 feet) in diameter were formed. The upper slopes of Poás were covered with mud and ash fell 35 km (22 miles) away from the crater.

Phreatic eruptions ejecting material high above the volcano also happened in 1914 and 1915, but were not as spectacular. Another period of particularly intense activity occurred between 1952 and 1954, when Poás spewed out large quantities of ash and the area was rocked by earthquakes. The crater lake dried out, and Strombolian activity took place in the dry crater, building a cone 40 m (130 feet) high – known simply as the "crater's interior cone."

Since the Poás Volcano National Park was formed in 1971, the volcano's activity has been recorded by the Park's staff. Several periods of phreatic eruptions took place between 1977 and 1979. On February 14, 1978, an eruption column reached 2,000 m (6,500 feet) in height. Phreatic eruptions continued through 1980 but, in 1981, the activity changed in style. There were no phreatic eruptions, but fumarolic activity increased in the "interior cone," and temperatures reached 940 °C (about 1725 °F). Gas columns were ejected from the interior cone and this type of activity continued until 1987, when phreatic eruptions from the hot crater lake started to take place again.

Poás has displayed four types of activity during the

The summit area of Poás volcano, located only about an hour's drive from Costa Rica's capital city, San José. The summit crater contains several explosion craters, such as the one shown here, which houses a small turquoise-colored lake of sulfurous steaming water. Note how little vegetation there is in the summit area, a sign of frequent eruptions. Occasionally, Poás has violent eruptions. (Photograph courtesy of Scott Rowland.)

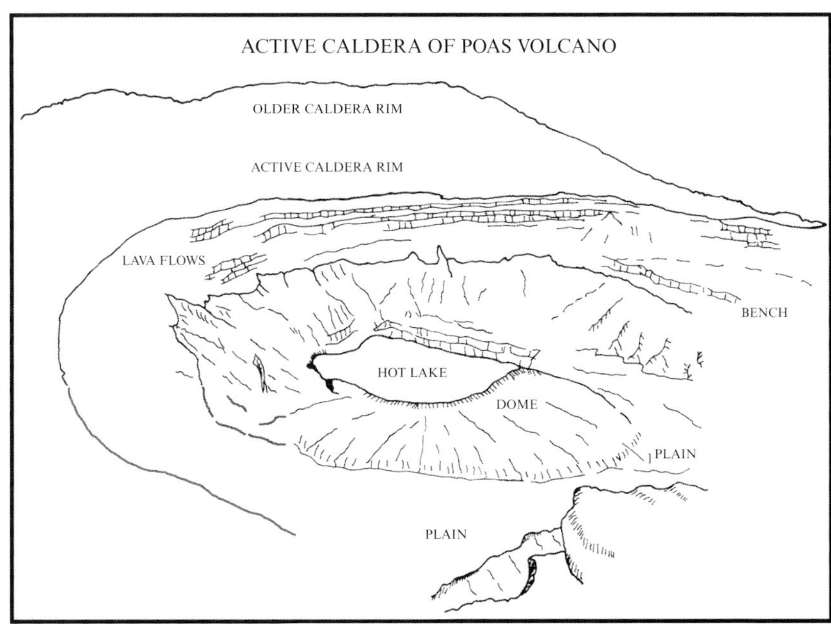

Sketch map of the summit of Poás volcano. The currently active caldera is located inside a much larger older caldera. The active crater is about 1.5 km (1 mile) in diameter. (After Casertano *et al.*, 1987.)

two centuries for which there are records. Geyser-like eruptions ejecting columns of muddy water and steam that go up meters or even hundreds of meters into the air have occurred at intervals from minutes to years. Violent phreatic eruptions, such as those in 1834 and 1910, cover large areas with ash and are the most dangerous. Strombolian activity, such as that during the 1953–5 period, does not present any great hazard for the local population. The most recent type of activity is also the most benign: quiet degassing. It is also the least spectacular to see. The type of activity one hopes to see at Poás is the first – the geyser-like eruptions that have earned the volcano the technically untrue but understandable reputation of being the largest geyser in the world.

The 1987–90 eruption and recent activity

This active phase began in June 1987 with several phreatic eruptions. These were not major events and all of the materials ejected fell inside the crater. It was noted that the temperature of the crater lake went up, from 58 to 70 °C (136 to 158 °F). The phreatic activity continued for many months, with some impressive eruption columns coming out of the crater. For example, a phreatic explosion on April 9, 1988, sent a column 1 km (3,000 feet) above the ground.

What made this series of eruptions particularly interesting was the fact that the lake level went down progressively between 1987 and 1989, eventually exposing the crater floor. Microgravity measurements indicated that there was a shallow injection of magma beneath the lake between 1987 and 1988. This may have exceeded the ability of the summit's hydrothermal system to buffer the volcano's heat output, resulting in a gradual decrease in the lake level by over 30 m (100 feet) in a little under 2 years. Fumaroles were exposed on the crater floor and their temperatures reached 85 °C (185 °F). Overall emission of gases from the volcano increased and noxious gases were carried by winds to the west and southwest.

In March 1989, a particularly large explosion took place and was even seen from San José, causing great concern to the residents. By April, the crater lake had totally disappeared. Gas jets lifted sediments from the crater floor high into the air, depositing them as far away as the town of Sarchi, 18 km (11 miles) from the crater. With the lake gone, a very interesting phenomenon was seen on the crater floor: the appearance of pools of liquid sulfur with temperatures as high as 140 °C (285 °F) – high enough to keep the sulfur molten and bubbling. The high temperatures were probably maintained by the effect of hot gases from below. The discovery of these molten pools of sulfur was important because this volcanic phenomenon had not been observed and documented before. Another unusual phenomenon was reddish flames coming out of some of the fumaroles – the flames were produced by combustion of sulfur in very hot fumaroles.

To make things even more interesting, sulfur and mud "mini-volcanoes" begun to erupt on the crater floor. A British team visiting the crater described several small yellow sulfur cones, about 1 to 3 m (3 to 10 feet) high, formed by fragmented sulfur and lithic particles. Occasionally, "chimneys" up to 1 m (3 feet) high formed at the top of the cones, but did not last long as the cones seemed to collapse often. A year later, small sulfur volcanoes were observed again on the dry lake bed and their formation became more clear. Vigorous gas pulses were seen to rip an apparently muddy surface, spraying mud and coagulated sulfur up to 5 m (16 feet) high. The falling tephra formed a concentric rampart around the vent, eventually building a cone. The sulfur cones tended to be short-lived; sometimes they collapsed in less than a day. The pits of collapsed cones often contained bubbling sulfur pools. It is likely that the cones were atop a molten sulfur "chamber" – a layer of liquid sulfur just below the dry lake surface. The eruption mechanism could have been similar to Strombolian except that, in this case, the gases were not coming out of the melt, but were instead fumarolic gases supplied by the volcano's hydrothermal system.

The strange and rather photogenic mini-volcanoes were of great interest to volcanologists, as sulfur eruptions are quite rare on Earth. Sulfur volcanism is really a by-product of silicate volcanic eruptions: the high temperatures of silicate magmas melt the sulfur accumulated at fumaroles, producing melted pools and flows such as the Mauna Loa sulfur flow described in Chapter 7.

The sulfur volcanism at Poás also served to confirm the existence of a molten layer of sulfur beneath the bottom of the crater lake, as had already been suggested. Back in 1977, greenish sulfur particles as large as 4 cm (1½ inches) were found around the crater, which had apparently been ejected during earlier eruptions. The particles looked like scoria in shape, suggesting that they were ejected while still molten.

The mini-volcanoes are, unfortunately, now gone, but visitors can again delight in the sight of the beautiful crater lake, which has returned to its usual level. However, the lake level fluctuated again during the 1990s and at times it dried up completely again.

This photo of the interior of the Poás crater was taken on April 11, 1989, during the 1987–90 activity. The lake level went down progressively during this period, eventually exposing the crater floor. By April 1989 the lake had disappeared and molten pools of sulfur were seen – a volcanic phenomenon that had not been documented before. (Photograph courtesy of G. Alvarado Induni.)

Poás occasionally has phreatic activity, explosions generated by contact of hot magma and water. When these explosions happen, Poás is usually closed to visitors. This photograph was taken on April 29, 1988 by Costa Rica volcanologist G. Alvarado Induni.

Between 1995 and 1997 the lake level went up, reaching about 50 m (160 feet) in depth. The activity of Poás during the mid-1990s has consisted of gas plumes and phreatic explosions, occasionally causing the Park to be closed to visitors.

A personal view: An ecological wonderland

Like most other visitors, I came to Poás to see the volcano, but left fascinated by the plants and the birds in the Park. It is worth allowing oneself plenty of time in the National Park to be able to appreciate these aspects of nature. There are four different habitats within the National Park – one with essentially no vegetation, an area of arrayans, one of stunted forest, and one of cloud forest.

No plants grow inside the active crater because of the lack of suitable soil, natural erosion, and, in particular, the effect of fumarolic gases. Some species of plants grow on the edge of the crater, clinging tenaciously to the hardened ash. These include the paddle fern (*Elaphoglossum lingua*), small *Pernettia coriacea* plants, mosses, and lichens. Further out from the crater is the arrayan (*Vaccinium consanguineum*) area, made up of dwarf plants. This zone includes the main crater lookout point and the beginning of the trail between the lookout and the Botos Crater. The plants in this zone can grow 2 to 3 m (6 to 10 feet) high, and include mountain mangrove (*Clusia odorata*), *Vaccinium poasanum*, *Didymopanax pittieri*, and small cypress (*Escallonia poasana*). As you walk farther towards the Botos crater, you will enter the zone of dwarf or stunted forest. This slow-growing forest of

The Poás crater lake as it usually looks to visitors: placid and bright turquoise blue. (Photograph taken by the author in April 1998.)

twisted branches is almost impenetrable. Species include the mountain mangrove, arrayan, tucuico (*Ardisia* sp.), and *Hesperomeles obovata*. Finally, you see the cloud forest surrounding the lake – very damp, misty, shady, with trees up to 20 m (65 feet) high and moss everywhere. Species include oak (*Quercus* spp.), small cedar (*Brunellia costaricensis*), and white cypress (*Podocarpus oleifolius*). On your way back to the Visitor Center and parking lot, look out for the "poor man's umbrella" plant (*Gunnera insignis*), which grows in clearings in the forest. Its leaves can be as large as 2 m (6 feet) in diameter and they would probably live up to their common name during a rainstorm.

Many years ago, the Poás area thrived with wildlife, but uncontrolled killing and deforestation led to a sharp decrease in the local fauna. A few species of mammals, such as ocelots and rabbits, are making a slow comeback. Near the Botos lagoon, one can often see frogs, toads, iguanas, and salamanders. Luckily, there is no shortage of birdlife: about 80 species of birds inhabit the Park and some of them are quite spectacular. Birdwatchers should look for the fiery-throated hummingbird (*Panterpe insignis*), the sooty robin (*Turdus nigrescens*), the black guan (*Chamaepetes unicolor*), and the emerald tucanet (*Aulacorhynchus prasinus*). The resplendent quetzal (*Pharomachrus mocinno*) has been reported here and lucky visitors may see one of these magnificent birds, considered by many to be one of the most beautiful in the American continent.

Visiting during repose

Poás National Park

The park, covering an area of 56 km^2 (22 square miles), is one of the oldest and best-run National Parks in Costa Rica. The highlight is the active crater, but don't miss the Laguna Botos and the well-run Visitor Center. There is essentially only one main path, divided into two parts – the first is a paved road that goes from the entrance to the active crater lookout (distance 0.75 km, ½ mile) and the second is a trail from the lookout to the Laguna Botos and back to about halfway through the first part of the path (total distance 3 km, 2 miles). The trail is quite easy to walk, with negligible elevation gain. Disabled visitors can ask permission to drive along the paved road that goes from the entrance to near the lookout.

The main crater (Cratera Principal)

This is an impressive feature and the highlight of the Park – a nearly circular hole with a diameter of 1,320 m (4,330 feet). The depth from the lookout point on the rim to the surface of the lake is about 300 m (1,000 feet). The crater can be a magnificent sight when the weather is clear – but this is largely a matter of luck. Going in the early morning during the dry season helps, but there are no guarantees. Improve your chances by walking straight to the crater as soon as you arrive. If it is clouded in, walk around the Park and keep checking back – the winds may blow the clouds away.

There are many faults in the summit region, which are indications of the eventful history of the volcano. The walls of the crater show layers of deposits from different eruptions, and are softened at the base by deposits from landslides. On the southern crater wall, gullies and ravines expose products from eruptions in the distant past. If you go to the highest platform of the tourist lookout and look behind you, away from the crater, you will notice poorly sorted phreatomagmatic deposits from a typical eruption of Poás.

The area west of the crater is barren because it is permanently exposed to volcanic gases and acid rain from the fumarolic emission of the crater. Plants near the exposed zone commonly show yellow spots due to acid rain and necrosis of leaf tips is very common. The affected area contrasts dramatically with the lush vegetation on the Botos cone.

Inside the active crater, on the eastern side, is a plateau. Between the tourist lookout and the lake is a reddish-colored ridge, part of the now almost completely destroyed dome from the 1953 eruption. Further down is the hot-water lake, often an exquisite turquoise in color but sometimes a dull muddy brown. The size and depth of the lake vary with rainfall; typically the lake is about 40 m (130 feet) deep and 300 m (1,000 feet) across. There are a number of active fumaroles inside the crater and yellowish sulfur deposits. During the nineteenth century, azufreros (sulfur-gatherers) used a trail to go inside the crater and extract the material. These days, it is not possible to descend into the crater without a permit, which is very difficult to obtain.

Laguna Botos crater lake

This is the volcano's cold-water, dark blue lake, about 400 m (1,300 feet) across. The name is thought to be that of a local tribe, called Botos by the Spanish. The crater has at times been known as Fria, because of its cold, sapphire-blue waters. Botos is really a cinder cone topped by a crater lake, but you cannot see any cinders now because the whole cone is covered by lush vegetation. The Botos crater was last active approximately 7,540 years ago, according to results from radiocarbon dating.

Visitor Center and Museum

This small but well laid-out museum is well worth a visit. Many of the exhibits have explanations in both English and Spanish. There are maps of Costa Rica showing the location of the volcanoes, an explanation of the tectonic setting, an impressive three-dimensional model of Poás, and a corresponding cross-section model explaining the different geological units. The previous activity of the volcano is illustrated in many photographs. The museum shop upstairs is worth visiting, it has a free showing of a video about Poás, in English, and offers books, slides and maps that are hard to find elsewhere.

Visiting during activity

Since Poás displays several different types of activity that vary in their degree of danger, it may or may not be possible to enter the National Park to view an eruption. For example, the Park was briefly closed in 1989 after a phreatic eruption sent a column of ash about 1 km (3,000 feet) high. The Park was closed intermittently during 1995, though the activity was relatively minor.

It is hard to predict whether the Park will be open or not when Poás displays a greater level of activity than the bubbling and steaming of the crater lake. Check with the tourist office in San José or with the Park itself before making the trip. On rare occasions, the Park closes even when the activity is minor, because wind conditions and gas emissions can make the Park unpleasant to visitors. At other times, visitors have been allowed in to see gas jets and geysers from the crater lake.

Officials at the Park are strict about keeping people away if the Park is closed and about restricting visitors to paths otherwise. Plumes from large phreatic explosions can be seen from as far away as San José. On a clear day, the summit of Irazú makes a good vantage point from which to see its neighbor Poás.

There are no known cases of persons killed at Poás

This dark blue lake is the volcano's cold-water lake. It sits atop a cinder cone that has not been active in over 7,000 years and has been covered by lush vegetation. The lake is about 400 m (1,300 feet) across. (Photograph by Frederico Chavarria, courtesy of Rodolfo Van der Laat.)

by volcanic activity; however, before the Park was established some 20 persons lost their lives from falls, drowning, or from exposure after losing their way in the fog. As is the case for many volcanoes, Poás can be treacherous for reasons other than volcanic activity.

Other local attractions

Catarata la Paz ("Peace Waterfall")

This is the best-known waterfall in Costa Rica, though it is certainly no Niagara. However, it is an easy side-trip from Poás and is located on the volcano's flanks. The Rio la Paz cascades almost 1,400 m (4,600 feet) down the flanks of Poás in less than 8 km (5 miles), resulting in the waterfall. It is located 8 km (5 miles) north of the town of Vara Blanca and can be seen from the road. A short trail leads to a viewpoint behind the falls, from where visitors can touch the water which, according to a local guidebook, is charged with "negative electricity" – apparently good for reducing stress.

View of the Alajuela Fault scarp

This major fault parallels the Cordillera Central and can be seen from San José's international airport as well as from several places along the road from San José to Poás. Streams have carved deep canyons along the trace of the fault.

Barva volcano

Located inside the famous Braulio Carrillo National Park, this dormant volcano, 2,906 m (9,500 feet) high, is a great place for hiking. Use the San José de la Montana park entrance if you plan to climb to the summit. The usual route is to go from the ranger station at Sacramento; from there it is about 4 km (2½ miles) one way. It is a good trail and most people can do the round trip in about 5 hours. The trail goes through the rainforest and up to three crater lakes at the top: Lagos Danta, Barva, and Copey. Danta is the largest, with a diameter of about 0.5 km (1650 feet), while Barva and Copey are only 70 m (230 feet) and 40 m (130 feet) across. If you are willing to backpack for a few days, there is the option of going north from

Barva into the lowlands, or to climb Barva's neighbor volcano, Volcán Cacho Negro, which is 2,150 m (7,000 feet) high. Visitors are allowed to camp inside the park, but the trails are not all as well maintained as the main trail up Barva and there are no camping facilities.

Visitors who only have a short time can drive through the highway that bisects the Park and stop for short hikes. There is an enormous variety of plant and animal life in this park, partly due to the range of altitude – from the summit of Barva down to about 50 m (160 feet) above sea level. The lush vegetation includes many types of orchids, palms, bromeliads, and ferns. Mammals such as peccaries and monkeys are common sights and lucky visitors may also see tapirs, sloths, or even the big cats – ocelots, pumas, and jaguars. Birdwatchers should definitely come here and hike the higher altitude trails, particularly up the slopes of Barva, for the best chance to see the resplendent quetzal.

Irazú

The volcano

Volcán Irazú is the highest active volcano in Costa Rica and one of the country's top tourist attractions. The volcano's highest point, 3,432 m (11,260 feet) above sea level, is one of the few places on Earth from where one can see both the Atlantic and the Pacific oceans – on clear days, of course. However, the view that draws the crowds is the volcano's chartreuse-green crater lake. As with Poás, there is always a chance that the weather will not cooperate and that the lake will be obscured by clouds. This happened during my first visit here but my companions and I decided to stay and walk around. About an hour later, the clouds cleared away, allowing us a breathtaking view.

Irazú is a famous National Park and the summit's ease of access and magnificent views bring hundreds of tourists here every day. The volcano's notoriety also adds to its popularity: Irazú poses a great potential hazard to local residents, including those from the capital, San José. Thanks to its history of explosive eruptions, Irazú has been dubbed "o santabarbara mortal de la naturaleza" – which translates as "nature's deadly powder keg". It is a colossal stratovolcano that erupts mostly basalts and basaltic andesites. The volcano's complex shape tells of a long and eventful history of frequent eruptions, which have been recorded since 1723, when the governor of the Province of Costa Rica, Diego de la Haya Fernández, described the event in writing. The 1723 eruption crater is now named after him. Irazú's most recently active crater at the summit is known simply as Cratera Principal ("Main Crater").

The alignment of craters and vents on Irazú, and the sequence of eruption products, tells us that the recent activity has migrated from east to west. The Diego de la Haya crater is located east of the Main Crater. Almost 4 km (2½ miles) to the northeast of the Main Crater are the remains of a pyroclastic cone, Cerro Alto Grande ("Big Tall Peak"). Several older vents are also known. South of the summit there are at least ten vents, among them the pyroclastic cone known as Cerro Nochebuena ("Christmas Peak") at about 3,200 m (10,500 feet) altitude, Cerro Gurdián at 3,066 m

Irazú volcano's summit has a distinct chartreuse-green crater lake. Irazú is Costa Rica's tallest active volcano, but the road up to the summit makes it an easy destination for visitors. (Photograph by the author.)

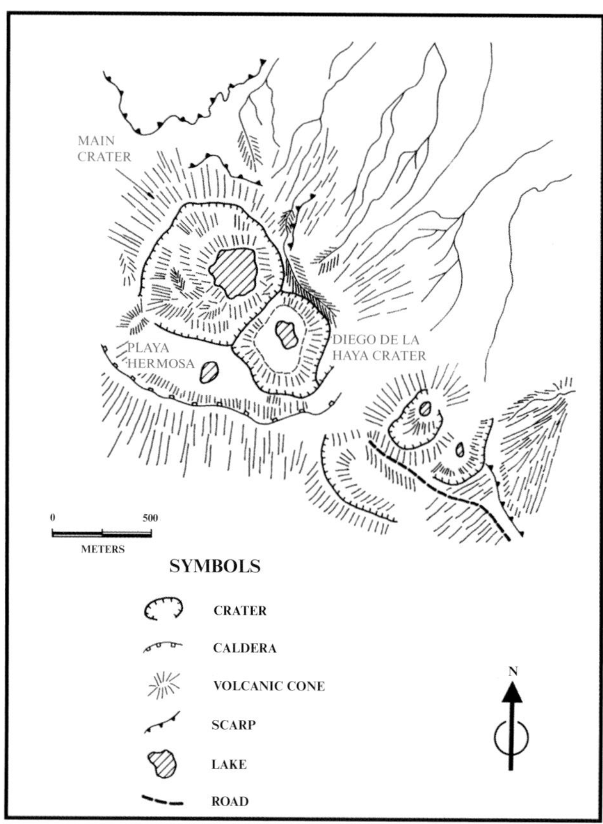

Sketch map of the summit of Irazú volcano. (After Alvarado Induni, 1989.)

(10,060 feet), and Cerro Pasquí at 2,554 m (8,380 feet). Five other vents, all eroded pyroclastic cones, are known as the Dussan–Quemados complex. At the southern base of two of these cones is the vent for the western part of the Cervantes flow, the youngest lava flow from Irazú that can be recognized on the volcano's flanks. The eastern part of the Cervantes originated from Cerro Pasquí. Radiocarbon dating gives an age of about 14,000 years for this flow.

The first recorded eruption on Irazú, that of 1723, was the first eruption in Costa Rica to be documented. The volcano may have erupted again in May 1726, and it certainly did in 1885, 1886, 1894, and 1899. There were two major eruptions during the twentieth century, one lasting from 1917 to 1920 and another from 1962 to 1965. Eruptions typically consist of a series of explosive events that eject large quantities of ash, scoria, and sometimes mud, and can cause great damage to the local agriculture. Ash from Irazú's November 30, 1918 eruption reached as far as the Nicoya Peninsula, some 120 km (75 miles) away. Eruption columns have reached 500 m (1,650 feet) high. The eruptions have been truly frightening to the local people, but not without some benefits. It is largely because of Irazú that the land of the Central Valley is so fertile.

The 1963–5 eruption

Irazú had been quiet for nearly 30 years; the last eruption had been in 1933 and consisted only of relatively minor explosions, sending ash as far as San José but causing no great concern. On August 9, 1962, Irazú stirred from its sleep with some mild explosions that again did not cause alarm. The activity became more intensive on March 12, 1963, just at the time when Ticos were preparing to receive a very distinguished visitor: US President J. F. Kennedy.

San José was all decorated for the historic event and the crowd that turned out to see Kennedy on March 18 was unprecedented in Costa Rica. As the people watched their own President and Kennedy on parade, a light "rain" of small particles began to fall – Irazú seemed to be celebrating as well, by spewing out great quantities of ash. Soon, great black clouds covered the normally blue sky and the residents of the Central Valley looked towards the volcano with great apprehension. Now and then the people saw lightning cutting through the dark mass and heard the rumblings of the volcano.

Prevailing winds eventually caused the column to drift towards San José and the effect of the ashfall on the city and its citizens was severe. The first ashfall, which lasted for only a few hours, left San José covered by a thin layer of ash and turned to gray the normal tropical greenery of trees and plants. Irazú seemed to stop belching out ash for a few days and residents hoped that was the end. Their optimism did not last, as soon the eruption resumed and ash rained on San José once again. A national emergency was declared on March 22 and the ashfall continued, varying in level from day to day. A contemporary article about the eruption describes some bizarre proposed efforts to stop the ashfall. One engineer suggested building a giant parasol over the volcano. A group of believers in the ancient religion engaged a shaman to perform various magical dances near the summit, but Irazú carried on unabated and the shaman contracted pneumonia. Radio stations took calls from many people who suggested throwing a variety of things into the crater, ranging from crucifixes and flowers to virgins and mothers-in-law. Finally, the plight of the citizens was somewhat alleviated when local officials decided to buy four great sweeping machines and the US Ambassador arranged for the US Army to fly them to San José for free. The long task of cleaning up the capital city got under way, even though the eruption continued.

Ashfall on San José was at its heaviest on July 17 and November 23, 1963. Bombs and lapilli fell nearer the volcano, fortunately not causing any deaths. The eruption lasted until 1965 and mostly consisted of periodic explosions that varied in duration and intensity. The largest explosions sent columns of steam, gases and ash up to 0.5 km (1,650 feet) above the crater.

Although the explosions were the most noticeable and spectacular aspect of the eruption, the volcano's activity also manifested itself in the rivers that have their source in the Irazú massif: Virilla, María Aguilar, Tiribí, Torres, Sucio, Toro Amarillo, Retes, Birrís, and particularly the Reventado. Throughout the eruption, the waters turned into rivers of mud and avalanches. The worst tragedy of the eruption happened before many people became aware of the volcano's renewed activity. During the night of March 9, 1963, a mudflow along the Reventado river destroyed at least 300 homes, most in the San Nicolás district west of Cartago, causing the deaths of more than 20 people. The mudflow also devastated farmland and killed numerous cattle. The disaster was not completely unexpected, as the Reventado river has long been known as a potential source of mudflows – several had happened before, not all related to Irazú's activity. It is because of this river that Cartago has the unappealing nickname La Ciudad del Lodo – Mud City.

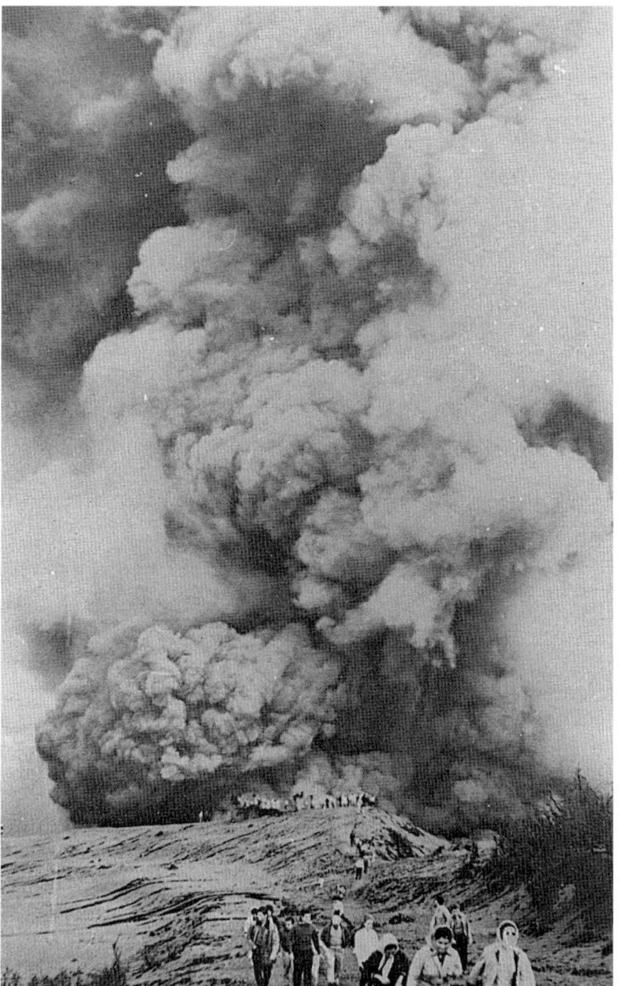

Irazú's eruption of 1963–5 became a great tourist attraction, as shown by this contemporary photograph. (Photograph, photographer unknown, courtesy of G. Alvarado Induni.)

A personal view: The historical importance

An active volcano usually exerts a great influence on the course of local history and Irazú is no exception. The volcano has made its mark on the development of Costa Rica, including the choice of a capital city. Irazú and its activity are mentioned in early historical records of the Spanish conquest. One historian relates that during the sixteenth century, when Spaniards came upon the bare trees and ash cover of the foothills of Irazú, they named the area "Valley of Desolation."

Costa Rica's first capital was the city of Cartago, founded in 1563 by Juan Vászquez de Coronado. For the next 150 years, Cartago remained the country's only city. It is located at the foot of Irazú, in the valley between the Cordillera Central and the Cordillera de Talamanca, at an altitude of 1,435 m (4,700 feet). The volcano's summit is about 2,000 m (6,600 feet) over the city and can be seen from there on clear days. Irazú was probably active a few years before Cartago was founded, possibly in 1559. In 1719, the Spanish governor of Costa Rica, Diego de la Haya Fernández, wrote that Cartago "is surrounded by very high mountains, the highest being one in which there is a water volcano." The name Volcán de Agua probably referred to a crater lake. Irazú has also been known as the Cartago volcano. The name Irazú is not mentioned on any old documents and was adopted after 1821, when Costa Rica became independent from Spain. Its meaning is not known; perhaps it is derived from the indigenous words "i" (earthquake) and "ara" (make noise or thunder).

The first major eruption of Irazú after the Spanish settlement occurred in 1723 and was documented by Diego de la Haya in his diary. The first sign of activity was a thick plume hovering over the volcano, which was seen from Cartago in the afternoon of February 16. Soon, "thunder" was heard and the people were terrorized, rushing to the church and praying. The governor himself rode out toward the volcano, but had to turn back "because of the dark covering most of the mountain, and because the sulphurous stench was most fatiguing." The volcano continued to thunder and rumble for the next few days and "flames were seen shooting up from the highest part of the mountain, and within the flames large balls of fire and other burning fragments, all accompanied by great blasts, thunder and rumbling." Earthquakes shook the city, causing great fears but no major damage. The citizens prayed, held processions and masses, but ash continued to fall and to cover trees, roofs, and fields. Irazú calmed down somewhat in mid March, though light quakes persisted and "smoke" continued to come out of the summit. Another major explosion occurred on April 3, and the governor reported that the volcano "threw up rocks and other burning fragments very high, so high that one could say the Apostles' Creed while they went up and came down." Irazú continued the show but with decreasing vigor, for several months. The governor's report, dated December 11 of that year, tells of continuing earthquakes, but no damage to houses, and of a continuation of the "fires, ashes, and sands."

The proximity of Cartago to Irazú and the numerous earthquakes that rocked the city during and between volcanic eruptions became a great concern for the citizens. Eventually, the safer location of San José was a deciding factor in moving the capital in 1823, just 2 years after Irazú had erupted again. The decision proved wise, as strong earthquakes struck Cartago in 1841 and again in 1910, ruining almost all of the old buildings. Another earthquake in 1926 destroyed the Basilica de Nuestra Señora de los Angeles, the most famous church in Costa Rica. The present-day Cartago has really nothing left to evoke the days of Diego de la Haya, but it remains a religious center. The Basilica has been rebuilt in the Byzantine style and is home to La Negrita, a statue of the Virgin Mary that was discovered on the site in 1635. Pilgrims come from all over South America in search of her alleged healing powers and they give to the church small metal ornaments, mostly representing body parts that were cured. Among the many arms, legs, and torsos, there is a model of Irazú, maybe given by someone who was somehow saved from the volcano's fury.

Visiting during repose

Irazú National Park

The National Park was established in 1955 and has an area of 23 km^2 (9 square miles) roughly in a circular form around the volcano. It is located only 54 km (33 miles) from San José and a paved road goes all the way up to the summit. To get there from San José, take the road to Cartago and follow signs to Irazú. The summit is a lunar-like landscape that contrasts with the yellow-green lake at the bottom of the Main Crater. The trails in the summit region are very limited. A path leads visitors around the southern rim of the two major craters, Diego de la Haya and the Main Crater. To the south of the Main Crater is a flat region called Playa Hermosa ("Pretty Beach") and the remains of an old crater rim. To increase your chances of seeing both the Atlantic and the Pacific oceans from the summit, as well as the crater lake, get to the summit as early as possible after dawn. Remember that it is usually cold and very windy there; the average annual temperature is 7.3 °C (45 °F).

Diego de la Haya Crater

This crater, located just east of the Main Crater, is the one that visitors come to first if they follow the path. The crater's shape is elongated, with a diameter of about 690 m (2,260 feet) and about 80 m (260 feet) deep, with a flat bottom. To the east of the Diego de la Haya crater is a small pyroclastic cone (3,364 m (11,100 feet) altitude) topped by an eroded crater.

Main Crater

This crater, active during the last eruption, is almost circular, 1050 m (3,445 feet) in diameter and about 300 m (980 feet) deep. Deposits from Irazú's main recent eruptions, those of 1723 and 1963–5 are exposed as layers on the crater's walls, which are dotted with active fumaroles. At the bottom is the yellow-green lake, remarkably photogenic in clear weather. Its peculiar color is due to dissolved iron and sulfur in the water.

The 1963–5 eruption caused some significant topographic changes in the summit region. There used to be two other, smaller craters on the summit, called "H" and "G", one located to northwest and the other to the southwest of the Main Crater. Prior to the eruption, landslides had almost incorporated crater "G" into the Main Crater. During the eruption, the Main Crater was enlarged and both "H" and the remains of "G" were completely swallowed up into the Main Crater, as was an area occupied by a small lagoon.

The highest point

It is possible to deviate from the tourist path by crossing a low wooden fence and climbing up to Irazú's

Irazú's highest point is fairly easy to reach, but watch your footing along those contorted paths. The view from 3,432 m (11,260 feet) above sea level is quite spectacular on clear days – one of the few places on Earth from where one can see two oceans. (Photograph by the author.)

highest point. Lots of tourists do this and no one seems to stop them. Note that climbing during low-visibility conditions can be hazardous, as you will have to walk along a narrow crest for the last part of the "unofficial" path. Do not attempt to go all the way around the crater, past the highest point, as the rim beyond is a knife-edge and definitely not safe. If you look away from the Main Crater from the highest point and down the northern slopes of the volcano, you may see an area rich in active fumaroles.

There is one particularly interesting place to stop along the "unofficial" path, just as you start along it. Go towards the western end of the tourist path, past the Main Crater, climb over the fence and you will see a notch to your right, going down towards the Main Crater. The exposed layers along the walls of the notch are surge deposits from the 1723 eruption. Some places show bomb sags, caused when ejected bombs from the crater landed in the soft deposits and partly sunk into the ash layers. In a few places you can see the 1723 surge deposits overlain by those from the 1963 eruption. There is evidence that some of the 1723 surges were wet, but not much work has been done yet to reconstruct the sequence of events.

Flora and fauna

Irazú's vegetation and animal life have been severely impacted by human activity on the volcano's lower slopes. Wildlife was once abundant, as a record from 1855 testifies: it tells of a ranch that was closed down because of predation by jaguars. These days, the animals you are most likely to see inside the Park are eastern cottontail rabbits (*Sylvilagus brasilensis*) and coyotes (*Canis latrans*). You may see a long-nosed armadillo (*Dasypus novemcinctus*) or a porcupine (*Coendou mexicanus*). The only cat left is the tiger cat (*Felis tigrina*). Birdlife is more diverse, even close to the summit, where several species of hummingbirds can be found. Other birds you may see on Irazú are the appropriately called volcano junco (*Junco vulcani*), the mountain robin (*Turdus plebejus*), ruddy woodcreeper (*Dendrocincla homochroa*), ant-eating woodpecker (*Melanerpes formicivorus*), and the unspotted saw-whet owl (*Aegolius ridgwayi*).

Plant life near the summit is scarce, consisting of a few species that are slow-growing and stunted, such as arrayan, *Agrostis tolucensis*, *Trisetum viride*, and the

Irazú's eruptions throw out large bombs that can cause the ash layers on the ground to "sag" as the bombs land. This is one of the bomb sags seen from the summit crater path. (Photograph by the author.)

shrub *Castelleja irazuensis*, which is typical of the region and can be recognized by its red flowers. At lower altitudes there are areas of open and stunted vegetation and forested areas. In the open areas, the vegetation includes arrayan, arrecachillo (*Myrrhidendron donnell-smithii*), and *Acaena elongata*. The most common trees in the forested areas are the black oak (*Quercus costaricensis*) and the miconia (*Miconia* sp.), though other types can be found, including the *Magnolia poasana*. The drive up to the summit provides good views of the changing vegetation.

Cervantes flow

Located outside the National Park, the remains of this flow are not easy to see, as they are covered by vegetation. The easiest way to spot the flow is by looking for its undulating topography from the 230 road south of the volcano, between the towns of Pacayas and Boquerón. The thickness of the Cervantes varies mostly between 7 and 25 m (23 and 82 feet), but is thicker where it ponded in valleys and depressions. Studies of this flow have been hindered by the thick vegetation cover, but it is known that it is a flow field made up of two major units, east and west, which came out of different vents and are slightly different in composition. Each of these two major flow units consists of a series of individual, relatively small flows. Radiometric dating of the flow gives its age at about 23,000 years.

Visiting during activity

The historic eruptions of Irazú have been moderate, with an estimated VEI of 2 to 3. Although fatalities have been few, the volcano is potentially very dangerous. During the eruption of 1963–5 many people went right up to the rim of the crater to see the spectacular columns of ash, but they were taking a considerable risk. Back then, there was no real control or restrictions on visitors, who were mostly local. It is hard to imagine that this would be allowed now. Although the ash columns are not particularly dangerous to watch from upwind, it is hard to predict if and when the volcano will eject bombs that could be lethal. Irazú's other major hazard is mudflows, so stay away from river valleys, particularly that of the Reventado river.

If the next eruption follows the usual historical pattern and is similar in character to the 1963–5 event, watching it will not be a particularly comfortable experience because of the ashfall. Airports may be closed and all services are likely to be disrupted to some extent. The activity may decline for days or weeks, but periodic explosions will produce a large, ominous-looking ash column above the volcano. The column will be seen from Cartago and, among other places, from the summit of Turrialba and from the Turrialba Volcano Lodge. Depending on the height of the column, it may also be seen from San José, where the effects of the ashfall may again be severe – in which case, one hopes the sweeping machines will still work.

Other local attractions

Turrialba volcano

This isolated stratovolcano, set in the midst of a yet unspoiled region of Costa Rica, is 3,328 m (10,918 feet) high and had its last major eruption in 1866. Turrialba has been quiet since then, except for fumarolic activity and an unusual increase in seismicity between May and June of 1996. The volcano's name comes from the Spanish Torre Alba ("White Tower") and supposedly reflects the columns of steam that early settlers saw above the summit.

Turrialba's reported style of activity is similar to that of its neighbor Irazú and the chemical composition of the magma ranges from pyroxenic andesites, basaltic andesites, to basalts. Access to Turrialba is not easy and the volcano is still off the beaten tourist path. The roads to the lower slopes of the volcano all require four-wheel-drive vehicles. The nearest town, also called Turrialba, is located only 15 km (9 miles) to the southeast, but getting from there to the lower slopes of the volcano requires driving over at least 30 km (19 miles) of dirt roads. There are several ways to get to the summit. The easiest is on horseback, either from the Turrialba Volcano Lodge or from the village of Pacayas, located halfway between Cartago and the town of Turrialba. If you don't have a four-wheel-drive vehicle and are not staying at the Lodge (they will pick you up with a vehicle), take a bus to the town of Santa Cruz (at 1,500 m (4,950 feet) altitude), from where a 21 km (13 miles) dirt road and track will take you to the summit. The lower slopes of Turrialba are very pleasant, with green fields and dairy farms, and you may be able to arrange cheap accommodation at farmhouses or permission to camp.

The summit of Turrialba consists of three craters, roughly aligned east to west and located inside a depression that probably was formed by older craters coalescing together. The depression measures about 2,200 m (7,200 feet) in a northeast–southwest

Turrialba is perhaps Costa Rica's most scenic but least visited volcano. It is easily reached on foot or horseback. This volcano last erupted in 1866 but its fumaroles still puff away. (Photograph by the author.)

direction and about 500 m (1,650 feet) from northwest to southeast. You can walk around the rim, which is made up of lavas and pyroclastics and in places reaches 70 m (230 feet) high. Inside the depression are the three craters, all of which are accessible. The central crater is the youngest, about 50 m (160 feet) deep. It is nested inside the largest of the three, which forms a wide basin. The westermost crater is deep and narrow and its interior is harder to get into, involving a scramble down. Be careful as you walk around this crater; near one of its walls is a hole leading to bottomless depths. Fumarolic activity dots the inside of all the craters, and yellow sulfur deposits are abundant in places.

The volcanic activity at Turrialba is largely unknown and even today the volcano is not well studied. Historical records tell of volcanic activity in 1723, 1847, 1853, 1855, 1859, and 1864–6. However, not all of these reports are thought to be accurate and descriptions of the eruptions are scarce. The report from 1723 is particularly suspicious, since Irazú was erupting at the time. Some fresh-looking lava flows are seen on Turrialba's flanks, indicating that they must be relatively recent, probably from activity in the nineteenth century.

Near Turrialba is the Guayabo National Monument, the only archeological park so far established in Costa Rica. Human occupation in the area goes back to around 500 BC. The interaction between Turrialba's pyroclastic deposits and the archeological remains has only begun to be studied. It is interesting that the Turrialba region, so remote and peaceful today compared to other areas of the country, has been inhabited for a long time and has had a colorful history. It is said that in 1666 some 800 enemy pirates marched up the Matina valley to sack Cartago, but were turned back from the town of Turrialba by the powerful statue of Our Lady of the Conception. Diego de la Haya refers to this incident in his description of the 1723 eruption of Irazú, during which the same statue was brought to try to shut down the volcano, unfortunately without effect.

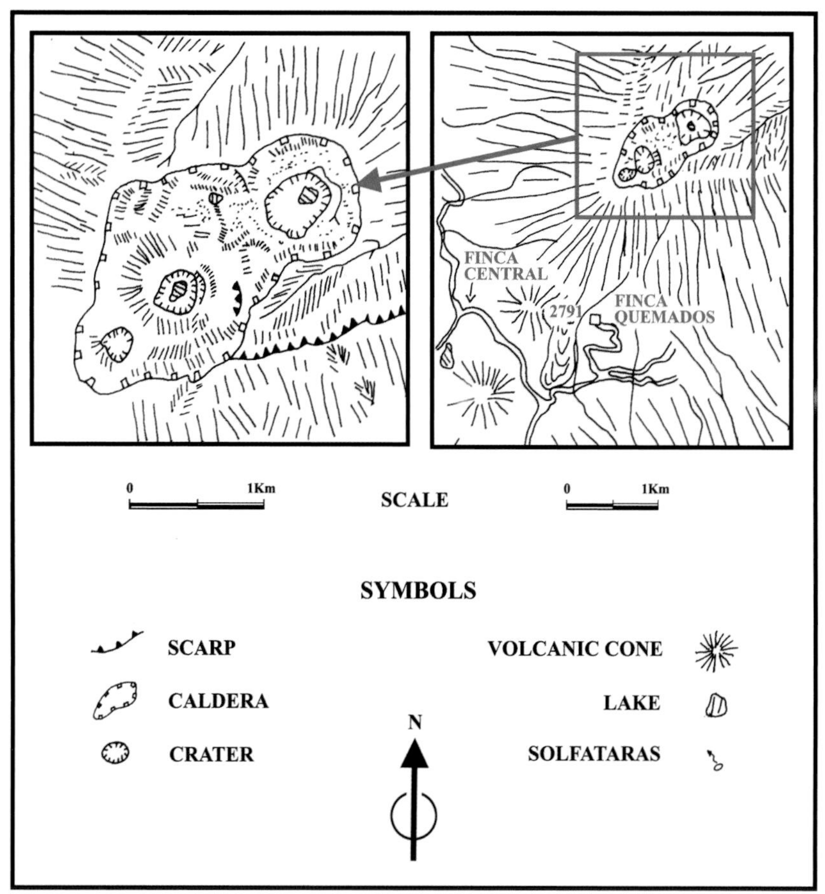

Sketch map of Turrialba's summit crater and surrounding areas. (After Alvarado Induni, 1989.)

The interior of Turrialba's summit crater is easy to go down into and explore. Regions of bright sulfur deposits contrast with the dark ash and tephra. (Photograph by the author.)

References

Alvarado Induni, G.E. (1989) *Los Volcanes de Costa Rica*. Editorial Universìdad Estatal a Distancia.

Casertano, L., A. Borgia, C. Cigolini, *et al.* (1987) An integrated dynamic model for the volcanic activity at Poás Volcano, Costa Rica. *Bulletin of Volcanology* **49**, 588–98.

Viramonte, J.G., M.N. Collado, and E.M. Rojas (1997) *Nicaragua–Costa Rica Quaternary Volcanic Chain*. International Association of Volcanology and Chemistry of the Earth's Interior.

13 Volcanoes in the West Indies

The West Indies

The West Indies' enchanting islands have some of the world's most interesting volcanoes, two of which have played a vital part in the development of volcanology as a science. The 1902 eruptions of Mt. Pelée, Martinique, and Soufrière, St. Vincent, attracted substantial scientific interest and led to the recognition of two volcanic processes. The first was the formation of deadly pyroclastic flows, also known as *nuées ardentes,* and second was the growth of lava domes or spines in the crater at the final stages of eruption. The most notable work was done by Frank A. Perret, a physicist who came to Martinique after the 1902 eruption of Mt. Pelée to study its deposits and subsequent activity. Perret devoted most of his life to the study of volcanoes and eventually became the first director of the Hawaiian Volcano Observatory.

Montagne Pelée is, with good reason, the most infamous of the West Indies volcanoes because of its catastrophic 1902 eruption. It is, however, only one of the potentially lethal volcanoes along the Lesser Antilles volcanic arc, which stretches from the Dutch island of Saba on the north down to the spice island of Grenada on the south. The volcanoes of the Lesser Antilles arc that have been active in historic times are: Saba, at the northern end of the arc, The Quill in St. Eustatius, Liamuiga in St. Kitts, Soufriere Hills in Montserrat, Soufrière in Guadeloupe, Microtin and Morne Patates in Dominica, Pelée in Martinique, Qualibou in St. Lucia, Soufrière in St. Vincent, and Kick-'em-Jenny, a submarine volcano just north of Grenada.

Recently, the volcano that has captured the news is Soufriere Hills in Montserrat, which begun erupting in 1995 and, at the time of writing, has not yet stopped. These two volcanoes – Mt. Pelée and Soufriere Hills – are highlighted in the following section as they illustrate the modus operandi of the Lesser Antilles volcanoes, in which pyroclastic flows play a major, and often deadly, role. Both volcanoes have caused untold destruction, but the nearly 100 years' worth of knowledge gained between their major eruptions meant that many thousands of lives could be saved in Montserrat, while 30,000 people perished in Martinique. In the case of Montserrat, it was the island's economy that took the brunt of the destruction, with the loss of the capital city, airport, tourism, and over half of the usable land.

Three other volcanic islands in the Lesser Antilles are described in the following section: Guadeloupe, St.

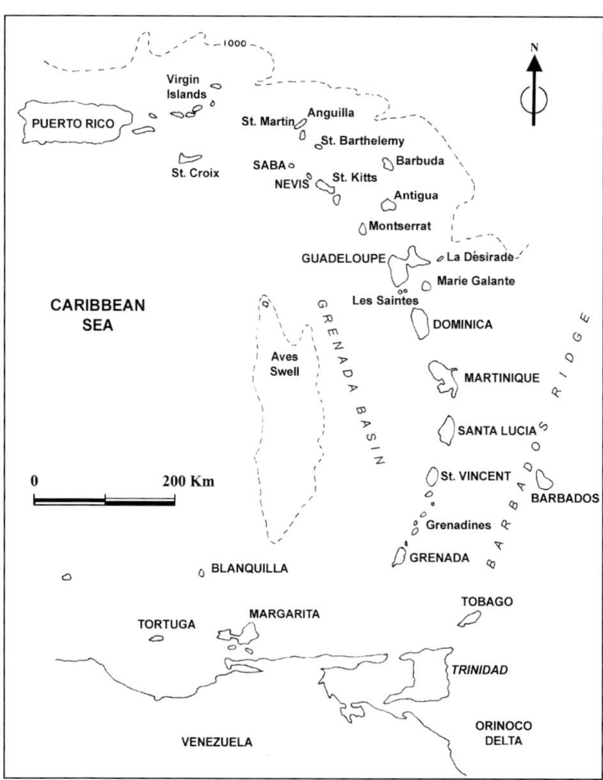

The Lesser Antilles island group. The volcanic arc goes from Saba in the north to Granada in the south. Dashed lines are isobaths in meters. (Modified from Westercamp and Tazieff, 1980.)

Vincent, and Dominica. The Morne Patates volcano in Dominica has not been active recently, but Soufrières in Guadeloupe and Soufrière St. Vincent have been frequently active in the recent past. Apart from active volcanoes, the Lesser Antilles offer much to visitors: scenery, beaches, cultural diversity, and historic heritage. Christopher Columbus discovered the islands during his second and third voyages, but they were later colonized by the Dutch, the British, and the French. Several islands still retain strong ties to Europe and even those that have become independent retain their European flavors. Martinique and Guadeloupe are strongly French, while Dominica, St. Lucia, St. Kitts, Montserrat, and St. Vincent and the Grenadines retain some of the ambience of their British colonial days. The Dutch influence is felt in Saba and St. Eustatius, which were settled by the Dutch West India Company. The Lesser Antilles islands therefore present a rich diversity in terms of language, architecture, cuisine, and local customs. The volcanoes are an integral part of life in these islands, though historic records of eruptions only begun after the islands were settled in the 1630s and, to date, fewer than 40 eruptions have been recorded. The volcanoes, however, have been intensively studied in the last few decades and the scarce historic records have been complemented by radiocarbon dates and field studies. Piece by piece, not only the activity of the volcanoes is being reconstructed, but clues are obtained about the underlying forces that are responsible for their existence.

Tectonic setting

The Lesser Antilles arc, stretching for about 740 km (460 miles), is a result of the subduction of the North Atlantic ocean floor beneath the Caribbean plate at a rate of about 2 cm/year (¾ inch/year). An important feature of this arc is its bifurcation north of Martinique. The northeastern branch (outer arc) is made up by the islands of Marie Galante, Grande-Terre de Guadeloupe, La Désirade, Antigua, Barbuda, St. Bartholemew, St. Maarten, Dog, and Sombrero. The northwestern branch (inner arc) comprises Dominica, Les Saintes, Basse-Terre de Guadeloupe, Montserrat, Redonda, Nevis, St. Kitts, St. Eustatius, and Saba. The islands of the northeastern branch, known as Limestone Caribbees, are all inactive, low-lying and partially or completely capped by limestones that date back to the mid-Miocene to Pleistocene. Underlying the limestone are volcanic rocks that make up the oldest arc. The islands of the northwestern branch are made up of younger volcanic rocks and include several recently active volcanoes. To the south of Martinique, there is a single arc that includes St. Lucia, St. Vincent, the Grenadines, and Grenada. The arc that contains all the active volcanoes, from Saba to Grenada, is named the Active Arc or Volcanic Caribbees.

The Lesser Antilles arc is the surface manifestation of a subduction zone, where the North American plate is underthrusting the Caribbean plate. Volcanism begun forming the older arc (from the Limestone Caribbees to Grenada) around 38 million years ago. The formation of this pre-Miocene arc was probably triggered by a change in the spreading rate of the North American plate. The volcanic activity shifted west in the northern part of the arc during post-Miocene times, probably due to variations in either the inclination of the subduction zone or else the depth at which the magma was tapped.

The present-day zone of underthrusting is located about 150 km (95 miles) east of the Active Arc and is inclined towards the west, where it reaches down to 200 km (125 miles) in depth. The details of the tectonic setting are quite complex. There are changes in the dip of the underthrusting plate along the length of the arc, and lower intensity of seismic activity at the southern part, perhaps because of a thick accumulation of sediments in the trench that may inhibit subduction and decrease seismicity. The presence of a number of faults transverse to the arc, which break up the plate boundary, has also been suggested.

The Active Arc shows five different types of volcanic centers. Mont Pelée and the Quill (St. Eustatius) are made up mostly of products from explosive activity. Saba, Soufriere Hills (Montserrat), and the Pitons du Carbet and Morne Conil (Martinique) are andesitic dome clusters surrounded by aprons of block and ash flow deposits. The various volcanic centers on Dominica and south-central St. Lucia are characterized by numerous dacitic pumice and ash flows. Soufrière (Guadeloupe), Soufrière (St. Vincent), and The Peak (Nevis) are stratovolcanoes in which about half of the material is from lava flows of basaltic to andesitic composition and half is from explosive eruptions. The South Soufriere Hills (Montserrat) and the Monts Caraïbes in Guadeloupe are composed of basalt–basaltic andesite lava flows and associated ashfall deposits.

The magmas erupted from the arc show more diversity in composition than is usual for volcanic arcs, suggesting that the Lesser Antilles have a particularly complex setting. Geochemical studies of the more recent magmas show that there are substantial quantities of sediments in the magma source regions. The forearm of the Lesser Antilles arc is, in fact, a region that receives sediment from a number of sources and thus several different types can be recognized. Some of the sediments show an unusual character in terms of radiogenic lead, which suggests an ancient age and continental source. Several studies show that these sediments come from the Archean Guiana Shield and are transported by the Orinoco River.

There is a strong similarity between the sediments and the arc magmas in terms of the lead isotope character. There are lavas from Martinique and St. Lucia that show high levels of radiogenic lead (around 3.5%) which suggests that the magma's source regions experienced high-level contamination by sediments. These lavas provide good evidence of the process of sediment subduction and incorporation in arc magmas. The Lesser Antilles have become one of the world's best locations for understanding the importance of sediments in the chemistry of subduction zone magmas. Another aspect of the uniqueness of the Lesser Antilles is its longevity – the arc has the longest continuous or semicontinuous history of activity of all the Earth's known volcanic arcs, spanning about 120 million years.

Practical information for the visitor

Tourism in the major industry of the Lesser Antilles and the islands offer a wide variety of facilities for visitors. At present, facilities in Montserrat are scarce, but are expected to improve as the eruption wanes. Visitors from far away may be tempted to visit more than one island during their stay and there is no doubt that one of the most attractive features of the Antilles is the variety of islands so close to one another. Not only do islands differ in terms of their volcanic scenery, but also in terms of history, culture, and heritage.

The islands were aptly described by colonists as "isles of perpetual June" – the climate is wonderful year-round. The difference between wet season (May–June and October–November) and dry season is very slight. If you can, avoid the month of September, when hurricanes are more likely to occur. A traditional rhyme says "June too soon, July stand by, September remember, October all over."

Transportation

Island-hopping in the Antilles is a great way to travel but can get expensive. Most travel between islands these days is done by air and the planes are known as the islands' bus service. The views during flights can be fantastic and the flights are a lot more interesting than those of more conventional airlines with

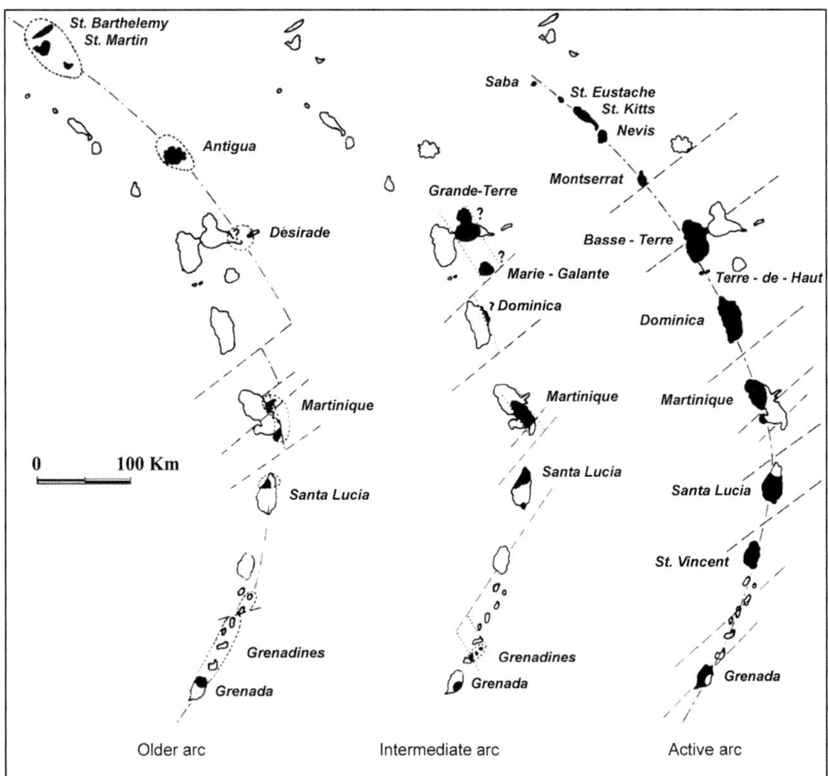

The Lesser Antilles arc through time. Black indicates activity at the time. The older arc's activity occurred more than 22.5 million years ago. The intermediate arc was active between 6.5 and 20 million years ago. The present-day active arc started its activity about 6 million years ago. Transverse faults are marked as dashed lines. Note how the present-day arc bifurcates north of Martinique. (Modified from Westercamp and Tazieff, 1980.)

jet planes. The "island buses" are small planes run somewhat informally – the pilot will often skip a landing if nobody wants to get on or off. Note that some of these local airlines do not take reservations very seriously, so reconfirm more than once and get to the airport early. The largest local airlines are LIAT and BWIA (pronounced "Beewee"). Air Martinique and Air Guadeloupe connect their home islands with Dominica and St. Vincent. At present, there is no air service to Montserrat, as the airport was destroyed by the recent eruption. To get there, take a flight to Antigua and then a helicopter or boat to Montserrat.

There are ferries between some of the islands, but services are more limited. There is a regular ferry service between Martinique and Guadeloupe (via Dominica) that is of particular interest to those with a volcano-hopping itinerary. It is also possible to travel by rented yacht or even to catch a ride on a freighter. The best way to find out is to go to the local harbor and ask around.

Rental cars are available on all the islands, though there is not much of a choice in Montserrat at present. Many of the major US rental car agencies have offices in the islands, but local offices may offer less expensive rentals. Martinique offers collective taxis ("taxis collectifs") from the waterfront in Fort-de-France to hotels all over the island, which are less expensive than regular taxis. Public buses tend to be rather crowded but are inexpensive, and bikes and motorcycles can be rented in Fort-de-France. In St. Vincent, the crowded and lively public buses are a cheap way of getting around. The driver will stop along the route for you and the buses are easy to spot: brightly colored mini-vans with names such as "Who to Blame." In Guadeloupe, biking is extremely popular and several places rent scooters, Vespas, and motorcycles. You can also get around by bus, the drivers here will also stop along the route if you are between bus stops (arretêbus). In Dominica, co-op cabs are offered from Melville Hall, the main airport, and individual taxis are also available. If you rent a car in Dominica, remember that driving is on the left and the mountainous roads are often steep, pot-holed, and have hairpin curves.

Tours

Tours by taxi can be arranged in all the islands, but more specialized options are available in some places. In Martinique, a wide variety of sightseeing tours by car, boat, and helicopter are available through hotels and agencies. Parc Naturel Régional de la Martinique (tel. 73-19-30) arranges guided hiking tours and Hélicaraibes (tel. 73-30-03) provides helicopter tours though at rather high prices. In Guadeloupe, guided hikes can be arranged through the Organization des Guides de Montagne de la Caraïbe (tel. 80-05-79). In St. Vincent, tours by air, sea, or land, can be arranged through Grenadine Tours (tel. 456-4176). In Dominica, hiking tours are offered by Dominica Tours (tel. 448-2638) and Land Rover tours by Rainbow Rover Tours (448-8650). In Montserrat, the tour business is expected to pick up again once the eruption is over.

Lodging

The variety can range from extremely luxurious to very plain and modest. The Antilles are a playground for the wealthy but most islands will have lodgings to suit a wide range of budgets. For those staying longer than a week in one place, rented villas and condominiums are a practical alternative to hotels. Many islands offer bed-and-breakfast places and camping. The French Antilles, in particular, have well-organized campgrounds. If you are unsure whether you need a permit for camping, ask at the police station.

Martinique

The island has many resort hotels, but these are not convenient for Mt. Pelée. The choice around St. Pierre is limited to mostly cheap but adequate rooms. The Auberge de la Montagne Pelée in Morne Rouge is a good choice as it faces the volcano and the views are spectacular. It only has nine rooms, so book well in advance (tel. 53-32-09). Residence Surcouf (tel. 78-13-86), located in Pécoul, is also a convenient choice. In St. Pierre itself, try the aging five-room La Nouvelle Vague (tel. 74-83-88). The best lodgings in the area are at the Plantation de Leyritz (tel. 78-53-08), located on the hillside above Basse Point. This eighteenth-century plantation house and associated slave quarters have been converted to a lovely 24-room hotel. Many visitors come here for lunch and to tour the plantation, where old sugar-cane crushing equipment is displayed. Visitors who prefer renting flats or villas should contact the Association des Gîtes Ruraux in Fort-de-France (tel. 73-67-92). Camping grounds are located in Anse à L'Ane, near Fort-de France, and in St. Anne on the south side of the island. Neither is convenient to Mt. Pelée and, although the distances are small (tens of kilometers), the roads are narrow and can be very slow.

Montserrat

It is hard at present to know what will be available in Montserrat even in the near future. The major resort hotel on the northern side, Vue Point, closed down during the eruption and it is not known if, or when, it might reopen. The best way to find lodgings is to consult the Montserrat Tourism home page (Appendix I), which will also have up-to-date information about rental cars and transportation to and from the island.

St. Vincent

This island remains relatively undeveloped and most of its hotels are owned and managed by locals. Kingston, the capital, and the nearby Villa are where most of the hotels are located. Since these are in the south side, they are not too convenient for the volcano, but there are no hotels in Georgetown or Richmond. A charming place is the Grand View Beach Hotel (tel. 458-4411), located on a hilltop within easy distance of Kingston. This is an original plantation house which was at one time used for drying cotton. Those wanting very exclusive lodgings should head for Young Island Resort (tel. 458-4826) on the private Young Island, located about 200 m (650 feet) off the south coast of St. Vincent.

Guadeloupe

There are plenty of hotels in Guadeloupe, most of them located along the south coast of Grande-Terre. Among these is the luxurious and well-known Hamak (tel. 88-59-99) which boasts not only a private white-sand beach but also a fleet of twin-engine planes for transporting guests. An inexpensive alternative, which is also better situated for the National Park, is the Auberge de la Distillerie (tel. 94-01-56), a small country-inn type of hotel located on Basse-Terre. Also on Basse-Terre is Le Rocher de Malendure (tel. 98-70-84), a well-known restaurant which rents out five bungalows for very reasonable prices. Le Rocher is set on a bluff above Malendure Bay and has a wonderful view. Family-run facilities are also available via the Gîtes de France Guadeloupe (tel. 82-09-30).

Dominica

Dominica is not a major destination for tourism and the choice of lodgings in more limited than for some of the other islands. The hotels on the south coast are the most conveniently located for the volcanic attractions. At the top end there is Fort Young Hotel (tel. 448-5000), a converted eighteenth-century fort which is set on a cliff in Roseau. Other high-end choices include Reigate Hall in Roseau (tel. 448-4031), a beautiful stone-and-wood building perched high on a steep cliff, and Springfield Plantation (tel. 449-1401), a colonial plantation house located close to Roseau. More moderate choices are the Layou River Hotel in Roseau (tel. 449-6281), a rambling estate, and the Layou Valley Inn (tel. 449-6203), a charming guest house under the peaks of the Morne Trois Pitons.

Safety and emergency services

The most severe natural disasters of the Lesser Antilles are not volcanic eruptions but hurricanes. If you are in the islands and hear that a hurricane is on the way, finder the strongest concrete shelter you can and don't come out when it suddenly goes quiet – this means that you are under the eye for a few minutes. If you are unfortunate enough to be on a boat and cannot get to land in time, head for a mangrove swamp for some shelter.

Crime can be a problem in the West Indies, though in the islands highlighted here, violence against visitors is relatively rare. You are more likely to encounter problems with poisonous trees or insects. Watch out for the manchineel (mancenillie) tree. These pretty trees, with small green fruits that look like apples, are so poisonous that even raindrops falling off them can cause skin to blister. In Martinique, look for red warning signs posted by the Forestry Commisssion and stay well away. On other islands, these trees may be marked with signs or red paint on the bark. It is a good idea to become familiar with its appearance, so you can avoid going near one in more remote areas where there are no warning signs. Insects can be a problem on all the islands; be sure to bring very strong repellent.

Emergency and rescue services are variable from island to island but are mostly reasonable, though without the sophistication of those in the USA.

Maps

It is not particularly easy to buy topographic maps on the islands, though road maps are widely available. Contact the Tourist Boards for the various islands to obtain maps showing roads, hotels, and the major attractions. They will also be able to inform you on how to purchase topographic maps. Martinique and Guadeloupe are departments of France and detailed maps of these islands are available from the Institut Géographique National in Paris (see Appendix I).

Mont Pelée and, in the foreground, the present-day town of St. Pierre. (Photograph courtesy of Charles Frankel.)

Mont Pelée

The volcano

Mont Pelée is one of the world's most famous and feared volcanoes. Its catastrophic 1902 eruption completely destroyed the city of St. Pierre, caused the deaths of some 30,000 people, and eventually gave its name to one of the major types of volcanic eruptions – Peléean. The lovely city of St. Pierre, known at the time as the "Paris of the Lesser Antilles," was the commercial and cultural center of Martinique. Its destruction shocked the world and became one of the most tragic proofs of the power of volcanoes – as well as one of the prime examples of gross mismanagement of a volcanic crisis.

Mont Pelée is located on the northwestern part of Martinique, where it rises to an elevation of 1,397 m (4,583 feet), dominating the landscape and forming an impressive backdrop to the present town of St. Pierre. Pelée is the only currently active volcano in Martinique, though two submarine centers may exist off the western coast, if some reports of "boiling seas" are credible. There are older volcanoes in Martinique – like other islands in the Lesser Antilles, Martinique was built by volcanic eruptions. The geology of the island is best described in terms of its eight major volcanic centers, which range in age from the pre-Miocene to the present. There are significant differences between these volcanoes, one of the most interesting being the large range in the isotopic ratios of the magmas, which implies that they came from different sources.

The oldest exposed rocks in the island are now exposed in the Ste. Anne peninsula in the southern part of the island, and in the Caravelle peninsula in the east. These rocks are older than 24 million years in age and mostly tholeiitic andesites in composition. They are the remains of lava flows from the older, outer Lesser Antilles arc (the Limestone Caribbees). As described earlier, the Lesser Antilles arc bifurcates in Martinique.

Martinique went through a period of quiescence of several million years, until activity restarted about 16 million years ago. During this second stage, there was a change in magma composition, from tholeiitic to relatively high-potassium calc-alkaline magmas (with a marked increase in radiogenic strontium and lead).

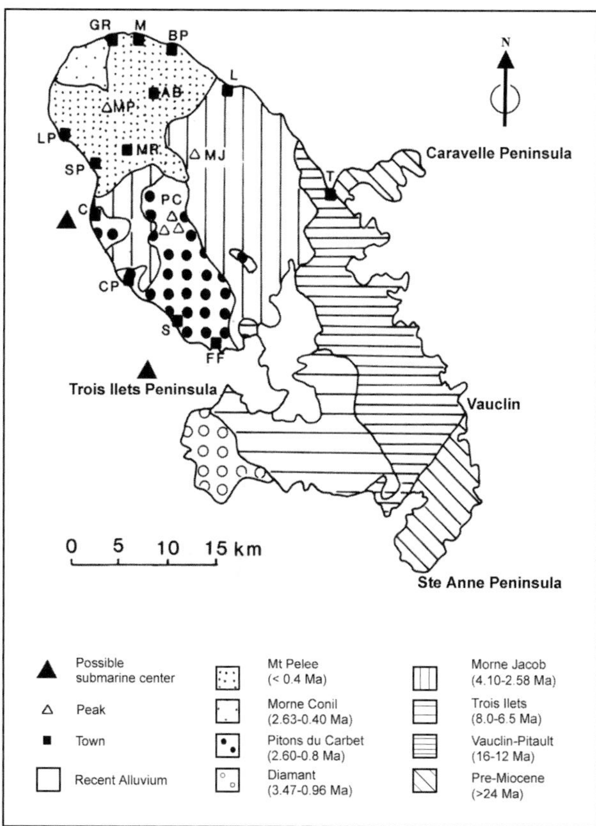

The different geological units, volcanic centers, and major towns of Martinique: Morne Jacob (MJ), Pitons du Carbet (PC), Mt. Pelée (MP), Trinité (T), Lorrain (L), Basse Pointe (BP), Macouba (M), Grande Rivière (GR), Le Prêcheur (LP), St. Pierre (SP), Carbet (C), Casse Pilote (CP), Schoelcher (S), Fort-de-France (FF), Ajoupa–Bouillon (AB), Morne Rouge (MR). (Modified from Smith and Roobol, 1990.)

The first volcanic center built during this time was Vauclin–Pitault, which now forms the southeastern part of the island, exposing basaltic lavas that were erupted under the sea. The next volcanic center was Trois Islets. Andesitic rocks from this time are now exposed in the Trois Islets peninsula in the southwestern part of the island.

During the Pliocene, the activity moved to the island's northeast, forming the Morne Jacob center, which is now deeply dissected by erosion. Morne Jacob erupted submarine basalts and, later, andesitic lavas on land. Before the activity ended, three other centers started up elsewhere: Diamant, Pitons du Carbet, and Morne Conil. The Diamant complex, on the southwest part of the island, is made up of several andesitic domes surrounded by pyroclastic deposits. Pitons de Carbet is a groups of seven, steep andesitic pitons that rise up to 1,196 m (3,924 feet) high. The steep summits of the pitons form a sharp contrast with the flanks, which have been deeply eroded and suffered from collapse. The Pitons de Carbet center is considered by some to be the precursor to Mt. Pelée. Morne Conil, active until about 400,000 years ago, is made up of overlapping breccia (broken-up lava) deposits that were formed by eruptions under shallow water, probably from many vents. These breccias now form the northwestern tip of the island and they seem to have been uplifted before activity started at Mt. Pelée.

Mont Pelée was born about 400,000 years ago and has continued active to the present time. The activity has been mostly explosive, though sometimes only mildly so. The evolution of the volcano has been well studied, and is usually described in terms of three stages. The first stage, which lasted until about 200,000 years ago, is the ancient Pelée. The volcano grew between Mont Conil and Morne Jacob, mostly by eruptions of andesitic lava flows. There is little evidence that violent explosive activity took place at that time. Some breccias from this First Stage can be seen close to the present summit – this tell us that the ancient volcano was about the same size as the present one. The breccias can also be found in exposures all around Mt. Pelée and overlapping Mont Conil and Morne Jacob. There are two main types of breccias: Tombeau Caraïbe and Macouba. The Tombeau Caraïbe breccias are found mainly on Mt. Pelée's western flanks and are probably pieces of lava that disintegrated as it flowed. This can happen if lava flows over steep slopes, the fronts of the flows break away and cascade down. The Macouba breccia is harder to interpret – it may also be broken-up lava flows, but appears to have been heavily eroded by fluvial activity.

During the second stage of Mt. Pelée's growth, between about 100,000 and 19,500 years ago, the intermediate cone was built. The volcano's activity became more explosive, depositing thick layers of pumice fall, as well as pumice-and-scoria flows and block-and-ash flows. Violent eruptions produced *nuées ardentes*, both of the Peléean type (associated with the growth of domes) and of the St. Vincent type (associated with the collapse of an eruption column). An important event during the second stage was the collapse of a large part of the mountain. The collapse formed a horseshoe-shaped structure on the southwestern flank, now marked by scarps that intersect Morne Macouba, Morne Plumé, Morne Calebasse, and Morne Essentes. The southern edge of the horseshoe can be traced to St. Pierre and then offshore for some 6 km (4 miles). The age of the collapse is uncertain, but we know that it is older than the voluminous *nuées ardentes* of the St. Vincent type that happened about 25,000 years ago. Similar collapse structures can be seen in Dominica, St. Lucia, and St. Vincent. In all of these islands, the slope failure happened on the Caribbean side of the arc, where slopes are considerably steeper than those on the Atlantic side.

Following a repose period of about 6,000 years, Mt. Pelée entered its third stage, which lasts to the present. There have been at least 34 eruptions during the last 13,500 years. Carbon dating of the deposits from this period and studies of the stratigraphy have enabled us to make a detailed reconstruction of the activity, particularly the events over the last 5,000 years. Mont Pelée has alternated between *nuées ardentes* and more violent, Plinian-type eruptions disgorging pumice-rich pyroclastic flows. Six of the eruptions – about once every 750 years – were quite large and produced significant quantities of pumice. Geologists know them as events P1 through P6. The deposits of the P1, P2, and P5 eruptions show that they begun with a thick ashfall and a Plinian-type pumice fall, which were followed by pumice-rich pyroclastic flows. The pyroclastic flows associated with the Plinian-type eruptions are mostly confined to valleys and some are up to 50 m (160 feet) thick. Deposits from surges or "ash hurricanes" – turbulent, low-density clouds that accompanied pyroclastic flows – have reached 10 km (6 miles) from the source. Surges are now recognized to be the major volcanic hazard for the region around Mt. Pelée.

Only two eruptions from Mt. Pelée have been documented, those of 1902–04 and of 1929–30. It is interesting that the last eruption before the catastrophic event of May 8, 1902, happened probably only a few years before the first European colonization of the island in 1635. This eruption denuded the mountain of vegetation, inspiring its name, Pelée, after the French word for bald. It also devastated the area near the coast where the city of St. Pierre was eventually built. If colonization had happened a few years earlier, the precarious position of St. Pierre relative to the volcano would have been noticed and the 1902 tragedy might have been prevented.

The 1902 eruption

The morning of May 8, 1902 marks one of the worse volcanic disasters of all time, with a death toll in excess of 30,000 people. Like so many other volcanic catastrophes, Mt. Pelée's did not come without warning signs – but a combination of ignorance about the

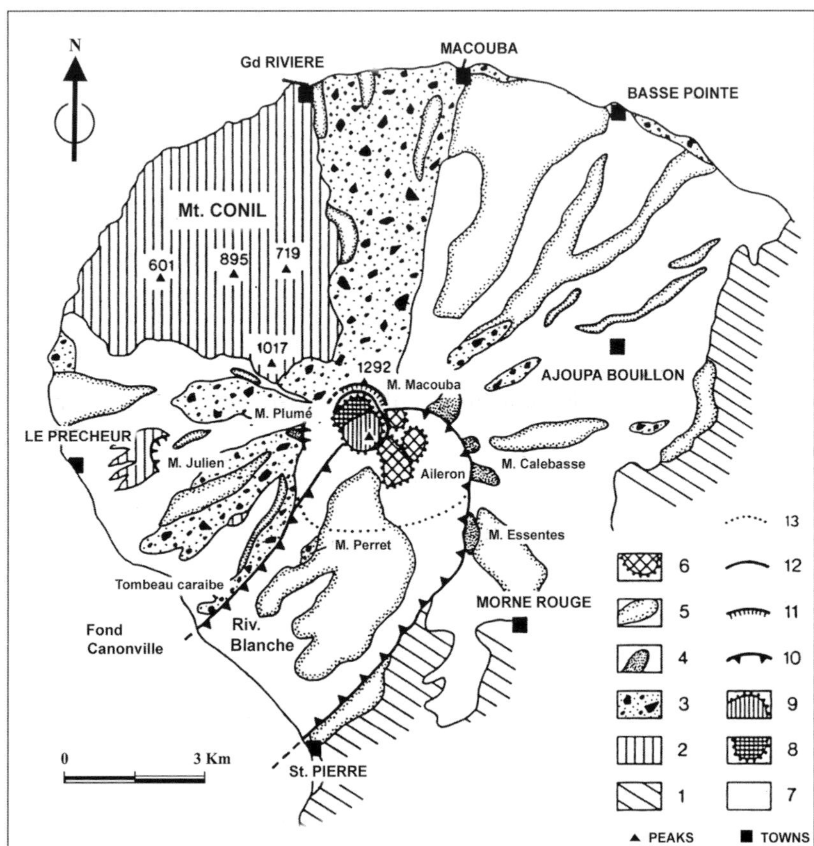

Geological sketch map of Mt. Pelée and Mt. Conil. The geological units are, from older to younger: (1) older rocks from Pitons du Carbet and Morne Jacob, (2) Mt. Conil formations, (3) paleo-Pelée breccias, (4) paleo-Pelée lava flows, (5) neo-Pelée scoria flows, (6) neo-Pelée prehistoric lava domes within the flank collapse structure, (7) neo-Pelée deposits of modern age (younger than 13,500 years), (8) 1902 lava dome, (9) 1929 lava dome. Rims of structures are: (10) rim of the flank-collapse structure, (11) rim of Macouba caldera, (12) rim of the present crater (Etang Sec), (13) southern rim of the paleo-Pelée caldera. (Modified from Vincent *et al.*, 1989.)

volcano and complacency meant that the tragedy was not prevented.

In 1902, Mt. Pelée was already known to be an active volcano. Some minor activity occurred in January of 1792 and a small eruption, between August and October of 1851, caused some minor ashfall in St. Pierre and left a steaming summit crater about 100 m (300 feet) in diameter, called L'Etang. This activity created enough concern at the time for a scientific investigation commission to be set up, but their report minimized the dangers: it called the eruption a "picturesque decoration" to St. Pierre.

The first known sign that Mt. Pelée was stirring again was some increased fumarolic emission in 1889, but that was largely ignored. In February of 1902, the emissions of sulfurous gases was large enough to be noticed in Le Prêcheur and St. Pierre, as the fumes killed birds and tarnished silver. April brought more ominous signs: on the 22nd, small earthquakes shook Le Prêcheur and, a day later, steam was seen rising from the volcano. The first explosion came in the morning of April 25 – a great noise was heard and an ash cloud rose high above the volcano, showering fine ash over Le Prêcheur.

The citizens of St. Pierre were naturally concerned, but life carried on very much as usual. On April 27, a significant event took place, but political rather than volcanic in nature: the first ballot in the elections for a representative for the legislature. The Progressive Party candidate, Fernand Clerc, the owner of a sugar mill, won by 348 votes over his opponent Louis Percin from the Radical Party, but that did not constitute an absolute majority. The election was rescheduled for May 11, which would be a Sunday. This turn of events was to prove fatal for St. Pierre.

Contrary to its name, the Progressive Party was conservative, colonial, and even reactionary. The party's wish to maintain white supremacy in Martinique begun to be threatened in 1899, when Amédée Knight, a black man, won a surprise victory on behalf of the Radical Party. Knight had high hopes of winning all the political seats in the island during the 1902 elections. The Progressive Party realized that it needed a more liberal image, and chose Clerc, who only paid lip-service to their policies, as a candidate. The governor of Martinique, Louis Mouttet, openly supported the Progressive Party, although his relations with the more liberal Clerc were somewhat strained.

Clerc was one of the few people who took Pelée's threat seriously. He had climbed to the summit in April and realized that the once-placid lake surface of L'Etang had become black, bubbling, and boiling. He then wrote a report to the governor, but his effort seems to have been ignored. Clerc continued to monitor the volcano using a telescope and to write what his observations were. On Friday, May 2, he heard a crash "like a broadside of cannon" and saw a column of ash shoot into the sky.

May 2 was probably the first day in which the volcano's activity was strong enough to worry the populace. There were rumblings and explosions heard, and a distinct glare seen over the volcano. Ashfall on Le Prêcheur was heavy, making life miserable for its residents, as well as for the country folks, whose crops and livestock began to die. People began fleeing into St. Pierre, where the ashfall was lighter. The captain of the steamer *Topaz* reported dead fish floating on the sea surface. The most tragic event of the day was the death of farmer Pierre Laveniѐre and several of his estate hands, when they were caught in a mudflow from the volcano.

Ironically, the May 2 edition of the local newspaper *Les Colonies* announced that a "leading authority" on volcanoes had told the paper that there was no reason for concern. The paper advertised an excursion to Mt. Pelée on Sunday, May 4. The volcano was both a nuisance and a tourist attraction, but its threat seems to have been largely overlooked. It turned out that the unnamed "authority" was no other than the paper's editor, Andréus Hurard, who concocted the story as a political favor to Governor Mouttet. The governor, residing in Fort-de-France, was doing all he could to stop concerns about the volcano, including intercepting a telegram by the American Council which was to be sent to Washington, expressing concern at the volcano's awakening. Mouttet decided to reassure the populace by planning to visit St. Pierre, arriving on the eve of Ascension Day.

By now the population of St. Pierre had swollen with fugitives from the countryside. Food was running scarce, drinking water was polluted by ashfalls, and animals were dying in the streets. The city was covered with fine ash. Schools and businesses closed, and life was downright unpleasant. One of the few people who fled St. Pierre was Madame Philomène Gerbault, a wealthy widow. On May 3, she left with her maid for Fort-de-France. She later recalled of those last few days that "one can hardly imagine a more hopeless scene of impending ruin." During the night, as her carriage traveled to Fort-de-France, she heard a series of explosions – ". . . frightful detonations. And then we observed one of the most extraordinary sights in nature – Mont Pelée awake at night. The glowing cone was soon hidden by an enormous black column of smoke traversed by flashes of lightning. A few moments later a rain of ash fell over the countryside." Unknown to her, ash reached as far as Fort-de-France.

Given what we know now about volcanoes, it may seem extraordinary that the people in the city did not flee, even if they were ignorant about volcanic activity. The truth is that politics played a major role in the deaths in St. Pierre. It is now widely believed that the governor wished to keep people in the city, so that the May 11 elections would go ahead as planned. The governor's only action about the volcano was to set up a Commission of Inquiry, which was to publish their findings on May 7.

The fireworks from the volcano on the night from May 3 to 4 actually caused excitement among many St. Pierre residents, who had been told there was no reason to fear the volcano. People leaned out of windows and stood in the streets to watch the spectacle. On May 4, the ashfall ceased and this was interpreted as a promising sign that the eruption was waning, but the respite did not last. That same evening, the eruption started up again and the village of Fond Corré, near St. Pierre, had to be evacuated. Events rapidly got worse the next day. At about noon, the Rivière Blanche, which had dried up, was suddenly flooded by a rapidly moving, hot mudflow, which was probably created by water flooding out of Etang Sec, Mt. Pelée's crater lake. A sugar mill on the river bank, Usine Guérin, was covered by some 3 m (10 feet) of mud, killing the owner and 22 workers. Mud and water rushed down to the sea and created a small tsunami. Horrified citizens in St. Pierre watched as the waterfront drew back some 30 m (100 feet), then rushed back towards the town, flooding its low-lying areas. The devastation along the lower half of the Rue Victor Hugo was almost complete. Rescuers recovered 68 bodies from the mulatto quarter alone. Twenty-eight children drowned when the wave hit the orphanage of Ste. Anne. A new horror came to light: rescue workers identified smallpox in the bodies of some victims dug out of the mulatto quarter. Previous outbreaks had decimated the black population of whole areas in the island. Meanwhile, in Fort-de-France, the governor studied a reassuring report of the Commission of Inquiry.

On Tuesday, May 6, the flooding in the Rivière Blanche continued and the telegraph between St. Pierre and the island of St. Lucia broke, probably because of the mudflows. So far, France knew little about the disaster happening on its colony. Amédée Knight, the black senator, breached protocol and managed to get a cable through to Pierre Decrais, France's Minister of the Colonies, pleading for assistance. Unfortunately for the citizens of St. Pierre, the power play between the black senator and the white governor was blatant and Decrais's reply was meant to stall, citing vacation of the French parliament. That same day, Louis Mouttet arrived in St. Pierre and so did troops of soldiers. Panic-stricken citizens who tried to leave St. Pierre were turned back by the soldiers. Fernand Clerc tried in vain to enlist support for an evacuation of the city. Early that evening, admitting defeat, he left for his estate outside the city, the soldiers letting him pass.

The British Consul, James Japp, ordered his faithful servant Bouverat to leave town by swimming along the Roxélane River. Bouverat carried a bundle of papers, with orders that he was not to open them until the drama in St. Pierre had resolved itself. These papers were Japp's valuable accounts of his last few days of the city. In the American Residency, Consul Thomas Prentiss finally persuaded his wife Clara to leave his side, impressing upon her that it was her duty to carry his report to the President of the United States, in which he described the terrible situation in St. Pierre. She agreed to sail on the Italian ship *Orsolina* in the morning of Thursday, May 8.

Shortly after midnight, drumming and voodoo chants begun in the mulatto quarter. Yvette de Voissons later described the horrifying night, as the voodoo wizards urged their people towards the cathedral. Carrying lighted torches, the voodoo followers passed under Madame de Voissons' window and she saw that many were drunk, singing and chanting in a wild frenzy. When they reached the cathedral, the wizards brought forward their sacrificial animals, a goat and two chickens, cut their heads off and flung them onto the heavy wooden door. The chanted words commanded the dead to leave their graves.

In the morning of Wednesday, May 7, a great cloud rose above the summit and then descended down towards Fond Corré. Reports tell of a second cloud that went down following the path of the first. Luckily, Fond Corré had been evacuated – these were either small *nuées ardentes* or rock avalanches of old fragmented materials. The clouds did not alarm the population – in those days the dangers of *nuées ardentes* were largely unknown. In spite of all the events, people in St. Pierre were more optimistic, because the mudflows were no longer reaching the sea and the height of the ash column above the summit had lowered. The Commission of Inquiry produced their report on schedule, concluding that St. Pierre was not threatened.

One person who was not calmed was Yvette de Voissons. The desecration of the cathedral the previous night made her decide to leave St. Pierre immediately – "I knew the thing had begun, and that this was no place for Christian people." She packed her bags and went to the waterfront, hoping somehow to get aboard the *Orsalina*, though she had no passage booked. Later she was to appreciate her amazing luck. The ship's captain, realizing the danger from the volcano, had decided to sail a day early. He sent a longboat ashore to pick up mail, not expecting any passengers to be ready a day early. The longboat carrying Yvette was chased by harbor officials, who boarded the ship and told Captain Leboffe that he could not change the sailing date. The captain reminded them that he was responsible for the safety of his ship, and ordered the officials to be escorted off the vessel. Shortly after 9 a.m., the *Orsalina* sailed away, leaving behind 18 passengers who had been booked to board, including Clara Prentiss.

By the afternoon explosions started up again and the Roxélane River that ran through downtown became flooded with muddy water. The unrest began to spread, with people demanding food. The mayor feared a riot but the populace was reassured by the governor, who even brought along his wife. The night brought worse fears, as incessant explosions made sleep impossible for many. Those who ventured a look at Mt. Pelée were rewarded with the sight of incandescent columns of ash and gas rising from the summit. Finally, at around 4 a.m., the volcano seemed to go to sleep. At 6 a.m. it woke up again, ejecting clouds of ash that came south and plunged St. Pierre into darkness.

May 8 was celebrated in Martinique as the Catholic Ascension Day. The fearful residents of St. Pierre were being called to church by tolling bells when they heard a loud explosion. Those who looked towards the mountain saw a great black cloud rapidly descending towards them. The moment of doom had finally come. At 7:50 a.m., as shown by the broken clock in the Military Hospital, St. Pierre was totally destroyed. Among the dead were the governor, his wife, the city's mayor, and most members of the Commission of Inquiry, who had concluded the previous day that the volcano posed no threat. Also dead were Thomas and Clara Prentiss, and James Japp, who had foreseen the disaster but had stayed out of a sense of duty.

The eruption left little standing in St. Pierre and devastated an area of about 58 km^2 (22 square miles) west and southwest of the volcano. The devastation also spread to vessels anchored in the harbor of St. Pierre; the passengers and crew of 18 ships were among the victims of the eruption. Only one ship was able to escape, the British steamer *Roddam*, but its captain and several crew members died from massive burns shortly after they arrived in St. Lucia. The total death toll from the eruption was about 29,000 people, most of whom were in St. Pierre. The sight that confronted rescuers as they arrived in the city was horrendous – burnt bodies of the victims were scattered in the streets amongst the rubble from destroyed buildings. Surprisingly, there were two survivors. One was a shoemaker named Leon Compère, who lived near Morne Abel and was able to run away on the road to St. Dennis.

Compère was lucky, but his fellow survivor was much more so. Auguste Cyparis, a local from Le Prêcheur, had been sentenced to prison for assault and battery and, shortly before the eruption, had been placed in the dungeon as a punishment for an attempted escape. On Sunday, May 11, rescuers who were combing the city looking for survivors heard faint cries from the prison. It was Cyparis, badly burnt but still alive. His underground dungeon, with only a small window near the top, had been his salvation. He quickly recovered, was pardoned by the government, and eventually came to the USA to become an attraction of the Barnum and Bailey Circus as "The Prisoner of St. Pierre." He died in 1929.

The eruption of May 8 became known around the world as one of the most tragic volcanic disasters of all time. Like the AD 79 eruption of Vesuvius, it has formed the background of several books, including the

View of the devastation caused by the 1902 eruption. (Photograph courtesy of US Geological Survey).

novels *The Violins of St. Jacques* by Patrick Leigh Fermor and *Texaco* by Patrick Chamoiseau, the latter winning the Prix Goncourt in France. The story of Cyparis has been turned into the play *The Prisoner of St. Pierre* by Pat Gabridge. The eruption and its actual events were described in detail (though not all of them accurate) by journalists Gordon Thomas and Max Morgan Witts in a book appropriately called *The Day the World Ended*.

The tragic events of May 8 were not the end of the eruption. An equally or even more violent explosion occurred on May 20, culminating with another *nuée ardente* that invaded St. Pierre, finishing its destruction but taking no lives. Other violent events happened on May 26, June 6, and July 9, and smaller explosions took place throughout August. On August 30, the eruption climaxed with an explosion and *nuée ardente* that were much more powerful than the tragic May 8 event. It destroyed an area some 115 km^2 (45 square miles), mostly to the south and east of the volcano. Several villages were destroyed, including Morne Rouge, and over 1,000 lives were lost. The activity subsided after that and ceased near the end of September, except for one curious phenomenon: a dome made up of pasty lava was being gradually extruded from the summit crater. The "Spine of Pelée," as it became known, started to form in July and, during November, it grew 230 m (750 feet) in 20 days. As it grew, it also crumbled, then grew some more. On May 30, 1903, the spine attained its greatest height: 340 m (1,122 feet) over the crater.

The eruption finally ended in 1904. The only subsequent activity to date was the extrusion of a lava dome from 1929 until 1932. Although some dome collapse produced small *nuées ardentes*, no lives were lost and the *nuées* were mostly confined to the Rivière Blanche valley. Since then, Mt. Pelée has settled down to a long period of repose. We know that it could wake up again anytime, so the volcano is monitored around the clock. St. Pierre and numerous villages have been rebuilt on its flanks and nobody wants to take chances again. Next time Mt. Pelée wakes up we can be sure that no political election will hold back an exodus, that no newspaper will state that the volcano is no threat, and that nobody will offer tourist excursions to the crater. Mont Pelée is best seen in repose.

A personal view: The significance for volcanology

The 1902–04 eruption, tragic though it was, represented a major step forward in our understanding of volcanoes. As is usually the case with a major catastrophe, the eruption attracted the world's attention and posed the question of whether a disaster of such magnitude could have been prevented. People from around the world compared the May 8 eruption with that of Vesuvius in AD 79, and St. Pierre became known as the Pompeii of the Caribbean. Scientists begun to study the deposits of the eruption as a clue to the mechanism which, when understood, could be the key to the prevention of future disasters. What, from a volcanologist's point of view, really happened on May 8? Ironically, although the deposits of the eruption have been studied in detail for a century, and the sequence of events is well known, the scientific interpretation remains controversial.

The first studies of the eruption were done by the French geologist Professor Alfred Lacroix. He amassed a lot of valuable information and concluded that the flow that destroyed St. Pierre had been the result of a lateral blast. In 1902, volcanology was still in relative infancy as a science, but Lacroix envisaged essentially the same mechanism that we now understand happened in the May 18, 1980 eruption of Mt. St. Helens. Other explanations for the Mt. Pelée eruption have been proposed, but it seems that Lacroix was not far off. We now think that a lateral blast similar to that of St. Helens occurred and that St. Pierre was destroyed by a thin, extensive, fast-moving and ground-hugging pyroclastic surge. A more conventional *nuée ardente* would have buried the town under rubble; and St. Pierre, although totally devastated, was left with only a few centimeters of ash on top. It is possible that a *nuée ardente* did come down during the fatal eruption but did not reach St. Pierre.

An alternative explanation is that St. Pierre was destroyed by a *nuée ardente* of the "tripartite" type. The definitions of various pyroclastic flows can get quite complicated but, regardless of semantics, a *nuée ardente* can be thought of as having three major zones. The first one is the core component, consisting of a dense avalanche of fast-moving, glowing debris. This part forms a ground-hugging pyroclastic flow that is channeled along topographic depressions and can have large boulders as well as fine ash inside. The second zone is the surge, which has a lower density and is not confined by topography – it can in fact run uphill. The third zone is the towering cloud that rises up into the air, sometimes for several kilometers. It is possible that the May 8 eruption produced a *nuée ardente* of this "tripartite" type. The dense core of the flow went down the Rivière Blanche where, indeed, one finds a thick accumulation of coarse deposits. This dense part did not pass through St. Pierre, but the *nuée*'s lateral component – the less dense surge part – rolled over the city.

The two competing interpretations – blast plus ground-hugging pyroclastic surge, or lateral, surge-like component of a *nuée ardente* – seem equally likely. Field studies and interpretation of the deposits have been quite difficult for a number of reasons. First, Martinique is in the tropics and vegetation has grown very fast over the area. Second, the *nuée ardente* started fires, and the products of the fire confuse the field interpretations. A third problem is that other *nuées* or surges tore through St. Pierre after May 8, including a major event on May 20, and sorting one from the other is very difficult.

We may never know exactly what happened on May 8, but there is no doubt that great advances in volcanology were made because of this eruption. Even the birth of the Hawaiian Volcano Observatory is in part due to Mt. Pelée. Frank A. Perret, an American electrical engineer who had worked with Edison, heard about the tragedy of St. Pierre and decided to make a career change. He spent the next 30 years of his life studying volcanoes all over the world and eventually became the first director of the Hawaiian Volcano Observatory. Perret observed many *nuées ardentes* during the 1929–1932 reawakening of Mt. Pelée, building himself a small observatory. On one occasion, he was engulfed by the clouds of a small *nuée ardente* and, ever the scientist, wrote down what happened, including that he experienced "burning of the air passages." Although volcanologists today disagree with some of Perret's interpretations, there is no doubt that his recording of the phenomena of *nuées ardentes* became a great contribution to science. The next major leap forward in the understanding of this type of phenomenon was to come many years later, triggered by another tragedy: the eruption of Mt. St. Helens in 1980.

Visiting during repose

There are two highlights to a visit to Mt. Pelée: the lively, charming town of St. Pierre, where one can still see ruins of the old city destroyed by the eruption, and the climb up Mt. Pelée, which is pleasant and not too strenuous. Those who are particularly interested in the deposits from the eruption will want to hike up the *nuée ardente* valley known as La Coulée Blanche. Keen hikers should not miss the trails in the Pitons du Carbet National Park.

The drive from the capital, Fort-de-France, to St. Pierre is an easy though slow one along the costal highway N2. Allow at least 45 minutes to get to St. Pierre, longer if possible, as there are some interesting stops along the way. The old village of Case-Pilote is worth the short detour: turn south off N2 at the Total gas station and you'll see a stone church, one of the oldest in the island. The village square and the brightly painted boats along the shore deserve a few pictures. The small boats are called gommiers, after the trees they are made from. The town of Carbet, also along the N2, has a sandy beach and a dive shop (see below for diving suggestion). About 1.5 km (1 mile) north of Carbet is Anse Turin, known for its sandy gray beach and its connection to Paul Gauguin. Another 0.5 km (⅓ mile) towards St. Pierre is La Vallée des Papillons, one of the island's oldest plantations. The plantation's stone buildings, now in ruins, make a lovely backdrop for lush tropical gardens. As the name indicates, the plantation is now a butterfly farm and it is open to the public.

An alternative drive from Fort-de-France to St. Pierre uses the slower inland road (N3, Route de la Trace) which winds its way through the lush mountainous interior of the Pitons de Carbet National Park. The Route was started by Jesuit priests in the seventeenth century and legend has it that its windy nature is due to the priests' love of island rum. It is a beautiful road, passing along the eastern flanks of the Pitons and the trailheads of several well-marked hiking trails leading to the rainforest and up the peaks. The Route ends in Morne Rouge, on Mt. Pelée's southern flank, which was partially destroyed by the 1902 eruption. To continue to St. Pierre, take road D1, another windy, scenic drive that is the way to reach the Volcano Observatory (see below).

St. Pierre

Located by the waterfront, some 7 km (4 miles) away from Mt. Pelée, St. Pierre is a rather fascinating town that has retained its old charm. The town has two districts, separated by the Roxélane River: on the north side is the Quartier du Fort and on the south side is the Mouillage (meaning anchorage). There are ruins from the old city throughout the town and many of the surviving stone walls were preserved and incorporated into the reconstruction. The major streets run along the waterfront south of the river. The waterfront park, next to the market, is the gathering spot for the 6,000 or so town residents – about one-fifth of the pre-1902 population.

Most visitors are day-trippers and lodgings are scarce, but this is still the best place from which to explore Mt. Pelée. The long black sand beach is an added attraction; the sand is soft and the beach generally uncrowded. Experienced scuba-divers should consider diving to see some of the dozen or so ships sunk by the 1902 eruption. The most spectacular is the Roraima steamer, which rests about 45 m (150 feet) underwater just off St. Pierre. There are several scuba-diving shops that organize dives and rent equipment.

Theater ruins and prison cell of Cyparis

The ruins of the old St. Pierre are scattered all over town, but in many places they are no more than stone foundations and walls. The most impressive ruins are those of the old theater, which had 800 seats and hosted troupes from France. The theater was widely regarded as the finest public building in the Lesser Antilles. A double staircase, leading to the partially-standing walls of the lower level, is still impressive enough to give a sense of the sophistication and grandeur that made St. Pierre the "Paris of the Caribbean." Another highlight is the prison cell of

The ruins of the prison cell that held Cyparis, allowing him to survive the eruption. (Photograph courtesy of Charles Frankel.)

Cyparis, located nearby. Make sure to walk along the Rue Bouillé, near the museum, where many ruins can still be seen.

Those who can understand French should consider taking a guided tour of the ruins on the Cyparis Express tram. The tour takes about 40 minutes and leaves from the Ruines de Figuier on Rue Isambert, near the museum. Tours in English are given occasionally; call their main office (55–50-92) for the schedule.

Musée Vulcanologique

Founded by Frank Perret in 1932, this small museum preserves some fascinating relics of the 1902 eruption. The items on display were found among the ruins after the eruption and they range from petrified rice to glasses fused together by the heat of the surge. One particularly impressive item is the cast iron bell from the cathedral's tower: after the eruption, it was found rather squashed. The museum (open daily) is located on Rue Victor Hugo and its parking lot offers a magnificent view of the volcano, the harbor and the ruins. Just across the street is the Museum of St. Pierre, worth a brief visit for its eruption-related exhibits.

Maison Coloniale de Santé

Located on Rue Levassor in the Quartier du Fort, the "health house" was actually a lunatic asylum run by the Sisters of St. Paul. Some 200 patients died here during the eruption, as well as 14 nurses and five nuns. A particularly gruesome sight is that of the steel chair bolted to the floor, where patients used to be strapped. The chair was bent forward by and away from the volcano by the force of the blast. Near the patients' cells there are some good outcrops of the eruption's deposits. The top of the deposits is a poorly stratified gray-colored layer some 50 cm (20 inches) thick, resulting from a pyroclastic flow and containing blocks up to 30 cm (12 inches) in diameter. Underneath this poorly sorted layer is a thin (2 cm, ¾ inch) pinkish-brown ash layer which contains some accretionary lapilli, the result of thin ash fall. At the bottom is a coarse layer about 15 cm (6 inches) thick, made up of dark gray ash with a significant amount of charcoal and building fragments, resulting from the surge during the earlier phase of the eruption.

Maison des Génies

Located near the Maison Coloniale de Santé, this building once belonged to the Army Corps of Engineers. It is worth stopping by to see an excellent exposure of the 1902 deposits. Look for the entrance to a ground-floor room in which rubble from the collapsed ceiling piled up on the floor (looking close, you can see roof tiles and the charred remains of roof beams). Across the entrance to this room is the layered deposit. From the top down, the first layer is a pyroclastic flow about 30 cm (12 inches) thick, gray in color and sandy ash in appearance, merging into the overlying soil. The second layer down is very thin (2 cm, ¾ inch), made up of yellowish brown ash resulting from ashfall. Below is a thick layer (50 cm, 20 inches) pinkish-gray in color, which is a pyroclastic flow deposit with boulders up to 30 cm (12 inches) in diameter. The fourth layer down is about 10 cm (4 inches) thick, made up mostly of blast material containing charcoal and fragments of dense andesite. The fifth layer is very thin (1 cm, ½ inch), yellowish-brown in color and made up of ash with some accretionary lapilli, the result of ashfall. At the bottom is the dark gray surge deposit layer, about 20 cm (8 inches) thick. This layer has a lot of building rubble in it, including roof tile and sheet metal fragments, plus charcoal and andesite blocks up to 10 cm (4 inches) in diameter. It rests on top of pre-1902 soil and is the remains of the surge that destroyed the city. Comparing this sequence with that from the Maison Coloniale is straightforward if you are aware that the top two layers cannot be seen in the Maison's deposits.

Church of the Fort

These ruins have special historical significance because the building, from 1640, is thought to have been the oldest French church in the New World. On the morning of May 8, Ascension Day, the church was full of worshippers, including many youngsters having their first communion. The power of the blast toppled the thick walls, and all inside were killed. No surge deposits remain in this location, but the ruins are some of the most evocative in St. Pierre.

Anse Tourin beach

Famous as one of the places that inspired artist Paul Gauguin, this location is worth a visit also because of some rather unusual horizontal cavities in the pyroclastic deposits, the formation of which is still a mystery. These deposits are not from the 1902 eruption, but from the early phase of the Pitons du Carbet activity in the island, which pre-dated the formation of Mt. Pelée. The best exposure of the pyroclastic deposits is located just south of the tunnel that gets you to the beach. Starting at the base, a conglomerate at road level, try to distinguish the different layers. Just above the base is a thick (3.5 m, 11 feet) layer resulting from a pumice fall.

This deposit has faint stratification and the color changes from light gray near the bottom to dark gray near the top. The next layer is thin (20 cm, 8 inches), sandy ash also resulting from ashfall. Next comes a 3 to 5 meters (10 to 16 feet) thick deposit from a pyroclastic flow, pale brown in color, becoming thinner ("pinching") towards the north. This is overlain by a 1 to 2 meters (3 to 6 feet) mudflow deposit, which is discontinuous, jumbled, and quite different in appearance from the other deposits. A layer of soil about 0.5 m (1½ feet) separates the mudflow deposit from a younger Plinian fall deposit, about 4 to 8 m (13 to 26 feet) thick. This layer ranges from pale yellow to gray in color, with a darker gray base. It thins to the south, away from the vent. Notice the "reverse grading" in this layer, with bigger pieces near the top. Those with keen eyes may be able to pick out four surge layers interbedded with the Plinian deposit near the top. Above this is a pumice-rich pyroclastic flow deposit, yellowish-gray in color, that is very thick (15 m, 50 feet) towards the north, near the tunnel, but thins to only 0.5 m (1½ feet) to the south. This sequence reflects some of the many events involved in the formation of Pitons de Carbet, over 1 million years ago.

While in Anse Tourin, stop by the Gauguin museum. Although it only houses reproductions of the artist's work, it has some interesting memorabilia, including letters from Gauguin to his wife. Gauguin and his artist friend Charles Lavalle lived in a shack near Anse Tourin for 5 months in 1887. Although poor and sick, Gauguin painted about a dozen masterpieces inspired by the island's people and landscape, including *Bord de Mer* and *L'Anse Tourin avec les raisiniers*.

Statue of the Virgin of the Mariners

This statue was erected in 1870 at a high point in the southern end of St. Pierre. There is nothing particularly special about it, except for its magnificent location. This is one of the best places from which to photograph St. Pierre and Mt. Pelée. The Rivière Blanche scarp, formed by the collapse of a sector of the volcano, is also prominent from this angle. To reach the statue, follow the D1 road from St. Pierre to Fond St. Denis and turn right directly across from the cemetery, up a road surfaced with concrete. There are places to park at the top of the road, and a cobblestone path leads to the statue.

Volcano Observatory

The Observatoire Volcanologique de la Montagne Pelée, run by the French Institut de Physique du Globe, is responsible for monitoring the volcano. Its operations include a network of seismographs and ground deformation stations. Although the observatory is not open to the public, it is worth coming here just for the view, another magnificent panorama of the volcano. It may be possible to arrange a visit by writing to the Observatory well ahead of time. If you are allowed to visit, make sure to ask to see the Quervain–Piccard seismograph, located in the basement of the building. This 20-ton instrument is one of the largest and oldest seismographs in existence.

To reach the Observatory, take route D1 out of St. Pierre and head east to Fond St. Denis. Take the sharp right turn onto the road to Morne des Cadets; the Observatory is located at the end of the road on the hilltop.

Climbing Mt. Pelée

The hike up Mt. Pelée is not hard, but it can be disappointing because the summit is often shrouded by clouds. Going during the dry season (December to April) increases your chances for a cloud-free day, but be aware that clouds can come and go fast, and can lift for a few hours, usually in the early morning or late afternoon. It is best to ask locally what the weather pattern has been over the last few days. On cloud-free days, the merciless sun can be a problem. As its name ("bald mountain") indicates, Mt. Pelée's summit is devoid of trees and shade, so sun protection is a must. It is best to set out early in the morning and avoid the midday hours at the summit. The round trip takes from 3 to 5 hours. If you plan to start the hike in the afternoon, bring a flashlight: night comes swiftly in the tropics.

There are two trails to the top: the most popular trail, which is described here, goes up the western flank from the village of Le Prêcheur. The other trail goes up the eastern flank, starting out from the village of Morne Rouge. To take the western route, drive out from St. Pierre to Morne Rouge on the N2 road, drive through the village and then take the N3 road towards Ajoupa–Bouillon for a short distance (about 500 m, ⅓ mile). There is a well-marked turn to the left onto the D39 road, also known as "Route de L'Aileron." This road climbs up to a parking lot by a radio tower, at an altitude of 830 m (2,700 feet). The trail starts from here. There is a posted map showing the route up the mountain, but hikers should have no difficulty finding their way, as the trail is clearly marked and well maintained. It is a popular hike, particularly on weekends, so don't expect solitude.

The summit area of Mt. Pelée: on the left is the 1929 dome, on the right is the 1902 dome. (Photograph courtesy of Charles Frankel.)

The trail begins with a steep climb up the first knob, called "L'Aileron," which is actually a lava dome. A few outcrops of Plinian ashfall from the AD 250 eruption can be seen in this first part of the climb, but the ground is mostly hidden by low vegetation. Stop at L'Aileron for magnificent views of the south and east flanks of the mountain. The next part of the climb, up to the rim of the crater, goes through some andesite flows and a few exposures can be seen, particularly in places where steps are carved in the bedrock. The path splits at the crater rim. Take the one to the left up to Monument Dufrénois (altitude 1,210 m, 3,990 feet) for an excellent view of the steep crater wall and the two andesite domes from the last eruption that grew inside the crater. To the north is the 1902 dome, rising to an altitude of 1,362 m (4,468 feet) and to the south is the 1929 dome, rising to 1,397 m (4,583 feet).

Go back to where the path splits and hike north along the crater rim, across the Plateau des Palmistes. The trail offers different perspectives of the crater and domes. The 1929 dome is in the foreground, the 1902 dome to the north; both are made up of a jumble of large andesitic blocks. The domes are already covered with mosses and low-standing vegetation, much like the rest of the mountain. At Morne La Croix (altitude 1,247 m, 4,115 feet), at the level of a little shelter ("Deuxième Refuge), the path descends into the crater moat between the steep crater wall and the 1902 dome. The path continues up the dome and then to the south up the 1929 dome, which is the current summit.

The 1902 dome grew after the explosive activity ended, and was the site of the "Spine of Pelée" described earlier. Disintegration and crumbling of the spine happened throughout the period of growth, which ended in October 1905. By that time, the dome's height was about 300 m (1,000 feet). The type of rock making up the dome is a hypersthene andesite, containing phenocrysts of plagioclase, hypersthene, and titanomagnetite.

The 1929 dome was built during the 1929–32 eruption. While not as spectacular, the new dome also produced a number of small spines that reached up to 50 m (160 feet) in height. This done eventually covered the south and eastern parts of the old dome. The summit of Mt. Pelée is now quiet, with not even a fumarole in sight, but it is not hard to imagine the spectacular spines and fiery explosions that will eventually happen again.

La Coulée Blanche

This hike up the Rivière Chaude valley provides excellent views of the deposits from the *nuées ardentes* of 1902–04 and 1929–32. The trail goes up the southwestern flank of Mt. Pelée and ends at a beautiful waterfall. It is not a hard trail, but can get rather grueling if the sun is out. It best to start early in the morning, as the first part of the trail offers no shade. The trailhead is located by road D10, on the right hand side, about 1 km (5/8 mile) north of St. Pierre. Drive out of St. Pierre and, after the road crosses the Roxélane River, the Rivière des Pères, and the Rivière Sèche, watch out for a big pumice quarry on the right-hand side of the road. You will see a small shelter and a poster map of the trail. Parking is on the opposite side of the road.

The trail starts at the quarry gate (hikers have to squeeze by on the side of the locked gate). The first part follows the quarry road, through blinding expanses of bright white pyroclastics and andesitic blocks. The trail, flagged in places by metal rods, proceeds through a tree grove (your last chance for shade) and then through low-standing vegetation and by the left-hand side of Morne Perret (also known as Morne Lenard), a hill that acted as a barrier to divert most of the pyroclastic flows down the axis of the valley. The trail then hugs the western edge of the valley and descends into the bed of the Rivière Chaude, at the foot of a beautiful cliff called Le Piton, which exposes sections of the 1902 and 1929 pyroclastic flows. The river bed is choked with andesite and pumice rocks in a wide range of sizes, shapes, and colors and the stream is deep enough in places to enable weary hikers to take a bath. Thoughtful hikers have raked the rocks into dams to make pools about 0.5 m (1½ feet) deep. The water is slightly warmer than what is typical for a mountain stream and rather pleasant in the local climate.

The trail continues upstream along the river bed for another half an hour or so of hiking. It ends by a waterfall about 5 m (16 feet) tall, which provides an even nicer bath. Daring hikers may want to climb up the waterfall using a leaning tree trunk and hike up to the source of the stream, one of Mt. Pelée's ring fractures. The water comes out at the source at about 30 to 40 °C (86 to 104 °F). The source area has sulfurous deposits and was the site of phreatic explosions in 1851 and 1902.

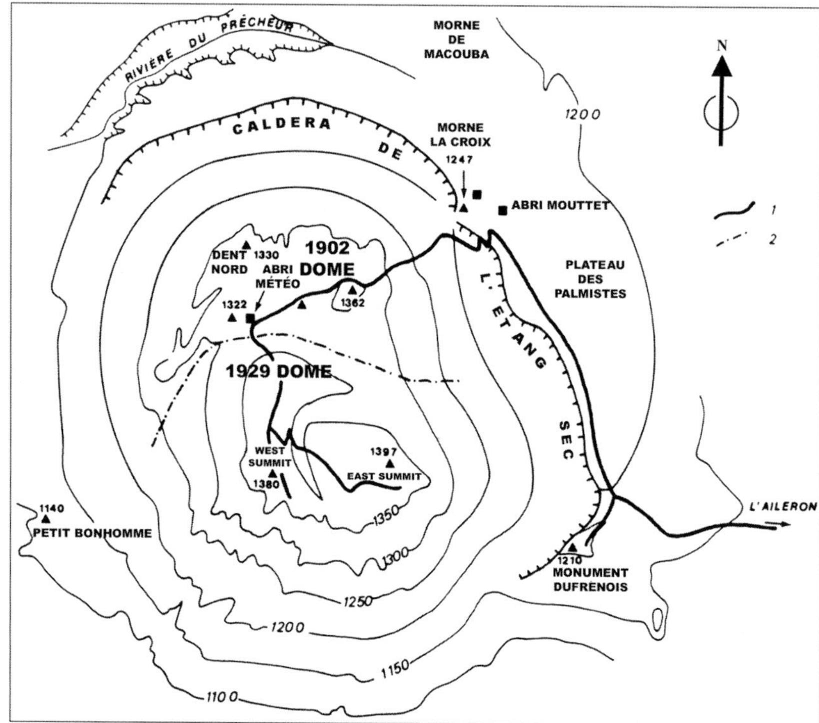

Map of the summit area of Mt. Pelée. Solid line marks the hiking trail, dashed line the boundary between the two historic domes of 1902 and 1929. Topographic contours are in meters. (Modified from Westercamp and Tomblin (1979.)

Visiting during activity

Mont Pelée is an extremely dangerous volcano and when it erupts again, it is likely that St. Pierre and other areas at risk will be evacuated. Pelée is one of those volcanoes best visited while in repose. However, depending on the type of activity, it may be possible for visitors to get close enough to see some action. Quiet dome growth, though potentially dangerous, could be viewed from the air and, depending on the assessed hazard level and permission to climb the mountain, even from the crater edge. Good vantage points away from Pelée that might be accessible are Morne Jacob and Pitons du Carbet.

Other local attractions

Dominica

It is said that when Christopher Columbus was asked by the king and queen of Spain to describe Dominica, he answered by crumpling a sheet of paper and tossing it onto a table. The analogy was a good one, as this unspoiled island is rather rugged, thanks to a number of young domes of dacitic andesite that form its distinct backbone. Starting from the south, there is Morne Patates (patates are sweet potatoes), where the last eruption in the island occurred some 500 years ago. The chain continues through Morne Anglais, Grande Soufrière, Morne Watt, Microtin, Morne Trois Pitons, and Morne Diablotin, and ends at Morne au Diable in the north. Morne Diablotin is Dominica's highest peak (1,421 m, 4,662 feet) and the second highest in the Lesser Antilles after Guadeloupe's Soufrière.

Dominica prides itself in not being a touristy destination. It is an ideal island for those seeking a rural, tranquil setting. It is also a prime destination for divers, because of its remarkable underwater volcanic arches, pinnacles, and caves. If you cannot dive, go snorkeling in Champagne, a submerged bubbly hot spring. On land, the island's major attraction is its mountainous interior dotted with waterfalls. The Trois Pitons National Park offers lots of hiking on volcanic terrain and beautiful areas such as Emerald Pool and Boiling Lake. Don't miss Sulfur Springs, the site of fumarolic activity where remarkable deposits of sulfur crystal can be seen. There is an excellent view of Morne Patates from the uppermost spring. The Patates's dome is surrounded by block-and-ash flows similar to those from Mt. Pelée's 1902 eruption.

The island's only volcanic activity in historic times was a phreatic explosion in January 1880, in the Valley of Desolation in the central region. This explosion was caused by the interaction of groundwater and hot magma within the crust and no fresh magma was erupted. This does not mean that Dominica's volcanoes are not likely to wake up. In fact, frequent shallow earthquakes under the volcanoes are considered a major sign that activity could start again at any time.

St. Vincent

This lush island is home of the very active Soufrière volcano, which presents a major hazard to the people living in the northern half. Soufrière is a restless stratovolcano 1,220 m (4,000 feet) high, with a large summit crater some 1.6 km (1 mile) in diameter and 300 m (1,000 feet) deep. Several eruptions happened in historic times, including a major event on May 7, 1902, one day before the eruption from Mt. Pelée. The climax of the Soufrière eruption was a violent explosion so loud that it was heard 170 km (106 miles) away in Barbados. The eruption column reached 15 km (9 miles) high, and deadly pyroclastic flows came down, killing nearly 1,600 people. Soufrière was active again in 1971, and more recently in 1979, but there were no fatalities from these eruptions. The 1971 event did not involve any explosive activity, only the growth of a lava dome in the crater lake, which formed a new island. The 1979 eruption had explosions that caused considerable amount of concern and resulted in the evacuation of 15,000 people. Pyroclastic block-and-ash flows came down the valleys and some entered the sea. Summit phreatomagmatic explosions destroyed the crater island formed in 1971, the crater lake was boiled off, and extrusions of basaltic andesite covered much of the remains of the island, with the sad consequence that the volcano's summit is not as attractive as it used to be.

Soufrière is an unusual stratovolcano because its pyroclastic flows are mainly made of basalt, while most of its lavas have a higher silica content and are basaltic andesites and andesites. This goes against the normal volcanic mode of operation and suggests the influence of factors other than composition. One of the major factors is probably the crater lake, as phreatomagmatic explosions can disrupt the basaltic lava, producing pyroclastic flows.

Soufrière is a popular attraction and the climb is relatively easy, on a good trail. The crater rim can be reached in about 2.5 hours. To get to the trail, take a boat from Kingston to the mouth of the Dry Wallibou, on the volcano's western coast. Hike up the gorge for about 500 m (1,600 feet) and you will see the volcano

Soufriere Hills volcano in Montserrat in a relatively quiet state during the author's June 1996 visit. (Photograph by the author.)

trailhead. The trail at first passes through cultivated farmland, then enters a secondary forest of tall trees, many draped with mosses and surrounded by a lush ground cover of ferns. As the trail leaves the forest, it enters a grassy terrain that was devastated by the 1979 eruption, and several outcrops of 1979 deposits can be seen along the trail. The deposits are from pyroclastic surges, flows, and ashfall.

The main crater of Soufrière is now largely covered by the 1979 lava dome, which is about 870 meters (2,850 feet) across and 130 meters (400 feet) high. Surrounding the dome are yellow-gray pyroclastic flow deposits from the explosive stage of the 1979 eruption. There are some small fumaroles in the southern part of the lava dome. If you look to the northeast across the crater, you can see on the horizon the lip of the crater formed by a large eruption in 1812. Far to the north is the amphitheater-like wall that was formed when a sector of the old volcano collapsed about 10,000 years ago.

There is a short trail going northwest along the crater rim that brings you to the ruins of the 1971 eruption observatory, which was destroyed during the 1979 eruption. The only sign of the building that is left is a steel rod sticking out of the pyroclastic deposit. Pyroclastic deposits from 1979 are exposed on the crater rim. Return on the trail the same way, down to the Dry Wallibou river mouth.

Soufriere Hills, Montserrat

The volcano

Soufriere Hills is the best recent example of a volcano that has totally changed the lives of thousands of people – in fact, of a whole island and country. From a scientific point of view, the continuing eruption has not been truly violent nor has it had a significant environmental impact worldwide. From the point of view of the people of Montserrat, it brought about the end of the world they had always known. Their lush "Emerald Island," often compared to Ireland, has been largely transformed into a barren desert. Its capital city and airport are now buried under pyroclastic flows, the once clean air poluted by ashfalls, and the lush forests destroyed. Soufriere Hills demonstrated to the world what havoc a volcano can cause, even without trying very hard. The explosions have been only moderatately violent, the warning signs plentiful, and the loss of life, though tragic, has been minor.

Montserrat is a tiny island shaped like a teardrop, only 106 km^2 (44 square miles) in area. It is a British Protectorate and was largely settled by Irish refugees fleeing religious prosecution in the early seventeenth century. Settlers remained loyal to Britain and fiercely held on to their Irish heritage: the island's flag shows Erin of Eire with cross and harp and St. Patrick's Day is celebrated with big festivities. The settlers kept the country's French-sounding name, which actually hails from Spain. When Columbus discovered the craggy island in 1493, he was reminded of the mountainous landscape around the monastry Santa Maria de Montserrat, near Barcelona.

Montserrat has volcanic mountains running along its length: in the north is Silver Hill, the oldest part of the island, reaching 403 m (1,320 feet) high. In the middle are the Centre Hills, about 4 million years old, where Katy Hill reaches 741 m (2,430 feet). Just to the south and west of the Centre Hills are St. George's Hill and Caribald Hills. In the south-central part of the island are the Soufriere Hills, a group of andesite domes where activity began in 1995. Further south are the South Soufriere Hills, which have basaltic lava flows and scoria fall deposits, as well as pyroclastic deposits of andesite. South Soufriere and St. George's Hill are considered active centers, but the possibility of eruptions from these locations is considered small. In contrast, an eruption from Soufriere Hills was considered likely years before the 1995 activity started.

The Soufriere Hills dome group is topped by Chances Peak, the island's highest point at 915 m (3,000 feet). The summit crater, English's Crater, is about 1 km across (⅝ mile) and shaped like an amphitheater. Its walls, about 100 to 150 m (300 to 500 feet) high, open to the east into the Tar River

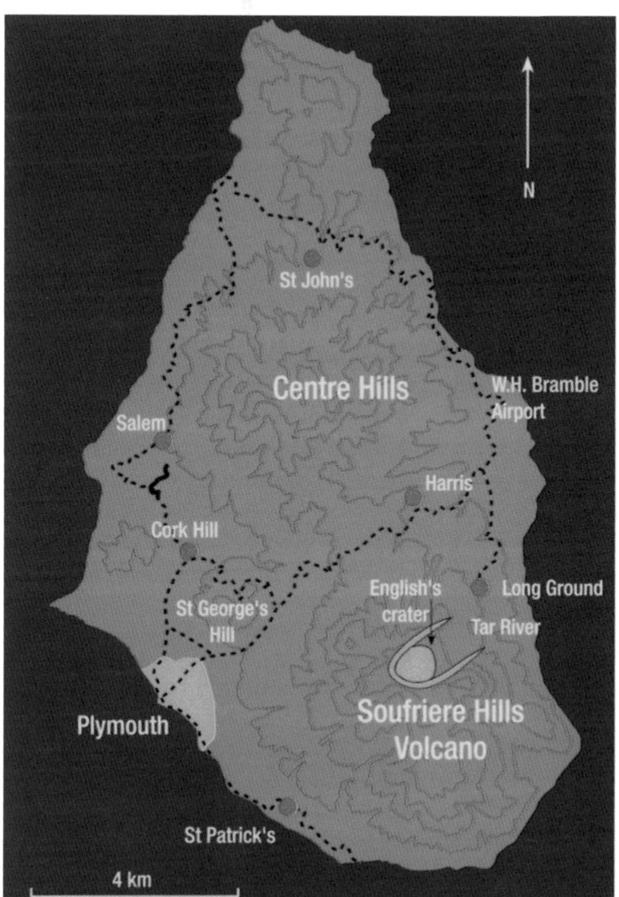

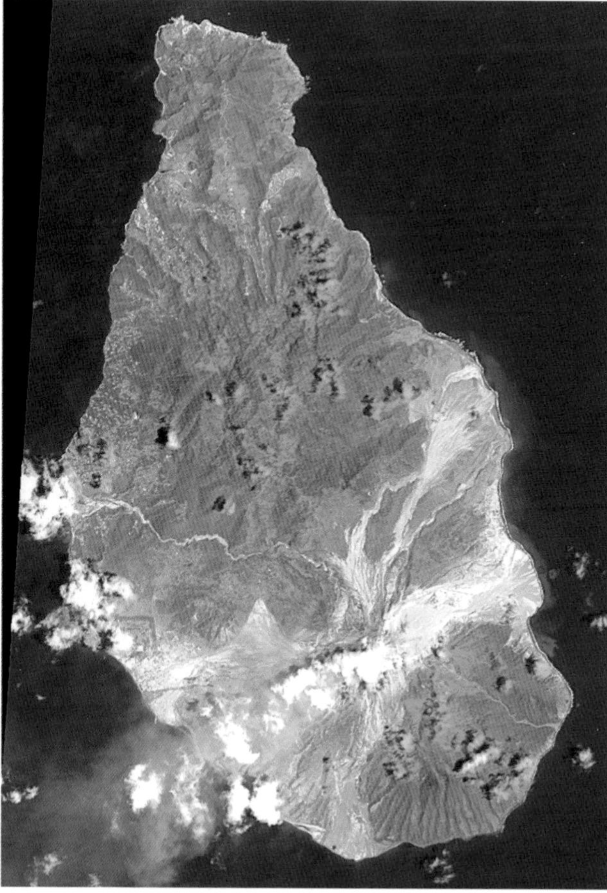

(left) Map of Montserrat, showing the locations of main population centers prior to the eruption. The old capital, Plymouth, was destroyed by the volcano, as was the airport. The dashed line represents the main road around the island. The island's highest point (just west of English's Crater) is at 915 m (3000 feet). (Modified from Young, 1998.) (right) Image of Montserrat taken by ASTER (Advanced Spaceborne Thermal Emission and Reflection Radiometer) on NASA's *Terra* satellite, showing pyroclastic flow deposits that covered the airport and Plymouth, and flowed down the Tar River Valley. Note the fans where various pyroclastic flows entered the sea. The image was taken on April 13, 2002 and shows simulated natural colors. (Image courtesy of Mike Abrams, NASA/JPL.)

Soufriere Hills during an eruption in August 1996, with a helicopter in the foreground. (Photograph courtesy of Christopher Kilburn.)

Valley. The crater was probably formed when a large sector of the mountain collapsed, probably some thousands of years ago. Soufriere Hills is about 26,000 years old, the youngest of the six volcanic centers in Montserrat. Radiocarbon dates show that many of the deposits are older than 16,000 years. There is no record of volcanic activity between 1632, when Montserrat was settled, and 1995. Ironically, radiocarbon dates show that pyroclastic flows were erupted in the mid seventeenth century. The settlers probably arrived just after the last eruption of Soufriere Hills ended. For the next three centuries, Soufriere's only activity consisted of fumaroles and hot springs. There were several signs that it was an active volcano, such as changes to the level of solfataric activity. Small earthquakes were reported on three occasions before the 1995 eruption started: in 1897–8, 1933–7, and 1966–7. Geologists studying the 1966–7 crisis took Soufriere's warning seriously, as Caribbean volcanoes typically have long periods of repose followed by violent eruptions. Because no eruption took place on any of these three crises, islanders were not particularly alarmed when seismic tremors begun in 1992. They were not to know that this time the volcano would get serious.

The 1995–present eruption

The eruption began on July 18, 1995 with an explosion within English's Crater. It was a surprise but not an unexpected one, as it came after 3 years of earthquake activity. During the first 4 months there were only steam (phreatic) explosions, caused by rapid heating of groundwater by the rising magma. No new magma reached the surface during these first few months, but it was clear that molten rock was moving up the conduit. Earthquakes continued to rock the island and, by mid November, the pasty andesitic magma finally reached the surface, piling around the vent to form a new dome. By this time, scientists from several countries had arrived to study the eruption and help local authorities in hazard management.

As more magma came up, the dome grew to hundreds of meters high and its steep, rubbly flanks eventually became unstable. Scientists monitoring Soufriere Hills were well aware that lava domes can suddenly disintegrate and avalanche down, forming deadly pyroclastic flows. In March, 1996, the dome had become large enough that parts of it begun to collapse. On April 3, 1996, the first substantial pyroclastic flow from Soufriere Hills rushed down the Tar River Valley. There were no deaths, as a large area of the southern part of the island, including its capital city Plymouth, had already been evacuated.

The dome continued to grow and in May – Mother's Day Sunday – it produced a major pyroclastic flow that came down the Tar River Valley and entered the sea, creating a delta of new land beyond the coastline. The flow happened just when scientists from the Volcano Observatory were overflying the volcano on a helicopter. Using a video camera, they were able to film the event, including a rare sequence of the fast-moving part of the pyroclastic flow – the surge – fanning out from the shore and flowing rapidly over the surface of the ocean. I was lucky enough to be able to examine the deposits from this flow only 2 weeks later.

Pyroclastic flow entering the sea at the mouth of the Tar River Valley, September 1996. (Photograph courtesy of Simon Young.)

Soufriere Hills went through a brief quiet period and did not send out another major pyroclastic flow until July, when activity begun to escalate. More pyroclastic flows came down in August and early September, dimming islanders' hopes that life might soon return to normal. September 17 saw the first explosive eruption, a frightening event that ejected a column of ash 14 km (9 miles) high. Rocks and ash showered the landscape; some of the bombs measuring 1 m (3 feet) across landed up to 2 km (1¼ miles) away. Volcanologists knew that magma was moving to the surface more rapidly than before. They concluded that the explosion was triggered by a large part of the dome, maybe as much as one-third, collapsing and avalanching down in the previous 12 hours. When the material was removed, it allowed the gas-rich magma inside of the volcano to decompress and rise rapidly to the surface. Sadly for Montserrat, the volcano had plenty more magma deep down. Barely 2 weeks after the big explosion, a dome was seen growing inside the crater, from the scar left by the big explosion.

Over the next few months, Soufriere Hills settled into a pattern of dome growth punctuated with pyroclastic flows. Eventually the dome became so large that it totally filled English's Crater. This was bad news as the 100 m (300 feet) high crater walls could no longer protect the southwestern, western, and northern flanks of the volcano from pyroclastic flows. The level of activity increased in late March of 1997 and the southern part of the dome collapsed, producing major pyroclastic flows to the southwest for the first time. In mid May, pyroclastic flows overtopped the northern crater wall. These were preludes to the worst event, which soon came: on June 25, 1997, 8 million m^3 (280 million cubic feet) of the dome avalanched down the northern flanks in less than 20 minutes, killing at least 19 people, all of whom had entered the Exclusion Zone against official advice. The pyroclastic flow destroyed some 200 houses in seven villages, devastated farmland, and nearly reached the airport, about 5.5 km (3½ miles) away.

The tragic event changed the topography of the dome, channeling new pyroclastic flows toward the west, where Plymouth lay waiting. In late July 1997, as Montserrat still mourned its dead, large pyroclastic flows rushed down the western valleys and engulfed the abandoned city. This was another major blow for the people of Montserrat. The capital that they knew and loved will probably never be rebuilt on the same precarious spot.

The behavior of the volcano changed slightly in early August. A major dome collapse was followed by a week in which explosive eruptions of Vulcanian type happened at fairly regular, 12-hour intervals. The explosions introduced a type of pyroclastic flow somewhat different from those formed by dome collapse. Pyroclastic flows formed by explosions tend to be less constrained by topography, because the explosions can eject materials in all directions around the volcano. Luckily, these flows turned out to be more predictable, because scientists monitoring the events detected a pattern of pressure build-up before the explosions happened. There was no further loss of life during this period, even though a pyroclastic larger than any before rushed down the flanks on September 21, reaching and destroying the airport terminal. This explosion tapped magma levels deeper than any of the previous events had – probably as deep as 4 km (2½ miles).

Montserrat was now only accessible by helicopter or boat. Life on the island continued to become increasingly difficult, while the pattern of activity showed itself to be slowly on the rise. By January 1998, about two-thirds of Montserrat's population had left and fewer than 4,000 people remained. Although these residents were considered to be safe in the northern part of the island, their life was not pleasant. Resources became scarcer and the island's main source of income – tourism – dried up. Refugees from the affected parts of the island crowded into tents and churches. Unemployment reached 60% and Montserrat became the most aid-dependent economy in the world. Scientists at the Volcano Observatory started to become concerned about the remaining residents, who tended to believe that by now they knew the volcano well. Warnings issued from the Volcano Observatory were not always taken seriously, unless the volcano itself gave clear signs of increased activity.

The dome continued to grow during late 1997, as Soufriere Hills warmed up for a special Christmas show. An intense earthquake swarm rocked the island on Christmas Day and, on December 26, the largest event of the whole eruption occurred. The dome, which had been growing over unstable rocks and hot springs, could no longer sustain itself: a massive landslide of volcanic rocks swept down the White River Valley, nearly reaching the sea. This landslide undermined the dome, causing it to collapse. Some 60 million m^3 (14 cubic miles) of dome failed, triggering a volcanic blast and violent pyroclastic surge that devastated 10 km^2 (4 square miles) of the southwestern part of the island in barely 15 minutes.

More dome growth happened after the collapse, but stopped quite suddenly in early March, 1998. However, the dome was still hot and unstable, and continued to collapse, producing new pyroclastic flows. The volcano also continued to vent ash and gases that escaped from the magma deep down. The high rainfall during the wet season of 1998 caused more problems, triggering mudflows, particularly when Hurricane George battered the island in September 1998. Flash floods and mudflows

Helicopter view of the dome growing inside Soufriere Hills, June 1996. Small pyroclastic flows were being emplaced during this time (note dark scars on dome where flows have come down). (Photograph by the author.)

New beach created by a pyroclastic flow of May 1996. (Photograph by the author.)

transported large amounts of debris down to the lower flanks, adding to the devastation.

Contrary to some optimistic reports, dome growth had only paused, not stopped. The crisis continued through 1999 and 2000, with rockfalls, pyroclastic flows, mudflows, ash venting, and seismic tremors. On March 20, 2000, part of the dome collapsed, triggering pyroclastic flows that ran down the Tar River Valley to the sea. A sudden explosion shot red lava hundreds of meters above the summit. The ash cloud reached at least 9 km (30,000 feet). Heavy rain may have triggered the dome collapse and the explosion, but there was no doubt that new lava was still being injected into the dome. By the end of March, a new dome could be seen growing in the scar left by the March 20 explosion.

By early 2004, the Montserrat crisis was still not over. The remaining residents still hung on to the hope that the volcano will soon be at rest. When that finally happens, the long task of rebuilding can begin. It will be fascinating to watch how nature and the tenacious islanders who refused to leave will transform the devastated landscape into a semblance of the paradise they lost.

A personal view: Visiting an island in crisis

I visited Montserrat in late May 1996, arriving shortly after the Mother's Day pyroclastic flows that reached the sea. I was lucky to be able to join scientists from the volcano observatory on a helicopter ride to see the lava dome and to land on the new beach formed by the flows. Given that lavas are my specialty and most of my field work has been done on effusive volcanoes, being able to see fresh pyroclastic flows was a rare opportunity for me. It was incredibly thrilling to be able to examine their deposits while they were still warm to the touch, and to watch small collapses from the lava dome making new flows which, thankfully, did not reach far.

I admit to feeling somewhat apprehensive when my colleagues and I arrived at the new pyroclastic beach, and even more so when our helicopter left. Our plan

Geologists on the new beach created by the May 1996 pyroclastic flow. (Photograph by the author.)

was to stay there for a couple of hours, studying various features of the flow deposits. Although I was soon engrossed by the fine details of the flows, I did not lose sight of the fact that we were in the preferred path for new pyroclastic flows. There had been no signs of impending dome collapse in the previous days and seismic events were at a relatively quiet level – but an awakened volcano is potentially a deadly one. I thought of the worst-case scenario and scanned the high cliffs for a place to escape to. It didn't take me long to realize that there would have been no escape if a pyroclastic flow reached our beach. I decided to turn my attention back to the deposits, which were quite fascinating. The surface of the flow had a dune-like structure, but on a very small scale: the dunes were only a few centimeters (under 1 inch) high, and were spaced at about 1 m (3 feet) intervals. This dune structure is typical of pyroclastic surges that reach topographically flat areas. Looking at the surge in cross-section, I could distinguish two units, which most likely represented pulses in the surge rather than separate events. The deposits were still warm at the surface and when I took a step that made my leg sink up to the calves in the soft ash, I realized that the flow's interior was unpleasantly hot. Luckily the burn was minor, but I was rather more careful after that.

We left the new beach without mishap and, later in the day drove up the Tar River Valley to see the pyroclastic flow deposits closer to the lava dome. This was an even more hazardous area than the beach, as the valley channeled pyroclastic flows, not all of which reached as far as the ocean. The small collapses from the dome, which happened frequently, had been very interesting to see from the helicopter, but I found them rather unnerving from the ground. I had no previous experience observing pyroclastic flows and I could not judge if the collapses would soon become larger and turn into deadly flows, following their predecessors down the valley. My more experienced colleagues reassured me that the collapses happening then were too small to be dangerous. However, they were very alert, keeping in radio contact with the observatory. We were all ready to get out fast: we had left our vehicles on the road pointing downhill, unlocked, with the keys in the ignition. I still consider the Tar River Valley to be the most dangerous place I have ever been in.

I learnt much about the volcano and pyroclastic flows during my visit but, with hindsight, I think the greatest lesson was to be in a small country that was trying desperately to cope with a major volcanic crisis. Montserrat taught me much about the human impact of eruptions. Not only were the people afraid, but they were extremely frustrated. Some eruptions are devastating but they are over quickly. Soufriere Hills was slowly squeezing the life out of Montserrat.

What I didn't know then was that the island was still in fairly good shape at that time. My short flight from Antigua landed at the W. H. Bramble airport, which later became one of the eruption's many casualties. I was able to rent a car there and drive through the evacuated city of Plymouth, which already looked like a ghost town but was still standing. I wondered then how much more destruction the eruption would cause and found it hard to imagine that it would get much worse. Life in the island was already difficult: most hotels and businesses had closed, residents from the southern part of the island were housed in refugee camps, and the main road between the airport and Plymouth was sealed off by police. The most obvious sign of the eruption was the ash – it was everywhere, making it impossible to keep anything clean. Montserrat was in a very sorry state indeed and seemed totally removed from the paradise-like descriptions that I'd read about. Yet the devastation was to become worse in the following years.

As I write, Montserrat continues to fight for its survival. The resilient few who remain wonder if the eruption can go on much longer – and how much longer? People have an uneasy relationship with the mountain that became the one thing the whole island revolves around. There is still fear and frustration, but also some appreciation for each time they can see Soufriere Hills erupting at night, perhaps for the last time. The paradox of locals having fun watching the eruption is well described in the song "Glowing" by the local group Zunky 'n' Dem, who give an insightful social commentary of the eruption in their *Seismic Glow* album. In the words of Randall "Zunky" Greenaway:

Sat in the moonlight
And we watched the mountain glow
In the moonlight
From Harry's Hill
Incandescent rocks broke away
Creating firework displays
It was really quite a thrill
Something that could be dangerous
So beautiful
It didn't seem right to make a fuss
Our joys were full
I told my mind – snap a picture,
I want to hold this scene forever.

Visiting during repose

Before the eruption started visitors to Montserrat used to comment on the poetic billboard erected by the island's Water Authority: "The sky is held up by trees. If the forest goes, the roof of the world collapses and Nature and Man will perish together." Montserrat was

The Tar River Valley during June 1996. The valley floor is blanketed by recent pyroclastic flows. (Photograph by the author.)

a lovely, quiet island where big hotel chains didn't venture and the few tourists who dared to drive soon found out that road signs weren't considered necessary and, to make driving more interesting, often the roads disappeared into overgrown paths. Montserrat's popularity as a tourist destination was on the rise, but mostly with independent travelers wanting to get away from the beach crowds. The Tourist Board cleverly promoted eco-tourism, advertising hiking and mountain biking, and claimed the island was "the way the Caribbean used to be." Ironically, the hike to Soufriere Hills was considered one of the major attractions.

Few guidebooks to the Caribbean mention Montserrat any more, as most do not consider that anyone will want to visit the island any time soon. The Tourist Board advises writers to describe Montserrat as it used to be and, hopefully, how it will be again one day. There is no doubt that once the eruption is over, Montserrat will need all the tourism it can get. For travelers seeking to see the deposits and effects of a recent volcanic eruption, Montserrat will be one of the world's prime destinations. If islanders are clever, they will turn Montserrat into the type of attraction that Mt. St. Helens became in the USA – the volcanic eruption that devastated tourism in St. Helens eventually turned the mountain into a more popular tourist destination than it had ever been. While Montserrat may not have the resources to build large visitor centers and IMAX theaters, it has the potential for unique attractions, such as beaches formed by recent pyroclastic flows. I can imagine diving into the blue waters of the Caribbean to see the flow deposits under water, climbing the volcano to look into the crater, and hiking along the Tar River Valley which is now lined by numerous pyroclastic flows. The ruins of Plymouth may become a tourist attraction and, perhaps in a century of so, they will become as fascinating to visitors as the ruins of St. Pierre in nearby Martinique are to us today.

The tourist center of the island will most likely be in the northern part, which has largely escaped devastation. Prices will probably initially be cheap and the island may for a while be the best bargain in the Caribbean. Once the eruption stops, it won't be long until the air is once again clean, forests start to grow and friendly guest houses open their doors to visitors.

Even after the eruption is over it will be necessary to take some precautions. Be aware that explosions could still happen due to water seeping down into the cooling dome. Mudflows will be a major hazard because of the enormous quantity of ash deposited by the eruption on the volcano's flanks. Ash will continue to be blown about and will remain a health hazard for some time. The volcano and dome may be unstable and landslides could occur. Even the lower flanks, near the sea, could be unstable and break away. Do not enter any area marked as no-go zones and check the information provided by the Montserrat Volcano Observatory or tourist authority.

Despite possible difficulties, it will be worth making the trip. Montserrat will be a fascinating place in which to learn not only about explosive volcanism but also about a country's struggle for survival.

Visiting during activity

Going to Montserrat while Soufriere Hills is still active is certainly possible and, from a volcanological point of view, rather desirable. However, it is extremely important to follow certain rules. Before making any plans, contact the Government of Montserrat Information Service (a web page is available) to find out the current state of the volcano and conditions in the island. Visitors on professional assignments, including geologists and journalists, should contact the Montserrat Volcano Observatory for guidance and permits to go into restricted areas.

If you are visiting as a tourist, respect the local restrictions and stay away from no-go zones. It is still possible to see the effects of the eruption without venturing into dangerous zones. For a start, the effects are impossible to miss, from the fine ash that constantly blows in the wind to the people who will tell you how the volcano has changed the island and everyone's lives. It is hard to predict what conditions will be like as long as the eruption continues, except that the southern part of the island will stay off-limits to most visitors. However, Soufriere Hills is an impressive volcano even from a distance and there are creative ways to see the products of the activity. Ask about renting a boat to circle the island and see the deltas created by pyroclastic flows. A more expensive alternative is to rent a helicopter or a light plane from nearby Antigua. Soufriere Hills and the Tar River Valley are spectacular seen from the air and, as long as no explosions are taking place, pilots may be willing to circle the crater.

How long will the eruption last? It is very hard to predict how long any eruption will last, but experts' opinions are that the total duration will be less than 15 years. This would mean that activity could last until 2010, though more likely it will end sooner. Those

Soufrière, Guadeloupe. The vehicle is parked on the road to the summit. (Photograph courtesy of Simon Young.)

wanting to visit the island should check the volcano's current status in advance and, once on the island, tune in to the local radio station ZJB for up-to-date information on the eruption.

Other local attractions

Guadeloupe

This archipelago is dominated by its two main islands, Grande-Terre and Basse-Terre, which together form a butterfly-like shape. Grande-Terre ("big land") forms the eastern wing and it is, in fact, a smaller and flatter island than Basse-Terre ("flat land"), the western wing, which is dominated by rugged hills and the Soufrière volcano. The names of the two islands seem rather strange unless you know that they refer to the winds that blow over them: they are stronger ("grande") over the gentle topography of the eastern island, but are stopped by the mountains to the west, ending up "basse."

La Soufrière is located at the southern tip of Basse-Terre and rises to an elevation of 1,467 m (4,813 feet). It is important to inquire about the state of the volcano's activity before hiking to the summit. La Soufrière's last activity, a phreatic eruption in 1976, caused great alarm and even injured some scientists. Even in its quiet state, this volcano poses some danger because of its many fumaroles. The fumes are extremely acidic, containing hydrochloric and sulfuric acids. Stay upwind from the fumaroles and consider carrying goggles and a gas mask, or even just a scarf to soak in water and put over your nose and mouth. Be sure to also bring appropriate clothing, as the summit is almost always shrouded by clouds and fog, and often rainy.

Hiking up La Soufrière is relatively easy, but the trail is steep. To get to the trailhead, drive to the town of St. Claude and follow signs to La Soufrière. The volcano is within a national park. The Maison du Volcan, located about 2 km (1¼ miles) after entering the park, has a small museum that includes displays about the 1976 eruption. Continue driving up to La Savanne à Mulet, a parking lot at altitude 1,142 m (3,770 feet). On the rare days when the weather is clear, there is a good view of the volcano from here. The hiking trail starts by the parking lot along a gravel bed and continues steeply uphill. Most people can hike up to the summit in about 1.5 hours. The trail goes up the volcano's western flank, which is covered by mosses and low shrubs, and crosses a deep gully marking a fissure that opened up during a phreatic explosion (Éboulement Faujas). At the top of the volcano there is a caldera plateau with knobs and deep fissures, and puffing fumaroles at the southern fissure zone (Cratère du Sud). The trail continues on the summit plateau and is well marked by wooden posts. It is important to remain on the trail, as there are steep fissures here that can be difficult to spot, particularly if you become enveloped by the steam from fumaroles. Use great caution in this area and remember that a repeat of the 1976 phreatic explosion could happen. Be aware of your location relative to the trail at all times and have an escape route planned.

References

Smith, A.L. and M.J. Roobol (1990) *Mont Pelée, Martinique: A Study of an Active Island-Arc Volcano*. Geological Society of America.

Vincent, P.M., J. Bourdier, and G. Boudon (1989) The primitive volcano of Mont Pelée: its construction and partial destruction by flank collapse. *Journal of Volcanology and Geothermal Research* **38**, 1–15.

Westercamp, D. and H. Tazieff (1980) *Martinique, Guadeloupe, Saint-Martin, La Désirade: Guides Géologiques Régionaux*. Masson.

Westercamp, D. and J.F. Tomblin (1979) Martinique. *Bulletin de Bureau de Recherches Géologiques et Minières* **4**.

Young, S. (1998) Monitoring on Montserrat: the course of an eruption. *Astronomy & Geophysics* **39**, (2), 2.18–2.21.

Publisher's Note

The information contained in Appendices I and II was correct at the time of going to press. Its inclusion here does not imply any endorsement by the publisher of the services listed, nor any guarantee of the ongoing currency of the addresses and websites cited.

Appendix I Useful information for preparing a volcano trip

1. Sources of information on volcanic activity

Volcano World is a website run by the University of North Dakota; its intent is to bring the excitement of volcanology to the general public. You can select a region of volcano and find a good summary of activity, pictures, and other information. The site includes links to many other volcano-related web sites and contains a list of frequently asked questions about volcanoes. URL: http://volcano.und.nodak.edu/

Smithsonian Global Volcanism Program's website includes copies of recent Global Volcanism Network, which contains reports on recent volcanic activity, links to other volcano websites, and a list of frequently asked questions. The URL is http://www.volcano.si.edu/gvp. The Smithsonian Institution's Bulletin of the Global Volcanism Network is also available in print by subscription from the American Geophysical Union, 2000 Florida Avenue NW, Washington DC 20009, USA.

Volcano Listserv: This is the professional volcanologists' email network run by Arizona State University. The network is an invaluable source of information and contacts. An administrator screens messages before sending them to subscribers. People interested in volcanoes may be included in the distribution list, at the discretion of the administrator. Send a message to volcano@asu.edu.

Michigan Technological University's website includes information on some current eruptions and detailed information on some volcanoes including Santa Maria and Fuego in Guatemala, and Pinatubo in the Philippines. The URL is http://www.geo.mtu.edu/volcanoes

The International Association of Volcanology and Chemistry of the Earth's Interior (IAVCEI) has a website that caters mainly to professionals, but has items of interest to all volcano enthusiasts. The URL is http://www.iavcei.org. For membership information, write to IAVCEI Secretariat, PO Box 185, Campbell, ACT 2612, Australia.

Dartmouth's Electronic Volcano website is another great source of information on active volcanoes worldwide, and includes maps, photographs, dissertation texts and a few elusive documents. The Electronic Volcano will also guide you to resources in libraries or resources on other information servers. URL: http://www.dartmouth.edu/~volcanoes

The Volcano System Center at the University of Washington has a website with volcano news and a very useful list of other volcano-related websites. The URL is http://www.vsc.washington.edu

The World Organization of Volcano Observatories (WOVO) has a website listing contact information for volcano observatories around the world. URL: http://volcano.ipgp.jussieu.fr:8080/wovo/intro.html

L.A.V.E. (L'Association Volcanologique Européenne). This amateur volcanologists' group is well organized, enthusiastic and offers a great website (in French) with information on several volcanoes their members travel to. Great photos. Mailing address is L.A.V.E., 7, rue de Guadeloupe, 75018 Paris, France. URL: http://membres.lycos.fr/lave/intro.html

2. Sources of information and useful addresses for specific volcanoes and countries

USA (including Hawaii)

US Geological Survey. This organization has numerous offices and offers a variety of brochures, photographs from archives, maps, and services. The US Geological Survey is the parent organization for the various volcano observatories in the US listed below. URL: http://www.usgs.gov. The public affairs office's mailing address is 119 National Center, Reston, VA 20192, USA.

US Geological Survey Volcano Hazards Program's website gives information on US active volcanoes and links to volcano observatories. The site posts a weekly newsletter from US Geological Survey scientists giving information on volcanic activity (http://volcanoes.usgs.gov). Mailing address: US Geological Survey, Information Services, P.O. Box 25286, Denver, CO 80225, USA.

Hawaiian Volcano Observatory: their website includes up to date information on the activity of Hawaiian volcanoes. If you are interested in volunteering at the Observatory, the website provided details on how to apply. (http://hvo.wr.usgs.gov). Mailing address is Box 51, Hawaii National Park, HI 96718, USA.

Cascades Volcano Observatory: their website includes descriptions of the studies carried out by the observatory, numerous fine images of the Cascades volcanoes, and a discussion on the Volcano Disaster Assistance Program. Their webserver index lists volcanoes all over the world, providing information and links. The URL is http://vulcan.wr.usgs.gov/home/html. Mailing address is 5400 MacArthur Blvd, Vancouver, WA 98661, USA.

Alaska Volcano Observatory is a joint program of the US Geological Survey, the University of Alaska Fairbanks, and the State of Alaska Division of Geological and Geophysical Surveys. The observatory monitors and studies Alaska's hazardous volcanoes. Their website includes maps, photos, video clips, and general information about volcanoes in Alaska and reports from the Kamchatkan Volcanic Eruption Response Team (KVERT) (http://www.avo.alaska.edu/). Mailing address: 4200 University Drive, Anchorage, AK 99508, USA.

University of Washington Volcano Systems Center website: includes reports on Mt. Rainier and other research activities conducted by the university. URL: http://www.vsc.washington.edu/

The Hawaii Center for Volcanology's Kilauea homepage includes up-to-date information on the volcano's activity, as well as maps and historical information. Check out the link to a "virtual field trip" over Hawaii homepage. The URL is http://www.soest.hawaii.edu/GG/HCV/kilauea.html

National Park Service: an invaluable source of information on volcanoes in the US National Parks, including maps, visiting information, and photographs. URL: http://www.nps.gov/parks.html. You can go to each Park's individual website from this main site. To write or call Parks, see below:

Hawaii Volcanoes National Park: PO Box 50, HI 96718, USA. Phone: (808) 967-7184.

Haleakala National Park: P.O. Box 537, Makawao, Maui, HI 96768, USA. Phone: (808) 572-9306.

Lassen Volcanic National Park, P.O. Box 100, Mineral, CA 96063, USA. Phone: (916) 595-4444.

Sunset Crater Volcano National Monument: Wupatki, Sunset Crater Volcano, and and Walnut Canyon National Monuments, 2717 N. Steves Blvd, Suite 3, Flagstaff, AZ 86004, USA. Phone: (520) 526–1157.

Mount St. Helens National Volcanic Monument, 42218 NE Yale Bridge Road, Amboy, WA 98601, USA. Phone for 24-hour information: (360) 247-3900. Climbing Hotline: (360) 247-3961.

Crater Lake National Park, P.O. Box 7, Crater Lake, OR 97604, USA. Phone: (503) 594-2211.

Mount Rainier National Park, Ashford, WA 98304, USA. Phone: (206) 569-2211.

Yellowstone National Park, P.O. Box 168, WY 82190, USA. Phone: (307) 344–7381.

Katmai National Park (Valley of Ten Thousand Smokes), King Salmon, AK 99613, USA. Phone: (907) 246-3305.

For lodging and camping within the parks, go to the respective websites for information, or use the contact information below:

Yellowstone: TW Recreational Services, Yellowstone National Park, WY 82190, USA. For cabins and lodges including the famous Old Faithful Snow Lodge. Phone: (307) 344-7311. For camping at Yellowsone, call MISTIX, (800) 365-2267 from within the US.

Mount Rainier: Guest Services, 55106 Kernahan Road East, P.O. Box 108, Ashford, WA 98304, USA. For campground reservations, and lodging at The National Park Inn, and the Paradise Inn. Phone: (206) 569-2275.

Mount St. Helens: There is no lodging in the National Monument. For camping and general traveling information, contact the Park Headquarters at (360) 750-3900, or the Randle Ranger Station at (360) 497-1100, or the Gifford Pinchot National Forest Headquarters at (360) 750-5009.

Lassen: For camping, contact the National Park Offices at (916) 595-4444. For reservations at the Drakesbad Guest Ranch, contact California Guest Services, Inc., Adobe Plaza, 2150 Main St., Suite 7, Red Bluff, CA 96080, USA. Phone: (916) 529-1512.

Hawaii (Big Island). Volcano House has a lodge on the rim of Kilauea crater and cabins in the Namakani Paio campground. For reservations, write to Volcano house, Hawaii Volcanoes National Park, HI 96718, USA, or call (808) 967-7321.For other campsites, contact the National Park Offices at (808) 967-7184. For a list of bed and breakfast places, contact Hawaii Best Bed and Breakfasts at (808) 885-4450.

Hawaii (Haleakala): There is no lodging in the park; for camping reservations contact the National Park offices at (808) 572-9306.

Crater Lake: For reservations at the Crater Lake Lodge, write to P.O. Box 128, Crater Lake, OR 97604, USA, or call (503) 594-2255. For Manzama Campground Cabins, write to the same address or call (503) 594-2511. For camping and general park information call (503) 594-2211.

Sunset Crater: There is no lodging inside the park, for campground information contact the Coconino National Forest, Peaks Ranger Station, 5010 N. Highway 89, Flagstaff, AZ 86004, USA. Phone: (928) 526-0866.

Mount Shasta: The Mt. Shasta Wilderness area is under the control of the US Forest Service. Information is available in the internet (http://www.wilderness.net/nwps, search for "Shasta"). Permits are required for entering the Mt. Shasta Wilderness area. For permits, camping, climbing, and general information, contact the Mt. Shasta Ranger District at 204 W. Alma Street, Mt. Shasta, CA 96067, USA, or call (916) 926-4511. You may also contact the McCloud Ranger District office, PO Box 1620, McCloud, CA 96057, USA, or call (916) 964-2184. Rangers can put you in touch with the outfitter/guide business under permit to guide climbing trips within the Wilderness area. Lodging is available in Mt. Shasta City and towns nearby.

Italy

Volcanologists Roberto Carniel and Juerg Alean created "Stromboli-on-Line" (http://educeth.ch/stromboli).

Although originally about Strómboli, this page has expanded to include news, information, and photos from volcanoes all over the world. Includes videos, expedition news, volcano livecams, teaching materials, and a whole lot more. In English, Italian, and German.

Volcanologist Boris Behncke runs "Italy's Volcanoes: The Cradle of Volcanology" webpage, dedicated to the volcanoes of Italy. Great source of news, photos, and maps of Italian volcanoes. URL: http://boris.vulcanoetna.com or http://stromboli.net/boris

The Istituto Internazionale di Vulcanologia in Catania maintains a homepage about Etna and the Aeolian Islands volcanoes, in Italian (http://www.ct.ingv.it/). Mailing address: Piazza Roma 2, 95123 Catania, Italy.

The Vesuvius Volcano Observatory maintains a website about Vesuvius and Campi Flegrei, in Italian (http://www.ov.ingv.it/italiano/home.htm). A version in English is under construction. For information about the Observatory itself, go to http://www.vesuvioinrete.it/e_osservatorio.htm. Mailing address for the Observatory is Via Diocleziano 328, 80124 Napoli, Italy.

Volcanologist Roberto Scandone maintains the "Explore Italian Volcanoes" website, which includes a discussion board, a glossary of geological terms, and historical information on the eruptions of Vesuvius. URL: http://193.204.162.114/indice.shtml. In English and Italian.

Club Alpino Italiano: This association publishes maps and guidebooks and is Italy's premier organization for mountaineering. They have a webpage in Italian (http://213.140.0.212:8080/index.jsp).

S.E.L.C.A., the Società Elaborazioni Cartografiche, publishes maps such as the Mt. Etna Carta Naturalistica e Turistica. Address: Via R. Giuliani n. 153, Firenze, phone: (055) 4379898.

Greece and Santorini

The Thera Foundation, 105–109 Bishopsgate, London EC2M 3UQ, UK. In Greece: 17–19 Akti Miaouli 185, 35 Piraeus, Greece. The Foundation promotes the study of ancient Thera and the Santorini eruption and has published several books (see Bibliography) and organized international conferences.

Institute for the Study and Monitoring of the Santorini Volcano. As the name says, this scientific institute monitors the volcano's activity. They also publish materials, including a guidebook to the volcano (see Bibliography). URL: http://www.santonet.gr/volcano

Santorini Volcano Live Camera. This website has cameras located on the terrace of Hotel Heliotopos, showing images every 3 minutes. Also plenty of information on tourism in the islands, including weather. URL: http://www.santorini.net

Santorini Decade Volcano's website has photos and information about the geology of the island. URL: http://www.decadevolcano.net

For more information on the internet, search for Santorini in Volcano World or one of the websites listed in section 1 above, or select Santorini from the Stromboli-on-line page.

Iceland

The Nordic Volcanological Institute has a website with descriptions of the volcanoes, photographs, and maps (http://www.norvol.hi.is). Mailing address is Nordic Volcanological Institute, University of Iceland, Grensásvegur 50, IS-108, Reykjavík, Iceland.

Icelandic Institute of Natural History's website has information on the geology and natural history of Iceland, including on rare birds. The URL is http://www.ni.is/english/. Mailing address is Hlemmur 3–5, Box 5320, IS-125, Reykjavík, Iceland.

Iceland news: For a variety of information on Iceland, including weather and lodging, try these websites: http://www.icelandnews.com/ and http://www.icelandreview.com/

Landmælingar Íslands, the Icelandic Geodetic Survey, sells maps and aerial photographs. Address: Stilholti 16–18, IS-300, Akranes, Iceland. You can also buy from their website (http://www.lmi.is).

Icelandair (offices in the USA and several European countries) provides a variety of useful information for visitors to Iceland. It is worth browsing their website (http://www.icelandair.net) and writing for information, even if you are not flying with them (see Appendix II for their tours to Iceland).

South Iceland Institute of Natural History has a website with information on volcanic activity in Iceland, as well as general information on the Westmann Islands. The URL is http://www.nattsud.is/nshomeuk.htm

Costa Rica

Volcanological and Seismological Observatory of Costa Rica, Universidad Nacional, Ovsicori-Una, Apartado 86-3000, Heredia, Costa Rica. The Observatory monitors the activity of volcanoes in Costa Rica.
URL: http://www.eco-web.com/register/05656.html

Instituto Costarricense de Electricidad (ICE). The Institute assesses natural hazards in Costa Rica and carries out research on volcanoes. Mailing address: Apartado 10032-1000, San José, Costa Rica. URL: http://www.ice.co.cr

Arenal Observatory Lodge. This wonderful lodge offers spectacular views of the volcano, a museum, and nature trails. For reservations, call (506) 257-9489 in Costa Rica or write to Box 321-1007, Paseo Colón, San José, Costa Rica. URL: http://www.arenal-observatory.co.cr

Poás Volcano Lodge: Ideal for those visiting Poás, this lodge is located between the Poás and Barva volcanoes, atop a ridge. Next to nature trails and neighboring a dairy farm. Address: P.O. Box 5723-1000, San José, Costa Rica. Phone in Costa Rica: (506) 482-2194.

Costa Rica Map. This website, in English and Spanish, gives lots of information about traveling in Costa Rica, including National Parks, the flora and fauna, lodging, hiking, climbing, and sports. Very useful maps. URL: http://www.costaricamap.com

COCORI. This "Complete Costa Rica" website offers maps, photos, airline and lodging reservations, and numerous articles. URL: http://www.cocori.com

Hacienda Lodge Guachipelín. This lodge's website has lots of information on the nearby volcano Rincón de la Vieja. URL: http://guachipelin.com. Mailing address: P.O. Box 636-4050, Alajuela, Costa Rica. Phone: (506) 442-2818.

Volcan Turrialba Lodge. The "hotel with a volcano in its backyard" has a website with photos and information about Turrialba volcano. Mailing address: P.O. Box 1632-2050, San José, Costa Rica. Phone: (506) 273-4335. URL: http://www.volcanturrialbalodge.com

West Indies

Guadeloupe Volcano Observatory (Observatoire Volcanologique Guadeloupe), Le Houëlmont, 97113 Gourbeyre, Guadeloupe. The observatory monitors the activity of La Soufrière volcano. Their website, in French, has photos and information on the volcano. URL: http://volcano.ipgp.jussieu.fr:8080/guadeloupe/stationgua.html

Observatory of Volcanology at Mt. Pelée (Observatoire Volcanologique de la Montagne Pelée), Fonds St. Denis, 97250 St. Pierre, Martinique. The observatory monitors the activity of Mt. Pelée volcano. Their website, in French, has photos and information on the volcano. URL: http://volcano.ipgp.jussieu.fr:8080/martinique/stationmar.html

Montserrat Volcano Observatory monitors the activity of Soufriere Hills volcano. Their website has frequent updates on the current activity. Mailing address: Montserrat Volcano Observatory, Mongo Hill, Montserrat, West Indies. URL: http://www.geo.mtu.edu/volcanoes/west.indies/soufrière/govt/

Volcano Island is the website of David Lea and his family, who run the Gingerbread Hill Inn. David is an avid volcano watcher and has made video documentaries about the eruption. This site has useful information on traveling to Montserrat, and you can book lodging at the Inn. URL: http://www.volcano-island.com

Other volcanoes: internet resources

To find information on volcanoes anywhere, go to Volcano World or one of the other sites in the internet listed in section 1 – you will find links to websites for other volcanoes. Below are a few of my favorites.

Japan

Volcano Research Center. Their webpage gives information in English and Japanese on volcanoes in Japan and some other locations in Asia. URL: http://hakone.eri.u-tokyo.ac.jp/vrc/VRC.html

New Zealand

Volcanoes in New Zealand. This website by the Institute of Geological and Nuclear Sciences of New Zealand has volcano cams, current status of volcanoes, and a whole lot more. URL: http://www.gns.cri.nz/

Réunion

Observatoire Volcanologique du Piton de La Fournaise. The volcano observatory monitors Réunion's very active volcano. In French. URL: http://volcano.ipgp.jussieu.fr:8080/reunion/stationreu2.html

Equipment

Most of the recommended equipment is easy to find in camping and climbing stores. Hard hats can be purchased at home remodeling stores. Gas masks are harder to find; look for suppliers of laboratory safety equipment and purchase a lightweight type that will protect you against "acid gases" such as sulfur dioxide and hydrogen chloride. In the USA, try Lab Safety Supply (P.O. Box 1368, Janesville, WI 53547, phone: (800) 356-0783). It is often best to call a company and ask for advice rather than to order from the internet or a catalog.

Unless you are planning on collecting fresh lava samples, an asbestos glove should not be necessary (and it is possible to collect samples without gloves, though one can end up with minor burns). If you really want these gloves, ask a laboratory safety supplier or search in the internet.

Appendix II Tours to volcanoes

The organizations listed below are well established and regularly offer tours. There are lesser-known companies that offer trips to volcanoes and often advertise in magazines that cater to adventure and natural history, such as *National Geographic Traveler*, *Outside*, and *Natural History*. When shopping for a tour, beware of small-budget companies that are likely to cancel tours at the last minute if not enough people sign up. Ask what their track record is, whether cancellation is a possibility, and when you would know for sure. If your plans are flexible and you want to go on a true adventure, seek out volcano enthusiasts who will offer small expeditions to far-flung volcanoes almost at cost, and are very clear about the possibility of cancellation. They usually advertise through their own web sites and the volcanologists's email network VOLCANO listserv (see Appendix I). Another alternative, if you can stay in the same place a couple of months, is to volunteer at a National Park or volcano observatory – check their websites for opportunities or write them a letter explaining your skills and availability.

Geological Society of America. This is the major US geological society and their GeoVentures department regularly runs a variety of field trips of geological interest. Most of their destinations are in the US, but have included Iceland and other volcanic locations. Non-members may join trips depending on availability. Further information from: Geological Society of America Geo Ventures, P.O. Box 9140, Boulder, CO 80301, USA (http://www.geosociety.org/)

Geological Association of Canada. This is Canada's national society for the geosciences and they organize field trips, sometimes to volcanoes. They sell some of their guidebooks, which are great sources of information. Further information from: Geological Association of Canada, Department of Earth Sciences, Memorial University of Newfoundland, St. John's, Newfoundland A1B 3X5, Canada. (http://sparky2.esd.mun.ca/~gac/gac.html)

Lindblad expeditions. A specialist in the Galápagos Islands, this award-winning company offers trips led by naturalists who are expert on various aspects of the islands' natural history. A few times a year, Lindblad offers family trips to the Galápagos which are an excellent opportunity to introduce children to the wonders of these islands. The company's other destination of volcanic interest is Costa Rica, one of the itineraries includes Arenal volcano. 720 Fifth Ave., New York, NY 10019, USA or http://www.lindbladexpeditions.com

Smithsonian Institution. The Smithsonian Study Tours present educational travel programs that reflect the vision and interests of the Smithsonian Institution. Travel programs provide a combination of study, discovery, and adventure at a variety of levels, including family study tours. The tours are offered to members; consult the web site (http://www.smithsonian.org) to find out how to join and what tours are available. Past tours have included Costa Rica, the Galápagos Islands, New Zealand, and Yellowstone.

Icelandair. Iceland's national airline offers many natural history tours of the country that include its magnificent volcanoes. They even offer horseback riding tours. Icelandair has offices in the USA and in many European cities. Their website for tours is http://www.icelandair.is/eurovac

Exodus calls itself "the different holiday." This award-winning adventure tour operator, based in the UK, has been in business for over 25 years, offering tours all over the world. Volcanic destinations have included the Aeolian Islands, the Galápagos Islands, the Andes, and climbing Mt. Kilimanjaro. Head office: 9 Weir Road, London SW12 0LT, UK. Phone: +44 (0)20 8675 5550. URL: http://www.exodus.co.uk

iExplore. This established adventure travel operator is associated with *National Geographic* magazine. They offer some 3,000 different adventure trips. Volcanic destinations have included Costa Rica (Arenal), the Galápagos Islands, Hawaii, and Rwanda (Parc des Volcans where the "gorillas in the mist" reside). Phone: (312) 492-9443. URL: http://www.iexplore.com

Earthwatch organizes expeditions in which volunteers help scientists in the field. Several past expeditions have been to volcanic locations, for example, helping topographic and geologic surveys of volcanic deposits on Santorini. For information in the USA, call (617) 926-8200, in the UK call (0865) 311-600, or consult their website at http://www.earthwatch.org

Yellowstone Institute. This is an educational offshoot of the Yellowstone National Park. They offer short family educational vacations and opportunities to explore the Park with guides. Contact the Yellowstone Institute, P.O. Box 117, Yellowstone National Park, WY 82190, USA. Phone: (307) 344-2294. URL: http://www.yellowstoneparknet.com

Bibliography

General texts on volcanoes

Volcanoes of the World (2nd edn), edited by T. Simkin and L. Siebert. Geoscience Press in association with the Smithsonian Institution, 1994. A regional directory, gazetteer, and chronology of volcanism during the last 10,000 years.

Volcanoes: A Planetary Perspective, by P. Francis. Oxford University Press, 1993. The late Peter Francis was a lively writer as well as one of the world's most respected volcanologists. This book is my top choice for a general introductory text to volcanoes and the science of volcanology.

Volcanoes of Europe, by A. Scarth and J.-C. Tanguy. Oxford University Press, 2001. An invaluable general text for those who wish to learn more about European volcanoes active or not, including those in Iceland, the Azores, and the Canary Islands.

Volcanoes of the Solar System, by C. Frankel. Cambridge University Press, 1996. Volcanoes and volcanic processes on the Earth and planets are described in layman's terms. A great introduction to the wonders of solar system volcanoes.

Encyclopedia of Volcanoes, edited by H. Sigurdsson, B. Houghton, S. R. McNutt, H. Rymer, and J. Stix. Academic Press, 2000. The articles in this encyclopedia, written by top volcanologists, cover just about every subject about volcanoes everywhere.

Volcanoes, by R.I. Tilling, US Geological Survey, US Government Printing Office, 1999. This booklet offers a concise introduction to volcanoes and the science of volcanology.

Volcano Watching, by R. and B. Decker, drawings by R. Hazlett. Hawaii Natural History Association and Hawaii Volcanoes National Park, 1984. Colorful introduction to volcanoes with focus on Hawaii.

Volcanoes, by S. Van Rose and I. Mercer. Her Majesty's Stationary Office for the Institute of Geological Sciences, 1977. Concise, colorful introduction to volcanoes.

Volcanoes: Fire from the Earth, by M. Krafft. Gallimard Press (in French), 1991; English translation, Harry N. Adams, New York, 1993. A lively, wonderfully illustrated, concise book on the history of volcanology, volcanic eruptions, and volcano stories. Maurice finished this book shortly before his death at Unzen.

Volcanoes and Society, by D. Chester. Edward Arnold, 1993. An introduction to volcanology with focus on how society responds to and makes use of active volcanoes.

Volcanoes: An Introduction, by A. Scarth. Texas A&M University Press, 1994. Good introduction to volcanoes and volcanology, covering all the basics.

Mountains of Fire: The Nature of Volcanoes, by R. and B. Decker. Cambridge University Press, 1991. Another book that introduces the subject of volcanology, written by the husband and wife "Double Decker" team. Now out of print.

Volcanoes, by G. Macdonald. Prentice-Hall, 1972. A classic book on volcanology, long out of print but worth looking for. Great descriptions of eruptions.

Volcanology, by H. Williams and A. McBirney. Freeman, Cooper, & Co., 1979. Another out-of-print classic, more technical and more focused on physical volcanology than Macdonald's book.

Volcanoes of the Earth, by F. Bullard. University of Texas Press, 1984. Another out-of-print classic, this very readable book is worth checking the library for. A lively introduction to volcanoes, with some personal accounts of field work on volcanoes.

Global Volcanism 1975–1985, edited by L. McClelland *et al.* Prentice-Hall with the American Geophysical Union, 1989. Compilation of reports on volcanic activity during a decade.

The Citizen' Guide to Geologic Hazards, prepared and published by the American Institute of Professional Geologists, 1993. This reference work covers hazards from geologic processes such as volcanoes and earthquakes, as well as from geologic materials and gasses.

Volcanic Hazards, edited by R.I. Tilling. American Geophysical Union, 1989. Collection of very readable papers on volcanic hazards and monitoring techniques. Although the techniques and technology have progressed significantly since this was published, it is still a very useful reference.

Monitoring Volcanoes: Techniques and Strategies Used by the Staff of the Cascades Volcano Observatory, 1980–90. US Geological Survey, US Department of the Interior, 1992. Collection of technical papers on volcano monitoring.

Monitoring Active Volcanoes, by R.I. Tilling. US Geological Survey, US Department of the Interior, 1983. Informative US Geological Service booklet about techniques.

"Monitoring active volcanoes," by R.I. Tilling. This popular-level article is a shorter version of the above. *Earthquake Information Bulletin,* **12**, no. 4, 1980.

"Volcano monitoring by satellite," by D.A. Rothery. *Geology Today,* July–August 1989. Useful description of satellite use for volcano monitoring.

"Remote sensing of volcanoes and volcanic terrain," by P.J. Mouginis-Mark *et al.* Technical review article on satellite and aircraft monitoring of volcanoes. *Eos, Transactions of the American Geophysical Union* **70**, no. 52, 1989.

"Lessons in reducing volcanic risk," by R.I. Tilling and P.W. Lipman. Discussion on how volcanological studies can assess risks from active volcanoes. *Nature* **364**, 277–80, 1993.

"Safety recommendations for volcanologists and the public," Report for the International Association of Volcanology and Chemistry of the Earth's Interior, by S. Aramaki, F. Barberi, T. Casadevall, and S. McNutt. *Bulletin of Volcanology* **56**, 151–54, 1994.

"Terrestrial volcanism in space and time," by T. Simkin. Technical review article. *Annual Reviews of Earth and Planetary Sciences* **21**, 427–52, 1993.

Volcanology and Geothermal Energy, by K.Wohletz and G. Heiken. University of California Press, 1992. This mostly technical book focuses on the potential of extracting geothermal energy from active volcanic areas.

"Liquid light delight," by G.B. Lewis. Short discussion on photographing volcanic eruptions by one of Hawaii's most prominent photographers (see cover photo). *Volcano Quarterly Newsletter* **1**, no.3, August 1992.

Hawaii

Volcanic and Seismic Hazards on the Island of Hawaii, by C. Heliker. US Geological Survey, US Department of the Interior, 1992. This booklet does a great job of introducing volcanism in Hawaii.

Geological Field Guide, Kilauea Volcano, by R. Hazlett. Hawaii Natural History Association, 1993. Useful text and maps, but non-geologists would find Hazlett and Hyndman's book (below) easier to use.

Eruptions of Hawaiian Volcanoes, Past, Present, and Future, by R.I. Tilling, C. Heliker, and T.L. Wright. US Geological Survey, US Department of the Interior, 1987. Well-illustrated, popular book on Hawaiian eruptions.

Volcanoes of the National Parks in Hawaii, by G. Macdonald and D. Hubbard. Hawaii Natural History Association and the Hawaii Volcanoes National Park, 1974. Although out of date with respect to the eruptions, this concise book is still a very good introduction to the volcanoes of Hawaii.

Roadside Geology of Hawaii, by R. Hazlett and D. Hyndman. Mountain Press, 1996. Part of the well-known series, this is an easy-to-use guide to all the Hawaiian Islands.

Hawaii Trails: Walks, Strolls and Treks on the Big Island, by K. Morey. Wilderness Press, 1992. A great resource for the hiker, this book is highly recommended for visitors to the Big Island. My copy is from 1992, but I have since met Kathy Morey at a local bed and breakfast, working on an update.

Volcanoes in the Sea: The Geology of Hawaii, by G. Macdonald and A. Abbott. University of Hawaii Press, 1979. A classic book on Hawaiian volcanism. Out of date with respect to the eruptions, but still a good reference.

Mauna Loa Revealed: Structure, Composition, History, and Hazards, edited by J.M. Rhodes and J.P. Lockwood. American Geophysical Union, 1995. Collection of technical papers on Mauna Loa.

Hawaii Legends of Volcanoes, by W.D. Westervelt. Charles E. Tuttle, 1979. Compilation of legends translated from the Hawaiian.

A Curious Life for a Lady, by P. Barr. Penguin Books, 1985. The story of Victorian traveler Isabella Bird; includes a wonderful account of watching eruptions in Hawaii.

Continental USA

Volcanoes of North America: United States and Canada, edited by C.A. Wood and J. Kienle. Cambridge University Press, 1990. An encyclopedia-type book of all volcanoes in North America. Includes instructions of how to get to each of them.

Volcanoes of the United States, by S.R. Brantley. US Geological Survey, US Department of the Interior. Very informative booklet on the volcanoes in the U.S.

Excursion 12B: South Cascades Arc Volcanism, California and Southern Oregon, by L.J.P. Muffler *et al.* New Mexico Bureau of Mines and Mineral Resources Memoir, 1989. Geological field trip guidebook/technical paper.

The Complete Guide to America's National Parks, published by the National Park Foundation, Washington, D.C. Comprehensive information on all American Parks. Updated regularly.

Fire Mountains of the West: The Cascade and Mono Lake Volcanoes, by S.L. Harris. Mountain Press, 1992. A very useful book for visitors to the Cascades.

Summit Guide to the Cascades Volcanoes, by J. Smoot. Chockstone Press, 1992. Describes climbing routes up 18 of the volcanoes, with emphasis on the easiest, most popular routes.

Adventure Guide to Mount Rainier: Hiking, Climbing, and Skiing in Mount Rainier National Park, by J. Smoot. Chockstone Press, 1991.

Mount Rainier, Active Cascade Volcano: Research Strategies for Mitigating the Risk from a High, Snow-Clad Volcano in a Populous Region, by the National Research Council. National Academy Press, 1994.

Pictorial History of the Lassen Volcano (3rd revd edn). This old classic was first published by California Press, San Francisco, in 1926 by B.F. Loomis. Loomis Museum Association (Mineral, CA) in cooperation with the National Park Service, 1971.

Through Vulcan's Eye, by P.S. Kane. Loomis Museum Association (Mineral, CA) in cooperation with the National Park Service, 1990. Describes the geology and geomorphology of Lassen Volcanic National Park in a well-illustrated book.

Lassen Trails, by S.H. Mattenson. Loomis Museum Association (Mineral, CA) in cooperation with the National Park Service, 1992. Useful booklet for hiking in the Park.

Hiking Trails of Lassen Volcanic National Park, by G.P. Perkins. George P. Perkins, 1989. More detailed and better illustrated than the above, includes information and photos of wildflowers found in the Park.

Lassen Place Names, by P.E. Schulz. Walker Lithograph, 1991. Written by a Park Naturalist, this interesting book discusses the origins and meanings of the place names in the park. (First published by the Loomis Museum Association (Mineral, CA) in 1949.)

"Pleistocene glaciation, Lassen Volcanic National Park," by P. Kane. *California Geology*, May 1982. Technical paper.

Crater Lake: The Story of its Origin, by H. Williams. University of California Press, 1941. A classic study, now out of print.

Road Guide to Crater Lake National Park, by R. and B. Decker, illustrated by R. Hazlett. Double Decker Press, 1995. Concise and useful guidebook to the Park.

The Mount Shasta Book, by A. Selters and M. Zanger. Wilderness Press, 1992. A guide to hiking, climbing, and exploring the region. Includes a topographic map.

Volcanic Hazards at Mount Shasta, California, by D.R. Crandell and D.R. Nichols. US Geological Survey, US Department of the Interior, 1987. Non-technical booklet about the volcano and its hazard potential.

Potential Hazards from Future Eruptions in the Vicinity of Mount Shasta Volcano, Northern California, by C. Dan Miller. Technical paper. Geological Survey Bulletin 1503, United Staes Government Printing Office, 1980.

Mount Shasta: History, Legends and Lore, by M. Zanger. Celestial Arts, 1992. Well-illustrated book with lots of historical information.

The 1980 Eruptions of Mount St. Helens, Washington. US Geological Survey, US Department of the Interior, 1981. Collection of technical papers on the activity. Lots of maps and photographs.

Eruptions of Mount St. Helens: Past, Present, and Future (revd edn), by R.I. Tilling, L. Topinka, and D.A. Swanson. US Geological Survey, US Department of the Interior, 1990. This booklet contains a very well-written summary of the 1980 eruption and discusses the mountain's future hazards.

Volcanic Eruptions of 1980 at Mount St. Helens: The First 100 Days, by B.L. Foxworthy and M. Hill. US Geological Survey, US Department of the Interior, 1982. Very readable and well-illustrated chronology of the eruption.

Volcano: The Eruption of Mount St. Helens, written and edited by the combined staffs of *The Daily News*, Longview, Washington, and *The Journal-American*, Bellevue, Washington. Longview Publishing Co., and Madrona Publishers. The story of the eruption told by local news staff. Wonderful photographs.

Roadside Geology of Mount St. Helens National Volcanic Monument and Vicinity, by P.T. Pringle. Washington Department of Natural Resources, Division of Geology and Earth Resources, 1993. Very informative geologic guide to the volcano.

Road Guide to Mount St. Helens, by R. and B. Decker, illustrated by R. Hazlett. Double Decker Press, 1993. Shorter and less detailed than the above, but still a very useful guide.

Mount St. Helens National Volcanic Monument Trail Guide, by L. Roberts. Northwest Interpretive Association in cooperation with Gifford Pinchot National Forest, this easy-to-use guide is invaluable to hikers.

Geologic Field Trips in the Pacific Northwest, vol. 2, edited by D.A. Swanson and R.A. Haugerud. Department of Geological Sciences, University of Washington, 1994. Collection of papers on the Cascades, focused on geologic field trips.

"Geologic nozzles," by S.W. Kieffer. *Reviews of Geophysics*, **27**, 1989. Technical paper.

Roadside Geology of Arizona, by H. Chronic. Mountain Press, 1983. Part of the *Roadside Geology* series, useful to those visiting Sunset Crater and other Arizona wonders.

Volcanoes of Northern Arizona: Sleeping Giants of the Grand Canyon Region, by W.A. Duffield, photographs by M. Collier. Grand Canyon Association, 1997. A concise, well-illustrated book on the volcanoes of the region, including Sunset Crater and the SP cone, with focus on geology and volcanology.

A Guide to Sunset Crater and Wupatki, by S. Thybony, photographs by G. H. H. Huey. Southwest Parks and Monuments Association, 1987. Another concise, well-illustrated guidebook to the region.

Of Men and Volcanoes: The Sinagua of Northern Arizona, by A.H. Schroeder, edited by Earl Jackson. Southwest Parks and Monuments Association, 1977. Concise history of the prehistoric natives who lived in the Sunset Crater region.

The Basaltic Cinder Cones and Lava Flows of the San Francisco Mountain Volcanic Field (revd edn), by H.S. Cohen. Museum of Northern Arizona, 1967. Classic technical paper on the geology of the region.

Yellowstone National Park: Its Exploration and Establishment, by A.L. Haines. US Department of the Interior, National Park Service, 1974. History of the exploration of Yellowstone, published to celebrate the 100th anniversary of the park. An important volume and a mine of information.

Geysers: What They Are and How They Work, by T. Scott Bryan. Roberts Rinehart, 1990. A concise explanation of geyser activity by a member of the Geyser Observation and Study Association.

A Field Guide to Yellowstone's Geysers, Hot Springs, and Fumaroles, by C. Schreier. Homestead Publishing, 1992. Well-illustrated, easy-to-use book describing all of the park's main attractions. Excellent book to carry around during a visit to Yellowstone.

Guardians of Yellowstone, by D. Sholly with S. Newman. William Morrow, 1991. A Park Ranger's account of the challenges of protecting Yellowstone.

Death in Yellowstone, by L.H. Whittlesey. Roberts Rinehart, 1995. Account of accidents and foolhardiness in the Park.

Thomas Moran, by N.K. Anderson. National Academy of Art, Washington, and Yale University Press, 1997. Catalog of the exhibition on Moran's work, which include his powerful paintings of Yellowstone.

"Inside Old Faithful," by S. Perkins. *Science News*, October 11, 1997. Excellent, non-technical account of the field work of J. Westphal and S. Kieffer on lowering a camera down Old Faithful.

"The Yellowstone hot spot," by R.B. Smith and L.W. Braile. *Journal of Volcanology and Geothermal Research* **61**, 121–87, 1994. Technical paper on Yellowstone and the Snake River Plains underlying hotspot.

"Yellowstone magmatic evolution: its bearing on understanding large-volume explosive volcanism," by R.L. Christiansen. In *Explosive Volcanism: Inception, Evolution, and Hazards*. National Academy Press, 1984. Technical paper.

Italy

Italian Volcanoes, by C. Kilburn and W. McGuire. Terra Publishing, 2001. Part of the *Classic Geology of Europe* series, this is a detailed, invaluable guide to Vesuvius, Campi Flegrei, the Aeolian Islands, and Etna, written by two top experts who I've had the pleasure to work with.

Mount Etna: Anatomy of a Volcano, by D.K. Chester, A.M. Duncan, J.E. Guest, and C.R.J. Kilburn. Chapman and Hall, 1985. A comprehensive, mostly technical book on Mt. Etna, focusing on research and history of eruptions (now out of print).

Guida Eolie: Escursionistico Vulcanologica delle Isole, by N. Calanchi, P.L. Rossi, F. Sanmarchi, and C.A. Tranne. Centro Studi e Ricerche di Storia e Problemi Eoliani, 1996. Even those who cannot read Italian will find this guidebook to the Aeolian Islands useful for its detailed maps and itineraries.

Somma–Vesuvius, edited by R. Santacroce. Consiglio Nazionale delle Ricerche, 1987. Collection of technical papers on the volcano, including maps. In English.

Mount Etna Volcano, edited by R. Romano. Società Geológica Italiana, 1982. Collection of technical papers on the volcano, including maps. In English.

"The Island of Stromboli," by M. Rosi. *Rendiconti Società Italiana di Mineralogia e Petrologia* **36**, 345–368, 1980. Technical paper.

"Volcanic hazard assessment at Strómboli based on review of historical data," by F. Barberi, M. Rosi, and A. Sodi. Technical paper. *Acta Vulcanologica* **3**, 173–187, 1993.

"Strómboli and its 1975 eruption," by G. Capaldi *et al. Bulletin of Volcanology* **41**, 1978. Technical paper.

"Geology, stratigraphy, and volcanological evolution of the island of Stromboli, Aeolian arc, Italy." *Acta Vulcanologica* **3**, 21–68, 1993. Technical paper.

"A statistical model for Vesuvius and its volcanological implications," by S. Carta *et al. Bulletin Volcanologique*, **44**, 1981. Technical paper.

"Mount Vesuvius: 2000 years of volcanological observations," by R. Scandone, L. Giacomelli, and P. Gasparini. Technical paper. *Journal of Volcanology and Geothermal Research* **58**, 5–25, 1993.

"The eruption of Vesuvius in AD 79," by H. Sigurdsson, S. Carey, W. Cornell, and T. Pescatore. Technical paper. *National Geographic Research* **1**, 332–87, 1985.

"Eruptions on Mount Etna during 1979," by J. Guest, J. Murray, C. Kilburn, and R. Lopes. *Earthquake Information Bulletin* **12**, 154–60, 1980. Non-technical eyewitness account of the fatal 1979 explosion.

"Etna erupts again: A VEST report of the March 1981 eruption of Mount Etna," by the UK's Volcanic Eruption Surveillance Team, *Earthquake Information Bulletin* **13**, 134–9, 1981. Sicily. Non-technical account of the scientific observations by the Volcanic Eruption Surveillance Team (the author was a team member at this time) of the eruption.

"Volcanism in Eastern Sicily and the Aeolian Islands," by H. Pichler. In *Geology and History of Sicily*, edited by W. Alvarez and K.H.A. Gohrbandt. Petroleum Exploration Society of Libya, 1970. Technical paper.

"Lava diversion proved in 1983 test at Etna," by J.P. Lockwood and R. Romano. *Geotimes* **30**, 10–12, 1985. An easy to read account of the lava diversion.

"Etna. 1. Eruptive history," by S. Calvari, M. Neri, M. Pompilio, and V. Scribano. *Acta Vulcanologica* **3**, 1993. Technical paper.

"Isle of fire," by L. Geddes-Brown. *Country Life*, August 4, 1988. Historical article on Vulcano.

"Evolution of the Fossa cone, Vulcano," by G. Frazzetta, L. La Volpe, and M.F. Sheridan. *Journal of Volcanology and Geothermal Research* **17**, 329–60, 1983. Technical paper.

"The Island of Vulcano," by J. Keller. *Rendiconti Società Italiana di Mineralogia e Petrologia* **36**, 369–414, 1980. Technical paper.

"Volcanic history and maximum expected eruption at 'La Fossa di Vulcano' (Aeolian Islands, Italy)," by G. Frazzetta and L. La Volpe. *Acta Vulcanologica* **1**, 107–13, 1991. Technical paper.

"Aeolian Islands (Southern Tyrrhenian Sea): Excursion B2." In *Excursions Guidebook*, by the International Association of Volcanology and Chemistry of the Earth's Interior. IAVCEI, 1985. Guidebook for the geologic field trips organized by the International Association of Volcanology and Chemistry of the Earth's Interior.

"1987–1990 unrest at Vulcano," by F. Barberi, G. Neri, M. Valenza, and L. Villari. *Acta Vulcanologica* **1**, 95–106, 1991. Technical paper.

"In the jaws of the volcano," by C. Kilburn. *New Scientist*, February 6, 1986. Article on the Phlegraen Fields unrest and its impact on the town of Pozzuoli.

"Historical activity at Campi Flegrei caldera, Southern Italy," by J. Dvorak and P. Gasparini. *Earthquakes and Volcanoes* **22**, 256–67, 1990. Informative article on the activity at the Phlegraen Fields.

Parco dell'Etna: Guida Turistica. Touring Club Italiano, 1993. Tourist guide (in Italian), contains a booklet and a map (1:50,000) showing topography and main cities and roads. Very useful, even if you cannot read Italian. Sold in towns and tourist stops around Etna.

Etna: Carta Naturalistica e Turistica, by R. Romano. Club Alpino Italiano and Società Elaborazioni Cartografiche, Geological map of Mt. Etna (first published in *Mount Etna Volcano*, see above), updated and including information on the vegetation and fauna. In Italian and English. Scale 1:60,000. Complements the above *Guida Turistica* and is sold in the Mt. Etna area. Very useful but, as with the above, unlikely to be up to date (Etna erupts far too often).

Greece and Santorini

Fire in the Sea, by W. L. Friedrich, translated by A. R. McBirney. Cambridge University Press, 1999. Subtitled *Santorini: Natural History and the Legend of Atlantis*, this is a beautifully illustrated and scientifically accurate book written at a popular level. Highly recommend background reading for Santorini.

Santorini Volcano, by T. H. Druitt, M. Davies, L. Edwards, R. S. J. Sparks, and R. Mellors. Geological Society, 1999. This is the most comprehensive recent work on scientific research at Santorini. Although quite expensive and aimed at professional scientists, it includes a large, folded geological map that many visitors to Santorini might like to have, particularly if they are geologists. The map can be purchased separately.

Santorini: Guide to "The Volcano," by G. Vougioukalakis. Institute for the Study and Monitoring of the Santorini Volcano. This very well-illustrated book, available in Santorini, is currently the best guidebook to Palea and Nea Kameni. Includes a volcanological map of the islands.

Santorini: A Guide to the Island and its Archeological Treasures, by C. Doumas. Ekdotike Athenon, 1996. Very well-illustrated book summarizing the archeological findings in the island, written by a leading archeologist.

The Wall Paintings of Thera, by C. Doumas. Thera Foundation, 1992. A large-format, beautifully illustrated book on the wall paintings found at Akrotiri. Worth purchasing if you are particularly interested in the paintings and their restoration.

Thera and the Aegean World, vol. 3. Thera Foundation, 1990. Part of a collection of scientific papers in three volumes, Proceedings of the Third International Congress.

"The end of the Minoan civilisation," by W. Downey and D. Tarling. *New Scientist*, September 13, 1984. Non-technical article on the eruption.

"Unsteady date of a big bang," by G. Cadogan. *Nature* **328**, 473, 1987. Commentary on dating the eruption and on the article below.

"The Minoan eruption of Santorini in Greece dated to 1645 BC?" by C. U. Hammer, H. B. Clausen, W. L. Friedrich, and H. Tauber. *Nature* **328**, 517–19, 1987. Technical paper.

Iceland

Iceland, Land of the Sagas, by D. Roberts and J. Krakauer. Harry N. Abrams, 1990. Informative text and beautiful photographs of Iceland.

Letters from Iceland, by W. H. Auden and L. MacNeice. Paragon House, 1990. A perceptive view of Iceland and its people by two young English poets who visited the country in 1937.

Guide to the Geology of Iceland, by A. T. Guðmundsson and H. Kjartansson. Bókaútgáfan Örn og Örlygur, 1986. Concise guide to Iceland's volcanoes.

Iceland, by M. Bamlett and J. F. Potter. Geologists' Association, 1994. Geologic field guide, part of a series edited by J. T. Greensmith.

The Control of Nature, by J. McPhee. Farrar Straus Giroux, 1989. Contains "Cooling the lava," a superb account of the efforts to stop the lava from reaching the port of Heimaey.

Ice and Fire: Contrasts of Icelandic Nature, by H. R. Bárðarson. Hjálmar R. Bárðarson, 1991. Well-illustrated book about Iceland, including descriptions and photos of various eruptions.

Iceland, Greenland, and the Faroe Islands: A Travel Survivor Kit, by D. Swaney. Lonely Planet, 1994. Unlike many travel guides, this one is strong on the geology. It also has good maps, and offers helpful advice on hikes, food, and a variety of other essentials.

Surtsey: The New Island in the North Atlantic, by S. Thorarinsson. Viking Press, 1967. Popular book on the eruption of Surtsey, written by Iceland's foremost volcanologist at the time. Beautiful pictures and insightful descriptions. (Out of print, but not hard to find second-hand).

Man against Volcano: The Eruption on Heimaey, Vestmannaeyjar, Iceland, by R.S. Williams and J.G. Moore. US Geological Survey, 1983. Non-technical account of the eruption by two U.S. volcanologists who studied it first hand.

"The 1991 eruption of Hekla, Iceland," by A. Gudmundsson *et al. Bulletin of Volcanology* **54**, 238–46, 1992. Technical paper.

The Eruption of Hekla, 1947–1948, 2 vols. Societas Scientiarum Islandica, 1978. Collection of papers on the eruption, also covers studies of previous historical eruptions of the volcano. The first volume, *The Eruptions of Hekla in Historic Times* by the late Sigurdur Thorarinsson, is very readable and recommended for those who wish to find out more about Hekla's eruptions and their relation to Icelandic history.

Costa Rica

Costa Rica, Land of Volcanoes, by G. Alvarado Induni, translated by O.L. Chavarría-Aguilar. Gallo Pinto Press, 1993. Highly recommended book for visitors to Costa Rica, written by a leading volcanologist. (Originally published as *Los Volcanes de Costa Rica*, Editorial Universidad Estatal a Distancia, 1989.)

Costa Rica: National Parks, by M.A. Boza. Editorial Heliconia, 1988. Well-illustrated book on the country's national parks.

Costa Rica's National Parks and Preserves, by J. Franke. The Mountaineers (Seattle, WA), 1993. Very good guide for the independent traveler.

Nicaragua–Costa Rica Quaternary Volcanic Chain: Field Trip 17 Guidebook. International Association of Volcanology and Chemistry of the Earth's Interior, 1997. Geologic field trip guide.

Volcán Poás, Costa Rica, by J. Barquero H. Published in Costa Rica, 1998. Booklet about the volcano in English and Spanish written by a local expert.

"Sulphur Eruptions at Poás volcano," by C. Oppenheimer. *Journal of Volcanology and Geothermal Research* **49** 1–21, 1992. Technical paper.

"An integrated dynamic model for the volcanic activity at Poás volcano, Costa Rica," by L. Casertano *et al. Bulletin of Volcanology* **49**, 588–98, 1987. Technical paper.

Mapa Geologico de Costa Rica, by J. Tournon and G. Alvarado Induni. Editorial Tecnologica de Costa Rica, 1997. Geologic map of Costa Rica (1:500,000) and explanatory booklet. In Spanish and French.

The West Indies

The Day the World Ended, by G. Thomas and M.M. Witts. Stein and Day, 1969. This out-of-print gem is worth looking for. Written by two journalists, this day-by-day description of the events leading to the eruption and on what followed is an eye opener. A few inaccuracies, but a wonderful read.

Martinique, Guadeloupe, St. Martin, La Désirade, by D. Westercamp and H. Tazieff. Masson, 1980. Guidebook to the geology of the islands, in French. Part of the *Guides Géologiques Régionaux* series.

Caribbean Volcanoes: A Field Guide – Martinique, Dominica, and St. Vincent, by H. Sigurdsson and S. Carey. Geological Association of Canada, 1991. A geological field guide to the islands.

Mont Pelée, Martinique: A Study of an Active Island-Arc Volcano, by A.L. Smith and M.J. Roobol. Geological Society of America, 1990. Technical work on the volcano, good summary of reseach.

La Catastrophe: The Eruption of Mount Pelée, the Worst Volcanic Disaster of the Twentieth Century, by A. Scarth. Oxford University Press, 2002. I received this book after finishing my manuscript, so it is strictly not part of the bibliography, but I highly recommend it. Like *The Day the World Ended*, it is a wonderful account of the events of 1902, but stronger on the science.

"The 1902–1905 eruptions of Montagne Pelée, Martinique: anatomy and retrospection," by J.C. Tanguy. *Bulletin of Volcanology* **60**, 87–107, 1994. Technical paper on the activity.

The Tower of Pelée: New Studies of the Great Volcano of Martinique, by A. Heilprin. J.B. Lippincott, 1904. Published shortly after the eruption, describes the evolution of the rock tower that grew out of the volcano's crater. Great photographs. Out of print but available at some academic libraries.

"Eruption of Soufriere Hills Volcano in Montserrat continues," by S. Young *et al.* Description of the first couple of years of the eruption. *Eos, Transactions of the American Geophysical Union*, **78**, 1997.

Monitoring on Montserrat: the course of an eruption, by Simon Young. Report on the first three years of the eruption and the monitoring techniques. Astronomy & Geophysics, vol. 39, issue 2, 1998. Published by the Royal Astronomical Society, U.K.

Montserrat's Andesite Volcano educational video produced by D. Lea and presented by S. Sparks. Geological Society Publishing House, 1999. This video, aimed at geology students, describes the volcano and the first few years of activity, and comes with a useful booklet. There is a lack of popular-level published material on the eruption, this video helps to fill that gap.

Other locations

Mexico's Volcanoes: A Climbing guide, by R.J Secor. The Mountaineers (Seattle, WA), 1988. A must for those planning to climb Colima and other volcanoes in Mexico.

Surviving Galeras, by S. Williams and F. Montaigne. Houghton Mifflin, 2001. A thrilling account of the tragic eruption of 1993 in which Williams nearly lost his life.

Krakatau 1883: The Volcanic Eruption and its Effects, by T. Simkin and R.S. Fiske. Smithsonian Institution Press, 1983. Compilation of eyewitness accounts and scientific interpretation of the famous eruption.

Lessons from a Major Eruption: Mount Pinatubo, Philippines, by the Pinatubo Volcano Observatory staff. American Geophysical Union, 1992. Popular-level discussion of predicting hazards from the Pinatubo eruption of 1991.

"In the volcano's shadow," by D. Shimozuru. *Nature*, **353**, 295–6, 1991. Discussion on the 1991 eruption of Unzen.

"Obituary: Harry Glicken (1958–1991)," by R.V. Fischer. *Bulletin of Volcanology* **53**, 514–516, 1991.

Glossary

aa lava The most common type of basaltic lava flow. Flows of aa typically move slowly, with a rubbly top surface gradually falling down in front of the flow and being overridden by it, in a caterpillar-like motion. Aa surfaces of cooled flows are a jumble of loose blocks, usually with razor-sharp protusions. Walking on one of these flows is hard work, hence the Hawaiian word for it: "a'a," for the sounds that usually escape someone's mouth when trying to cross this type of flow.

Andesite Name given to volcanic rocks that have chemical compositions containing 53–63% silica (SiO_2), intermediate between basalt and rhyolite. Andesite is the most common type of lava on island arc volcanoes. When in a molten state, andesite has moderate viscosity, generally higher than basalt but lower than rhyolite.

Ash Fragments of lava or rock expelled by volcanic eruptions. Ash is strictly the name given to fragments that are less than 4 mm in diameter.

Ashfall Volcanic ash falling to the ground after being transported by volcanic plumes.

Basalt Name given to volcanic rocks that have chemical composition containing less than 53% silica (SiO_2). When in a molten state, basalt has low viscosity and forms lava flows of pahoehoe or aa type. Basalts are the most common type of volcanic rock on Earth.

Base surge Density currents made up of volcanic ash and gases mixed with air. Base surge currents flow over the ground, and were named after currents observed spreading along the ground following nuclear explosions.

Blast A high-energy, fast-moving pyroclastic flow, extremely destructive.

Blocky lava A type of lava flow with fractured surfaces resembling a jumble of blocks, some up to 1 m (~3 feet) across. Unlike aa flows, the surfaces of blocky lavas have fragments that are angular with smooth sheer faces. These type of flows move slowly, usually less than 1 km (⅝ mile) per day. Lavas forming blocky lavas usually have high silica content.

Bocca A volcanic vent, the opening where lava comes out. (From the Italian meaning for "mouth".)

Bomb A volcanic fragment larger than 6 cm (2½ inches) across. When bombs are expelled during eruptions, they cool and become rounded. During flight, they may develop distinctive exteriors (e.g., fusiform or spindle bombs). The surfaces of more viscous bombs may form cracks that look like crusty bread, and these are named breadcrust bombs.

Breccia Volcanic breccias are broken volcanic rocks. A breccia deposit consists of angular blocks within a matrix of fine particles.

Caldera A large volcanic crater, usually larger than 1 km (⅝ mile) in diameter. Calderas are formed either by collapse (most often) or explosion. The name comes from the Portuguese, meaning cauldron, and seems to have first been used to name Caldera Taburiente on La Palma, Canary Islands.

Composite volcano A steep-sided volcano composed of many layers of rock, some resulting from lava flows, some from fragments from volcanic explosions. Examples are Mt. Fuji in Japan and many of the volcanoes in the Cascades Range, USA.

Dacite Name given to volcanic rocks that have chemical composition containing 63–68% silica (SiO_2). Dacitic lavas have high vicosity and flow slowly.

Debris avalanche The type of avalanche that occurs when part of a volcanic edifice collapses.

Dome A steep-sided mound that forms when viscous lava piles up near a volcanic vent. Domes are formed by andesite, dacite, and rhyolite lavas.

Fallout Term to describe volcanic particles and fragments that rise high into the atmosphere and then fall out over the surrounding areas.

Fumarole An opening that releases volcanic gases, including steam.

Hornito A small cone formed by weak explosions, usually formed of layers of spatter. (From the Spanish, meaning "small oven.")

Hydrovolcanic eruption *See* **phreatomagmatic eruption**.

Ignimbrite A pumice-rich pyroclastic flow, normally associated with large-volume explosive eruptions.

Island arc Destructive boundaries of plates in ocean areas occur where one plate subducts back into the mantle. Island chains of active volcanoes form at island arcs such as the Lesser Antilles in the Caribbean.

Lahar A flowing mixture of water and rock debris that forms on the slope of a volcano. Also called mudflow or debris flow. The name lahar comes from Indonesia, where this type of flow has often happened. Lahars can be triggered by heavy rains or snowmelt and can be very destructive.

Lapilli Magmatic fragments 2–6.5 cm (½ to 2½ inches) across. Accretionary lapilli are formed in eruption clouds when coatings of ash form concentric layers around a tiny nucleus.

Lava Molten rock that erupts from a volcanic vent. *See also* **magma**.

Magma Molten rock that contains dissolved gases and crystals. Magma is formed deep within the Earth; when it reaches the surface, it is called **lava**.

Magma chamber Region of magma which supplies magma to the volcano.

Mudflow *See* **lahar**.

Nuée ardente The term generally used to mean a **pyroclastic flow**; however, the strict definition is a pyroclastic flow of poorly vesiculated magma. (From the French, meaning "glowing cloud.")

Pahoehoe lava Hawaiian word for lava flows with smooth, easy to walk on surfaces. Pahoehoe lava usually flows through tubes and oozes out as a series of small overlapping "toes" up to a few meters across. In some places, tumuli form, where lava pushes up the surface causing it to break out. In others, flowing lava collects into tight folds called ropes (ropy pahoehoe).

Phreatic eruption (phreatic explosion) A type of volcanic explosion that happens when water comes into contact with hot rocks or ash, causing a steam explosion. Phreatic explosions do not eject juvenile (new) lava fragments.

Phreatomagmatic eruption Same as phreatic eruption, but involves New lava. Also called hydrovolcanic eruption.

Plume Vertical turbulent flow of volcanic particles, volcanic gas, and air. The plume rises in the atmosphere because its density is lower than that of air. Plumes form both from explosions, producing an eruption column above the volcano, and by fine ash particles and gases escaping from pyroclastic flows.

Pumice A light-colored volcanic rock ejected during explosive volcanic eruptions. Pumice contains abundant bubbles formed by trapped gases and, because of this, it is very light and commonly floats on water.

Pumice flow A coarse-grained pyroclastic flow containing substantial amounts of pumice blocks and pumice fragments as well as volcanic ash.

Pyroclastic Term used to refer to fragments of lava. (From the Greek, meaning "fire-broken.")

Pyroclastic fall The rain-out of particles through the atmosphere from an eruption jet or plume during an explosive eruption.

Pyroclastic flow A hot, fast-moving, high-density mixture of ash, pumice, rock fragments, and gas formed during some volcanic eruptions. Pyroclastic flows are one of the greatest hazards from volcanic eruptions. *See also* ***Nuee Ardente***.

Pyroclastic surge The same process as pyroclastic flow (above) but having lower density. *See also* **base surge**.

Rhyolite A volcanic rock containing more than 68% silica (SiO_2). It has a very high viscosity when in a molten state.

Scoria Generic term for broken volcanic rock; it usually refers to aa lava fragments and the ejected products of Strombolian-type eruptions that are smaller than bombs. (From the Greek, meaning "dung.")

Shelly lava A type of pahoehoe lava that forms a thin, glassy surface which easily breaks when stepped upon.

Shield volcano A volcano shaped like an inverted warrior's shield, with gentle slopes. Shield volcanoes are formed by eruptions of low-viscosity, basaltic lavas.

Silica The molecule formed of silicon and oxygen (SiO_2) that is the basic building-block of volcanic rocks and the most important factor controlling the fluidity of magma. The higher the magma's silica content, the higher its viscosity, or the more "sticky" it is.

Spatter Generic term for large volcanic fragments (usually bombs) ejected by Strombolian eruptions and fire fountains, that become flattened during impact.

Spine A protusion of semi-solid lava that forms at the surface of lava flows. Large spines, such as occurred at Mt. Pelée, are pushed out at the vent along ductile shear zones.

Steam vent Vent formed when superheated water reaches the surface and boils explosively, forming a jet of steam. This type of activity usually happens during the early stages of an eruption, when rising magma heats up groundwater.

Tephra Generic term for all volcanic fragments that are explosively ejected. (From the Greek, meaning "ashes.")

Tuff Consolidated ash and pumice deposit, usually associated with pyroclastic flows. The deposit becomes consolidated over time by circulating water, which deposit minerals that cement the volcanic fragments.

Vent The opening at the Earth's surface through which volcanic materials escape. A volcano is the vent and also the landform that is constructed by the erupted material.

Vesicles Gas bubbles within magma.

Volcanic gases Gases (volatiles) dissolved in magma at depth that are released when pressure decreases as the magma comes to the surface. The main volcanic gas is usually water (steam), with small amounts of sulfur dioxide (SO_2), carbon dioxide (CO_2), sometimes hydrogen sulfide (H_2S), and halogen gases such as chlorine and fluorine.

Volcanology The science dedicated to the study of volcanoes and their products.

Xenolith Fragments of old rock trapped inside magma. (From the Greek, meaning "foreign stone.")

Index